DYNAMICS OF STAR CLUSTERS

INTERNATIONAL ASTRONOMICAL UNION
UNION ASTRONOMIQUE INTERNATIONALE

DYNAMICS OF STAR CLUSTERS

PROCEEDINGS OF THE 113th SYMPOSIUM OF THE INTERNATIONAL ASTRONOMICAL UNION, HELD IN PRINCETON, NEW JERSEY, U.S.A, 29 MAY – 1 JUNE, 1984

EDITED BY

JEREMY GOODMAN
California Institute of Technology, Pasadena, California, U.S.A.

and

PIET HUT
The Institute for Advanced Study, Princeton, New Jersey, U.S.A.

D. REIDEL PUBLISHING COMPANY
A MEMBER OF THE KLUWER ACADEMIC PUBLISHERS GROUP
DORDRECHT / BOSTON / LANCASTER

Library of Congress Cataloging in Publication Data

CIP

International Astronomical Union. Symposium (113th: 1984: Princeton, N.J.)
Dynamics of star clusters.

Includes index.
1. Stars–Clusters–Congresses. 2. Stars–Globular clusters–Congresses.
3. Stars–Open clusters–Congresses. I. Goodman, Jeremy, 1956–
II. Hut, Piet, 1952– III. Title.
QB853.I58 1984 523.8′5 85-1870

ISBN-13: 978-90-277-1965-2 e-ISBN-13: 978-94-009-5335-2
DOI: 10.1007/978-94-009-5335-2

Published on behalf of
the International Astronomical Union
by
D. Reidel Publishing Company, P. O. Box 17, 3300 AA Dordrecht, Holland

Softcover reprint of the hardcover 1st edition 1985

Sold and distributed in the U.S.A. and Canada
by Kluwer Academic Publishers,
190 Old Derby Street, Hingham, MA 02043, U.S.A.

In all other countries, sold and distributed
by Kluwer Academic Publishers Group,
P. O. Box 322, 3300 AH Dordrecht, Holland

TABLE OF CONTENTS

MAY 31

EXTERNAL FIELDS AND FINITE-STAR-SIZE EFFECTS

OPEN CLUSTERS

JUNE 1
THE DYNAMICS OF STAR CLUSTERS: FUTURE PROSPECTS

PREFACE

The emphasis in these proceedings of IAU Symposium No. 113, *Dynamics of Star Clusters*, and the main reason for organizing the symposium in the spring of 1984, was the rapid increase during the preceeding year in our understanding of core collapse. The last I.A.U. Symposium to discuss the dynamics of star clusters at length was No.69, *Dynamics of Stellar Systems*, held in Besançon in 1974. For a few years afterwards, globular clusters received much attention due to the discovery of X-ray bursters and the mounting evidence that X-ray sources in globular clusters were formed in completely different ways than those within our galaxy. Globular clusters, which until this time had a reputation for sedate old age, turned out to lead violent private lives at high energies.

However, in the early 80's globular clusters seemed to lose some of the glamor of the 70's. The grand speculations of heavy black holes lurking in their centers had to make way for a variety of observational evidence which indicated that the X-ray sources are low-mass close binaries instead. But, though dynamical fashion turned to heavy galactic halos and so on, some of the unsolved theoretical problems regarding the evolution of star clusters kept their fascination for a number of relatively isolated workers. After several years of inconspicuous labor, a number of preprints suddenly appeared in the spring of 1983 that studied the evolution of globular clusters *after* core collapse. This problem had been recognized clearly in the heyday of X-ray enthusiasm, but with the notable exception of early work by Hénon, all previous attempts to model post-collapse evolution posited the existence of a massive central black hole.

With the appearance of detailed models of the post-collapse regime, a renaissance in globular cluster evolutionary calculations occurred, comparable to the rapid progress in stellar evolution in the late 50's. It was this development which led us to propose this symposium. The international scientific organizing committee decided to organize the symposium in a slightly unusual fashion. Recognizing the many developments which had been taking place during the ten years since the previous IAU symposium in Besançon, the committee expressed their intention that the new symposium play a double role: to review those past developments; and to provide a forum for discussion of future directions of research, both theoretical and observational. The feeling of the organizers was that these goals could best be reached by scheduling only a limited number of invited speakers, together with a larger number of participants in poster sessions. In this way two objectives could be realized: the many new lines of research could be reviewed in a less hectic and more coherent way than would be possible if time were divided equally among all participants, and more time would be available between the talks for informal discussion. We will leave it to the participants and to the readers of these proceedings to judge the success of this plan.

We regret that Dr. L. M. Ozernoy was not able to attend the symposium and deliver his invited lecture, due to circumstances beyond his control. Although he informed us that he intended to submit the invited paper for the proceedings, we unfortunately had not yet received his contribution by the time these proceedings went to press.

We editors do appreciate the many positive reactions we received at the end of the symposium; the enthusiasm of the participants was all the more gratifying because of the nearly incessant rain that they suffered and that obscured a near-total eclipse. (The group photograph included here gives an utterly false impression of the usual weather, as any participant will admit.)

Those of you who actually read prefaces, and especially those who work on open clusters, may have noticed a bias in favor of globular clusters. The main reason for this is that much recent progress has been made in understanding the latter objects, whereas the dynamics of open clusters have been less well explored, both theoretically and observationally. In many ways the dynamics of open clusters are more difficult to study than those of globular clusters: theoretically, because the combination of the lower number of stars and the younger age of open clusters make a statistical description much more difficult and its validity more questionable; and observationally, because the velocity dispersion in an open cluster is typically an order of magnitude lower than that in a globular cluster. These difficulties notwithstanding, significant progress has been made recently and is reported in these proceedings, and it is possible that in another ten years the study of open clusters may dominate a future star cluster symposium, in part because this study promises to increase our understanding of star formation in the galactic disk.

We express our hope that in the coming years the proceedings of this symposium will prove to be as valuable as those of its predecessor, I.A.U. Symposium 69, as a reference work for research on the dynamics of star clusters. To this end we urged all invited speakers to make a broad survey of their topics. In addition, on the final day of the symposium a discussion was held among a panel of six wise men to assess the scientific progress revealed by the conference and to identify the most promising directions for future research, a summary of which has been included here.

Four appendices have been added. The first contains an English translation of a classic paper by Ambartsumian, in which he showed for the first time that the evolution of an isolated self-gravitating star cluster under the influence of two-body relaxation does not cease with the establishment of a maxwellian velocity distribution. The second appendix is a translation of that seminal paper by V.A. Antonov which led eventually to our present understanding of the gravothermal instability. We would have liked to include a translation of Michel Hénon's (1961) Ph.D. thesis, which has been much neglected but predicted many of the essential phenomena of cluster evolution after core collapse; we felt, however, that the paper was too long to include in this already lengthy volume. To honor Hénon's fundamental contributions, we have instead reproduced one of the illustrations which appeared in his thesis, side to side with a new calculation (reported in these proceedings by Stodolkiewicz) confirming many of the results obtained by Hénon nearly a quarter century ago. We hope that his thesis will soon appear in translation elsewhere and thereby become accessible to a larger fraction of those working presently on the dynamics of star clusters.

The third appendix contains a list of observed parameters of globular clusters with the original references, compiled by Webbink, and a similar list for open clusters, by van Leeuwen, forms the fourth appendix. These appendices should increase the long-term value of this book sufficiently, even with the inevitable increase of the volume and thereby the price of the present proceedings. A detailed index of names and subjects follows. This index covers all papers and discussions included in these proceedings with the exception of the appendices which are already written in the form of compilations.

We must apologize to the reader for our failure to correct many small errors of grammar and spelling (sometimes even in the spelling of names) in the camera-ready manuscripts; we felt that since none of these errors is likely to cause confusion, our limited resources and the importance of publishing the proceedings quickly required that we let them pass. For errors in the organization of the book and in the indices we are, of course, wholly responsible.

We acknowledge the assistance which we received from many people, without whose help this symposium would not have been possible. We thank John Bahcall, Jeremiah Ostriker and Lyman Spitzer for their enthusiastic support of our early plans to organize this symposium. We thank Harry Woolf and the Institute for Advanced Study for their hospitality. We thank Mary Wisnovsky, the conference coordinator, for her administrative expertise and enthusiastic support. We thank all the participants for their contributions, and especially the panel members who were rounded up at the last moment to give their impromptu comments. We thank Jens Verner Villumsen and Robert Lupton for their efficient and smooth organization of local affairs, scientifically and culturally. We thank Nels Anderson, Web Ewell, Steve Ratcliff and Tom Statler for their untiring operation of audiovisual aids. We thank Donald Lynden-Bell, Stuart Shapiro and Lyman Spitzer for supplying us with copies of originals and translations of the material in the first two appendices. The first appendix has been translated by Ludmilla Wightman, whose cheerful help we happily acknowledge. The second appendix is a revision by Ludmilla Wightman of a translation by the British Admirality Translation Service. We thank Ron Webbink and Floor van Leeuwen for delivering custom-made appendices, number three and four, respectively, on short notice. We thank Grace Rapp, Marge Barbosa and Barbara Pinkham for their indefatigable assistance in the various stages of organizing the symposium and preparing these proceedings. Finally we express our thanks to the California Institute for Technology and especially to the Institute for Advanced Study, for bearing the expenses of both computer time and secretarial assistance needed for the preparation of these proceedings.

The members of the Scientific and Local Organizing Committees are listed below. We are grateful to them for their contribution to the success of this symposium.

Scientific Organizing Committee:

T.A. Agekian, K.C. Freeman, D.C. Heggie, P. Hut (Chairman), S. Inagaki, I.R. King, L.M. Ozernoy, L. Spitzer, J. Stodolkiewicz, R. Wielen.

Local Organizing Committee:

J.N. Bahcall, R. Lupton, J.V. Villumsen (Chairman), M. Wisnovsky.

The Editors,

Jeremy Goodman & Piet Hut

1. Cudworth
2. H. Smith
3. Lake
4. N. Bahcall
5. Jernigan
6. Aarseth
7. Hanes
8. van Leeuwen
9. Freeman
10. Dones
11. Colin
12. Meylan
13. Wielen
14. Casertano
15. van Albada
16. G. Smith
17. Da Costa
18. Mayor
19. Davoust
20. Luwel
21. Severne
22. Dagan
23. Mikkola
24. Inagaki
25. Larson
26. Aurière
27. Semenzato
28. Applegate
29. Statler
30. Kandrup
31. Ipser
32. Lachieze-Rey
33. Webbink
34. Des Forets
35. Harris
36. Seitzer
37. Peterson
38. Adler
39. Jarvis
40. Jog
41. Bettwieser
42. Sygnet
43. Giannone
44. McMillan
45. Margulis
46. Pryor
47. Sugimoto
48. Shara
49. Latham
50. Moffat
51. Stodółkiewicz
52. King
53. Grindlay
54. J. Bahcall
55. White
56. Barnes
57. Godwin
58. Djorgovski
59. Frenk
60. Terlevich
61. Leonard
62. Heggie
63. Ratcliff
64. Cohn
65. Lugger
66. Duncan
67. Lightman
68. Goodman
69. Wandel
70. Tremaine
71. Lupton
72. Spitzer
73. Hut
74. Villumsen
75. Shapiro
76. Mamon
77. Ostriker
78. Radhakrishnan
79. Mathieu

LIST OF PARTICIPANTS

Aarseth, S. J.	Cambridge, England, U.K.
Adler, D. S.	Charlottesville, VA, U.S.A.
Aguilar, L. A.	Berkeley, CA, U.S.A.
Anderson, N.	Princeton, NJ, U.S.A.
Applegate, J.	Pasadena, CA, U.S.A.
Auriere, M.	Observatorie du Pic du Midi, France
Bahcall, J. N.	Princeton, NJ, U.S.A.
Bahcall, N. A.	Baltimore, MD, U.S.A.
Barnes, J.	Berkeley, CA, U.S.A.
Beichman, C.	Pasadena, CA, U.S.A.
Bettwieser, E.	Göttingen, Germany
Casertano, S.	Princeton, NJ, U.S.A.
Cohn, H.	Bloomington, IN, U.S.A.
Colin, J.	Besancon, France
Cudworth, K.	Williams Bay, WI, U.S.A.
Da Costa, G. S.	New Haven, CT, U.S.A.
Dagan, E.	Jerusalem, Israel
Davoust, E	Besançon, France
Des Forets, G.	Paris, France
Djorgovski, S.	Berkeley, CA, U.S.A.
Dones, L.	Berkeley, CA, U.S.A.
Duncan, M. J.	Scarborough, Ontario, Canada
Ewell, M.	Princeton, NJ, U.S.A.
Fall, S. M.	Cambridge, England, U.K.
Freeman, K. C.	Canberra, Australia
Frenk, C. S.	Santa Barbara, CA, U.S.A.
Giannone, G.	Palermo, Italy
Gilden, D.	Princeton, NJ, U.S.A.
Godwin, P.	Edinburgh, Scotland, U.K.
Goodman, J.	Pasadena, CA, U.S.A.
Grindlay, J.	Cambridge, MA, U.S.A.
Gunn, J. E.	Princeton, NJ, U.S.A.
Hanes, D.A.	Kingston, Ontario, Canada
Harris, W. E.	Hamilton, Ontario, Canada
Heggie, D.C.	Edinburgh, Scotland, U.K.

Hut, P.	Princeton, NJ, U.S.A.
Inagaki, S.	Kyoto, Japan
Ipser, J. R.	Gainesville, FL, U.S.A.
Jarvis, B.	La Serena, Chile
Jernigan, J. G.	Berkeley, CA, U.S.A.
Jog, C. J.	Princeton, NJ, U.S.A.
Kandrup, H. E.	Austin, TX, U.S.A.
King, I. R.	Berkeley, CA, U.S.A.
Kylafis, N.	Princeton, NJ, U.S.A.
Lachieze-Rey, M.	Gif-sur-Yvette, France
Lake, G. R.	Murray Hill, NJ, U.S.A.
Larson, R. B.	New Haven, CT, U.S.A.
Latham, D.	Cambridge, MA, U.S.A.
Leonard, P. J. T.	Toronto, Ontario, Canada
Lightman, A. P.	Cambridge, MA, U.S.A.
Lo Cascio, L.	Palermo, Italy
Lugger, P.	Columbia, MO, U.S.A.
Lupton, R.	Princeton, NJ, U.S.A.
Luwel, M.	Brussel, Belgium
Mamon, G.	Princeton, NJ, U.S.A.
Margulis, M.	Tucson, AZ, U.S.A.
Mathieu, R. D.	Cambridge, MA, U.S.A.
Mayor, M.	Sauverny, Switzerland
McMillan, S.	Urbana, IL, U.S.A.
Meylan,G.	Sauverny, Switzerland
Mikkola, S.	Turku, Finland
Moffat, A.	Montreal, Quebec, Canada
Ostriker, J. P.	Princeton, NJ, U.S.A.
Peterson, C. J.	Columbia, MO, U.S.A.
Phinney, E. S.	Princeton, NJ, U.S.A.
Pryor, C. P.	Victoria, BC, Canada
Radhakrishnan, V.	Bangalore, India
Ratcliff, S.	Princeton, NJ, U.S.A.
Ratnatunga, K.	Princeton, NJ, U.S.A.
Schommer, R. A.	New Brunswick, NJ, U.S.A.
Seitzer, P.	Tucson, AZ, U.S.A.
Semenzato, R.	Padova, Italy
Severne, G.	Brussel, Belgium
Shapiro, S. L.	Ithaca, NY, U.S.A.
Shara, M.	Baltimore, MD, U.S.A.

Smith, G.	Cambridge, MA, U.S.A.
Smith, Jr., H.	Gainesville, FL, U.S.A.
Spitzer, L.	Princeton, NJ, U.S.A.
Statler, T.	Princeton, NJ, U.S.A.
Stodólkiewicz, J.	Warszawa, Poland
Sugimoto, D.	Tokyo, Japan
Sygnet, J. F.	Paris, France
Terlevich, E.	Hailsham, East Sussex, England, U.K.
Toomre, A.	Cambridge, MA, U.S.A.
Tremaine, S. D.	Cambridge, MA, U.S.A.
van Albada, T.	Groningen, The Netherlands
van Leeuwen, F.	Hailsham, East Sussex, England, U.K.
Verter, F.	Troy, NY, U.S.A.
Villumsen, J. V.	Princeton, NJ, U.S.A.
Wandel, A.	Princeton, NJ, U.S.A.
Webbink, R. F.	Urbana, IL, U.S.A.
White, S.	Santa Barbara, CA, U.S.A.
Wielen, R.	Berlin, Germany

OBSERVED SURFACE DENSITIES IN GLOBULAR CLUSTERS

Ivan R. King
Astronomy Department
University of California, Berkeley

ABSTRACT. This review of observations is oriented toward the problem of core collapse in globular clusters. After a brief discussion of methods of determining surface densities, recent results are cited which show that the central brightness peak seen in M15 also appears in several other clusters. The central peaks of brightness are interpreted as the result of core collapse, which theories have repeatedly predicted. If binaries stabilize the collapsed core, the cluster should follow a Hénon model. In two of the three cases tested this model is a reasonable fit. Remaining theoretical questions are listed, as are observational needs.

1. INTRODUCTION

I am glad to be able to open this Symposium by presenting globular clusters from an observational point of view. There are several reasons for saying this. First, the symposium is indeed devoted largely to globular clusters rather than open clusters, because the problems of the globulars are more straightforward and well defined. Second, in a conference where theoretical talks outnumber observational ones by three to one, it is a good thing to be able to start by emphasizing the essential role that observations play.

Not that I wish to set observations against theory; they are two parts of the same story, and neither is the center. When Robert Benchley, the humorist, was an undergraduate at Harvard, he was faced with an exam question requiring a discussion of a U.S.-Canadian fisheries dispute, as seen from either point of view. Benchley chose the point of view of the fish. In this vein, but more seriously, I ask you to consider the point of view of the globular clusters. Our central effort should be to understand what their reality is. The name for this task is interpretation, and both theory and observation merely offer tools for it. Computational experts, too, should recognize that the heavens have more than a hundred analog computers that always get the right answer, by definition. Observation tries to read those answers,

J. Goodman and P. Hut (eds.), Dynamics of Star Clusters, 1–17.

while theory tells us what to look for. The interaction of the two shows us what the observations mean, and what the theory should interpret.

Of the questions concerning globular clusters, the one that I think will receive the most attention at this meeting is core collapse. Instead of introducing it directly, however, I would rather give a little background, in order to place the core-collapse question in its setting.

The basic dynamical equilibrium of globular clusters is fairly well understood. (For a summary, see King 1981.) Relaxation, through stellar encounters, drives the velocity distribution as close to Gaussian as it can get, given the existence of a finite escape velocity. In a steady, self-gravitating system this velocity distribution implies a corresponding density distribution, which can be found from a straightforward algorithm. Except for scale factors, the only free parameter is the tightness of binding of the cluster, relative to the galactic tidal field. The cluster models calculated in this way form a one-parameter family (Fig. 1), and they fit actual clusters (as, for example, in Fig. 2). This picture is a satisfying one, because given their different binding energies, clusters are as simple as they can possibly be.

Unfortunately the picture that I have just sketched is badly oversimplified. Difficulties arise, it is true, from anisotropic velocity distributions and from the stratification of stars of different mass; but by far the worst problem is the dynamical evolution of star clusters, which turns out to be unstable. This phenomenon is discussed lucidly by Spitzer in this volume; here I will merely note two ways of looking at the problem. One is that relaxation ejects stars and causes the cluster to contract (since its binding energy is shared among fewer stars); the contraction speeds up the relaxation, and the whole thing runs away. A second point of view is that beyond a certain central concentration the core of the cluster can no longer stay in equilibrium with the envelope. It makes a thermodynamic turnabout, energy flows the wrong way, and the center of the cluster collapses. I will not try to reconcile these two points of view, or even to discuss whether they are in some way equivalent. Either way, we should expect the core of a cluster eventually to collapse. This really should happen, in actual clusters. The central concentrations of many clusters are on the verge of instability, and the relaxation times in these and other clusters are short enough to drive them over the brink before long -- so much so that we wonder why they have not collapsed long since.

Much of this volume is devoted to description and discussion of the theories and calculations that predict core collapse. Here I will merely reproduce a single example (Fig. 3), which is typical of the Monte Carlo calculations of a decade ago. Since the seminal work of Hénon (1961) and Lynden-Bell and Wood (1968), theoreticians have had little doubt about core collapse.

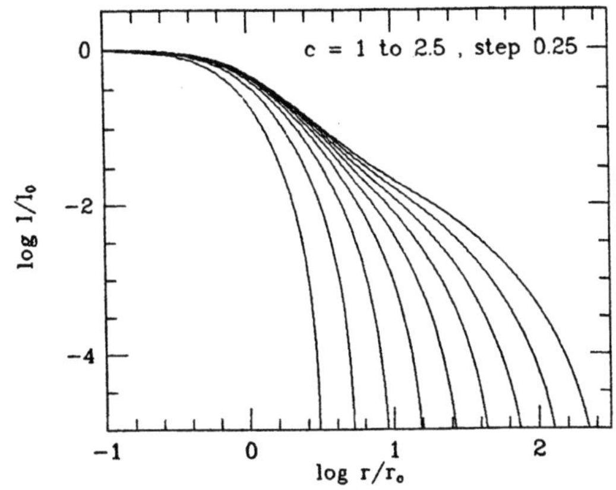

Fig. 1. Projected densities for a family of models based on a steady-state relaxing velocity distribution (King 1966a).

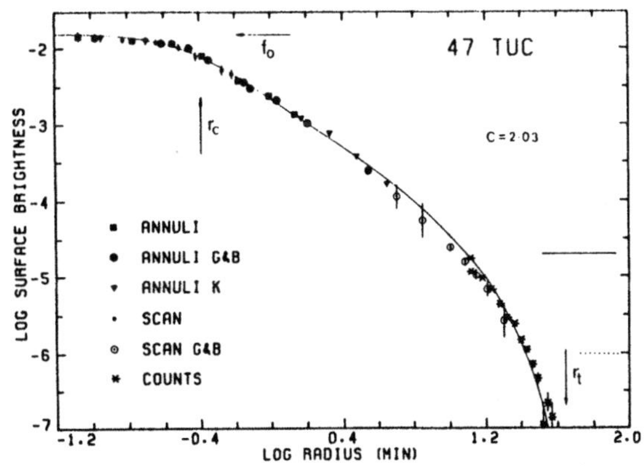

Fig. 2. Surface densities in 47 Tucanae, fitted to a model from the family shown in Fig. 1 (Illingworth and Illingworth 197).

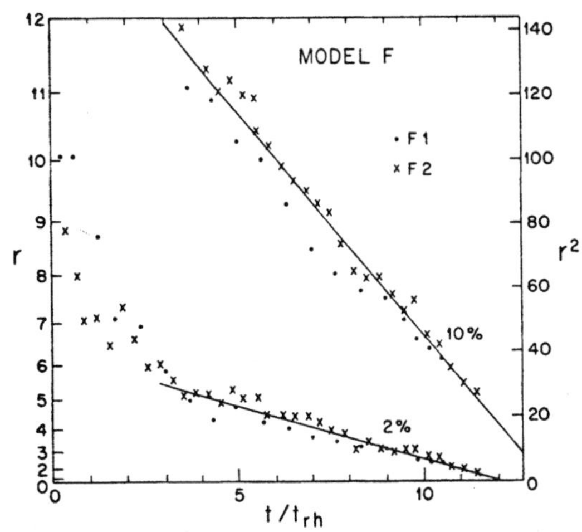

Fig. 3. Final contraction of cluster core in a Monte Carlo calculation (Spitzer 1976). The two curves show the radii containing 2% and 10% of the mass, respectively.

Observationally, however, the situation has been less clear. It appeared that nearly all clusters lived on the tranquil sequence of models that I have described, without any inkling of the doom that lurked in the theoretical crystal ball. Only M15 refused to behave simply; its brightness rises continuously in the center, instead of flattening into a typical cluster core. This anomaly has been commented on frequently; suffice it to reproduce some data of 18 years ago (Fig. 4):

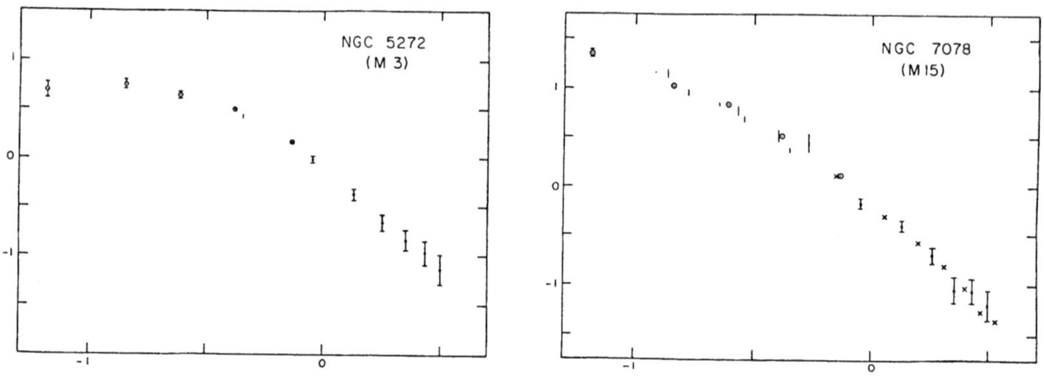

Fig. 4. Photoelectric surface brightnesses in M3 and M15 (King 1966b).

Is this core collapse? It has the right appearance indeed, but why did this type of profile seem to be confined to this one cluster? The answer has turned out to be that there are indeed other such clusters but that the observations had been inadequate to detect them -- or else marginally sufficient observations had been inadequately interpreted. It was not the crystal ball that was cloudy, but the telescope -- or at least those who were looking through it.

2. OBSERVATIONAL DATA

2.1 Methods of Observation

To explain the observational situation I shall have to fill in a little background about the observation of density distributions in globular clusters. Again I refer to a more detailed discussion elsewhere (King 1980) and will merely mention the factors that are directly relevant here.

Even in globular clusters the limitation in density studies is generally the small-number statistics of the stars, when they are taken region by region. The smoothest density studies involve counts of the numerous faint stars. In the central regions, however, the faint stars are too crowded, and only the less numerous bright stars can be counted. Close to the center of a dense cluster, even the bright stars are too crowded to be counted reliably. The images of many of them can be pried apart by heroic efforts (see, for example, Aurière and Cordoni 1981), but this will yield only colors and magnitudes of individual stars. Densities cannot be determined reliably in this way, because the ability to separate stars depends on the density itself.

In the central regions of a dense cluster, densities can be measured only by surface photometry -- measuring the amount of integrated starlight per unit area. The traditional method of doing this has been by photoelectric photometry. The procedure is indirect. It would be of little use to trail a measuring aperture across a cluster, because the light in a small aperture tends to be dominated by so few individual stars that it is statistically quite unstable. Far better is to measure through a set of apertures of graduated size, all centered on the cluster center, and to difference successive apertures to get the total amount of light in each annulus. It does not matter that this procedure violates the stricture against differencing data; the observational errors are of little consequence compared with the statistical uncertainty in the amount of light in the small areas involved.

Photoelectric observations of this sort have two drawbacks, however. First, the observer usually centers on the cluster by a somewhat coarse eye-estimate. Second, and more fundamental, photoelectric observations are unable to work accurately with apertures smaller than 5 or 10 seconds in diameter, because the seeing is time-variable and too much light flickers in and out of the edges of such a small area. The next observation has a different sequence of jiggles, and the photometric accuracy of a sequence of observations is poor.

For the present problem high angular resolution is important. At the distances of even nearby globular clusters, the phenomena that we are discussing subtend only a few seconds of arc. Thus we need good telescope scale, reasonably good seeing, and the ability to make effective use of them. The way to achieve accurate high-resolution surface photometry is to make measurements on a recorded image, in which all parts have been observed simultaneously over the same interval of time. Seeing then blurs the image, but it does not introduce photometric distortions.

In principle, any imaging device will do, as long as it is capable of reliable photometry. CCD's are in many respects ideal, because of their speed, linearity, and dynamic range. Traditional photography can also be used reliably, however; and it actually offers some advantages, particularly in ultraviolet observations, where most CCD's are rather insensitive.

Why should one prefer the ultraviolet? The answer is the statistical fluctuations in the integrated surface brightness, which is dominated very much by a small number of bright stars. At longer wavelengths the major contributors are the few red giants, and the brightness distributions are quite ragged. As we go to shorter wavelengths, however, the importance of the red giants is reduced; more individual stars contribute effectively to the light, and the accuracy improves. The nature of the improvement is clear if we examine HR diagrams of the same globular cluster in V and in U (Fig. 5). Moving to the ultraviolet has flattened out the red-giant branch, so that a much larger number of individual stars makes effective contributions to the light. Also, clusters that have a strong horizontal branch (like M3, for which these diagrams were made) have an even larger number of effective contributors.

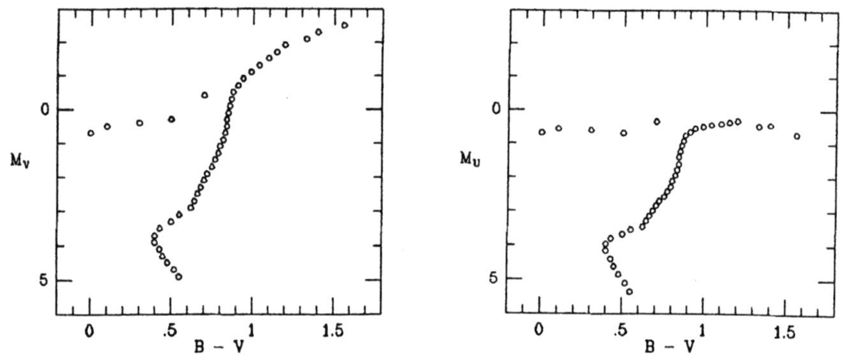

Fig. 5. Schematic HR diagrams for M3 in V and in U.

2.2 Recent Results

The foregoing reasoning was the starting point of a recently published study (Djorgovski and King 1984) of the central regions of nine high-concentration southern globulars. U-sensitive photographs were taken at large scale, so that photographic grain had no effect and small areas could be measured well. Sensitometer spots provided reliable calibration. PDS tracings at high resolution were converted to intensity, point by point, and the light was added up in each annulus, just as a photoelectric photometer would do, if it could work with such a high resolution.

The improvement in resolving power is dramatic. Figure 6 shows, for one cluster that turned out to be particularly interesting, the previously available density data, from star counts, and the region to which our new observations give access.

The results were also exciting (see Figure 7). Of the nine clusters, three clearly show the M15 effect: NGC 6624, 6681, and 7099. To emphasize the difference between these and normal clusters, Figure 8 shows a fitting of models from Fig. 1 to a normal cluster and to a condensed-core cluster. In NGC 5824 the fit is reasonably good, but in NGC 6624 it is hopeless; the brightness never flattens into a normal core but continues to rise toward the center until it is lost in the

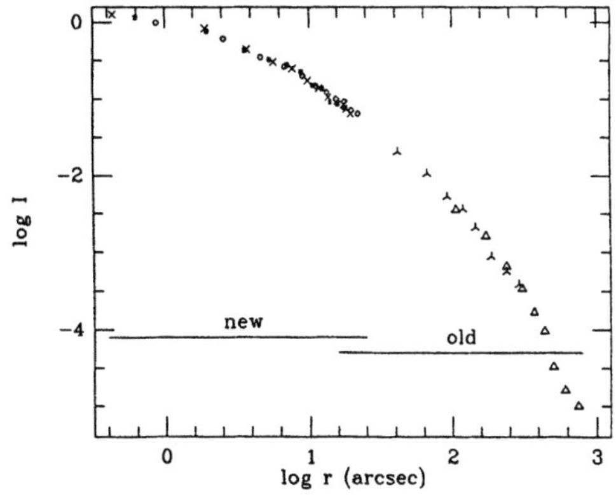

Fig. 6. Surface densities in NGC 7099, showing the radial range covered by previous and by new observations.

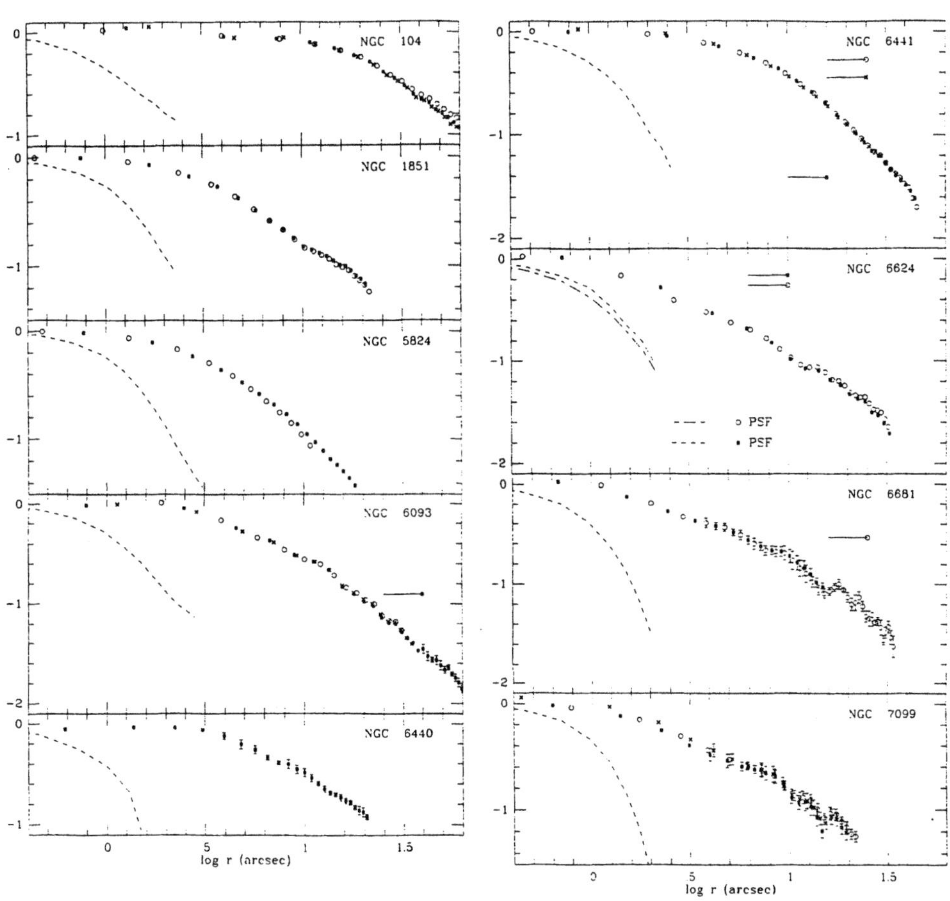

Fig. 7. Surface densities in 9 high-central-density southern globular clusters (Djorgovski and King 1984).

seeing disc. Instead, it seems to fit much better to a −1 power of radius, which agrees with a singular isothermal sphere in projection. (I will return to this question in greater detail later.)

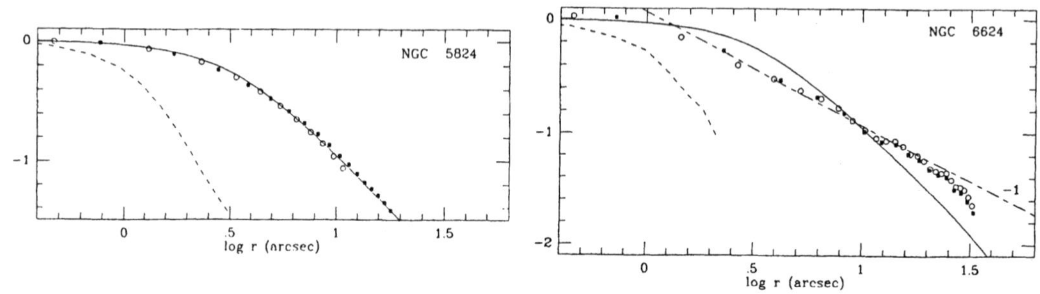

Fig. 3. Normal-cluster models fitted to the normal cluster NGC 5824 and to the abnormal cluster NGC 6624.

Another provocative phenomenon in Fig. 7 is the indication of bumps in the brightness curves. These may or may not be real. In NGC 6681 the bump is almost certainly not real; by going back to the original images I have been able to pick out the individual bright stars that are responsible for the prominent bump at ordinate 1.2. In NGC 1851 the bump at 1.2 (or dip at 1.0) might conceivably be real -- but personally I doubt it.

To give a better feel for the roughness or smoothness of the images, I have reproduced in Figure 9 the central contours of a very smooth cluster and a moderately ragged one.

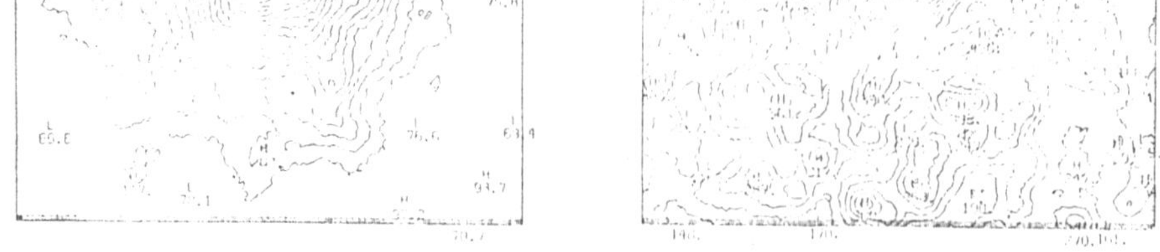

Fig. 9. Contours of the central regions of UV images of NGC 5824 and 6093.

3. COMPARISON WITH THEORY

3.1 Immediate Conclusions

Including M15, we now have four clear cases of what I will hereafter refer to as collapsed cores. The first question to ask is, what other characteristics do they have to distinguish them from normal clusters? Unfortunately there is no simple answer. One characteristic that they all share is high central density. That may indeed be physically significant, but the fact is that the clusters in this study were chosen largely for that characteristic in the first place. They also have a short relaxation time, but they were chosen for that characteristic too. And in neither case is the correlation with collapsed cores good, even within this selected sample, as is easily seen from the values of central densities and relaxation times given by Peterson and King (1975). I will return later to the implications of this, but for the time being let us simply note that the correlation is poor.

A very interesting question is whether the presence of a collapsed core correlates with the emission of X-rays, as might have been suspected from the existence of an X-ray source in M15. The answer is clearly negative. NGC 6624 is indeed an X-ray source, but NGC 6681 and 7099 are not, and the well-known X-ray sources NGC 6440 and 6441 have normal cores. X-rays from globular clusters are spectacular fireworks, but they are no more responsible for collapsed cores than ordinary fireworks are responsible for the holiday that they celebrate. The connection might in fact go in the opposite direction; a collapsed core may well provide dynamical surroundings that are particularly conducive to the formation of the close binaries that are believed to be responsible for X-rays in globular clusters.

3.2 Tentative Theoretical Picture

It may be, however, that a coherent picture is now beginning to emerge from the observations and the theory. Evolution, due to stellar encounters, causes the cores of some clusters to undergo a runaway collapse. In its final stages the collapse leads to a spatial density law of −2.2 power (−1.2 power in projection), but this phase is so brief in time that it is unlikely to be observed. The collapse is finally halted by the formation of one or more binaries, a phenomenon that becomes more and more inevitable as the density increases and the number of stars in the central peak decreases. The binaries stabilize the core, supplying energy at the same rate at which relaxation removes it. Relaxation keeps the stabilized core isothermal, with a velocity dispersion that is the same as that in the surrounding part of the envelope. Small changes occur as binaries are ejected and new binaries form, but for the most part the cluster remains for a long time in this state, with a core profile that is for all practical observational purposes a singular isothermal sphere.

A very similar model was predicted in detail in Hénon's thesis (Hénon 1961). He imposed the condition that a cluster remain self-similar while relaxation takes place at its appropriate rate at all points. Since, except for self-similarity, his assumptions were nearly the same as mine (King 1966a), it is not surprising that his model is very close to the limit of my model sequence at high central concentration. This is effectively a singular isothermal model with a tidal cutoff. There are small physical differences, which lead to density differences that are less than the thickness of the line in a graph; but in what follows I shall use the high-concentration limit of the so-called King models, and shall refer to it as a Hénon model.

3.3 Comparison with Observation

Thus, according to the tentative theoretical interpretation that I have sketched, a post-collapse cluster should look like a Hénon model. What do the observations say about this? The answer is very incomplete, but I have been able to assemble some data for three of the collapsed-core clusters. In Figure 10 is shown a fit of NGC 6624 to a Hénon model. The data are taken from Djorgovski and King (1984), from photoelectric photometry by van den Bergh (1977) and by Canizares et al. (1978), and from star counts by Bahcall and Hausman (1977). The fit is reasonably good, except for a central depression of the observations that can clearly be attributed to the seeing-limited resolution. In Figure 11 is shown a similar fit for NGC 7099, where the outer data are taken from star counts by King et al. (1968). Again the fit is reasonable.

For M15 I have combined the central photoelectric photometry shown in Fig. 4 (King 1966b) with star counts in the envelope (King et al. 1968). The result is shown, fitted to a Hénon model, in Figure 12. (Better data are available in the center [Newell and O'Neil 1978], but I have not yet had a chance to plot them on the necessary scale.) The fit is not very good, especially in the envelope; I would find this discouraging were it not for the better fit in the other two cases. The difficulty in M15 may be due to the fact that the outermost densities come from star counts of low-luminosity stars ($M_V \sim +8$), which should have a distribution different from that of the giant stars on which the central part of the curve depends. Indeed, the deviation is in this sense -- a flatter distribution for the stars of the outer envelope.

For NGC 6681 I have unfortunately been unable to find any density data other than those in Fig. 7.

This is all that the observations can say at the present time about my theoretical surmises. It appears that the interpretation that I have suggested is by no means excluded, but it certainly has not yet received a ringing endorsement.

Fig. 10. Fit of a Hénon model to combined surface densities in NGC 6624.

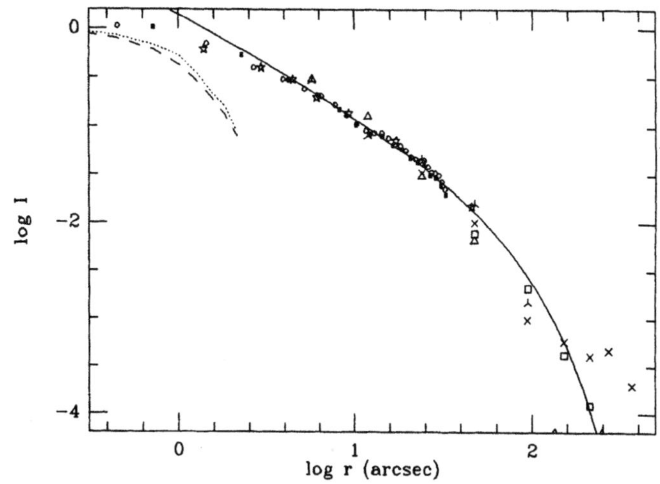

Fig. 11. Fit of a Hénon model to combined surface densities in NGC 7099.

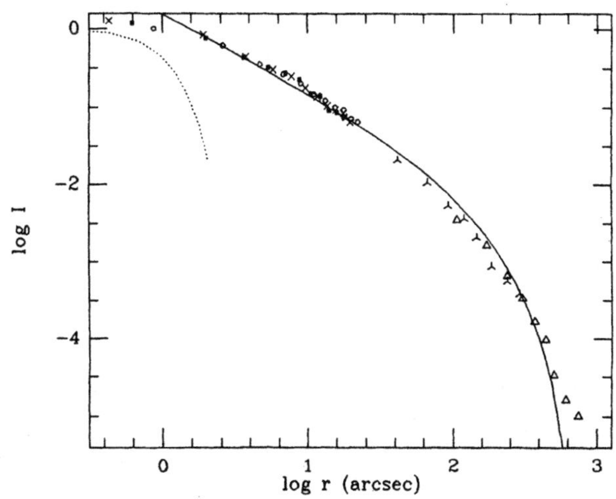

Fig. 12. Fit of a Hénon model to combined surface densities in M15.

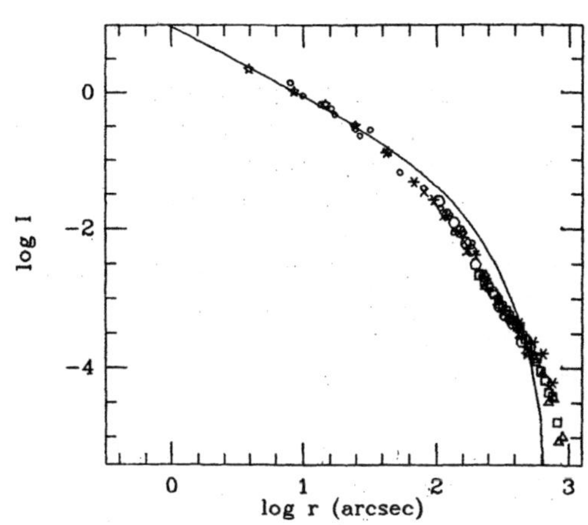

3.4 Frequency of Collapsed Cores

Besides the radial profile in an individual cluster, there is another way in which the theory can be confronted with observation. How many collapsed-core clusters are there? And are they the right ones -- that is, are they the ones that theory predicts will go the fastest? Here we run into a real contradiction: there are not enough collapsed-core clusters. In a survey that is reported briefly at this symposium, Djorgovski and Penner have looked at the centers of more than 30 additional clusters, including all those north of declination -20 degrees that are worth looking at, and they have turned up only two more probable cases and two others that are "possible."

Relaxation times of only two of the collapsed-core clusters are given by Peterson and King (1975). The values at the centers are 121 Myr for M15 and 15 Myr for NGC 7099. Their Table IV includes 16 other clusters with central relaxation times less than 100 Myr, and many of these have already been observed, with negative results. To be sure, the relaxation times given by Peterson and King have considerable uncertainties, especially for collapsed-core clusters, which do not fit the models that they used; but the numbers must mean something. Even worse, all relaxation times in this range imply a collapse time that is less than a Hubble time. Even though such a prediction refers to the future rather than to the present state of the clusters, it seems highly improbable that so many clusters should now be close to disaster while so few have yet reached it.

This anomaly has frequently been noted, and attempts have been made to interpret it. Lightman, Press, and Odenwald (1978) have suggested that core collapse is a disaster indeed, so much so that the cluster is destroyed by it; they assert that our remaining globular clusters are only the fortunate few that have been able to avoid core collapse until now. But this seems a most improbable picture, even aside from my previous suggestion that we do see some real post-collapse clusters. What seems fatal to their picture is that Lightman et al. fail to indicate how the collapse of a small core could cause the dissipation of the much more massive envelope of a cluster.

Another interpretation is given by Cohn and Hut (1984). They suggest that many clusters have already been through collapse and that they are simply mixed into the general distribution of present-day relaxation times. But this picture also seems implausible to me, because all the post-collapse clusters should show the characteristic profile that I have just claimed is so rare. A way out would be that post-collapse cores return, in a reasonably short time, to their previous undisturbed profile; but from what I understand of the theory, this is unlikely.

But the worst anomaly, in my view, is not in the globular clusters but in the open clusters, nearly all of which have relaxation times short enough to bring on core collapse in much less than a lifetime. As

Wielen will explain later in this symposium, some of them are saved from collapse by expansive effects that come from outside, but many others are not. Observations, as far as they go, have not shown any indication of collapsed cores in open clusters, however. Perhaps they are all saved by the formation of binaries, which is much easier when the number of stars is so much smaller; but until this is demonstrated I will regard the absence of collapsed cores in open clusters as an outstanding anomaly -- or else as an indication that post-collapse cores really do expand back to normal again.

Before leaving the comparison with theory, I should acknowledge not having considered any black-hole theories. These have their basis in the idea that core collapse might lead to a coalescence of material into a black hole, which thereafter dominates the central dynamics of the cluster. Black-hole theories, which will be reviewed later in this symposium by Shapiro, have had rather little attention in recent years. This is due largely, I believe, to the demonstration (Grindlay *et al.* 1984) that the X-ray sources in globular clusters are much more likely to be close binary stars. But some of the current disfavor of black-hole theories must certainly be due to their inherent improbability. If this remark shows prejudice, I submit to you that it is less prejudiced than the wish to see black holes everywhere. Such a view seems to me like the reasoning of Aquinas, who began by assuming the existence of God and then assiduously sought logical proofs of it. If black holes are to be supported by scholastic argument, then I would cite against them that other stalwart medieval, William of Ockham, whose razor excises all that is not necessary. At the same time, I must acknowledge that the quest for black holes did make the positive contribution, in the hands of Lightman, Shapiro, Bahcall, and others, of promoting interest in core collapse theory at a time when many of us observers were saying that it wasn't true.

4. OUTSTANDING PROBLEMS, AND FUTURE NEEDS

4.1 Theoretical Questions

Where does all of this leave us? As an observer, I would ask the theoreticians a number of pressing questions. What is the profile of a post-collapse cluster core (not during the all-too-rapid stages of the collapse itself, but a long time afterward)? Does it retain some characteristic shape by which it can be recognized? Is the profile that of a singular isothermal sphere? Or will a collapsed core re-expand so that it shows no signs of a past collapse? Should we expect intermediate cases between collapsed cores and normal ones? For the sake of future higher-resolution observations with Space Telescope, how small should the core radius of a collapsed cluster be, after binaries stabilize it (if indeed they do)? How should the profile of a collapsed core tie onto that of its presumably undisturbed envelope? Should a collapsed core show a radial variation of velocity dispersion? Will core collapse produce a mass-segregation that might have observable consequences? What does theory predict for open clusters, and what should we look for?

With these questions I wish also to plead for one condition on the answers: give them to us in the observational domain. We know that the real universe has three dimensions, but please give us results projected into two dimensions, on the plane of the sky. You can do this easily while making the theoretical calculations, while it is hard for us to do it, especially if we have to do it by starting with astrometry on published graphs.

4.2 Observations Needed

It is of course easy to make demands on others, but what should the observers be doing meanwhile? Here also, there are some clear needs. The detailed surface photometry of cluster cores has only begun. The Djorgovski-Penner survey is only a preliminary one; it needs particularly to be followed up with a study of the smoother ultraviolet images of the interesting clusters. In addition, the southern clusters still need to be looked at. Fortunately, this work is already under way. Grindlay, Cohn, and Lugger have Cerro Tololo observing time in which they will work hard to get UV CCD images of a number of southern clusters. Djorgovski and I also have observing time, unfortunately with a CCD that does not work effectively in the UV. But with our shorter exposures we can survey a more complete list of clusters.

When a cluster is found to have an unusual core, we often lack adequate data for its envelope. Surface photometry needs to be extended farther out; and, more important, star counts to faint magnitudes are still lacking for a number of interesting clusters. Many of these are in low galactic latitude, with sometimes-appalling densities of field stars superimposed. To remove the field stars selectively, one should probably place each individual star in a color-magnitude array, so as to select cluster stars by the criterion of their having the right color for the main sequence.

When available, Space Telescope observations will be invaluable, because only then will we be able to see farther into the centers of the clusters (with about 20 times ground-based resolution). Not only will we see better central density profiles; for the first time we will have really good data on the relative distribution of stars of different mass. But ST time is very scarce; we must know beforehand which clusters are most profitable to look at.

In addition to the surface densities, we need to look much more carefully at the velocity dispersions. There are reports at this symposium by Da Costa and Freeman and by Mayor and Meylan, but they merely whet our appetites. Surface densities sample two of the six dimensions of phase space; but the addition of radial velocities will give us a third component, and the confrontation of theory with observation will be that much more effective.

5. CONCLUSIONS

It appears now that globular clusters are more complicated than the previous simplified picture had suggested. The phenomenon of a central brightness peak, previously known only in M15, occurs in a number of other clusters, in very similar form. Existing observations, which are still far from adequate, are consistent with a picture of post-collapse cores that are singular-isothermal, perhaps with binaries stabilizing against further collapse. The profiles of such clusters can probably be fitted by a Hénon model, which for them is the analog of the King models for normal clusters. A serious problem, however, is why more clusters do not show these collapsed cores. Both for theory and for observation, core collapse is a challenge for the future.

REFERENCES

Aurière, M., and Cordoni, J.-P. 1981, Astron. and Ap. Suppl. 46, 347.
Bahcall, N. A., and Hausman, M. A. 1977, Astrophys. J. 213, 93.
Canizares, C. R., Grindlay, J. E., Hiltner, W. A., Liller, W., and McClintock, J. E. 1978, Astrophys. J. 224, 39.
Cohn, H., and Hut, P. 1984, Astrophys. J. Letters 277, L45.
Djorgovski, S., and King, I. R. 1984, Astrophys. J. Letters 277, L49.
Grindlay, J. E., Hertz, P., Steiner, J. E., Murray, S. S., and Lightman, A. P. 1984, Astrophys. J. Letters, 282, L13.
Hénon, M. 1961, Ann. d'Astrophys. 24, 369.
Illingworth, G., and Illingworth, W. 1976, Astrophys. J. Suppl. 30, 227.
King, I. R. 1966a, Astron. J. 71, 64.
King, I. R. 1966b, Astron. J. 71, 276.
King, I. R. 1980, in Globular Clusters, ed. Hanes, D. and Madore, B. (Cambridge Univ. Press), p. 249.
King, I. R. 1981, Quart. J. Roy. Astron. Soc. 22, 227.
King, I. R., Hedemann, E., Hodge, S. M., and White, R. E. 1968, Astron. J. 73, 456.
Lightman, A. P., Press, W. H., and Odenwald, S. 1978, Astrophys. J. 219, 629.
Lynden-Bell, D., and Wood, R. 1968, Mon. Not. Roy. Astron. Soc. 138, 495.
Newell, E. B., and O'Neil, E. J. 1978, Astrophys. J. Suppl. 37, 27.
Peterson, C. J., and King, I. R. 1975, Astron. J. 80, 427.
Spitzer, L. 1975, in Dynamics of Stellar Systems, ed. Hayli, A. (Dordrecht: Reidel), p. 3.
van den Bergh, S. 1977, Astron. J. 82, 796.

DISCUSSION

COHN: I address two questions raised by King: (1) what should a cluster look like after core collapse? (2) How long after core collapse should it look "funny?" I noted that calculations of cluster evolution including binaries that I've carried out with Hut and Goodman indicate that a central cusp with slope close to a singular isothermal forms during core collapse and persists long after core collapse. These calculations predict the evolution of the detailed density and velocity dispersion profiles.

OSTRIKER: The inner parts of M15 appear to be more centrally concentrated than the singular isothermal ($I \propto r^{-1}$) model. What is your best estimate for the logarithmic slope of the surface brightness in the inner parts?

KING: I don't know why you call it more concentrated, but I'll try to answer. The logarithmic slope is approximately -1, but I don't think that these observations could distinguish that from -0.75 or -1.25, for example. Nor is the slope constant; seeing flattens the center, and the slope gets steeper in the envelope.

LACHIEZE-REY: Three clusters out of nine in your sample show a possibility of core collapse. Do you expect this proportion to be universal (for all globular clusters) or was your sample selected to exhibit specially this effect?

KING: My clusters were carefully chosen to be those that were most likely to have this phenomenon. It is likely that rather few other clusters will show it.

BAHCALL, N.A.: Why does the M15 surface brightness profile not show the flattening due to seeing as seen in NGC 6624 and 7099? Has it been taken out?

KING: I didn't show the center of M15 in as much detail. If I had, it would have looked similar to the other two.

TREMAINE: A few years ago, singular density profiles were usually interpreted as evidence for massive black holes in the cluster centers. Is there any direct observational evidence that favors core collapse over balck holes as the explanation?

KING: No, I don't know of direct observational evidence; but the two hypotheses would presumably predict different profiles, which might be distinguishable. Against central black holes I would cite (1) Ockham's razor, (2) the fact that a black hole should have infalling material that emits X-rays - which some core-peak clusters don't, and (3) Grindlay's observation that X-ray sources are not at the centers of their clusters. But I am not asserting that the density peak is due to core collapse - I'm asking the theoreticians to tell us.

SUGIMOTO: How many stars do you think are contained within the smallest collapsed core that is recognizable by present-day ground based observations?

KING: Assuming some very rough typical numbers, the answer for a 1-arcsec seeing limitation would be about 100 stars.

INAGAKI: Some clusters have bumps in their luminosity profiles. Is there some evidence for mass segregation in such clusters?

KING: It is too hard to see mass segregation in the central regions of clusters; especially in these high-density clusters, the crowding is hopeless. This is one area in which Space Telescope will make a large breakthrough, however.

RADIAL VELOCITIES AND PROPER MOTIONS OF GLOBULAR CLUSTER STARS

Robert Lupton
James E. Gunn
Princeton University Observatory, Princeton
Roger F. Griffin
Institute of Astronomy, Cambridge

ABSTRACT
We describe techniques and results for obtaining radial velocities and proper motions for individual stars in globular clusters. One set of these results for the nearby northern clusters M92 and M13 are discussed in terms of new dynamical models and distance estimators.

1. Introduction

In order to make realistic models of globular clusters, one needs to know something about both the stellar velocities and the light profile. Most earlier studies of clusters have only used the photometry and in a few cases, central velocity dispersions (see, for example, Illingworth 1976) derived from the width of the lines in integrated spectra either by Fourier techniques or direct least-squares fitting in wavelength space. Gunn and Griffin (1979, hereinafter GG) first made dyamical models of a cluster (M3) using radial velocity data for stars throughout the cluster. The fact that those data badly disagreed with the most sophisticated model constructed till that time (Da Costa and Freeman 1976) and demanded both significant anisotropy in the outer parts of the cluster and a significantly steeper mass function (all within the framework of the King (1966)-Mitchie (1963) models, about which we will say more later) says quite certainly that without such data one cannot even approximately understand the dynamics of globular clusters. It is clear that if one had in addition proper motions, essentially all of the ambiguity of reconstructing the three-dimensional velocity distribution from the observations would disappear, and one would have in addition a very powerful handle on the distance. The proper motions available are not quite sufficiently accurate to allow this to be done straightforwardly, as we shall see, but some considerable leverage on the models and quite good distance estimates can still be obtained.

2. Radial velocities.

Radial velocities of stars at these brightness levels ($B < 15$, say) and to the requisite accuracy (about 1 km/sec) are now being obtained routinely by two techniques which yield comparable efficiencies. The

J. Goodman and P. Hut (eds.), Dynamics of Star Clusters, 19–31.

original Griffin cross-correlation technique, applied at the coude of the 200-inch Hale telescope (Griffin and Gunn 1974) has been used in all our work. This technique involves the use of an infinite-contrast mask at the focal plane of a high-dispersion spectrograph, the mask having transparent slots which correspond in location to absorption lines in the stellar spectrum. The mask is scanned across the spectrum (or, in our and other cases, the spectrum across the mask) and the total flux through the mask recorded. The position of the mask which corresponds to a minimum in the flux represents the best agreement between the mask and spectrum and directly yields a radial velocity relative to those of a set of standard stars. In practice the scanning is done rapidly and periodically and the scans accumulated in a real or software-simulated multichannel analyzer; the position of the minimum is found in later analysis. Possible problems resulting from spectral-type mismatch between the template star and the object observed, the fact that the shift to match is constant in wavelength and the doppler shift proportional to the wavelength, zero-point and scale drifts, are all unimportant in practice in properly designed equipment, and are discussed extensively in Griffin and Gunn (1974). This technique is some thousands of times faster than photographic spectroscopy, and probably yields more accurate results, if for no other reason than that the zero point of the instrument can be monitored on essentially arbitrarily short time scales. In a modification incorporating an echelle as the dispersing element, Mayor (see Mayor, et al. 1983) in the Coravel machine has solved at least the problem of the shift mismatch and has gained the high luminosity efficiency of echelle spectrographs. This instrument has been used to obtain high-quality velocities in 47 Tuc, as reported by Da Costa and Freeman in this volume. The advantages of the technique are speed and stability; the photomultiplier is used only as a flux collector and even large magnetic perturbations to its electron optics have negligible effects on the results. The disadvantage is that time spent scanning away from the cross-correlation dip is time wasted; this can be minimized to some extent by clever control of the scanning, and this is done to advantage in Coravel. It need not be a major loss for the globular cluster problem, since the line widths in the stars alone demand instrumental profiles some twenty km/s wide and the velocity dispersions are smaller than this--thus not much "continuum" need be scanned once the cluster velocity is found. One looks fondly today at CCD quantum efficiencies, but the rapid time response required and the large photocathode areas required by the realities of optics make their use in this configuration impossible.

It would seem that it would be much more efficient to record all the spectrum at once and do the analysis later in the computer. Since there are many lines, one need obtain only very poor signal-to-noise in the spectrum to obtain very good velocities. The requirement is clearly for a detector with no noise of its own, high quantum efficiency, and excellent dimensional stability. The first requirement today means very high-gain intensification, whatever the ultimate detector. This approach was used by the Stromlo group with an intensified SEC vidicon to obtain a few velocities in NGC6397 (Da Costa et al. 1977). It has since been applied by the Latham and his collaborators at Harvard to a large number of problems, most notable in this context to a followup of Gunn and Griffin's velocity measurements in M3. The original measurements strongly suggested that the incidence of spectroscopic binaries in M3 was very small compared to the field, but the statistics were limited and the result has been called into question by Harris and McClure (1983). Latham and collaborators (1983,1984) have measured all the stars in the GG sample again, and have in fact found one binary; this confirms the greatly reduced numbers but is at clear variance with GG's suggestion that there may be none. Latham et al. have used an echelle spectrograph with a pulse-counting intensified Reticon detector of the Shectograph (Shectman and Hiltner 1974) type. The efficiency, as expected, is somewhat higher than that of the cross-correlation technique, but it is not clear whether the ultimate accuracy is as high. The latter question is irrelevant for globular cluster work in any case, since the errors are small compared to the cluster velocity dispersions.

The ideal machine for this problem has not been built yet but will doubtless be soon. An echelle spectrograph with a resultion of about 0.3 Angstroms with coatings for high throughput in the 4000-6000 A range with a low-noise violet-sensitive CCD as a detector would be at least an order of magnitude more sensitive than any instrument now being applied to the problem. It should not be difficult to build an instrument of this sort with 30 percent quantum efficiency on the sky, and with some effort even higher efficiencies might be achieved. At 15th magnitude, such a spectrograph would yield 20 electrons per resolution element per second; with a thousand stellar absorption lines of about 20 percent average depth at this resolution, one km/sec accuracy would be obtained in of the order of ten seconds, longer for very metal-weak objects just inversely as the line strengths. A CCD with very good low-level transfer efficiency, very good geometric properties, backside illumination with control of the surface potential profile for efficient accululation of electrons generated very close to the surface by violet photons, and very low noise (clearly 2-3 electrons would be desirable) is required. While such chips so not exist today, we can look forward confidently to their existence within a couple of years. Such a machine would allow easy extension of the techniques discussed here to the most distant globular clusters and to dwarf spheriodals.

3. Proper Motions

Proper motions of individual stars in globular clusters measured by the techniques available today require very long time baselines. The field has been completely dominated by Cudworth and his collaborators, who have made use of the Yerkes plate collection, which contains first-epoch plates for the bright northern clusters taken in the first few years of this century. Since one-dimensional velocity despersions for clusters are of the order of five kilometers per second and the nearest clusters are about ten kpc distant, the relative proper motions are of order 10 milliarcseconds per century. Cudworth and Monet (1979) have measured proper motions in M13 with an accuracy of 13 mas/century,

Cudworth in M3 (1979a) and M5 (1979b) to 20 mas, and in M15 (1976a) and M92(1976b) to 30 mas; all quoted errors are one-component standard deviations.

None of these results are sufficiently accurate to use the inferred motion of individual stars from the proper motions in dynamical analyses, though in the case of M13 its small distance and the relatively great accuracy of its proper motions allows one to approach this goal. What one can do is to use the proper motions in a statistical analysis, if the errors are well enough understood. Since globular clusters are pressure-supported structures, the measurement errors simply add in quadrature to the random velocities, and any systematics such as anisotropy should show in large enough samples. Since the fields are relatively small, it has not been possible to tie to the the background well enough to measure any rotation in the plane of the sky (though it is in principle possible to detect it as an induced anisotropy parallel to the equatorial plane in the random proper motions) or indeed even well enough to measure the absolute proper motion of the cluster, which is expected to be huge on the scale of the relative internal velocities.

If one can convice oneself that the errors are well understood, it is possible, of course, to derive a tangential velocity dispersion up to the distance scale factor. With accurate radial velocities one can derive a radial velocity disperison, and the ratio of these, calculated (with not very much model sensitivity) from a dynamical model allows a distance to be determined. This technique was first applied by Cudworth to his M3 proper motions and the model of GG constructed using their radial velocities; the derived distance of 9.6 kpc compared well with accepted values between 9.2 and 10.5, but the errors were large enough not to be able to choose within the favored range. We will discuss a model of M13 below for which a quite good distance estimate can be obtained.

4. Models for M13 and M92

The two nearest bright northern clusters, M13 and M92, both show dynamically significant rotation in the radial velocity data which we have obtained for them, a fact which significantly complicates the construction of dynamical models--particularly as for both the effects of anisotropy are also important in the outer parts, as was the case for M3. One must use models with three integrals to describe such systems, and we discuss such models in this and the following sections.

The models used are three component modified King(1966a) models, with distribution functions of the form

$$f_j\ (E,J,J_z) = e^{-\beta J^2}\ e^{-A_j \Omega J_z}\ (e^{-A_j E}-1)\ ,$$

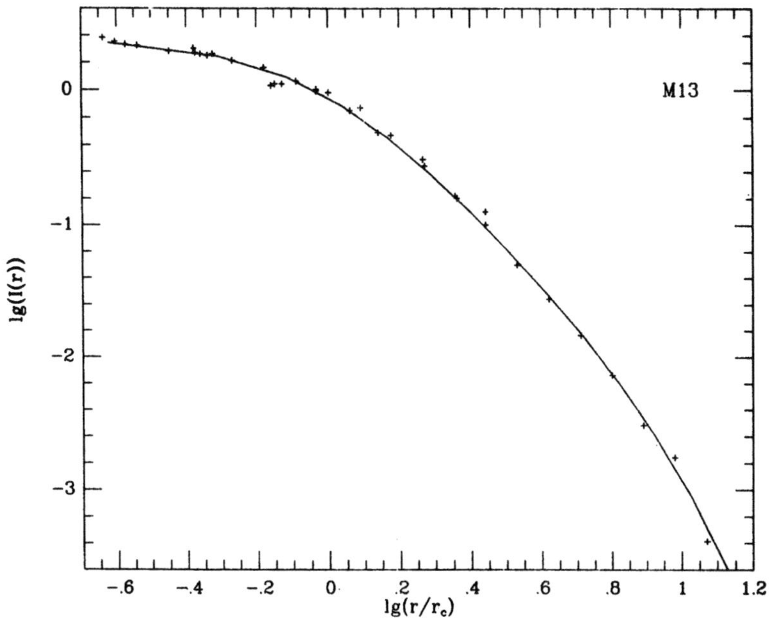

Figure 1a.
The M13 radial light profile. Crosses are the observational points, and the solid line represents the best fitting M13 model.

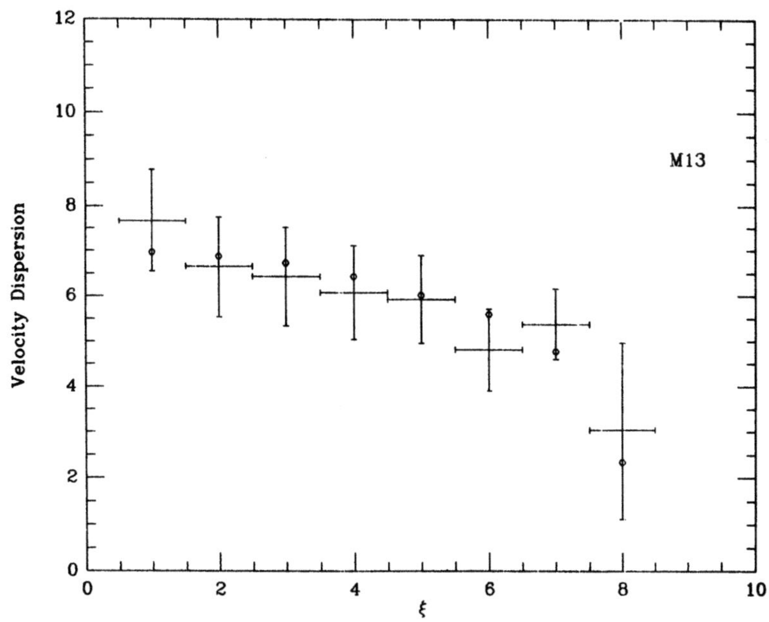

Figure 1b.
Velocity dispersion profile for the M13 model, with the stars binned in radius with 20 stars per bin. The bins are centered at 0.35, 0.77, 1.16, 1.74, 2.42, 3.08, 4.62 and 11.82 core radii.

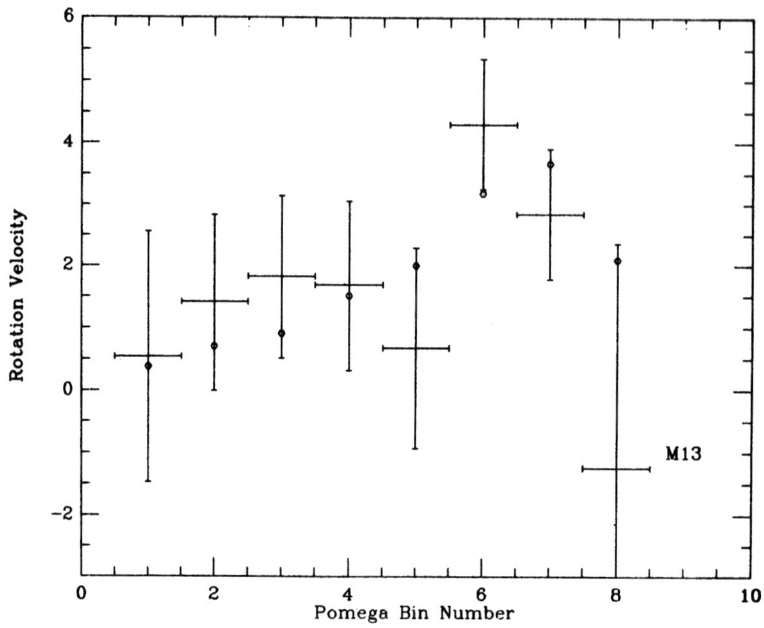

Figure 1c.
Rotation curve for the M13 model, with the stars binned in polar radius with 20 stars per bin. The bins are centered at 0.115, 0.327, 0.517, 0.803, 1.26, 2.13, 3.77 and 9.16 core radii.

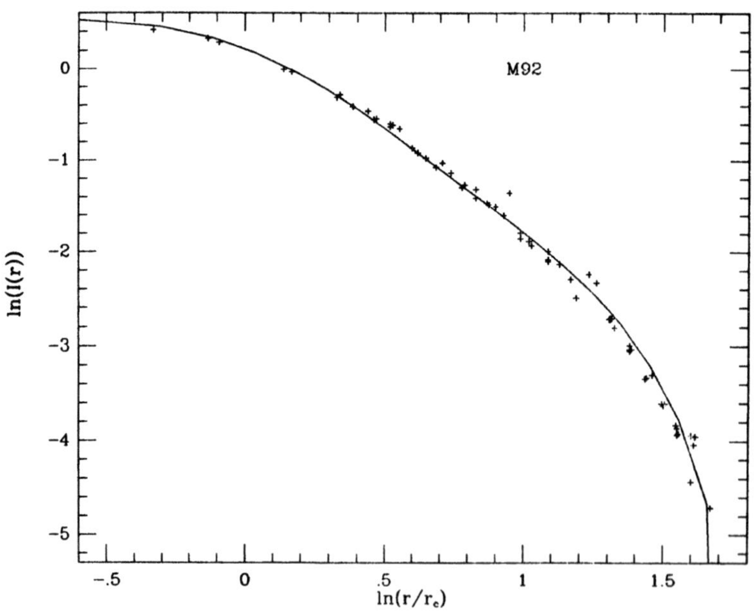

Figure 2a.
The M92 radial light profile. Crosses are the observational points, and the solid line represents best-fitting M92 model.

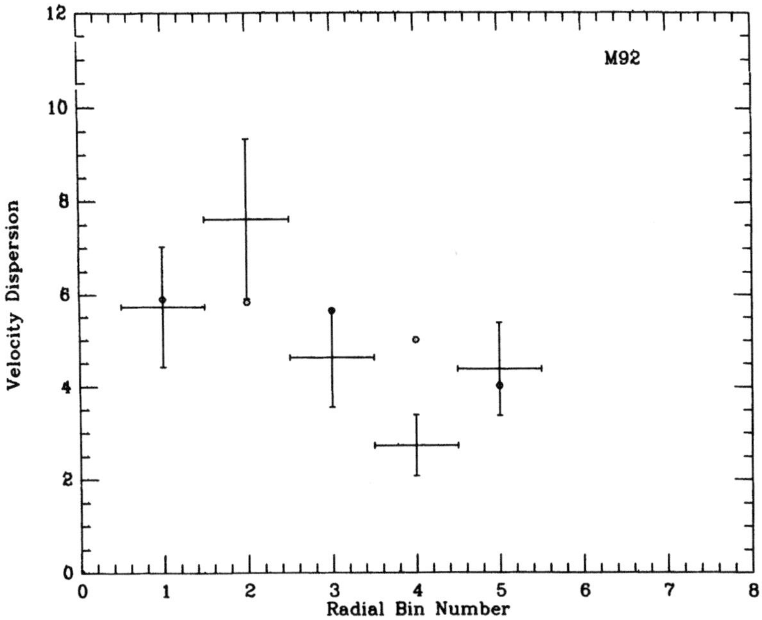

Figure 2b.
Velocity dispersion profile for the M92 model, with the stars binned in radius with 10 stars per bin. The bins are centered at 0.62, 1.97, 3.63, 7.47 and 13.9 core radii.

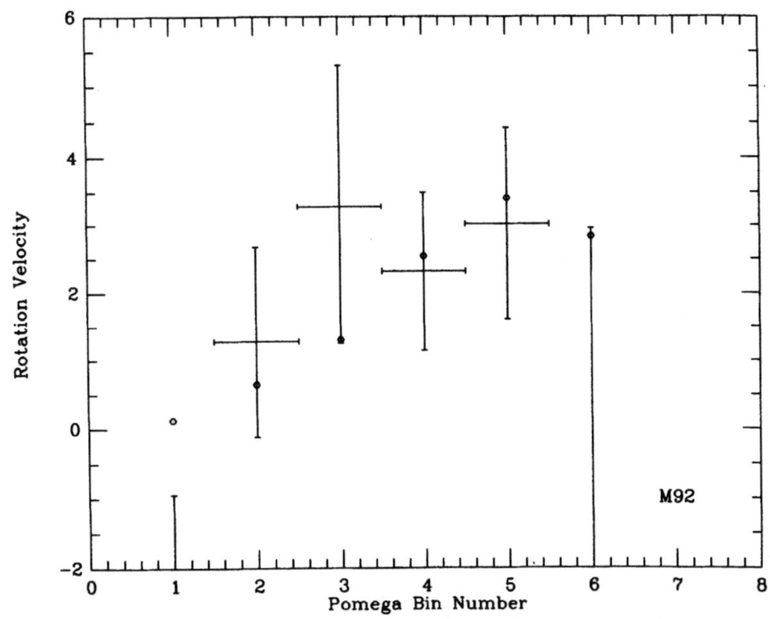

Figure 2c.
Rotation curve for the M92 model, with the stars binned in polar radius with 10 stars per bin. The bins are centered at 0.356, 0.928, 1.79, 3.40 and 7.67 core radii.

Or, in dimensionless form,

$$\tilde{f}\ (\vec{\xi}, \vec{u}) = \alpha_j C_j \exp(-\frac{1}{2}\mu_j u^2 \xi^2 \xi_t^{-2})\ \exp(\mu_j \tilde{\Omega}\tilde{\xi}+u\phi)\ [\exp(\frac{1}{2}\mu_j u^2+\mu_j W)-1]$$

for each of the three mass classes, labeled with j. The notation is the same as that of GG; the new quantities Ω and $\tilde{\Omega}$ are the central angular velocity and its dimensionless counterpart (the "units" of the latter are scale velocities per core radius). It is assumed that the three mass classes are in thermal equilibrium in the central regions, so that the central velocity dispersions are inversely proportional to the square roots of the masses. The models are rotating, with a linear rotation curve in the core, and are anisotropic in the outer parts. These models, and the techniques used to construct them, are described in Lupton and Gunn (1984), following the notation and precepts outlined in GG. In particular, the use of the total angular momentum as an approximate integral is shown to introduce errors of less than about 1% in the density distributions.

5. The Observations

a). Light Distribution. In globular clusters, almost all of the light comes from the stars on the giant branch, with a smaller contribution from the stars on the horizontal branch and near the turnoff. This means that essentially all the light comes from a relatively small number of stars, and that the light distribution is seriously affected by small number statistics. The problem is especially serious in the core, where for example in M3 the integrated light is uncertain to about 6% (GG). Seriously affected as well are the measurements of the small ellipticities of the relatively slowly-rotating systems discussed here, though the flattening is obvious on the POSS prints. We have used the ellipticity data of Kadla et al (1976) and our own photographic photometry from out-of focus 1.2-meter Schmidt plates to determine the ellipticities.

The radial photometry is in better shape, consisting of photometry and star counts by King (1966b) and King et al. (1968), and of our own data. The brightness may be integrated to give an estimate of the total magnitude of the clusters being studied.

b). Velocities. The radial velocities used in this paper were measured with the Griffin-Gunn (1974) cross-correlation coude' spectrometer on the Hale telescope. Typical accuracies for a single measurement are about 1.2 km/s, to which must be added (as was the case for M3 as well) a "jitter" of about the same size to bring the cumulative chi-squared of the measurements down to an acceptable value. This extra error is believed to be due to motions in the atmospheres of the high luminosity giants studied, but its origin is immaterial to the dynamical analysis. Only stars brighter than about V = 14.0 can be studied easily, so these radial velocities are restricted to a relatively small number of stars in each cluster. 142 velocities are available for M13 while for M92, where velocities are very difficult to measure owing to the extreme weakness of the metal lines, we have only 49. These velocities are presented and discussed in detail in Lupton, Griffin, and Gunn (1984) (hereinafter LGG).

The proper motion data are those of Cudworth and Monet for M13 and Cudworth for M92 discussed above. Since the errors in these motions are large compared to the expected velocities we did not use them in the model-making procedure. We did, however, use our models in conjunction with the proper motions to estimate the distance to the cluster, and to compare a postiori the velocity ellipsoids observed with those predicted by the models.

6. Fitting Procedures

We have adopted fitting procedures which are analogous to those presented in GG, though the added complexity of the rotation and subsequent lack of circular symmetry both in the brightness and in the kinematics complicates things somewhat. As was done there, we have restricted ourselves to models which fit the azimuthally averaged brightness profiles essentially perfectly. We have used here a maximum-likelihood technique with an approximate distribution for the likelihood along with realistic estimates of the errors in the brightness profiles to judge the significance of deviations from the observed profile; we had no such tool for the M3 work, and it is possible that those models were overconstrained by the observed brightness distribution. The ellipticities played no role in the fitting, but were examined afterward.

The scale of the fit to the brightness gives the angular core radius and the total flux, which with an assumed distance gives the physical core radius r and the total luminosity. The shape parameters are essentially those described in GG, the inner core shape being determined by the total mass in massive remnants (taken here, as there, to be 1.2 MO white dwarfs), the slope in the "body" of the cluster, determined mostly by the depth of the dimensionless central potential, and the cutoff radius, determined here by a combination of the central potential, the anisotropy, and the rotation.

The radial velocity data was likewise fit to the model by a maximum likelihood technique, here a generalization of the one described in GG. Here we have three parameters to fit, the central velocity dispersion (the velocity scale in the models), the mean cluster velocity, and the position angle of the rotation axis on the sky. As in the spherical models, the run of dispersion with radius is a critical diagnostic for mass in low-mass stars and for anisotropy. The slope of the mean rotation curve in units of central velocity dispersion per core radius determines the dimensionless central angular velocity for any assumed inclination angle, but the inclination angle itself is very poorly determined, even if one uses the ellipticity data, as has already been noted by Wilson (1975) . That this should be so is an immediate consequence of the fact that the ellipticity is quadratic in the angular velocity, the observed velocity is proportional to the sine of the inclination angle, and the observed ellipticity to the square of that quantity. The isotropic component of the centrifugal potential affects the cluster density distribution, so one has a weak handle, but this near-degeneracy makes for a good deal of uncertainty in the final models.

The list of parameters to be fit is the one for the spherical three-component ones enumerated in GG: Three dimensional ones, the total light, the core radius, and the central velocity dispersion; three dimensionless parameters for the mass function, here taken to be the mass fraction in 1.2 MO white dwarfs, the mass fraction in lower main sequence stars, and the mean mass of those objects (these two parameters manifest themselves as the slope and lower mass cutoff for continuous mass functions), and two dimensionless model parameters, the depth of the central potential in scale velocity units and the anisotropy radius in core radii. To this list are added three parameters associated with the rotation: the dimensionless central angular velocity, and the two angles, the inclination and the position angle of the apparent axis. A daunting list, but GG outlined how the first set are quite unambiguously determined by the behavior of the data; the position angle is well determined, but as noted above, the combination Ω sin i is all that is determined well by the observed rotation curve.

The final test of the models is the agreement or lack thereof between the dynamical mass, $Av_o r_c/G$, where the dimensionless number A is calculated with the model, and the population mass, $L^*(M/L)$, where M/L is the mass-to-light ratio for the assumed mass function. It is important to note that this comparison is affected by errors in the distance, since the dynamical mass is proportional to the first power of the distance and the population mass to the square; hence claims of the presence or absence of "missing mass" in globular clusters based on these techniques should be viewed with requisite caution.

One last physical check on the models is to determine whether the anisotropy radius is in fact outside the radius where the local deflection relaxation time is of order 10 billion years; if it is not, it is difficult to understand how the anisotropy has survived till the present.

7. Results

a). M13. With an inclination angle of 60 degrees, near the most probable value, we easily obtained good fits to both the radial velocity and brightness data. The fits are shown in figures 1 and 2. The derived mass function parameters correspond to a main sequence slope of about 2.2, compared to a value of 2.0 for the best M3 model in GG and a value less than 1 for the solar neighborhood. This large value is driven largely by forcing agreement between the dynamical and population masses; adding remnants at the massive end or very low-mass stars at the other end to achieve the same end is precluded by the core shape in the first instance and the observed rapid fall of velocity dispersion with radius in the second.

The parameters are summarized in the Table 1; a surprising result is the very small value of the anisotropy radius, only five core radii.

We also built M13 models with an inclination angle of 30 degrees, which, of course, almost doubles the central angular velocity. These models, though not near the classical ratio of 0.14 for ordered kinetic to gravitational energy for the onset of instability, represent the most rapidly rotating models we can easily construct. The centrifugal potential pulls in the wings of the cluster, and to fit the observed light profile the anisotropy radius must be very small, only 2.2 core radii, at which radius the deflection relaxation time is of order one billion years. We thus regard this model as unlikely to represent the cluster, though the fit is very nearly as good as the best 60 degree model.

The scale velocity of M13 is quite high, 7.8 km/sec, and the distance fairly small, the favored values clustering about 6.5 kpc. The characteristic proper motions are thus pretty large on the scale of such motions, being about 25 millarcseconds/century. With the phenomenally low errors achieved for the proper motions in this cluster by Cudworth and Monet, only about 12 mas/century, very good comparisons can be made for the tangential motions and an excellent distance can be obtained. The anisotropy is obvious in the data, and appears to set in at about the place predicted by the models, but the amplitude of the anisotropy appears too large in the proper motion data, as if the errors had been slightly underestimated resulting in too large a ratio of the corrected radial to tangential motions. These points are discussed in detail in LGG, as are the schemes used for deriving the distance from various "best fit" criteria to the observed proper motions. We obtain a distance of 6.5 (-0.4,+0.6) kpc, in excellent agreement with the accepted value.

b). M92. We have radial velocities for only 49 stars in this cluster, and so the fitting is much less certain. The kinematics of the outer parts is particularly poorly constrained, so the radius of the onset of anisotropy is very uncertain; in any case, very nearly isotropic models seem to fit well. The rotation is not as important dynamically as is the case in M13, and the rotation curve is poorly determined, so we have build models only for an inclination of 60 degrees.

The best-fitting model has an anisotropy radius of about twenty core radii, less extreme than M3 and much less so than M13, and in agreement with Cudworth's proper motion data, which indicates that the motions are isotropic over the area covered by his data, almost all of which come from within this radius. The mass function of the best model again has a slope of 2.2, and the comments made about that value for M13 we could repeat here, though the larger uncertainties must be kept in mind. The modulus of M92 is commonly thought to be slightly greater than that of M13, and our derived scale velocity smaller; these, combined with the vastly larger proper motion errors for this cluster make the dynamically derived distance estimate very uncertain indeed. We obtain a value of 6.1 (+1.0,-0.9) kpc, much smaller than the accepted 8.3 kpc value. Cudworth has indicated to us that the errors of the proper motions, which dominate this estimate, are likely to have been underestimated, and that a reanalysis using more modern techniques is underway.

The parameters again are summarized in the table.

Table 1. M13 and M92 models: properties of best fit models

cl	r_c(pc)	R(pc)	v_o(km/s)	M/L_V	M_5	Td_7	W_o	ζ_T	$\tilde{\Omega}$
M13	2.17	22.9	7.858	2.39	5.84	17.	7.5	5.0	0.34
M92	0.74	61.0	6.077	2.39	3.92	1.6	11.5	20.	0.14

r_c is the core radius in pc, R the mean limiting radius, v_o the velocity scale (very nearly the central one-dimensional velocity dispersion), M/L_V the visual mass-to-light ratio, M_5 the total mass in units of 10^5 solar masses, Td_7 the reference central deflection relaxation time for the mean mass star, W_o the dimensionless central potential, X_t the dimensionless anisotropy radius, and OM the dimensionless angular velocity.

8. Discussion

We have discussed techniques for obtaining radial velocities and proper motions for globular cluster stars and the application of these data to the construction of dynamical models of the generalized King type, in particular the application of approximate three-integral models to photometric and kinematic observations of the rotating globular clusters M13 and M92. Excellent fits are obtained, and relatively unambiguous determination of the large number of parameters which characterize the models is achieved with the notable exception of the inclination of the rotation axis to the line of sight. The inclination is strictly unobtainable in the limit of small rotation but does not affect the other parameters (except, of course, the angular velocity) and is in practice unobtainable for more rapid rotation but confuses the fitting of other parameters.

The reliability of the models themselves is questionable on two fundamental points. First, the relatively arbitrary form of the distribution function, and, second, the assumption of thermal equilibrium among the mass classes. The points are related, because thermal equilibrium in a single mass class dictates a distribution function not qualitatively different from the one we are using, though one may quibble with the exact form of the energy cutoff and such. Many of our conclusions about anisotropy and the lower end of the mass function depend quantitatively but not qualitatively on thermal equilibrium and the concommitant mass segregation. The models, though complex, are probably the simplest ones which include the physics one knows to be necessary to model real clusters; what is in question is not whether there is enough, but whether it is correct. Only real evolutionary models (and real data on mass functions and hopefully someday kinematic data for stars of radically different mass) will finally tell. These models in any case represent probably the farthest one wishes to proceed with the formalism and philosophy originated by King(1966a); the next step will of necessity be very much more complicated.

Acknowledgements
Financial support for this work has come from the U.S. National Science Foundation. RHL would like to thank the Fulbright-Hays commission and a Charlotte Proctor Fellowship for support during this project.

References

Cramer H., Mathematical methods of Statistics, Princeton University Press (1946)
Cudworth, K.M., A.J. 81, 519 (1976a).
Cudworth K. M., A.J. 81, 975 (1976b).
Cudworth, K. M., A.J. 84, 1312 (1979a).
Cudworth, K. M., A.J. 84, 1866 (1979b).
Cudworth, K.M. and Monet, D.G., A.J. 84, 774 (1979).
da Costa, G.S. and Freeman, K.C., Ap.J. 206, 128 (1976).
da Costa, G.S., Freeman, K.C., Kalnajs, A.J., Rodgers, A.W., and Stapinski, T.E., A.J. 82, 810, 1977.
Griffin, R.F. and Gunn, J.E., Ap.J. 191, 545 (1974).
Gunn J.E. and Griffin R.F., A.J. 84, 752 (1979) (GG).
Harris, H.C. and McClure, R.D., Ap.J. 265, L77 (1983).
Illingworth, G., Ap.J. 204, 73 (1976).
Kadler Z.I., Richter N., Stugatskaya A.A. and Hogner W., Sov Astron 20, no. 149 (1976).
King, I.R., A.J. 71, 64 (1966a).
King I.R. A.J. 71, 276 (1966b).
King, I.R., Hedeman, E., Hodge, S.M., and White, R.E., A.J. 73, 456 (1968).
Latham, D.W., BAAS 15, 924 (1983).
Latham, D.W., and Stefanik, R.P., BAAS 14, 612, (1982).
Lupton R.H. and Gunn J.E. A.J. submitted (1984).
Lupton R.H., Griffin, R.F., and Gunn J.E. in preparation (1984) (LGG).
Mathieu, R. D., and Latham, D. W., BAAS 16, 494 (1984).
Mayor, M., Imbert, M., Andersen, J., Ardeberg, A., Baranne, A., Benz, W., Ischi, E., Lindgren, H., Martin, N., Maurice, E., Nordstrom, B., Prevot, L, A.A. Suppl. 54, 495 (1983).
Peterson C.J., and King, I.R., A.J. 80 427 (1975).
Shectman, S.A., and Hiltner, W.A., BAAS 6, 324 (1974).
Wilson, C.H., A.J. 80, 175, (1975)

CHEMICAL GRADIENTS IN GLOBULAR CLUSTERS

K.C. Freeman
Mount Stromlo and Siding Spring Observatories
Research School of Physical Sciences
The Australian National University

INTRODUCTION

First I will describe briefly some older work on chemical gradients in globular clusters. Then I will discuss some more recent data on the chemical and kinematical properties of 47 Tuc and Omega Cen, that leave us in little doubt about the reality of such chemical gradients.

REVIEW OF OLDER WORK

Many years ago, Martin (1937) demonstrated that the radial distributions of RR Lyrae stars and red giants in Omega Cen are clearly different: the RR Lyrae stars are less centrally concentrated. This was originally believed to result from dynamical relaxation, because the masses of the RR Lyrae stars are less than those of the red giants. However we now know that the relaxation time at the center of this cluster is significantly longer than the total lifetime on the giant branch and the horizontal branch, so this explanation seems unlikely. Another possibility suggested by this observation is that the cluster has a radial gradient of chemical abundance, because the morphology of the horizontal branch is sensitive to abundance. This suggestion led to much of the recent work on abundance gradients in clusters. We will see later that the direct evidence for a chemical gradient in this cluster is very strong indeed.

In external galaxies, chemical gradients are seen most easily in the form of color gradients. Such color gradients have also been observed in some globular clusters. Gascoigne and Burr (1956) first reported the color gradient in 47 Tuc. This was confirmed by Chun and Freeman (1979); they also made photoelectric UBV concentric aperture of several other clusters, which showed that some clusters have color gradients and some do not. They noticed that the clusters with color gradients are those whose central relaxation times are longer than about 10^8 years. This makes sense if the color gradients are set up early in the life of the clusters; the gradients could then survive to

J. Goodman and P. Hut (eds.), Dynamics of Star Clusters, 33–42.

the present time only in the clusters with relatively long relaxation times.

In globular clusters, most of the visible light comes from the giants, so the color gradients must be associated with some change in the bright end of the color-magnitude diagram or of the luminosity function. Chun (1976) studied the color-magnitude diagram for 47 Tuc in several radial zones; he showed that the CMD for the giant branch is very similar at all radii. However the bright end of the luminosity function does change with radius. There are more bright giants ($M_V < -1$) per unit luminosity in the inner regions. A similar radial change in the luminosity function has been found in Omega Cen (Freeman, unpublished). Again, these luminosity function changes are clearly not dynamical in origin, because the masses of all the giants are almost the same. Chun was able to show that, for 47 Tuc, the radial changes in the bright end of the luminosity function are sufficient to produce the observed color gradient.

It seems, then, that the color gradients are associated with luminosity function gradients. However the reason for these gradients is not yet understood. Probably, it has to do with chemical gradients (Freeman, 1978; d'Antona et al, 1984) or possibly with radial changes in some other quantity, like stellar rotation (Norris, 1981). I have suggested (Freeman, 1978) that the excess bright giants could be AGB stars; their numbers are sensitive to the distribution of stars on the HB, which in turn could be affected by radial changes in abundance. To get more AGB stars near the tip of the giant branch, towards the cluster center, we would need a redder HB towards the center. From Chun's photometry in 47 Tuc, it seems that the HB is indeed redder in the inner parts where the excess of bright giants is found.

There is some direct older evidence for chemical gradients in globular clusters. Firstly, many clusters have C and N inhomogeneities which are seen most readily as star to star variations in the strength of the CN bands in the stellar spectra. The distribution of CN band strengths is frequently bimodal (Norris and Smith, 1981). One such cluster is 47 Tuc. From a sample of about 140 giants with measured CN band strengths, Norris and Freeman (1979) found that the relative strength of the two CN modes changes markedly with radius in the cluster. (The CN-strong mode is most prominent near the cluster center). Smith (1979) has also detected a radial CN gradient from integrated spectra of the inner parts of 47 Tuc. In Omega Cen, Norris et al (unpublished) have a sample of about 200 stars with measured CN strengths, all the way from the cluster center out to the tidal radius. This sample shows a spectacular CN gradient: the inner parts have stars covering a wide range of CN strengths, while the outer parts have only CN-weak stars (see Figure 4). I will discuss the CN gradients in 47 Tuc and Omega Cen again, later. However, it should be said here that there is still disagreement about whether the CN spread (and therefore the CN gradients) come from mixing processes within these low mass stars, or from accretion of CN-rich material on to the surfaces of these stars early in the life of the cluster, or from CN inhomogeneities in the gas from which the cluster formed. See Freeman and Norris (1981) for a discussion of this problem.

Secondly, Omega Cen shows star to star inhomogeneities in the heavy element abundances (see Freeman and Norris 1981). Only Omega Cen and M22 (to a lesser degree) have this property. This inhomogeneity is seen very well in the sample of 58 RR Lyrae stars with measured [Ca/H] values (Butler et al, 1978). These RR Lyrae stars show a clear radial gradient in [Ca/H]; again, the inner stars are more Ca-strong (Freeman, 1978; Smith, 1981). The heavy element inhomogeneity in Omega Cen is particularly interesting, because these elements cannot be produced in the low mass stars that we are now observing in this cluster. We do not yet understand the origin of the heavy element inhomogeneity and radial gradient in Omega Cen, but it seems very likely that they were established early in the life of the cluster. Maybe they were produced by self-enrichment at the time of cluster formation, while the massive stars were evolving, or come from the merger of pre-cluster fragments of different abundances (Searle and Zinn, 1978). In Omega Cen, the relaxation time is so long that a radial abundance gradient would readily survive to the present time.

THE RADIAL DISTRIBUTION OF CN IN 47 TUC

Figure 1 shows the radial distribution of CN strength for Norris and Freeman's (1979) sample of giants in 47 Tuc. This sample extends out from the center of the cluster to about half the tidal radius. The CN gradient is not immediately obvious, because we see stars with strong and weak CN over the whole range of radius. However the gradient is seen clearly when we compare the frequency distributions of CN strength at different radii. Generalised histograms of the CN distribution in three radial intervals are shown in Figure 2. The relative population of the two CN modes changes with radius: the inner parts of the cluster have a much larger proportion of CN-strong stars than do the outer parts.

Da Costa and I have accurate radial velocities for this sample of stars in 47 Tuc (see our paper in this volume). We have examined this velocity data, to see whether the CN gradient has any associated kinematical effects. For example, the CN-strong stars are more centrally concentrated, so one might expect that their velocity dispersion would be smaller, particularly at larger radii. However, we were unable to find any significant dependence of velocity dispersion on CN, at any radius. The lack of an observable effect probably results from the limited radial range of our sample (few of the stars have $R > R_t/3$). We also estimated the mean rotation of the CN-weak and CN-strong subsamples: both rotate, but there is again no significant difference.

In summary, 47 Tuc has a well-defined CN gradient. Whatever produces the two modes in the CN distribution (whether abundance bimodality or some mixing process) is itself radially varying.

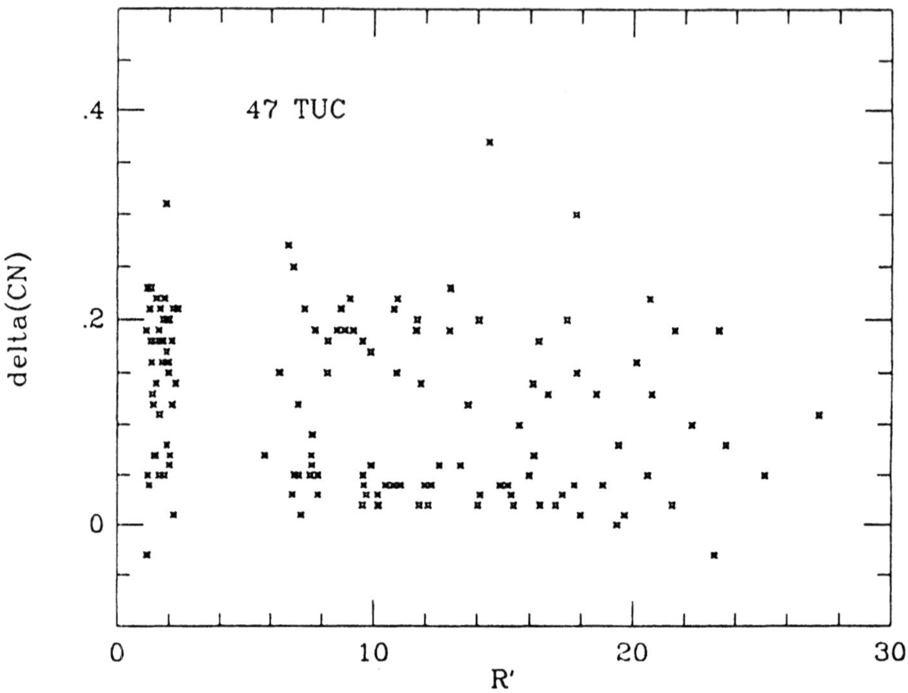

Figure 1. The radial distribution of CN strength delta(CN) for stars in 47 Tuc. High values of delta(CN) correspond to strong CN band strengths. Note the wide range of delta(CN) at all radii.

A comment on the persistence of the CN gradient in 47 Tuc: in the inner parts of the cluster (R < 2'.5), no gradient is visible (see Figure 3). In this region, $R < 5R_c$, where R_c is the core radius, and the relaxation time is less than about 10^9 years. At the next radial zone shown in Figure 2, where $R > 12\ R_c$ and the relaxation time is longer than the age of the cluster, we see that the CN gradient has clearly set in. This is all consistent with the CN gradient having been established early in the cluster's life and now diffusing away through dynamical relaxation: the inner parts of the cluster have already been homogenised by relaxation. It is unfortunate that there are no stars in our sample in the radial zone $5R_c < R < 12R_c$, because in this region we should see the CN gradient gradually appearing. This would be a nice test of the above picture. We will observe a sample of giants in this region in the coming season.

THE RADIAL CN DISTRIBUTION AND KINEMATICS OF OMEGA CEN

Omega Cen is <u>the</u> chemically inhomogeneous cluster. It also has a very long relaxation time, so any initial chemical gradients would have an excellent chance of surviving to the present time, almost unmodified by

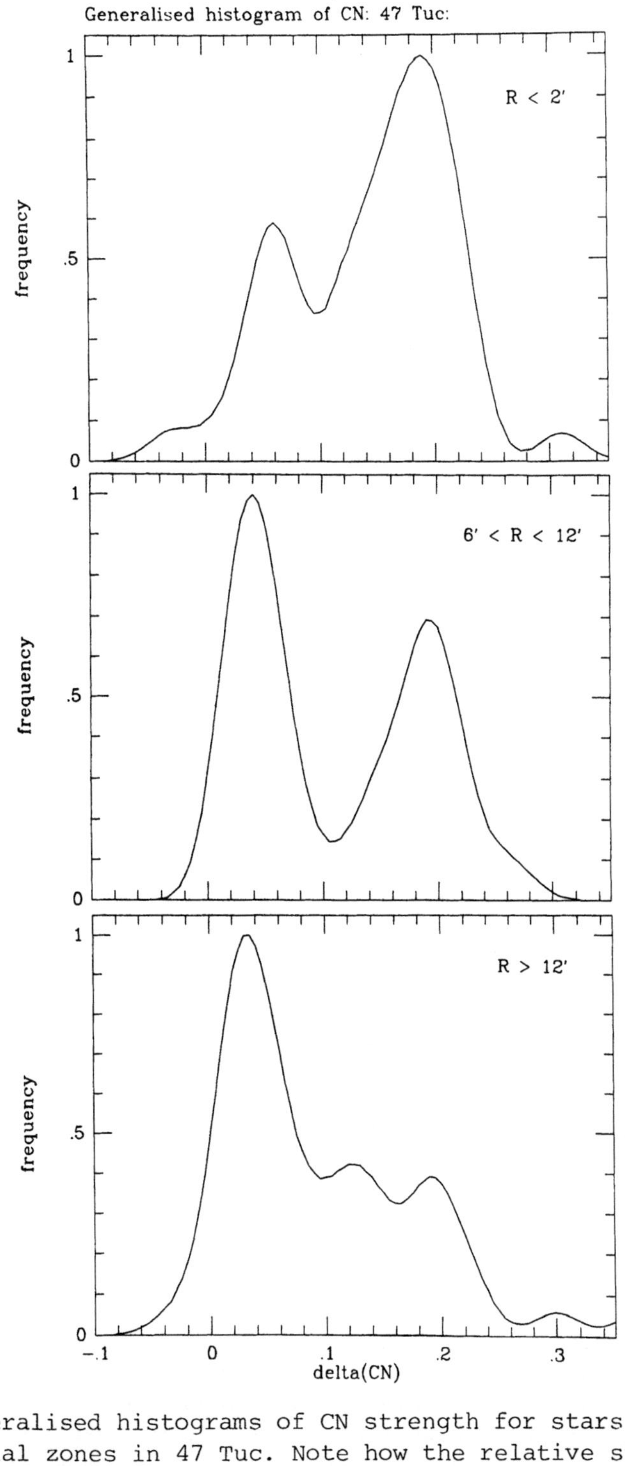

Figure 2. Generalised histograms of CN strength for stars in three radial zones in 47 Tuc. Note how the relative strength of the two CN modes changes with radius.

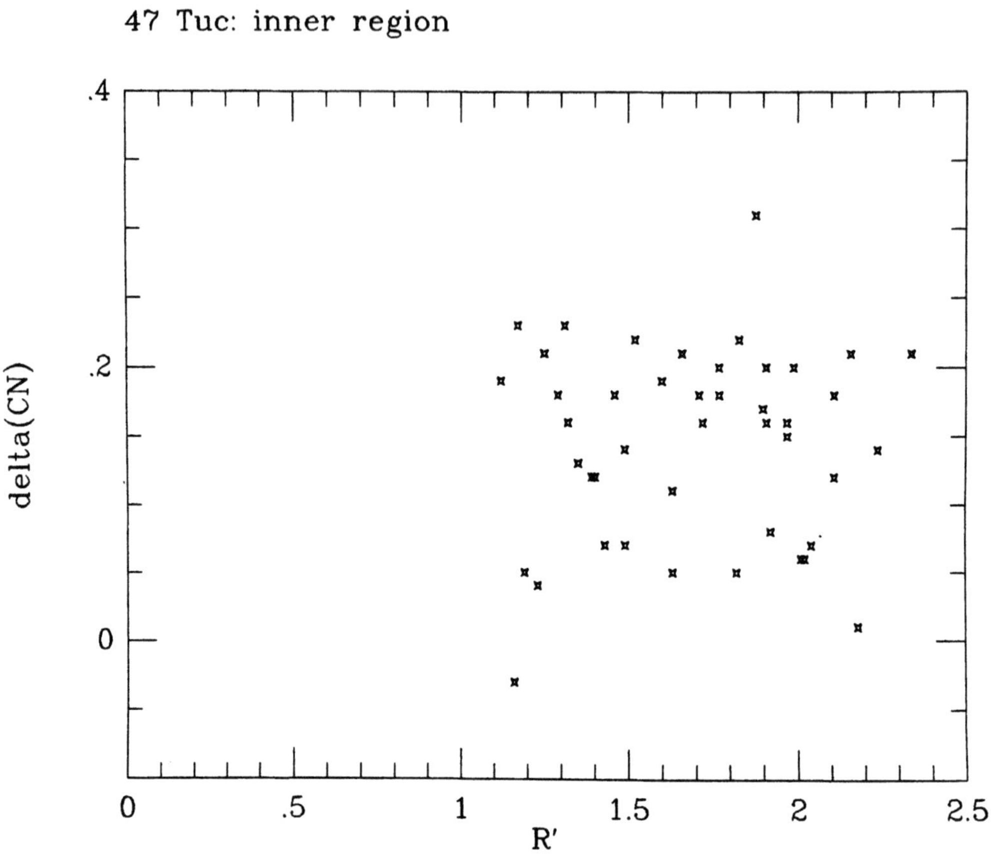

Figure 3. The radial distribution of CN strength for stars in the inner region of 47 Tuc. Note the absence of any CN gradient in this region.

relaxation effects. Furthermore, Omega Cen is a very massive and not very highly concentrated cluster, and it has a large systemic radial velocity: together, these properties make it relatively easy to find member stars in the outermost parts of the cluster. Following a program by Seitzer and Freeman to find such stars, we now have a sample of 115 red giants, with measured CN strengths and accurate radial velocities, from the cluster center out to the tidal radius.

Figure 4 shows the CN-radius diagram for these stars. The CN gradient in the outer half of the cluster is unmistakable. The outer parts of the cluster are CN-weak, while the inner parts have stars with a wide range of CN strengths. Such an obvious CN gradient should have a kinematical counterpart. This is seen in the velocity dispersion-radius relation for stars with weak, intermediate and strong CN, shown in Figure 5. The CN-strong stars are found only in the inner half of the cluster. Near the maximum radius at which these stars are found,

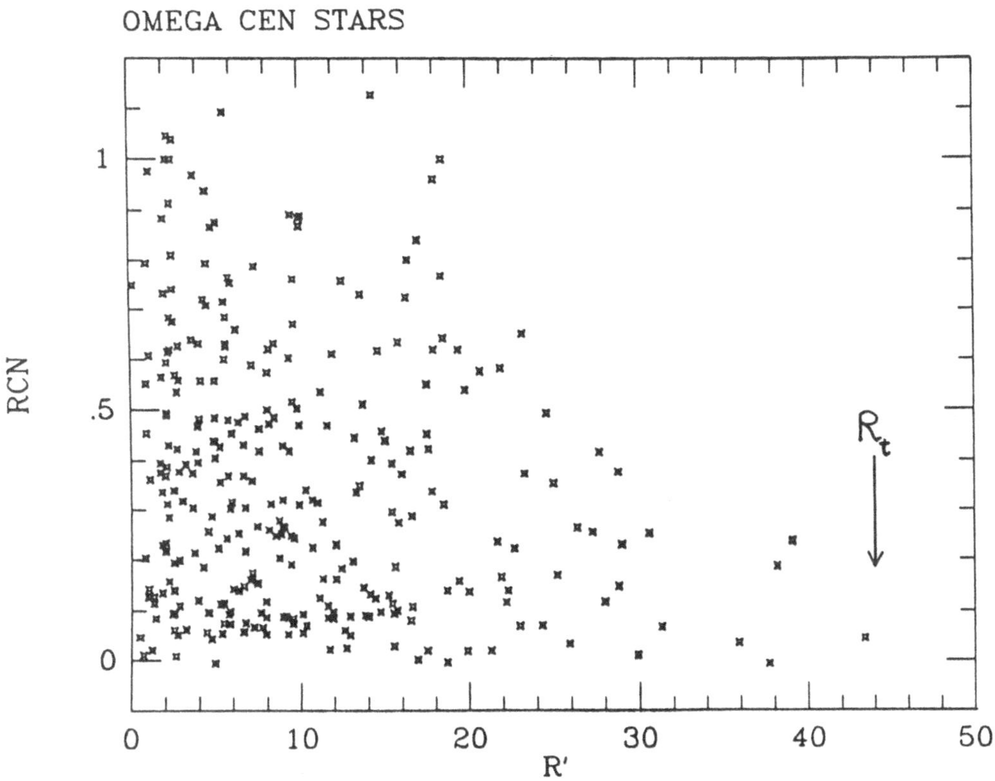

Figure 4. Radial distribution of the CN strength RCN (RCN increases with increasing CN strength) for stars in Omega Cen. The CN gradient is obvious. The arrow marked R_t is at the tidal radius.

their velocity dispersion becomes low. The stars near this maximum radius are the most energetic stars of their CN class, seen now near their maximum distance from the cluster center and therefore with low velocities. Similarly, the velocity dispersion of the intermediate and weak CN subsamples becomes low towards the maximum radius at which such stars are found. So, in Omega Cen, the kinematics and the radial distribution of CN give us a consistent picture of the chemical gradient, in which the CN-strong stars are the ones most bound to the cluster center.

Omega Cen is a relatively flattened cluster.Analysis of the stellar velocities shows that the cluster is rotating. The position angle of the kinematic major axis agrees very well with that for the major axis determined from the star count distribution. In the $(V/\sigma)-\varepsilon$ plane (V is the maximum rotational velocity, σ is the central velocity dispersion, and $\varepsilon = 1 - b/a$, where a and b are the apparent

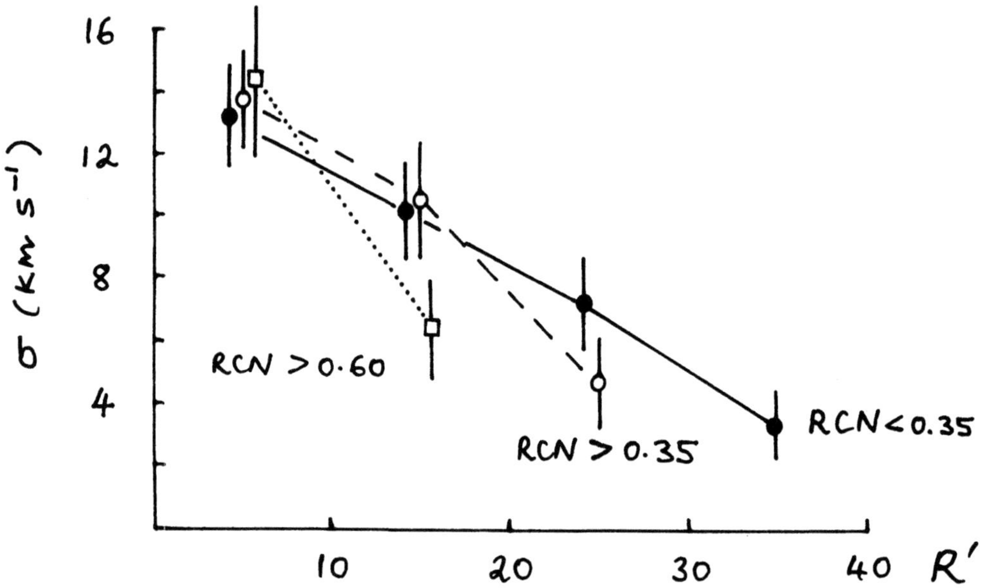

Figure 5. The radial variation of stellar velocity dispersion in Omega Cen, for the three CN-strength subsamples shown by the labels. The velocity dispersion decreases most rapidly with radius for the CN-strongest subsample, which is the most concentrated to the cluster center.

major and minor axes of the star count distribution at the count level corresponding to the maximum observed flattening), the cluster lies close to the oblate isotropic line (see Seitzer et al, 1984) so it seems very likely that Omega Cen is flattened by its rotation.

It seemed worth comparing the rotation for the CN-weak (RCN < 0.35) and CN-strong (RCN > 0.35) stars in Omega Cen. From our sample, it appears that the subsystem of CN-weak stars is rotating significantly more rapidly than the CN-strong subsystem. This result is somewhat unexpected, and more stars with good velocities and CN measures are needed in the inner parts of the cluster to check its validity. We know that, in galaxies, the dependence of rotation on chemical abundance is the other way around: the metal rich component is the more dissipated and rapidly rotating. However Omega Cen is unique among the clusters with its heavy element inhomogeneity. If this inhomogeneity results from the coalescence of two or more fragments of different abundance, then the chemical gradient and the difference in rotation of the CN-weak and CN-strong components probably reflects the different dynamical properties of these fragments before coalescence. On the other hand, if self pollution produces the inhomogeneity and the chemical gradient, then the rapid rotation of the CN-weak component becomes an interesting part of the formation picture. We recall again that the relaxation time in Omega Cen is very long, so the dynamical properties of the cluster are probably much as they were immediately after the time of formation.

REFERENCES

Butler, D., Dickens, R.J. and Epps, E. 1978. Ap.J. 225, 148.
Chun, M.S. 1976. *Australian National University Thesis.*
Chun, M.S. and Freeman, K.C. 1979. Ap.J. 227, 93.
d'Antona, F., Gratton, R. and Chieffi, A. 1983. Mem.Soc.Astron. Ital., 54, 173.
Freeman, K.C. 1978. *Nato Advanced Study Institute: Globular Clusters* ed. D. Hanes and B. Madore, p103. Cambridge University Press.
Freeman, K.C. and Norris, J.E. 1981. *Annual Reviews of Astr.and Astrophys.* 19, 319.
Gascoigne, S.C.B. and Burr, E.J. 1956. Mon.Not.Roy.Astron.Soc. 116, 570.
Martin, W.Ch. 1937. *Leiden University Thesis.*
Norris, J.E. 1981. Ap.J. 248, 177.
Norris, J.E. and Freeman, K.C. 1979. Ap.J. 230, L179.
Norris, J.E. and Smith, G.H. 1981. *Astrophysical Parameters for Globular Clusters*, ed. A.G. Davis Philip and D.S. Hayes, p109. L. Davis Press, Schenectady.
Searle, L. and Zinn, R. 1978. Ap.J. 225, 357.
Seitzer, P., Freeman, K.C. and Norris, J.E. 1984. To be published.
Smith, H.A. 1979. A.J. 84, 176.
Smith, H.A. 1981. A.J. 86, 538

DISCUSSION

KING: Most of these abundance gradients are mystifying, but one result is quite astounding. Your high-CN stars have just as great a velocity dispersion at the center, but both their density distribution and their velocity dispersion drop off much faster. Is there any rational model that will fit this?

FREEMAN: Yes. If we write your lowered Maxwellian as $f(E) \approx (e^{-E/\sigma^2} - e^{-E_0/\sigma^2})$, then the CN-rich stars would need a somewhat higher value of the scale dispersion σ and a somewhat more bound value of the cutoff energy E_0 to produce the observed density and velocity dispersion behavior.

LARSON: Does any of the evidence rule out the possibility that ω Cen is just a superposition of two clusters, a metal-rich one and a metal-poor one, that merged, and that all of the observed radial trends just represent varying relative proportions of the two populations of stars?

FREEMAN: The CN-distribution in ω Cen is not bimodal, as it is in many other clusters. It is a broad and fairly continuous distribution, which would argue against your suggestion.

HANES: How strictly correlated, star-by-star, are the CN and Ca variations? Obviously, if the Ca bimodality is primordial, its correlation with CN tells you something about the origin of variations in the latter.

FREEMAN: The correlation is indeed fairly well defined, which does support a primordial origin for the chemical inhomogeneities, as you suggest.

OSTRIKER: A variety of calculations has shown that angular momentum diffuses outwards in related systems. Thus the metal rich components would be expected to "spin down" somewhat. Can this effect quantitatively account for a lower rotational velocity of the CN rich component in ω Cen?

FREEMAN: It seems unlikely, unfortunately. In ω Cen, the relaxation time is close to the Hubble time, even in the center of the cluster.

LACHIEZE-REY: A correlation also exists between color gradient and rotation in elliptical galaxies. As a result of a multicolor CCD photometry of elliptical galaxies we found that color (B-R) gradients exist which cannot be reduced to metallicity gradients and that they correlate well to the quantities (v/σ) or $(v/\sigma)^*$.

X-RAYING THE DYNAMICS OF GLOBULAR CLUSTERS

Jonathan E. Grindlay
Center for Astrophysics
60 Garden St.
Cambridge, Mass. 02138

ABSTRACT

Recent studies of the x-ray sources in globular clusters have provided important new clues for both the dynamical processes in clusters and the stellar content and evolution of globular clusters. Very deep x-ray images of several globular clusters show evidence for diffuse x-ray emission from hot gas which may be related by a simple shock model to properties of both the cluster, such as its orbit in the Galaxy, and the interstellar medium in the halo of the Galaxy. The x-ray surveys conducted with the Einstein Observatory are reviewed and the results derived for the luminosity function, masses and nature of the compact x-ray sources are discussed. The evidence for the compact binary nature of the sources is now overwhelming, but long-term x-ray variability studies previously reported may suggest that some of the systems are in fact triple systems with distant companions. Possible relationships between the initial mass function, stellar density and cluster evolution are discussed, and our arguments that the ostensibly similar compact x-ray sources in the galactic bulge are remnants of a population of globular clusters disrupted by giant molecular clouds are updated.

1. INTRODUCTION

Globular clusters may be regarded as a laboratory for the study of many of the most interesting current problems in stellar dynamics and evolution. The origin, evolution and internal structure of globular clusters continue to provide challenges for a number of the most currently active areas of astrophysics. It is becoming increasingly clear that these major problems are not isolated but are related in their solutions. For example, the core collapse which must inevitably occur in a globular cluster of high central density will apparently be strongly affected by the stellar content, and particularly the binary content, of the stars in the cluster core (Cohn and Hut 1984, Hut 1985, and other papers in these proceedings). The binary stars, in turn, are formed predominantly by capture processes within the cluster cores but

J. Goodman and P. Hut (eds.), Dynamics of Star Clusters, 43–61.

survive and evolve predominantly by processes dependent on their initial mass and composition. Thus a complete understanding of globular clusters, or even their internal stellar dynamics, must involve a synergistic study of a number of cluster properties. X-ray observations of globular clusters offer a relatively new and powerful tool for achieving new insights into a number of these problems.

In this paper we review the major investigations of cluster content, internal dynamical structure and long-term cluster evolution which x-ray observations have stimulated. In accordance with the primary topic of this Symposium, most of the studies to be discussed are related to the dynamical structure and properties of globular clusters. We begin with a discussion of the large scale structure and motions of globulars as revealed by the recently discovered diffuse x-ray emission which may arise from their motion through the interstellar medium of the Galaxy. Cluster proper motions and total mass loss rates may be inferred from more sensitive studies of this emission in the future. We then discuss the results of the surveys for point sources in globulars. The derived luminosity function for these sources out to 10 core radii suggest they are compact binaries containing either degenerate dwarf or neutron star remnants and were formed by tidal capture within the cluster cores. The long-term x-ray variability of several of these compact binaries suggests they may have an appreciably probability having a distant third-body companion which modulates the mass-transfer rate in the compact binary during each periastron passage of the third body. We consider the possibility that the very formation and existence of compact binary x-ray sources contributes to the rapid demise of the cluster by first heating the core and causing the cluster to expand when it can then be more easily disrupted by tidal forces and giant molecular clouds in the Galaxy. The so called galactic bulge x-ray sources, with x-ray properties very similar to those for sources in globular clusters, may have both luminosity and spatial distributions consistent with their previously suggested (Grindlay and Hertz 1985, Grindlay 1984) origin in globular clusters now disrupted. Additional tests of this hypothesis are described.

2. CLUSTER WINDS AND ORBITS

The discovery of diffuse x-ray emission from globular clusters by Hartwick, Cowley and Grindlay (1982) was interpreted as being due to hot gas since it appears to be asymmetric with respect to the cluster cores. The gas may arise from cluster giants which lose mass in a wind. Such a wind could be shock-heated to x-ray temperatures by the ram pressure of a diffuse interstellar medium through which the cluster is moving. From the observed surface brightness and approximate temperature of the x-ray emission, we may derive constraints on the interstellar medium density density and cluster mass loss rate for an assumed cluster velocity. Detailed measurements of the emission temperature would eventually allow the cluster velocity and thus orbit to be measured.

We consider the case for the diffuse emission from the globular cluster ω-Cen. The hot gas (or diffuse emission) is detected primarily on the south side of the cluster at about 20 arcmin from the cluster center. Hartwick et al. note that this is where a bow shock might be expected from hot gas flowing out of the cluster and encountering the ISM since the direction of the cluster proper motion (Murray et al. 1965) is also to the south. The apparent angular size of the emission (approximately a shell with angular dimensions 5 x 20 arcmin) and the total diffuse emission luminosity (approximately 5×10^{33} erg sec^{-1}) allow the density of the radiating gas to be derived for an assumed temperature (or cluster velocity; see discussion above). Unfortunately, the counting statistics of the detected emission and the energy response of the IPC are such that the x-ray temperature cannot be measured directly. However the emission appears soft as it is detected only in the lowest energy channels of the IPC detector in the Einstein observations. Thus the temperature is probably not greater than 10^7 K, and is in fact only 1.4×10^6 K if the emission is due to an adiabatic shock with temperature proportional to the ram-pressure of the moving gas. For an assumed cluster velocity v, the temperature is (Spitzer 1978)

$$T = (\text{const})(\rho/\rho_0)v^2$$

where ρ/ρ_0 is the compression factor in density in the shock. This yields a temperature of 1.4×10^6 K for an assumed cluster velocity of 300 km sec^{-1} and a compression factor $(\rho/\rho_0) = 4$ behind the shock. The corresponding density in the shock is 0.016 cm^{-3} or a reasonable 4×10^{-3} cm^{-3} in the ambient ISM/galactic halo at the 1.3 kpc height above the galactic plane for ω-Cen. This density is compatible with recent estimates for the density and scale height of gas in the galactic halo as derived from LMC absorption line studies (Savage and deBoer 1979); the globular cluster diffuse x-ray emission would provide a new tool for measuring the density and distribution of this halo material. In deriving the density, we have assumed a optically thin bremsstrahlung emission process and cooling function given by Dalgarno and McCray (1972). This implies a cooling time for the shocked gas which, at 3×10^7 years, is much longer than the 1.7×10^5 year sonic crossing time across the shocked shell, so that the shock is indeed adiabatic and not isothermal.

The total mass of radiating gas in the shocked shell is about 12 $M_\odot$, which must be re-supplied at a rate $\dot{m}$ = Mgas/Tcool = 4×10^{-7} $M_\odot$/yr. Allowing for the approximate solid angle subtended by the radiating shell, this requires some 300 giants in the cluster to be supplying the mass loss at a rate of 3 x 10-9 $M_\odot$/yr per star. This is comparable with the mass loss rates found for extreme giants in clusters in the spectroscopic studies of Cohen (1976) and thus seems reasonable. The implied density of pre-shocked gas in the cluster core is about 0.01 cm-3, or a total gas mass of 0.1 $M_\odot$ if most of the giants are in the cluster core with core radius 4 pc. This mass is well below the typical upper limits for cool gas in cluster cores set by various searches (e.g., Scott and Durisen 1978).

More sensitive x-ray observations of ω-Cen and other relatively large or nearby globular clusters in which the diffuse emission from winds lost by the cluster giants could be detected (e.g., 47 Tuc and M22, as also detected by Hartwick et al.) could be expected to open up new dynamical studies of globular clusters. The tangential velocities and thus cluster velocity vectors could be derived or at least constrained for the first time for a large sample of clusters, and cluster orbits in the Galaxy finally studied. Gas dynamics within the clusters could also be investigated, and questions such as the degree of gas stripping in clusters just exiting from passage through the galactic plane could be studied. These studies should be possible to carry out with the permanent x-ray telescope, AXAF, to be launched in about 1992; preliminary studies may be possible with the German x-ray satellite ROSAT set for launch in 1987.

3. X-RAY LUMINOSITY FUNCTION: WHITE DWARFS VS. NEUTRON STARS IN GLOBULARS

The x-ray survey of some 71 globular clusters in the Galaxy reported by Hertz and Grindlay (1983) (hereafter HG) resulted in the first luminosity function for compact galactic x-ray sources in general and globular clusters in particular. Since the same deep exposures on ω-Cen, 47 Tuc and M22 which showed the diffuse x-ray emission discussed above also showed multiple point sources within several core radii (typically 10) of each cluster center, the globular cluster x-ray luminosity function actually derived is that for the brightest source in each cluster. Compact x-ray sources were detected in 16 clusters: 8 with high luminosity ($\gtrsim 10^{36}$ erg/sec) sources (including 7 clusters previously known and 1 discovered in the IR — Grindlay and Hertz 1981), and 8 with low luminosity ($\lesssim 10^{34.5}$ erg/sec) sources. The most striking result of the distribution of detected luminosities and the 56 upper limits derived in the Einstein survey is the gap in x-ray luminosity between approximately $10^{34.5}$–10^{36} erg/sec. The existence of this gap, as well as the approximate shape of the x-ray luminosity function derived by HG, lead HG to suggest that the low luminosity sources (below the gap) were predominately accreting white dwarf systems (i.e. cataclysmic variables) whereas those above the gap were compact binaries containing neutron stars. Support for the neutron star binary identification of the high luminosity sources comes from a wide variety of arguments although primarily from the study of x-ray bursters (often found in globular clusters) as reviewed by Lewin and Joss (1983) and the direct measurement of the masses of globular cluster x-ray sources (see discussion below) recently reported by Grindlay et al. (1984).

The existence of the gap and the shape of the overall x-ray luminosity function have now been even better constrained by Hertz and Wood (1984) with the addition of a large number of upper limits derived from the all-sky survey carried out with the HEAO-A1 experiment. Some 134 globulars were observed, or nearly double the sample studied with

Einstein, but no new sources were detected despite sensitivities for many clusters below the 10^{34} erg/sec level. Thus the probability that the gap is real is increased to be at greater than the 99% confidence level, and the total number of globulars in the Galaxy that could be expected to have their brightest source with luminosity in the gap is less than 0.5. The luminosity function derived by Hertz and Wood for the brightest source in each globular is shown in Figure 1. In deriving this brightest source luminosity function (BSLF), most of the detected clusters contained only one x-ray source. However, as reported by the Einstein survey of HG, the several clusters observed in very deep exposures contained several sources each although their total luminosity is less than a factor of 2 times that of the brightest single source. Thus the luminosity function shown in Figure 1 is equivalently the luminosity function for the total x-ray luminosity from all compact sources in the globular cluster. The additional contribution from the diffuse emission from hot gas as discussed in the section above, is relatively small: in the extreme example (in cluster mass) ω-Cen it is comparable with the lowest bin in the point source luminosity function.

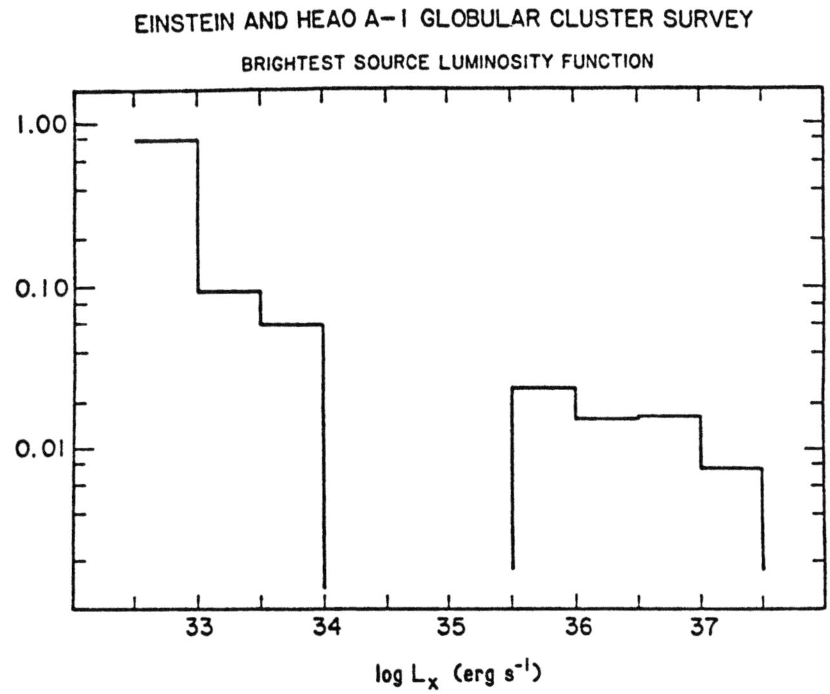

Figure 1 X-ray luminosity function for the brightest sources in globular clusters as derived by Hertz and Wood (1984).

The BSLF may be related to the actual luminosity function of the sources within a given globular cluster provided a model for the formation and nature of the sources is specified. Hertz and Wood (1984) have carried out this exercise for the tidal capture formation mechanism for the low luminosity sources and the simplifying assumptions for the accretion x-ray emission process, the cluster mass spectrum and the white dwarf mass-radius relation. The resulting luminosity function is a power law with index 1.3 and cutoffs at approximately 10^{31} and 10^{34} erg/sec corresponding to the range in accretion rates ($\sim 10^{-11}$–10^{-8} $M_\odot$/yr), white dwarf masses and binary parameters expected from the capture model. The total number of low luminosity x-ray sources in a typical globular cluster is derived by Hertz and Wood to be

$$N_{tot} = 9.2(n_x/8.3\text{x}10^3\ \text{pc}^{-3})(n_x/1.3\text{x}10^3\ \text{pc-3})(V/3.2\ \text{pc}^{-3})$$
$$\text{x}\ (v_o/5.7\ \text{km s}^{-1})^{-1.2}\ (t/10^9\ \text{yr})$$

by integrating this luminosity function. The capture formation volume V is approximately the cluster core radius cubed, and the stellar density n_s and velocity dispersion v_o are mean values for galactic globular clusters (Madore 1980). Thus for a fractional density of white dwarf stars in the cluster, or n_x/n_s, of about 15%, the typical cluster parameters given in the expression yield $N_{tot} \simeq 9$ low luminosity x-ray binaries per cluster. These parameters are such that the predicted BSLF agrees with that observed. Note, however, that the total number of capture binaries expected in the cluster core from this analysis is greater than the number of currently emitting x-ray binaries N_{tot} by the ratio of cluster age to x-ray emitting lifetime t (which is approximately 10^9 yrs, cf. Lightman and Grindlay 1982) so that the total number of compact binaries in the cluster is expected to be of order 100.

Most of these binaries will be in the cluster core (or at most a few core radii out) where they will have been "hardened" by successive interactions with stars in the cluster (e.g., Hut 1983). Their typical binding energies will thus be much greater than that of the cluster and thus their orbital periods should be less than a year. In fact, since these binaries are inferred entirely on the basis of their producing the low luminosity x-ray sources, their orbital semi-major axes must be typically less than a few (perhaps 10, at most) stellar radii for them to have been produced by tidal capture. The mass ratios of these binaries will be very nearly unity since the white dwarf masses are expected to cluster tightly around 0.5 $M_\odot$ (Renzini et al 1984) and the typical companion star mass should also be about 0.5 $M_\odot$. Thus their periods will be typically less than about 10 days (assuming binary companions with radii 1 $R_\odot$) and their orbital velocities will be typically greater than 70 km/sec. Allowing for random orbital inclinations, the expected orbital velocities of the companion (i.e. light emitting) star are still in excess of 50 km/sec. These expected orbital velocities are then sufficiently far from the cluster mean velocity that these stars would be rejected as cluster members in many searches.

A search for these expected compact binaries in the cores of clusters with the Hubble Space Telescope, therefore, need "only" be a spectroscopic survey of several hundred stars, each observed once with only moderate resolution (3-5 km/sec would be adequate). We have embarked on a pilot version of such a search in the diffuse globular cluster NGC 6712, which also contains one of the high luminosity x-ray sources within only 0.2 core radius of the cluster center (Grindlay et al 1984). These observations will be carried out with the FLWO 1.5-m and the MMTO MMT telescopes using Echelle spectrographs with reticon detectors for a velocity resolution of about 1 km/sec in this low velocity dispersion cluster. Further motivation for these observations is discussed in Section 7 below where we discuss the possible evolution of globular clusters containing large numbers of compact binaries (i.e. those clusters most likely to have a high luminosity source now detectable).

4. MASSES AND NATURE OF THE HIGH LUMINOSITY SOURCES

We have recently published (Grindlay et al. 1984) the final results of the analysis of the precise positions of the high luminosity globular cluster sources (measured with the Einstein X-ray Observatory) as well as the cluster centers and core radii (measured with optical data from CTIO) and have derived the constraints on the masses of the sources using the methods of Lightman, Hertz and Grindlay (1980). In this analysis, several simplifying assumptions were made about the sources and the gravitational potentials of the clusters they are found in. First, the cluster potentials were assumed to be isothermal spheres described by King (1966) models of single mass stars. Second, the source masses were assumed to be all the same, and the best estimate of that mass was derived from the data and the cluster model. Both of these assumptions were justified by first testing the observed distribution of source offsets against that expected for a model cluster potential containing single-mass sources; a two-sided Kolmogorov Smirnov test indicated the observed distribution would be obtained in more than 90% of the realizations of the model for sources with mass ratio $q = m_x/m = 2.6$ (where m_x is the total mass of the x-ray binary and m is the mean mass of the cluster field star). Thus, given the measurement errors for both the x-ray and optical position measurements, our data were completely consistent with both a single-q, single-core model.

Although we must await future x-ray observatory telescopes such as AXAF with both higher sensitivity and higher angular resolution to improve upon the x-ray data, more constraints may be inferred from the current observations. First, the fact that one of the eight sources used in the analysis is the source in 47 Tuc, which is actually a low-luminosity source ($L_X = 2 \times 10^{34}$ erg/sec) means that the single-q assumption may be invalid if the low luminosity sources are white dwarf binaries with total mass 1 $M_\odot$, as argued by HG and the discussion above. Of course it is also possible, as discussed by HG and also by Verbunt et al. (1984), that the low luminosity sources are actually a mixed sample with mostly white dwarf binaries but also some neutron star binaries in a very low accretion state (i.e. the low states observed in the so-called transient x-ray sources, at least one of which may in fact be the "low luminosity" source in the globular cluster NGC 6440--cf. HG). If we remove 47 Tuc from the KS test analysis for the source mass, the remaining 7 globulars yield an absolute likelihood distribution for L(q) of the definite high luminosity sources that is shown below. This distribution is somewhat broadened and shifted to lower q than that for all 8 sources originally presented. However, the original range in q (with 90% confidence interval 1.8-3.8) derived from the relative likelihood analysis would be only slightly changed to 1.5-4.3 in this modified sample.

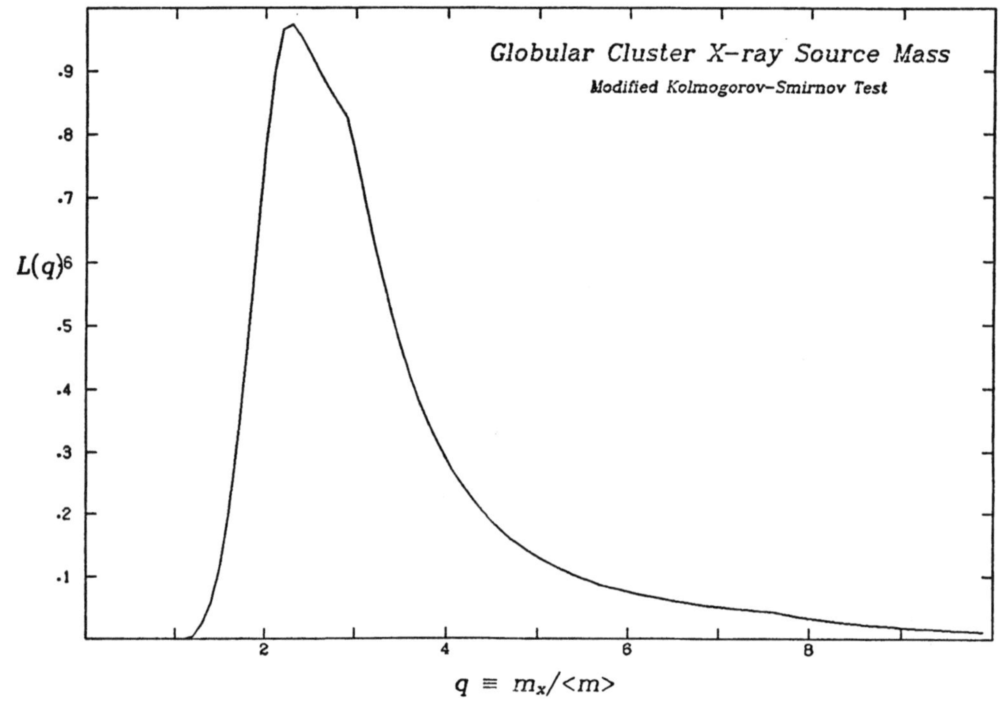

Figure 2. Absolute likelihood distribution for L(q) with 47 Tuc omitted from the sample originally reported by Grindlay et al. (1984).

A second possible modification which could now be made to our original analysis for the mass of the high luminosity sources comes from using the more precise x-ray source positions which can be measured by detecting the radio counterparts of these sources and measuring their positions with the VLA (Grindlay and Seaquist 1984, hereafter GS). In a survey of globular cluster and (the closely related--see discussion below) galactic bulge x-ray sources, all of the high luminosity globular cluster sources except NGC 1851 were observed with the VLA in short (15 min) observations at 6 cm and (for several fields) 20 cm wavelengths.

Only NGC 6624 was detected with a significance of 5 sigma at 20 cm; our upper limit at 6 cm suggests that the source may have a non-thermal spectrum as was also derived for the brighter galactic bulge sources in these same observations. A similar percentage of the galactic bulge x-ray sources not apparently in globular clusters was also detected with significance greater than 5 sigma (and 3 sigma effects, probably consistent with the fluctuations expected from the background in the maps produced, were found within the Einstein HRI error circles [90% confidence radius 3.2 arcsec] for the source in the globular Liller 1 as well as one galactic bulge source). The VLA position for the source in NGC 6624 has an uncertainty of only 0.1 arcsec as opposed to the 0.9 arcsec uncertainty we derived for the mean of a number of separate x-ray observations. The radio and x-ray positions agree but the radio position is closer to the cluster center by 1.8 arcsec. Using this refined source position for the NGC 6624 source but with all the other source positions as used in Grindlay et al. or in Figure 2 above, we derive the L(q) distributions shown below.

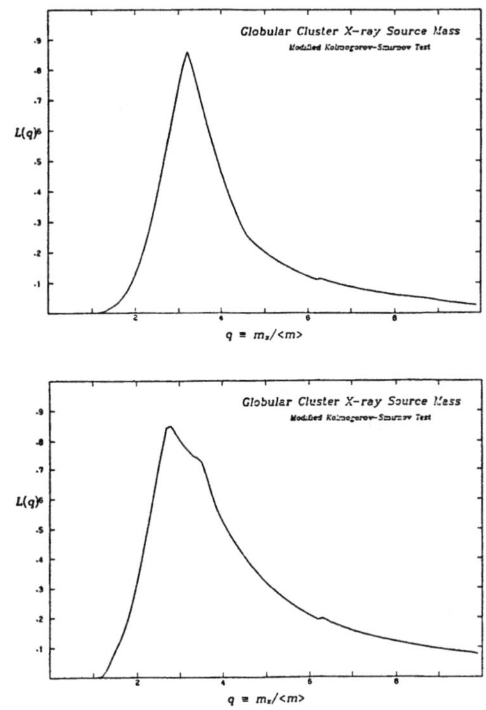

(a)

(b)

Figure 3. L(q) distributions using the VLA radio position for the source in NGC 6624 and a) the 7 other sources measured with the HRI (i.e., the total sample reported by Grindlay et al. 1984) and b) the 6 other sources known also be high luminosity or neutron star binary sources (i.e. 47 Tuc omitted as in Figure 2).

The more precise position for the source in NGC 6624 moves the peak in the L(q) distribution to slightly larger values of q, as would be expected.

The maximum value for L(q) in any of the three absolute likelihood test results shown here (in Figures 2 and 3) is comparably large (approximately 0.9) and essentially indistinguishable when compared to our original results indicating the single-q (albeit different values) and single-core models are equally well described by the different data sets. Thus either more precise x-ray source position offsets (perhaps from further radio studies) or more high luminosity cluster sources are needed before the single-q, single-core models can be further tested. There is immediate hope for both possibilities: more sensitive VLA observations are planned, and the heavily reddened globular clusters containing high luminosity x-ray burst sources will be studied at both IR and x-ray wavelengths. For example, the IR globular Grindlay 1 which contains the burster MXB1728-34 (Grindlay and Hertz 1981) will be mapped in very deep I-band CCD observations to derive a precise cluster center and core radius, whereas the bursters apparently in the globulars Terzan 1 and Terzan 5 (Makishima et al. 1981) might be located precisely in future ROSAT (and certainly AXAF) observations.

The existing data may nevertheless be used to derive a revised range for the source mass ratio q, or the source mass itself, which includes the uncertainty of whether 47 Tuc is really a high luminosity source and also includes the more accurate radio position for the one cluster source now detected. A maximum likelihood analysis for the 90% confidence interval for the value of q which is consistent with any of the three modified data sets discussed above gives the range 1.5-4.9 for q. That is, for a mean cluster star mass of approximately 0.5 $M_\odot$ (cf. Grindlay et al. 1984), the total mass of the x-ray source in these 7 or 8 high luminosity-source clusters is in the range 0.8-2.5 $M_\odot$ at the 90% confidence level provided (again) that the sources all have the same mass and that the cluster cores are isothermal. This somewhat expanded range of allowed q-values does not weaken our firm conclusion that the high luminosity x-ray sources in globular clusters must be compact binaries containing neutron stars (since the maximum mass allowed still does not require a black hole) but it does weaken somewhat our inference that these neutron stars are born in globulars with masses probably below the 1.4 $M_\odot$ value inferred for neutron stars in the galactic plane. Our previous conclusions (cf. HG and Lightman and Grindlay 1982) for the source formation mechanism also remain valid: namely, that the observed rate of 0.1 high luminosity source per cluster implies each typical cluster contains 1-3 neutron star binaries (each with x-ray emission phase of duration ~10^9 years) formed by tidal capture from a reservoir of about 1% of the cluster mass in neutron stars. However, we now present new evidence (culled from existing data or published results) that some of these compact binaries in cluster cores may in fact have distant companions and thus be members of hierarchical triple systems.

5. GLOBULAR CLUSTER AND GALACTIC BULGE SOURCES: BOUND TRIPLES ?

The compact binary systems containing either white dwarfs or neutron stars that we infer to be in globular cluster cores must themselves have a significant probability of capturing a distant third body companion. This is because these basically double-mass compact binary systems will both be more likely found in the high density cluster core, where capture of a distant companion is enhanced by the increased stellar density, and will have reduced isothermal velocity such that again (for the same reason) capture is more likely. Indeed the recent models of the evolution of cluster cores in which binaries are forming show that these systems often have a third body distant companion (cf. McMillan 1983, and Lightman, this volume). The orbits of these third body companions will be stable if the ratio of the third-body period to the compact binary period is large enough. Period ratios greater than about 100 are stable over a wide range of relative orbital inclinations (cf. Soderhjelm 1982). Thus if the compact x-ray binaries themselves have periods of order 3 hrs-1 day, as is now observed for several x-ray bursters/galactic bulge x-ray source systems, the third-body period expected would be in the 100-day range. Since the third body will itself be most efficiently captured by tidal encounters with passing field stars in the cluster, the surviving systems (with long enough periods for stability) will most likely have cluster giants as the third body. We might then expect third-body periods as long as 1-3 years for capture of cluster giants with radii as large as 0.3 AU at typical capture radii (orbital semi-major axes) of 3 stellar radii.

Such long-term periods may have been detected in 2 or 3 of the 8 high luminosity globular cluster x-ray sources as well as several of the closely related galactic bulge x-ray sources. Perhaps the best case, and the one first made for a possible long-term periodicity in the x-ray emission of a globular cluster x-ray source, is the occurrence of burst activity from the so-called rapid burster in the globular cluster Liller 1. Grindlay and Gursky (1977) pointed out shortly after the discovery of this remarkable source by Lewin et al. (1976) that the times of bursting activity seemed to cluster around the February-March and August-September time periods despite historical observations throughout the year with the Uhuru satellite as far back as 1971. Subsequent observations with the x-ray satellites Ariel 5 and SAS-3 (1975-1979), HEAO-1 (1978-1980) and the Japanese satellites Hakucho and Tenma (1980-1984) have remained consistent with these apparent "windows" for burst activity. The source became burst-active again in August 1983 for the first detected activity since August 1979 despite a number of extended observations with the Japanese experiments (Inoue et al. 1985). The overall record of burst activity is shown in Figure 4, which is from Lewin and Joss (1983) and does not include the 1981-1983 Japanese observations. The occurrence times of burst activity suggest, but do not prove, that the bursts may be triggered with a ~6 month period. Bursts are not always observed during these times, so that the trigger is not totally reliable. We propose that this trigger may be the tidal influence of a third-body companion with a ~6 month orbital period and small eccentricity in its orbit. Burst activity is initiated each periastron passage by a change in the accretion rate in the compact binary induced by the tidal shock of the passing third-body companion. The increased mass transfer rate and burst activity is sustained partially by the x-ray heating effect of the Type II (or rapid) bursts themselves but may not occur if the phase of the compact binary is not optimally aligned (i.e. approximately pointing towards the distant companion) at periastron passage.

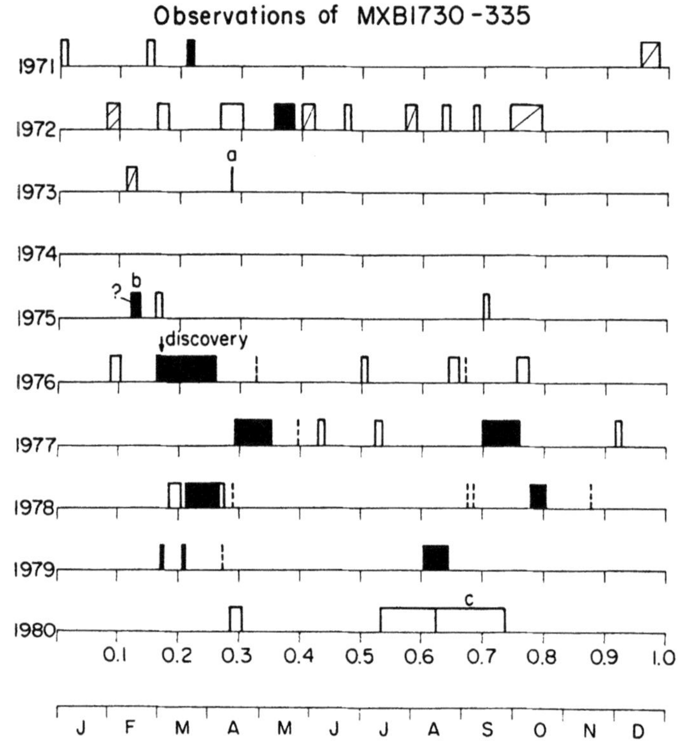

Figure 4. Long-term burst activity record for the burster MXB1730-33 in the globular cluster Liller 1 (from Lewin and Joss 1983). Burst activity periods are denoted by the filled rectangles, quiescent periods with the open rectangles, and ambiguous periods (in which probably only Type I bursts from a nearby burster were detected) are marked with the diagonals in the open rectangles. Short periods (1 day) of observation in which no burst activity was observed are indicated with the vertical dashed lines.

Another high luminosity globular cluster source which has recently been reported (Priedhorsky and Terrell 1983) to have a 176.4±1.3 day period in its overall x-ray intensity is the burster in NGC 6624. Long-term variations of the sources were observed with the 3-12 keV x-ray detector on the Vela 5B satellite over the interval 1969-1976. The apparent periodicity discovered is in the persistent emission, not the burst activity, of the x-ray source. The x-ray flux varies by a factor of 2 over the 176 day cycle, and the light curve shape can be described qualitatively as having a relatively rapid rise and longer decay. This relatively rapid onset and slower decay may suggest modulation of the mass transfer rate by a distant third body followed by x-ray heating effects as mentioned above but with a still larger period ratio or smaller eccentricity so that the variations are more regular and smooth. Alternatively, of course, the 176 day modulation could be simply a binary period in which case the binary is relatively wide separation and the companion (to the neutron star) must be a cluster giant. Such systems have in fact been proposed by Webbink, Rappaport, and Savonije (1983) as possible counterparts for galactic bulge x-ray sources.

The major difficulty with this interpretation for the bursters in globular clusters is that it is in conflict with what is becoming increasingly clear for bursters in the galactic bulge — namely, that their binary periods are in the range ~1-5 hours with perhaps a few in the range ~1-5 days. The most striking case is the galactic bulge burster 4U1915-05 which shows a 50 min period of x-ray dips (Walter et al. 1982, White and Swank 1982) which is almost certainly due to a binary, not spin, period (such dips, with similar periods in the ~1-5 hour range, have now been detected in several other bulge sources). What is remarkable about 4U1916-05, though, is that Priedhorsky and Terrell (1983) also find it to have a long period of 199 days. Thus the ~200 day periods found with the long-term Vela satellite data cannot all be binary periods and we consider it simplest to assume that none of them are. The obvious test is to carefully search sources with possible long term (~100-300 day) periods for short (~1-5 hours) periods; this is being undertaken with EXOSAT observations which can finally search this most interesting period range which was previously very difficult for x-ray satellites in low-earth orbit. Alternatively and in addition, more examples of long-term periods in either known or unknown short period binaries might be sought. Priedhorsky (private communication) has evidence for a 0.93 year period for the high luminosity x-ray source in NGC 6441 and a 0.80 year period for the galactic bulge source/burster GX17+2. We shall return to the implications of this possible GX17+2 period in the next section.

The ~200-400 day periods discussed here could naturally be, therefore, third-body periods for the globular cluster sources. What about the galactic bulge sources apparently not in globular clusters such as 4U1915-05 or GX17+2? The long period (199 days) for 4U1915-05 is such a large multiple of the 50 min binary period that we consider it unlikely this long period is due to precession. This is because the period ratios do not scale from systems with comparable long periods, which may be due to precession (e.g., Her X-1, SS433, Cyg X-1), unless the mass of the companion star is implausibly small (see, however, discussion by Priedhorsky and Terrell 1984). Thus we propose that the galactic bulge sources with long-term periods are also bound triple systems. General arguments that such systems might be relatively common for x-ray binaries were in fact made some time ago by Bahcall et al. (1974) with specific application to Cyg X-1. The question now is, how do such systems arise (if they indeed exist) for apparently isolated galactic bulge sources?

6. GLOBULAR CLUSTER DISRUPTION PICTURE FOR GALACTIC BULGE SOURCES

We have suggested (Grindlay and Hertz 1985, Grindlay 1984) that the so-called galactic bulge or galactic x-ray burst sources (or GXRBs, within ~30° of the galactic center) are compact binaries which formed by tidal capture in globular clusters before the clusters themselves were (rapidly) disrupted by encounters with giant molecular clouds (GMCs) in the galaxy. This was proposed as a hypothesis to explain a variety of facts including the high probability of optically identified bursters being in globular clusters, the basic similarities of bursters in and out of globulars, and the apparently significant coincidence in the spatial positions of 4 bursters in the field and G-K giants, which themselves could not be the binary companion (e.g., one of these 4 sources is 4U1915-05 with its 50 min period). The G-K giants could be either tracers of the remnant core to which the compact x-ray binary is still bound or, more likely (cf. Section 5 above), bound triple companions. Since both 4U1915-05 and GX17+2 now show evidence for long-term periods (see discussion above), this interpretation now seems more likely. The most basic argument for this cluster-disruption origin of GXRBs, however, is that a simple application of the Spitzer (1958) impulse approximation theory of cluster disruption to globular clusters and the relatively recently recognized large scale distribution of GMCs in the Galaxy (e.g., Soloman, Sanders, and Scoville 1979) suggests that GMCs will indeed disrupt globulars.

Several points have emerged since our last discussion (Grindlay 1984) of this hypothesis. First, Van Paradijs and Lewin (1984) have pointed out a numerical error in our (Grindlay and Hertz 1985) probability for the G-K star alignments, which we interpreted as remnant cluster giants gravitationally bound to the GXRBs as triple systems or remnant cluster cores. The chance alignment probability is ~0.03, not ~3 x 10^{-3}, for the assumptions given in our paper and is therefore only marginally significant. However, this probability is that <u>any</u> star of $m_V \lesssim 19$ is within 3 arcsec of the GXRB; the chance probability is still further reduced if all of the stars are G-K giants (rather than more numerous K or M dwarfs) as at least two appear to be (Grindlay 1984). Second, Verbunt et al. (1984) have argued that the luminosity distribution of GXRBs in the Galaxy appears to extend to significantly larger luminosities than does the x-ray luminosity function (HG) for globular cluster sources. However, this difference is complicated by several factors. Most notably, the difference is derived by comparing average burst fluxes, which are assumed to be standard candles, without allowing for possible differences in absorption. Since GXRBs are often more absorbed (by an unknown amount) than the globular cluster bursters (which have known optical extinctions in most cases), associating their burst fluxes with a standard candle luminosity will put them farther away than if absorption were included. Hence, the GXRBs would appear to have larger quiescent luminosity than the globular cluster bursters. Note that this "reddening" effect does not affect the burst fluxes the same as the persistent fluxes (and thus cancel out) since the burst spectra are generally softer, and thus more absorbed, than the persistent emission spectra. Indeed there seems to be more of a (positive) correlation between the L_X and distance estimates derived by Verbunt et al. for the GXRBs than for the globular clusters. Thus the difference in the high luminosity end of the L_X distributions for the two types of sources, which was estimated by Verbunt et al. to have only a 0.03 probability of being the same, may not be significant.

A third development relevant to our globular cluster disruption and GXRB origin hypothesis are the updated CO survey results of Sanders, Scoville, and Solomon (1985). These suggest an even larger number (~6000) of massive ($\gtrsim 10^5$ $M_\odot$) GMCs are present in the inner Galaxy (R $\gtrsim$ 2 kpc) than we assumed (in Grindlay 1984). Their results continue to show the GMCs are concentrated in a ring-like feature between 4-8 kpc (for an assumed galactic center distance R_0 = 10 kpc). Thus the probabilities for disruptive collisions between globular clusters with orbits near (i $\lesssim 10°$) the galactic plane and semi-major axes which extend out to the GMC ring are even further enhanced. Although it appears that only a few globular cluster-GMC encounters (with impact parameters $\lesssim$150 pc) are needed to disrupt a typical globular, detailed calculations are still needed. The (open) cluster disruption calculations presented by R. Wielen and by E. Terlevich at this meeting are a good starting point for the globular cluster case.

We conclude that the globular cluster and GXRB sources are fundamentally the same and that GXRBs could have been formed by capture processes in globular clusters now (largely) disrupted. The statistical significance of the GXRB/G-star associations is marginal, however, and very deep optical/IR/radio studies of these fields are needed to confirm physical associations. Such observations were planned (again) for 1984 but were not carried out due to bad weather; they are now planned for 1985. If the physical associations are confirmed, and the G-K giants are isolated but bound to the compact binary GXRB as triple systems, then important clues are available to guide studies of the final evolution and dissolution of globular clusters.

7. CONCLUSIONS: ROLE OF X-RAY BINARIES IN CLUSTER EVOLUTION

We conclude with a few remarks on the possible role of compact x-ray binaries in the evolution of globular clusters. It is clear from the large number of papers at this Symposium that both pre- and post-core collapse evolution of the cluster is strongly affected by its binary content. It therefore seems likely that clusters which were able to form the largest number of capture binaries (by virtue of their initial density distributions) should have evolved most rapidly. If the central cusp features found in two of the eight high luminosity x-ray cluster surface brightness profiles (Hertz and Grindlay 1984) as well as in several non-x-ray clusters (Djorgovski and King 1984) are indicative of post core-collapse in clusters, then these features may be dominated by binaries. The fact that the cusps are seen in clusters not detected as high luminosity x-ray sources does not argue the cusps are not related to the binary production or content of the core since only ~10% of the neutron star binaries are expected to be x-ray active (cf. Section 4). Thus we have initiated a massive observational project to search for cusps in most globulars in the Galaxy; preliminary results are presented by Lugger, Cohn, and Grindlay (these proceedings).

If binaries are driving the post-core collapse evolution of globulars, then it may be that it is precisely those clusters with the largest binary content which have also disrupted most rapidly. If post-core collapse evolution is accompanied by cluster core expansion, as Sugimoto (these proceedings) suggests, then these clusters may be the first to be disrupted by the tidal field of the Galaxy and individual GMCs as discussed above. Thus the disrupted population of globulars would have contained the largest concentration of compact binaries including the (~10%) presently x-ray active ones which we now detect as GXRBs and CVs. Since the number of binaries formed will also be sensitive to the number of massive stars or their remnants in the cluster (as these are most easily captured), there is also probably a connection between the cluster IMF and its subsequent evolution. If the disk globulars were formed initially with more massive stars (than the halo globulars), perhaps that accounts for their higher metallicity and propensity to self-destruct by formation of compact binaries. Clearly these various connections are ripe for considerable future study.

ACKNOWLEDGEMENTS

I thank P. Hertz and W. Priedhorsky for discussions of their results in advance of publication. This work was partially supported by grant NAS8-30751.

REFERENCES

Bahcall, J., Dyson,F., Katz, J., and Paczynski, B. 1974, Ap.J.(Letters), 189, L17.
Cohen, J. 1976, Ap.J.(Letters), 203, L127.
Cohn, H. and Hut, P. 1984, Ap.J.(Letters), 277, L45.
Dalgarno, A. and McCray, R. 1972, Ann.Rev.Astron. and Astrophys., 10, 375.
Djorgovsky, G. and King, I. 1984, Ap.J.(Letters), 277, L49.
Grindlay, J. and Gursky, H. 1977, Ap.J.(Letters), 218, L117.
Grindlay, J. and Seaquist, E. 1984, Ap.J., submitted.
Grindlay, J. and Hertz, P. 1981, Ap.J.(Letters), 247, L17.
Grindlay, J., Hertz, P., Steiner, J., Murray, S., and Lightman, A. 1984, Ap.J.(Letters), 282, L13.
Grindlay, J.E. and Hertz, P. 1985, in Proc. 7th North Am. Workshop on CVs (D. Lamb and J. Patterson, eds.), Reidel.
Grindlay, J.E. 1984, Adv.Sp.Res., 3, No. 10, 19.
Hartwick, F., Cowley, A., and Grindlay, J. 1982, Ap.J.(Letters), 254, L11.

Hertz, P. and Wood, K. 1984, Ap.J., submitted.
Hertz, P. and Grindlay, J. 1983, Ap.J., 275, 105 (paper HG).
Hertz, P. and Grindlay, J. 1984, Ap.J., submitted.
Hut, P. 1983, Ap.J.(Letters), 272, L29.
Hut, P. 1985, these proceedings.
Inoue, H., et al. 1985, in Proc. 7th North Am. Workshop on CVs (D. Lamb and J. Patterson, eds.), Reidel.
King, I.R. 1966, A.J., 71, 64.
Lewin, W.H.G., et al. 1976, Ap.J.(Letters), 207, L95.
Lewin, W.H.G. and Joss, P. 1983, in Accretion Driven Stellar X-ray Sources (W.H.G. Lewin and E.P.J. Van den Heuvel, eds.), Cambridge Univ. Press, p.41.
Lightman, A., Hertz, P., and Grindlay, J. 1980, Ap.J., 241, 367.
Lightman, A. and Grindlay, J. 1982, Ap.J., 262, 145.
Madore, B. 1980, In Globular Clusters (D. Hanes and B. Madore, eds.), Cambridge Univ. Press, p.21.
Makishima, K., et al. 1981, Ap.J.(Letters), 247, L27.
Murray, C., Jones, D., and Candy, M. 1965, Roy.Obs.Bull. #100.
McMillan, S. 1983, Ph.D. Thesis, Harvard University.
Priedhorsky, W. and Terrell, J. 1983, BAAS, 15, 980.
Priedhorsky, W. and Terrell, J. 1984, Ap.J.(Letters), in press.
Renzini, A., Buzzoni, A., and Fusi Pecci, F. 1984, Mem.Soc.It., in press.
Sanders, D., Scoville, N., and Solomon, P. 1985, Ap.J., in press.
Savage, B. and deBoer, K.S. 1979, Ap.J.(Letters), 230, L77.
Scott, E. and Durisen, R. 1978, Ap.J., 222, 612.
Soderhjelm, S. 1982, Astron. and Astrophys., 107, 54.
Solomon, P., Sanders, D., and Scoville, N. 1979, in IAU Symp. 84, The Large Scale Characteristics of the Galaxy, (W. Burton, ed.), Dordrecht: Reidel.
Spitzer, L. 1958, Ap.J., 127, 17.
Spitzer, L. 1978, Physical Processes in the Interstellar Medium, Wiley: New York.
Van Paradijs, J. and Lewin, W.H.G. 1984, Astron. and Ast., in press.
Verbunt, F., Van Paradijs, J., and Elson, R. 1984, MNRAS, in press.
Walter, F., et al. 1982, Ap.J.(Letters), 253, L67.
Webbink, R., Rappaport, S., and Savonije, G. 1983, Ap.J., 276, 678.
White, N. and Swank, J. 1982, Ap.J.(Letters), 253, L61.

DISCUSSION

SPITZER: Could you compare your estimate that 1% of cluster stars are neutron stars with theoretical expectations? What fraction of the neutron stars produced would need to be retained in the cluster?

GRINDLAY: The estimated fraction of ∿1% neutron stars in the cluster was derived by Lightman and Grindlay (1982, *Ap.J.*, 262, 145) to account for the frequency of the bright X-ray sources in globulars for the 2-body tidal capture-formation mechanism. This estimate is similar to that proposed on other grounds -- e.g., by Illingworth and King 1977, *Ap.J.Lett.* 218, L109 -- to account for the central cusp in M15. I have not made a detailed comparison of this required neutron star number with the numbers expected for various initial mass functions. However a back of the envelope estimate would suggest that for a Salpeter mass function and neutron star progenitors of $\gtrsim 8M_\odot$, most of them would have to be created in relatively "quiet" collapse (or at least symmetric collapse) so that they are not kicked out of the cluster.

SHAPIRO: One might expect binaries to be ejected from clusters in (at least) two different ways. One way is by tidal disruption of an entire post-collapsed cluster. In this case one might expect to observe both "low" luminosity (e.g. white-dwarf binaries) *and* "high" luminosity (e.g. neutron star binaries) for galactic bulge sources which you claim may originate from clusters. Alternatively, one might get binaries ejected from the cluster core directly, during the supernova formation of the neutron star or by small-N large-angle scattering effects. In this case one expects to observe "high" luminosity sources only. Which mode is suggested by the X-ray luminosity function of the Galactic bulge sources?

GRINDLAY: Unfortunately, the luminosity function for the other compact X-ray binaries in the galactic bulge is very poorly determined (if at all) since the distance to these sources, not apparently in globular clusters, is in general not possible to derive. However, it can be stated that the luminosity function for both the "dim" and "bright" X-ray sources in globulars is not inconsistent with the available constraints for the corresponding sources in the bulge. The Einstein Galactic Plane Survey (Hertz and Grindlay 1984, *Ap.J.* 278, 137) demonstrated this consistency.

APPLEGATE: What fraction of the high liminosity X-ray sources in globular clusters burst? What fraction pulse?

GRINDLAY: Essentially, *all* of the high luminosity cluster sources burst (only the source in M15 has not been observed to burst), whereas *none* of the cluster sources have been observed to pulse despite very intensive searches (see, e.g., Leahey *et al.* 1983, *Ap.J.* 266, 160)

KING: You have more than once referred to globular clusters as "fragile." I don't believe that they are. A high-concentration globular cluster is one of the most durable things that you can find in the Galaxy. Before speculating about products of disrupted clusters, one should have a quantitatively effective mechanism for disrupting them.

GRINDLAY: I couldn't agree more that a detailed theoretical treatment of cluster disruption is needed, as I urged in my talk. My point about the "fragility" of clusters here is this discussion was actually with reference to their relatively low binding energies (escape velocities) which would seem to require a "quiet" formation mechanism for the neutron stars we know must be present in cluster cores. As for cluster disruption by GMCs, as I have discussed in a recent paper it appears that a simple application of Spitzer's (1958) theory shows that a typical globular will be disrupted by GMCs in the $\sim$4-6 kpc ring within several cluster orbits if the cluster orbit is near the disk. However, once again, the detailed appearance and time scale(s) for a cluster in the process of disruption is a topic ripe for theoretical treatment so that comparisons with some of the observations I discussed may be made.

OSTRIKER: Perhaps extra destruction of globular clusters could occur if our galaxy had a bar. Then orbits would be box-like and clusters would occasionally come very close to the center and suffer extreme tidal forces.

BAHCALL, N.A.: 1) If the galactic X-ray burst sources are due to disrupted globular clusters, they should have a similar luminosity function to that observed for the cluster sources. Do they? 2) How significant is the gap in the liminosity function of the globular cluster sources?

GRINDLAY: 1) Unfortunately, as I remarked in response to Shapiro's question, the luminosity function for the galactic bulge sources (i.e. bursters apparently outside of globulars) is essentially unknown due to their very uncertain distances. To the extent bulge source luminosity estimates are available, however, they are consistent with the luminosity function derived for the cluster sources although the bulge sources *may* extend to somewhat higher luminosities than the globular sources, as recently discussed Verbunt *et al.*(1984, preprint). 2) The gap is now extremely significant as a result of the added upper limits from the HEAD-1 survey as recently reported by Hertz and Wood (1984, preprint). With these additional upper limites, $\leq$0.5 globular clusters in the Galaxy (at the 90% confidence level) are expected to have their brightest sources in the gap.

DJORGOVSKI: A comment on NGC 6712: Our surface photometry shows no indication of post-core-collapse morphology.

GRINDLAY: I agree NGC 6712 does not show an obvious central cusp (as a result of the photometry Hertz and I have done). However, for a cluster sufficiently past the post-collapse phase, any central cusp may have long since expanded (by binary heating) provided clusters do not oscillate.

A HIGH SPATIAL RESOLUTION INVESTIGATION OF THE CORE OF SOME DYNAMICALLY EVOLVED GLOBULAR CLUSTERS.

M. Aurière[1], J.P. Cordoni[2], O. Le Fèvre[1], A. Terzan[3].
1 Observatoires du Pic-du-Midi et de Toulouse, 65200 Bagnères de Bigorre, France; 2 Laboratoire d'Astronomie,Université de Montpellier, France; 3 Observatoire de Lyon, France.

We present a high spatial resolution photographic study of the core of some concentrated globular clusters (F.W.H.M. of averaged star intensity profiles reaches 0''5 on the best plates). Good plates of M 79, M 53, M 3, M 5, M 13, M 92, NGC 6397, M 15, M 2 and M 30 were obtained (Aurière and Cordoni, 1983, Astron. Astrophys. Suppl. Ser. 52, 383 and references therein). The observations are mainly obtained at the 2 m and 1 m Cassegrain telescopes of Pic-du-Midi Observatory. The high spatial resolution of our plates enables us to achieve stellar photometry, at least for the bright stars, in these overcrowded regions. Surface photometry is also obtained.
We now present two new results concerning M 15 then M 30.

A possible optical counterpart for the M 15 globular cluster X-Ray source.

We have achieved a new high spatial resolution photographic study of 23 stars resolved in the error circle of the M 15 X-ray source. Our observations were obtained at the 3.6 m Canada-France-Hawaï telescope (2U plates; F.W.H.M. of star profiles : 0''85) and at the 1 m and 2 m Pic-du-Midi telescopes (4B plates, 2V plates, 19 unfiltered image-tube plates, F.W.H.M. of star profiles : 0''5 - 0''85).
We propose star AC 211 of the Aurière-Cordoni (1981) catalogue to be the optical counterpart of the X-ray source since :
- It is located 2''7 east from the X-ray source position (Hertz and Grindlay, 1983, Astrophys. J. 275, 105).
- It is a UV star ($B \sim 16$; $U - B \leq -0.8$).
- It is a variable star (it almost disappears on one observation)
- Its optical properties are consistent with those of some low mass X-ray binaries (including Her X - 1) scaled to the X-ray flux of M 15 rather than with those of classical cluster variable stars.

The central brightness excess in M 30 (NGC 7099)

M 30 is found to show a central brightness excess over theoretical King's profiles (Djorgovski and King, 1984, Astrophys. J. 277, L 49.

J. Goodman and P. Hut (eds.), Dynamics of Star Clusters, 63–64.

Kron G.E. et al. 1984, Publ. Astr. Soc. Pac. 96, 198). We have made a B and V photographic study of the central part of M 30 with the 1 m Pic-du-Midi Observatory and 1.5 m Danish La Silla (ESO) telescopes. If a King's model distribution with $r_c = 15''.3$ is fitted to our data, a brightness excess occurs at radial distances smaller than 10". We resolve 46 stars in a circle centred on M 30 and of 10" radius. The light of these 46 stars corresponds to 75 % of the total brightness observed in the B and V passbands in the 10" radius circle. Their V magnitude distribution is quite consistent with that is expected from Sandage (1957, Astrophys. J. 125, 422) M 3 luminosity function. We have computed the brightness which arises from random fluctuations ($\sqrt{N}$ stars, for N stars in the sample under study) in the number of the resolved stars of the region $0 < r < 10''$ and counted in the B and V magnitudes intervals : 13-14...16-17. We find that such a statistical fluctuation may explain only about 50 % of the observed brightness excess in the B and V passbands.

We conclude that if the core radius value of $r_c = 15''.3$ is significant for M 30, our observations and calculations show that the central brightness excess is unlikely to be due to a random fluctuation in the bright stars distribution.

DYNAMICAL MODELING OF M13 PROPER MOTIONS

Kyle Cudworth*
Yerkes Observatory
University of Chicago

D.N.C. Lin and Kap-Soo Oh
Lick Observatory
University of California, Santa Cruz

ABSTRACT

Proper motion data of M13 cluster members are very useful for the investigation of the cluster's internal dynamics. We introduce a maximum likelihood method to fit these data in terms of conventional King-Michie models.

Cudworth and Monet (1979) obtained high precision proper motions for some 440 stars in the region of the globular cluster M13 using many plates from the Yerkes 40-inch refractor taken from 1900 to 1977. They derived cluster membership probabilities for individual stars and concluded that about 330 of these stars are highly probable cluster members. Most of these proper motion determinations have errors less than or comparable to the expected cluster internal motion dispersion and are thus useful for an investigation of the cluster's internal dynamics.

In preparation for a dynamical study, these proper motion data have been carefully analysed for any systematic effects. A small net radial motion of the cluster members was found and removed. The most likely cause for this motion is due to the details of the original reduction method. No other significant systematic trends in the proper motion were found. The errors of the motions have also been studied carefully, with only one significant trend noted: there are relatively fewer stars with very accurate motions in the outer parts of the cluster. This effect was expected, since some of the old plates were not full-sized, leading to proper motions based on fewer plates for the outlying stars.

When these proper motion data are fully analysed, they are particularly useful for determining the velocity dispersion ellipsoid and the

*Alfred P. Sloan Foundation Fellow

J. Goodman and P. Hut (eds.), Dynamics of Star Clusters, 65–67.

potential of the cluster as functions of distance from the cluster center. They can also be used to determine the evolutionary history of the cluster. Traditionally, the "velocity dispersion" of the cluster stars is determined by the variance of a Gaussian distribution function which is fitted to the motions of a collection of stars. These stars are binned according to their distance from the center of the cluster. When this method is applied to the M13 data, we find that the "velocity dispersion" of the cluster stars is essentially isotropic near the cluster center and is strongly anisotropic beyond about 7 core radii (300"). These data suggest that the outlying stars are primarily moving in highly eccentric orbits about the cluster core. This behavior is probably associated with two body relaxation process in the cluster.

Although the binning method is very informative, its statistical significance can be questioned. For example, since it is the variance of a Gaussian distribution function, the "velocity dispersions" deduced from the binning analysis may not necessarily have a Gaussian distribution about the values deduced from any theoretical dynamic models. Thus, direct comparison between these "velocity dispersions" and the conventional theoretical models, such as the King-Michie model, may be inappropriate and misleading. Furthermore, the "observed" anisotropy radius could depend upon the choice of bin boundaries. Some of these problems can be overcome if the proper motion data for individual stars are analysed with a maximum likelihood method. The basic principle of this method is to fit the observed proper motion of every member star directly to the stellar distribution function of an ab initio chosen theoretical model such as the King-Michie model. For any given theoretical model, the stellar distribution function at any location in the cluster is essentially determined by a set of structural parameters such as the central potential, anisotropy radius, and stellar mass function. The primary objective for modelling the proper motion data is to determine these structural parameters. When it is properly normalized, the six dimensional phase space distribution function can be used as a probability function for individual stars with particular values of velocity and position. For a given set of structural parameters, the total probability of the model can be evaluated from the product of the individual probability functions. In reality, stellar position in the cluster is measured in terms of projected distance from the cluster center. The spatial projection effect can be accounted for with appropriate integration of the probability function along the line of sight. In addition, if the one or two velocity component data are available, we can still make use of the maximum likelihood method after integration over the missing phase space element. Finally, the measurement error can be accounted for by deconvolution. The major advantage of this method is that the statistical significance of the model can be directly measured from the total probability function. The most probable structural parameters for the theoretical model are selected from the maximum total probability function. At this time we have constructed a numerical code for this purpose. As a test for the method, we generated a set of pseudo data for 1,000 stars and we have successfully fitted relevent theoretical models to these pseudo data. With the testing of the code essentially completed, we expect to run it with the real M13 data in the near future.

This research has been partially supported by the California Space Institute, the National Science Foundation, Korean Government Fellowship, and by an Alfred P. Sloan Foundation Fellowship to KC.

REFERENCE

Cudworth, K.M. and Monet, D.G.: 1979, Astron. J., 84, 774.

THE DYNAMICS OF 47 TUCANAE

G. S. Da Costa
Yale University Observatory
Box 6666, New Haven, CT 06511

K. C. Freeman
Mt. Stromlo and Siding Spring Observatory
Private Bag, P.O. Woden, ACT 2606
Australia

Observations made at Las Campanas Observatory and at the Anglo-Australian Observatory have been used to determine line-of-sight velocities with an average accuracy of 3 kms^{-1} for 135 member stars in the globular cluster 47 Tucanae. The velocities were derived from cross-correlation techniques applied to 30 A/mm spectra obtained with digital sky-subtracting detectors. The spectra themselves have been used to analyze the cyanogen anomalies on the red giant branch in this cluster (Norris et al., 1984). When combined with the velocities published by the CORAVEL group (Mayor et al., 1983), these observations yield velocities for 212 stars with projected distances from the cluster center ranging from 3 to 68 core radii. After radial binning and analysis these observations yield the following results:

(i) The inner parts of the cluster show appreciable differential rotation with a maximum projected rotation velocity of approximately 6 kms^{-1} in the region 6 - 18 core radii. However, at larger radii the rotation declines rapidly and is essentially zero for radii greater than 30 core radii. This result is illustrated in Figure 1. To within the errors of the determinations, the position angle of the maximum rotation and that of the major axis of the stellar density distribution coincide.

(ii) In contrast to M3 (Gunn and Griffin 1979), "thermal equilibrium" multimass models (c.f. Da Costa and Freeman 1976) can ONLY reproduce the observed velocity dispersion values by including a substantial amount of "dark matter"; i.e. unlike M3, there is "missing mass" in 47 Tuc. In order to retain a fit to the surface brightness profile of the cluster, this "dark mass" (which provides perhaps 30 to 40 percent of the total cluster mass) cannot have a distribution much different from that of the cluster giants if it is in the form of stars and "thermal equilibrium" is maintained. In this case the obvious candidates for the dark matter are the white dwarf remnants of the stars more massive than the current turnoff mass, though many more such remnants are required than the number expected from extrapolating the present mass function. The difference between M3 and 47 Tuc in this case then

J. Goodman and P. Hut (eds.), Dynamics of Star Clusters, 69–72.

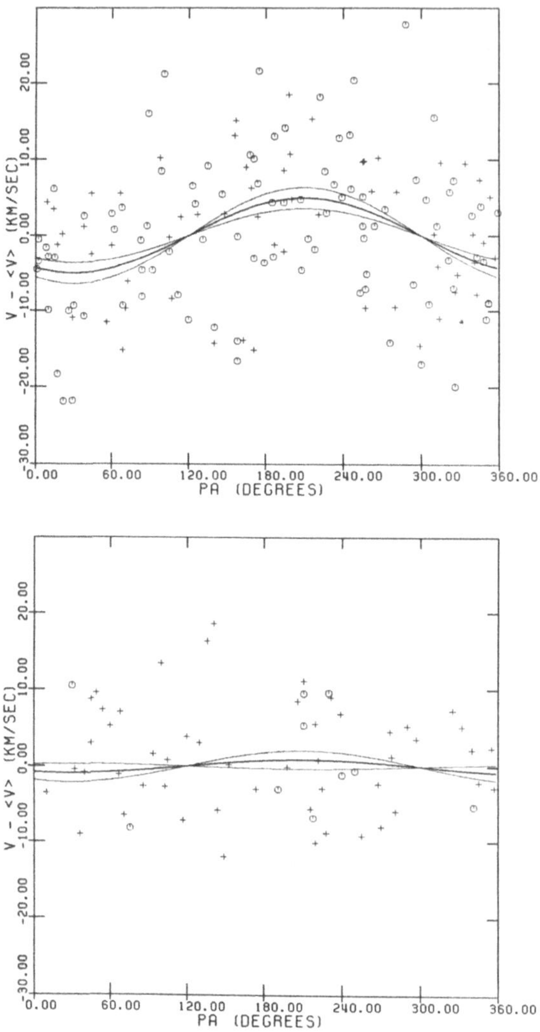

Figure 1. In the upper panel velocity difference (observed-mean for the entire sample) is plotted against position angle (increasing from North to East) for stars in the inner 5 radial bins (3 to 25 core radii). Open circles represent stars observed by the CORAVEL group including the stars in common with the LCO/AAO observations. The velocity error for these stars is approximately 1 kms^{-1}. Plus signs represent stars observed at Las Campanas or at the AAO (or both) only. These stars have errors of approximately 3 kms^{-1}. The thick line is the fitted curve $v - \langle v \rangle = A \cos(PA - PA_o)$ with $A = -5.0$ kms^{-1} and $PA_o = 30°$. The thin lines are $\pm 1\sigma$ (1.4 kms^{-1}) for A with PA_o fixed. The $\pm 1\sigma$ uncertainty in PA_o is $\pm$ 15°. The lower panel is a similar plot but for the outer 2 radial bins (25 to 65 core radii). For these stars $A = -0.9 \pm 1.2$ kms^{-1} at $PA_o = 30°$.

implies that the 47 Tuc initial mass function had many more massive stars than did that for M3. The work of Freeman (1977), who demonstrated large IMF variations in the 8 - 1.5 solar mass range in young Magellanic Cloud clusters, provides observational support for this interpretation.

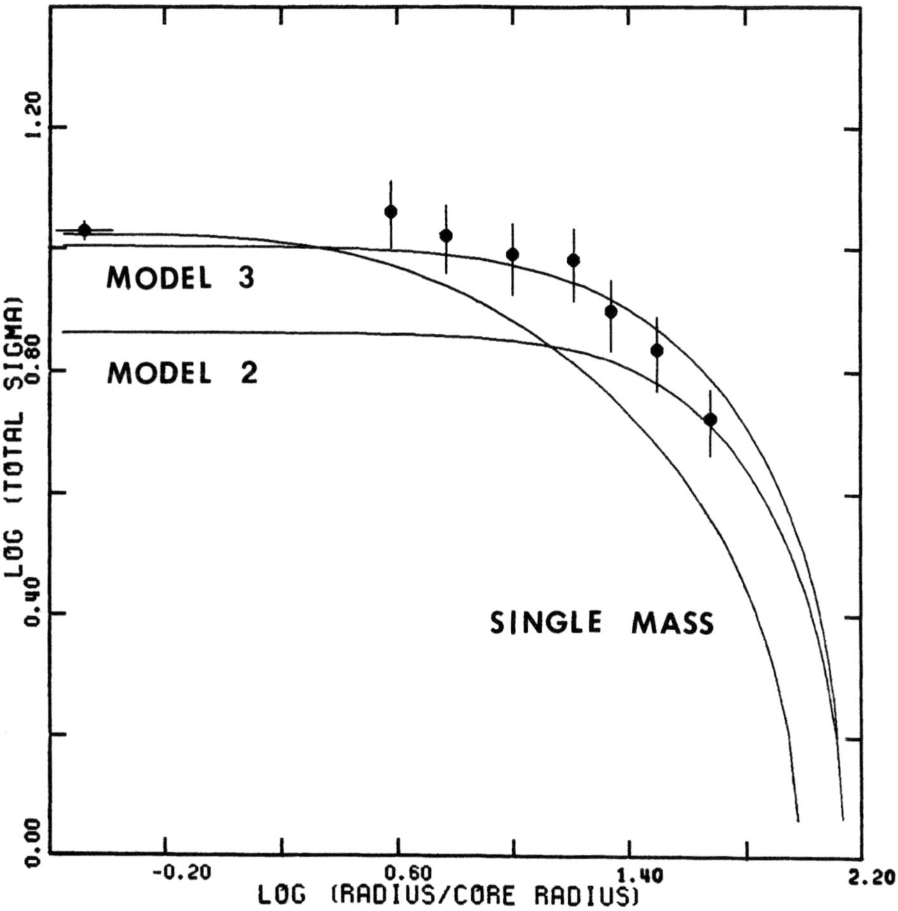

Figure 2. The logarithm of the total line-of-sight velocity dispersion (i.e. no correction for the rotation) is plotted against the logarithm of the radius in units of the core radius. The error bars are $\pm 1\sigma$ errors that result principally from the finite number (typically 30) of stars in each radial bin. With the exception of the curve labeled "single mass", the model curves are multi-mass isotropic King models from Da Costa (1977); a description of their properties is given in the text. The "single mass" curve is the dispersion predicted by the single mass isotropic King (1966) model which fits best the surface brightness profile of the cluster, scaled to fit the observed central velocity dispersion (Illingworth 1976). Note that this model has a smaller limiting radius than the multi-mass models which give superior fits to the surface brightness profile.

The requirement for additional mass in 47 Tuc is illustrated graphically in Fig. 2 where the model velocity dispersion profiles are compared with the observations. Model 2 is the model which fits best the surface brightness and star count data for the cluster (Da Costa 1979, 1982). The number of remnant stars in the model was set by assuming that the stars more massive than the current turnoff mass followed a Salpeter initial luminosity function and that all such stars have produced a remnant that remains within the cluster. The model M/L_V is 2.6 and the remnants contribute 9% of the total cluster mass. In these respects the model is similar to those of Gunn and Griffin (1979) for M3. However, as is obvious from Fig. 2, this model fails completely to reproduce the observed line-of-sight velocity dispersion measurements.

For model 3, the number of white dwarf remnants was increased until the model central velocity dispersion agreed with the observed value. For this model the number of remnants is approximately 7 times the number for model 2 and these stars make up 35% of the total cluster mass. The model M/L_V is 4.0 and it provides acceptable fits to both the star count and surface brightness data. Illingworth and King (1977) have reached similar conclusions from similar models fitted only to the surface brightness profile and the central velocity dispersion.

The difference between M3 and 47 Tuc can also be seen in a consideration of their central M/L_V values. These can be calculated directly from the observed central surface brightness, core radius and central velocity dispersion with little dependence on model parameters. For M3 using the data of Gunn and Griffin (1979), $(M/L_V)_o = 0.7 \pm 0.1$ but for 47 Tuc the value is almost a factor of two higher; $(M/L_V)_o = 1.3 \pm 0.1$.

REFERENCES

Da Costa, G. S. 1977, Ph. D. Thesis, Australian National University.
Da Costa. G. S. 1979, Astron. J., 84, 505.
Da Costa, G. S. 1982, Astron. J., 87, 990.
Da Costa, G. S. and Freeman, K. C. 1976, Astrophys. J., 206, 128.
Freeman, K. C. 1977, In "The Evolution of Galaxies and Stellar Populations", eds. B. M. Tinsley and R. B. Larson (New Haven: Yale University Observatory)
Gunn, J. E. and Griffin, R. G. 1979, Astron. J., 84, 752.
Illingworth, G. 1976, Astrophys. J., 204, 73.
Illingworth, G. and King, I. R. 1977, Astrophys. J. (Letters), 218, L109.
King, I. R. 1966, Astron. J., 71, 64.
Mayor, M., Imbert, M., Andersen, J., Ardeberg, A., Baranne, A., Benz, W., Ischi, E., Lindgren, H., Martin, N., Maurice, E., Nordstrom, B. and Prevot, L. 1983, Astron. Astrophys. Suppl., 54, 495.

A SEARCH FOR POST-COLLAPSE CORES

S. Djorgovski & H. Penner
Astronomy Department
University of California
Berkeley, CA 94720

ABSTRACT. We report the preliminary results of a surface photometry survey of globular cluster cores. Two new cores with post-collapse morphology have been found, and two possible candidates. The estimated fraction of clusters with this core morphology in the Galaxy is only a few percent. Most clusters do not show a morphology similar to the predictions of the central black hole models.

1. THE SURVEY

The problem of globular cluster core collapse is currently receiving much attention. Recent theoretical developments are described in the reviews by Heggie, Spitzer and Cohn elsewhere in this volume, whereas a review of relevant observations to date is given by King.

We have completed a CCD surface photometry survey of 36 compact globular clusters at Lick Observatory during 1982-1983. Some 190 CCD frames were obtained in a wide red bandpass. This is a very preliminary report; the data reductions are now almost complete, and the full details and further analysis will be presented in the upcoming paper.

The use of a red bandpass is not optimal: discreetness and the small number of red giants (which contribute most of the light) are the principal cause of observational difficulties and errors in this work. The error bars are determined by azimuthal variations of surface brightness in the concentric annular apertures used for photometry. Thus, if there is a positive "bump" in the profile, with large error bars associated with it, it is due to a statistical fluctuation in the spatial distribution of red giants. Some of our surface brightness profiles shown in Fig. 1.

2. THE PRELIMINARY RESULTS

The main scientific goal of this survey was to search for more globular clusters with a post-collapse morphology, following the survey by Djorgovski & King (1984). The characteristic signature of a post-core-

J. Goodman and P. Hut (eds.), Dynamics of Star Clusters, 73–76.

-collapse (PCC) cluster is a power law surface brightness profile with a slope ~ -1 to -1.2, going all the way into the seeing disk. The King (1966) models are all steeper in the outer parts, and have flat cores. We have found two new PCC clusters, NGC 6642 and NGC 6342, and two possible ones, NGC 6235 and NGC 6717. Our sample also contains the two "classical" PCC clusters, NGC 7078 (M 15) and NGC 7099 (M 30). The statistics of the known PCC clusters so far is given in the Table 1 below.

TABLE 1

Cluster	Best fit pow. slope	Fit range (arcsec)	Bandpass	Reference
NGC 6342	-0.94	2 - 30	Red	This work.
NGC 6624	-1.09	3 - 20	UV	Djorgovski & King (1984)
NGC 6642	-1.11	2 - 30	Red	This work.
NGC 6681 (M 70)	-1.03	3 - 20	UV	Djorgovski & King (1984)
NGC 7078 (M 15)	-0.91	3 - 30	Red	This work. See also Newell & O'Neill (1978)
NGC 7099 (M 30)	-0.85	3 - 30	Red	This work.
	-1.13	3 - 20	UV	Djorgovski & King (1984)

The estimated errors for the power law slopes are ~ 0.05 . Note that the distribution of slopes is in the range predicted by the PCC models (approaching the singular isothermal sphere, with the slope -1), and that the central black hole model prediction (slope -0.75) seems to be ruled out.

There are also four more possible PCC clusters, viz. NGC 6235 and NGC 6717 (this work), NGC 1851 (Bahcall, Lasker & Wamsteker 1977) and NGC 6397 (Auriere 1982). It should be noted that the surveys so far have been biased in the favor of finding possible PCC clusters, and we estimate that the fraction of clusters with PCC morphology in the Galaxy is only several percent. This may be in conflict with the timing estimates of what now emerges as a "standard model" - viz., many more should be found. One possibility is that the effects of the core collapse may be hidden in the distribution of dark stellar remnants (Goodman 1984). If this is true, then more complete statistics of PCC clusters can teach us something about the IMF variations among the globulars. Note in this respect the models of Larson (1984). However, this apparent discrepancy may also be an indication of a defficiency in our understanding of the PCC phenomenon. Another interesting fact is that several of the "established" and possible PCC clusters are very sparse, e.g., NGC 6642, NGC 6235 or NGC 6717. This may be additional evidence

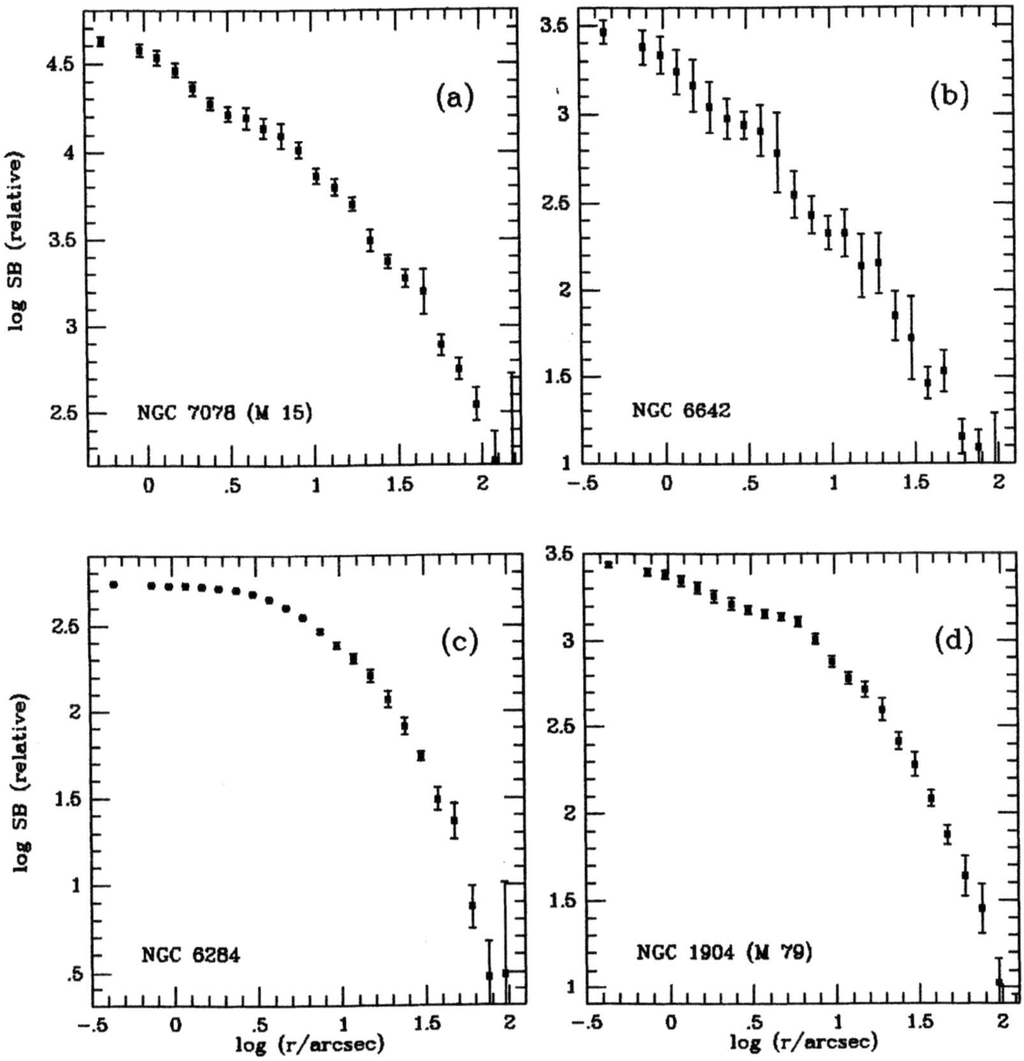

FIGURE 1. Selected surface brightness profiles from our CCD survey. (a) M 15 is the classical example of PCC morphology. (b) NGC 6642 is one of the newly found PCC clusters. (c) NGC 6284 is an example of a highly concentrated, but purely King model-like cluster. (d) M 79 is one of our "strange" clusters, which exhibit a shoulder in their light profiles, and cannot be fitted by either a King model or a simple power law.

for their advanced state of dynamical evolution. The central binary which stops the collapse and stabilizes the cluster also serves as a source of energy which accelerates the dissolution of the cluster.

Some clusters, of which NGC 1904 (M 79) may be the best example in our survey, exhibit a prominent shoulder in their suface brightness profiles, which appears to be statistically significant. They cannot be fitted either by King models, or seeing-convolved power laws. However, they may be fitted with a King model with a seeing-smeared shallow power law cone added in the center. The only model which predicted a profile similar to this is the central black hole model of Bahcall & Wolf (1976). The reality of such features should be confirmed with the data taken in the UV, and in better seeing.

This work was partly supported by the NASA contract NAS5-28086 and the NSF grant AST81-18557.

REFERENCES:

Auriere, M. 1982, Astron. Astrophys. **109**, 301.

Bahcall, J. N. and Wolf, R. A. 1977, Astrophys. J. **216**, 883.

Bahcall, N. A., Lasker, B. M. and Wamsteker, W. 1977, Astrophys. J. (Lett.) **213**, L105.

Djorgovski, S. and King, I. R. 1984, Astrophys. J. (Lett.) **277**, L49.

Goodman, J. 1984, Astrophys. J. **280**, 298.

King, I. R. 1966, Astron. J. **71**, 64.

Larson, R. 1984, M.N.R.A.S. (in press).

Newell, E. B. and O'Neill, E. J. 1978, Astrophys. J. Suppl. **37**, 27.

CORE AND TIDAL RADII OF THE CARINA DWARF SPHEROIDAL GALAXY FROM UK SCHMIDT TELESCOPE PLATES

P.J. Godwin
Department of Astronomy, University of Edinburgh
Blackford Hill, Edinburgh EH9 3HJ, Scotland, U.K.

ABSTRACT. Core and tidal radii of the Carina dwarf galaxy are determined by fitting King dynamical models to number count radial profiles, derived from COSMOS data. These values are compared with those of the other six known Local Group dwarf spheroidals.

PREPARATION AND REDUCTION OF DATA

Six good quality UK Schmidt Telescope plates (five IIIaJ, and one IIIaF), were mapped by COSMOS and then thresholded at 10% above the local sky background level in intensity space. This level was chosen as being a good compromise between (a) enabling faint images to be detected down to the horizontal branch with a relatively high S/N and (b) reducing the effects of crowding and hence making the blending of images less severe.

All the data were paired and calibrated enabling reliable J magnitudes to be determined and a (J-R, R) colour magnitude diagram drawn so that only the giant and horizontal branch stars could be used for the following analysis, thus eliminating to some extent the very high field star contamination ($l = 260^{\circ}.1$, $b = -22^{\circ}.2$). After being binned in 2′ bins, this data was contoured and finally smoothed using a 5 x 5 gaussian filter in order that the centre of the dwarf galaxy (XC, YC) could then be defined as the (x, y) coordinate of the bin with the maximum count. The (x, y) positions of the bins with a specified count starting at a value slightly higher than that of the background, could then be fitted (after transforming them to polar coordinates with origin at (XC, YC)), using a non-linear least squares procedure, to the generalised functional form of an ellipse, with the orientation and semi-major/minor axes left as the free fitting parameters. No significant variation of ellipticity with radius was found, but the effect of the bright central star SAO 234657 in the COSMOS data makes this result hard to confirm inconclusively with the present plate material. Work is now in progress (Godwin 1984), using the larger plate scale capacity of AAT prime focus plates to try and resolve this.

Hence a mean ellipticity (0.31) and orientation (72°) could then be found for all the fitted ellipses and the original data binned

J. Goodman and P. Hut (eds.), Dynamics of Star Clusters, 77–79.

in ellipses with these parameters. The background count was found by finding the value at which the counts summed over the ellipses levelled out. This was then subtracted from the original counts and the resulting mean radial profile with associated errors fitted to the empirical and theoretical models of King (1962, 1965, 1966). Values of $r_t = 47'$, $r_c = 9.6'$ were found. Taking the distance to the Carina dwarf galaxy as 91 kpc (Mould and Aaronson 1983), r_t comes out to be 1.24 kpc.

The effects of anisotropy in the outer regions of the Carina dwarf galaxy are now being investigated by fitting Michie models (Hodge and Michie 1969) to this radial profile (Godwin 1984).

DISCUSSION

The predicted limiting radius at perigalacticon of a stellar system moving in an elliptical orbit around the Galactic centre (King 1962) is

$$r_{lim} = R_p \left[\frac{M_d}{3.5M_g} \right]^{1/3} \tag{1}$$

assuming elongated orbits ($e \sim 0.5$). R_p is taken to be the present distance when calculating this quantity, and hence the value derived is an upper bound for the actual r_{lim}. If the observed tidal radius is larger than this upper bound, then the Galaxy must be seriously disrupting the system. Such is the case with Ursa-Minor and to a lesser degree with Draco and from the data here (see table I) Carina.

Hodge and Michie (1969) found that the ratio of the Galaxy's tidal, to the dwarf galaxy's internal gravitational force could be expressed in the approximate form

$$\left| \frac{T_z}{V_z} \right| \sim \frac{2\,(b/a)\, M_g\, r_t^3}{D^3 M_d} \tag{2}$$

where D is the distance to the stellar system. Using $M_g = 3 \times 10^{11} M_\odot$, these values are recomputed from the data in table I, and plotted against the mean ellipticity of the isopleths (figure 1a). As Hodge and Michie (1969) noted, and as would be expected, the closer the dwarf galaxy to the Galactic centre, the greater the role that the Galactic tidal force plays in determining the structure of its outer regions. Figure 1b shows $|T_z/V_z|$ plotted against r_t/r_c. A large value of this former ratio, should be indicative of a large Galactic influence on the dwarf galaxy, producing a low stellar density envelope at large distances from its centre. As can be seen from these two figures, the Carina dwarf galaxy joins Draco and Ursa-Minor as satellites of our Galaxy seriously affected by the Galactic tidal field.

ACKNOWLEDGEMENTS

My thanks go to Drs. M.T. Brück, R.D. Cannon, D.C. Heggie, H.T. MacGillivray (COSMOS Unit, ROE), S.R. Heathcote and Mr S.M. Beard for helpful suggestions and comments.

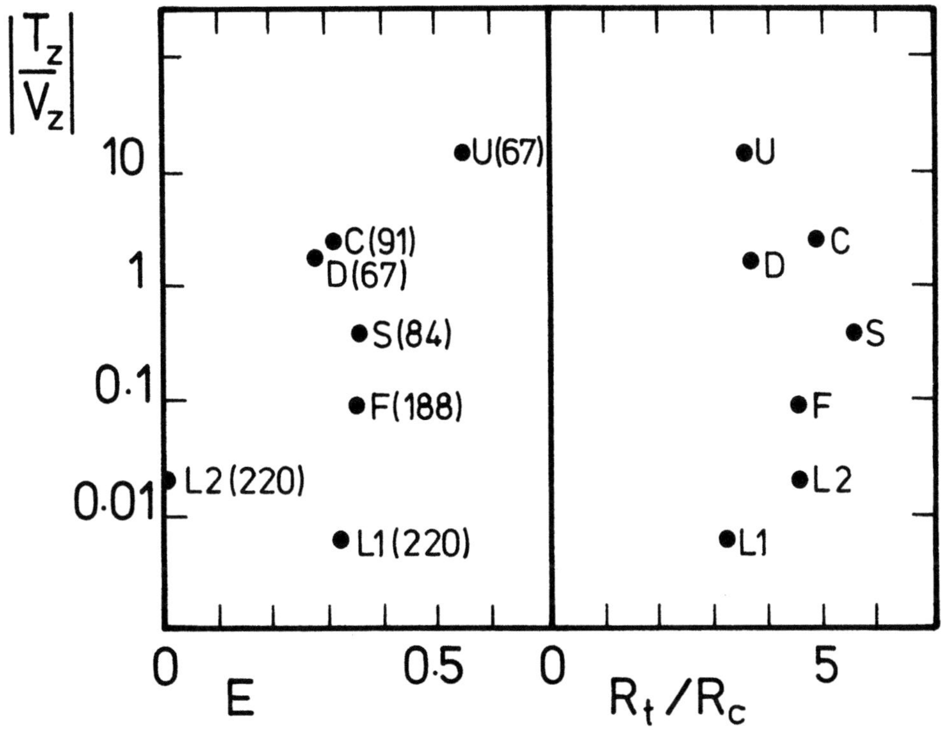

Fig. 1a,b. $|T_z/V_z|$ - ellipticity/concentration parameter relations for the Local Group dwarf spheroidals (distances in brackets).

Table I. Local Group dwarf spheroidal data.

Name	r'_t	r_t(kpc)	r_{lim}	r'_c	r_t/r_c	E	$M/M_\odot$	D(kpc)	T_z/V_z
Draco	35	0.51	0.33	9.27	3.85	0.29	1.2E+5	67	1.57
Leo I	13	0.91	3.44	4.33	3.13	0.31	4.0E+6	220	0.007
Leo II	10	0.65	2.16	2.28	4.43	0.01	1.0E+6	220	0.02
Sculptor	57	1.20	1.19	10.31	5.53	0.35	3.0E+6	84	0.38
Ursa-Minor	47	1.20	0.31	13.29	3.57	0.55	1.0E+5	67	15.51
Fornax	57	3.10	5.02	12.71	4.49	0.35	2.0E+7	188	0.09
Carina	47	1.24	0.71	9.60	4.90	0.31	5.0E+5	91	2.09

REFERENCES

Godwin, P.J. (1984) PhD Thesis, University of Edinburgh (in prep.)
Hodge, P.W., Michie, R.W. (1969) Astron. J. 74 p587
King, I.R. (1962) Astron. J. 67 p471
King, I.R. (1965) Astron. J. 70 p376
King, I.R. (1966) Astron. J 71 p64
Mould, J.R., Aaronson, M. (1983) Ap.J. 273 p530.

A DEEP LUMINOSITY FUNCTION FOR 47 TUCANAE

William E. Harris*
Department of Physics, McMaster University
Hamilton, Ontario, L8S 4M1 Canada

James E. Hesser*
Dominion Astrophysical Observatory
Herzberg Institute of Astrophysics
Victoria, B.C. V8X 4M6 Canada

CCD photometry in B and V reaching B(lim) $\simeq$ 25 has been employed to obtain the luminosity function and color-magnitude diagram for the main sequence of 47 Tuc. For $5<M_V<10$ we find that its LF is essentially flat ($\Delta\log n/\Delta m \sim 0$). The CMD is successfully matched by isochrones with [Fe/H] = -0.5 and t $\simeq$ 15 x 10^9y.

In October 1983 at CTIO, we obtained CCD exposures in B and V for a series of fields in and near the prototypical "metal-rich" globular cluster 47 Tucanae (NGC 104 = C0021-723). Two fields on the west side of the cluster field (at r $\simeq$ 19 arc min, or about 40% of the tidal radius), observed on consecutive nights, have been used to construct the color-magnitude diagram shown in Figure 1. Each pair of CCD frames is based on (5x800 sec) exposures in B and (10x500 sec) in V. Photometric calibration of the CMD was achieved by direct tie-in to fundamental E-region UBV standards and is believed to be systematically correct to $\pm$ 0.01 in V and B-V. All the final CCD photometry and reduction was completed with the Stetson DAOPHOT program, which employs automatic star finding, empirical PSF definition, and profile fitting.

Field-star contamination is remarkably low for V<21. Fainter than that, the contribution of the "background" SMC halo becomes a dominant feature (lower left of CMD). To derive a preliminary luminosity function for the 47 Tuc main sequence, we have simply assumed all stars falling within $\simeq 2\sigma$ of the main sequence ridge line in the CM diagram to be cluster members. The resulting differential LF is shown in Figure 2. Note this preliminary LF has not been sufficiently corrected for field stars in the same <u>color range</u> as the cluster main sequence; taking these into account would slightly lower the points (for V > 22 particularly) and make the overall LF even flatter than it appears here.

---

*Visiting Astronomer, Cerro Tololo Interamerican Observatory, operated by AURA, Inc. under contract to the National Science Foundation.

J. Goodman and P. Hut (eds.), Dynamics of Star Clusters, 81–83.

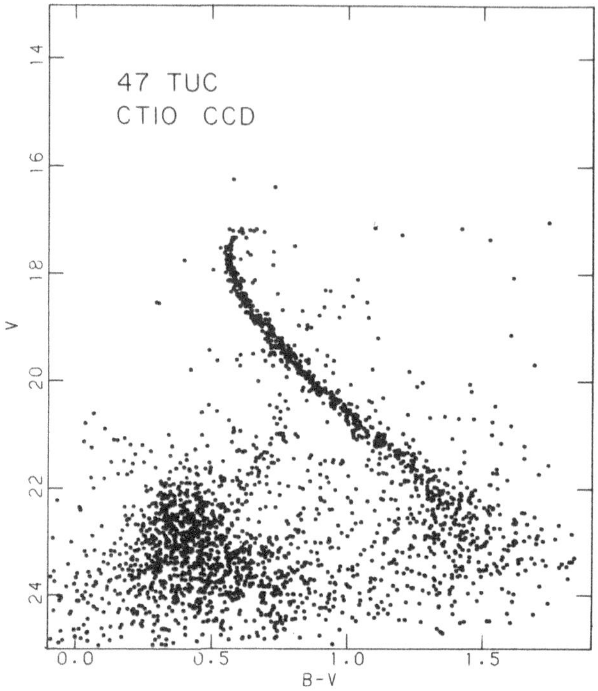

Figure 1. Color-magnitude diagram from CCD photometry of two 3' x 5' fields in 47 Tuc, obtained with the PF/CCD system at the CTIO 4-meter telescope.

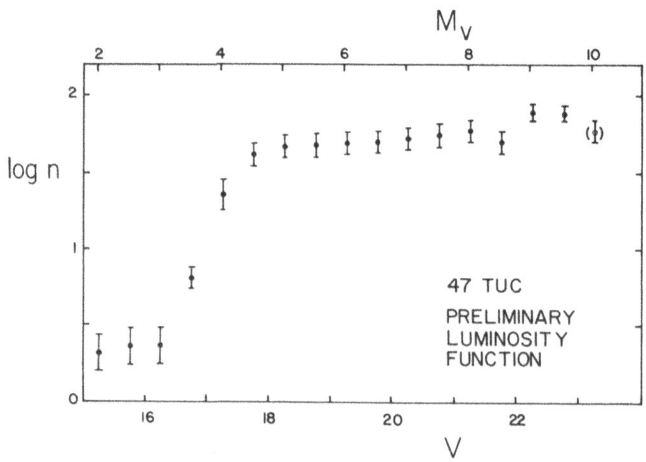

Figure 2. Differential luminosity function for the 47 Tuc main sequence, derived from Fig. 1 as described in the text. Here n is the number of stars per half-magnitude interval. The portion of the LF for $V \lesssim 17$ was obtained from Hesser and Hartwick (1977), appropriately scaled to match the upper main sequence. The absolute magnitude scale at top assumes $(m - M)_V = 13.2$.

We are currently measuring two additional CCD "background" fields taken just outside the cluster tidal radius. These will eventually allow the construction of an improved LF in which point-by-point field star subtraction has been made and incompleteness at faint levels has been specifically estimated. However, we believe completeness to be essentially 100% for $B<23$ because of the rigorous automatic star-finding and image subtraction features in the DAOPHOT program. For $B>24$, completeness drops severely.

Da Costa (1982) used a series of V photographic plates to obtain the LF for 47 Tuc to $V(\mathrm{lim}) \simeq 22.5$. We find our results agree substantially with his for $V < 20$, but deviate significantly as we go further down the main sequence (our LF is flatter). If the Da Costa LF is correct (scaling it to match ours for $V < 20$) we should expect to see $\simeq 1350$ main sequence stars brighter than $V = 22.5$ in our frames. In fact, we find $\simeq 570$ such stars, i.e. a factor 2.4 fewer.

The CM diagram itself provides a precise definition of almost 6 magnitudes of the cluster main sequence. Note particularly the absence of any obvious binary-star sequence sitting above the ZAMS, or of any blue stragglers. We have carried out a preliminary fit of the diagram to Vanden Berg's (1984) most recently calculated isochrones. After a reddening correction of $E(B-V) = 0.04$, these models produce an excellent match to the observed main sequence for $(m-M)_V = 13.2$ (or $M_V(\mathrm{HB}) = 0.9$) and an age of $(15 \pm 1) \times 10^9$y. The adopted model composition is $Y = 0.2$, $Z = 0.006$ ($[\mathrm{Fe/H}] = -0.5$), with $\alpha = 1.6$.

Full discussions of this work are being prepared and will appear in the journal literature. For much practical help and stimulation we are indebted to Peter Stetson and Don Vanden Berg.

REFERENCES

Da Costa, G.S.: 1982, Astron. J. 87, 990.
Hesser, J.E., and Hartwick, F.D.A.: 1977, Ap. J. Suppl. 33, 361.
Vanden Berg, D.A.: 1984, in preparation. See also Ap. J. Suppl. 51, 29 (1983).

OBSERVED DYNAMICAL PARAMETERS OF STAR CLUSTERS IN THE SMC

E. Kontizas and M. Kontizas

Observatory of Athens, University of Athens

The dynamical study of the SMC clusters is very important for many reasons. (a) The disk "blue" clusters are more massive and older than the galactic open clusters whereas the "halo" red clusters are less massive and younger than the galactic globulars (Kontizas, Danezis, Kontizas, 1982; Kontizas and Kontizas, 1983). (b) All clusters show density profiles that fit the available theoretical models (King, 1962) giving evidence of relaxed systems, even for the young clusters with evolutionary ages 10^6-10^8 yr. The relaxation mechanism has to be reconsidered for the SMC and possibly in connection to the metal abundance. (c) The ellipticities of young and old SMC clusters have shown that there is no obvious age-ellipticity dependance and that the SMC clusters are statistically more elliptical than the clusters of the LMC and our Galaxy (Frenk and Fall, 1982, Kontizas et al, 1984).

The tidal radii, masses, relaxation times and the $^M/L$ ratios of 43 clusters have been derived by means of star counts on plates taken with the 1.2 m U.K. Schmidt Telescope in Australia (Kontizas, 1984). The "red" halo clusters have tidal radii and concentration parameters comparable to those of our galaxy whereas the young "blue" clusters have diameters much larger than those of the mean values of the galactic open clusters.

The derived ellipticities of 25 clusters have shown that there is not statistically significant correlation with age or integrated magnitude (Van den Bergh, 1981 ; Kron and Kron, 1983).

For the LMC clusters Van den Bergh (1984) has suggested that the correlation age-ellipticity, reported previously, results from the fact that the blue LMC clusters are, in the mean, brighter than the old red LMC clusters.

The ellipticities of the disk SMC clusters are plotted versus their masses in Fig. 1. From this diagram it can be seen that there is a linear correlation of mass-ellipticity. The halo clusters are not considered since their derived masses have been found within two values

J. Goodman and P. Hut (eds.), Dynamics of Star Clusters, 85–87.

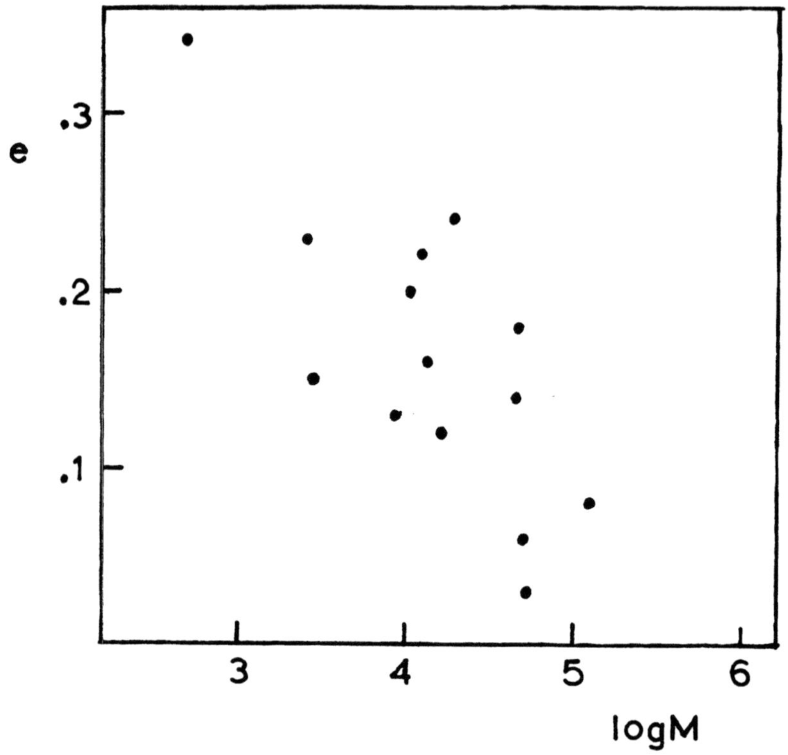

Fig. 1. Ellipticity-mass relation for the disk SMC clusters.

M_{max} and M_{min} (Kontizas, 1984).

A preliminary investigation of the orientation of the major axes of all the 25 studied clusters has shown that there is a preferable direction within 30°. The orientation of the major axes and the mass-ellipticity relation permit us to suggest that rotation should be considered as an important mechanism of the clusters' flatness.

REFERENCES

Frenk, C.S. and Fall, S.M., 1982. M.N.R.A.S. 199, 565.
Gordon, K. and Kron, G. : 1983, PASP, 95, 461.
King, I.R. : 1962, Astron. J., 67, 471.
Kontizas, M., Danezis E., Kontizas, E. : 1982, Astron. Astroph. Suppl. 49, 1.
Kontizas, E., Kontizas, M., : 1983, Astron. Astroph. Suppl. 52, 143.
Kontizas M., : 1984, Astron. Astroph. 131, 58.

Kontizas E., Prokakis Th., Dialetis D. and Kontizas M. : 1984, IAU Colloquium No. 78, in press.
Van den Bergh, S. : 1981, Astron. Astroph. Suppl. 46, 79.
Van den Bergh, S. : 1984, IAU Symp. No. 108, p. 1.

WHERE'S THE CUSP? (OR CCD PHOTOMETRY OF GLOBULAR CLUSTER CORES)

Phyllis M. Lugger and Haldan Cohn
Department of Astronomy
Indiana University
Swain Hall West 319
Bloomington, IN 47405

Jonathan E. Grindlay
Harvard-Smithsonian Center for Astrophysics
60 Garden Street
Cambridge, MA 02138

ABSTRACT. UBVR CCD surface photometry for a sample of 11 globular clusters has been obtained at the Kitt Peak National Observatory (KPNO) #1 0.9 m telescope as part of a major study of globular cluster core structure. These data are being used to test for the presence of central cusps in the surface brightness profiles, which are expected to be present in clusters which have already undergone core collapse (Cohn and Hut 1984).

1. INTRODUCTION

Cohn and Hut (1984) have recently predicted that a significant fraction of high central concentration globular clusters have already undergone core collapse. While massive black holes can produce post-collapse expansion, it now appears rather unlikely that clusters do in fact harbor massive central black holes (Hertz and Grindlay 1984). Post-collapse expansion is instead expected to be driven by energy release from hard binaries (Inagaki 1984). Recent theoretical studies of binary driven post-collapse evolution (see e.g. Heggie elsewhere in this volume) predict that the central mass density profile should be characterized by a singular isothermal $\rho \propto r^{-2}$ or surface brightness $\sigma \propto r^{-1}$. Thus, the surface brightness in a post-collapse cluster should continue to increase with decreasing radius, in contrast to the behavior of a King profile which flattens out inside of the core radius. Therefore, central surface brightness cusps are predicted even in the absence of black holes.

The electronographic surface photometry of the cluster M15 reported by Newell and O'Neil (1978) indicates the presence of a central cusp that has been confirmed by the photographic photometry of Hertz and Grindlay (1984), the CCD photometry of Djorgovski and Penner (elsewhere in this volume), and the CCD photometry reported here. A

J. Goodman and P. Hut (eds.), Dynamics of Star Clusters, 89–92.

cusp has also been detected in NGC 6624 in these latter three studies. Djorgovski and King (1984) and Djorgovski and Penner (elsewhere in this volume) report cusps in NGC 6342, NGC 6642, NGC 6681, and NGC 7099.

2. THE SAMPLE

To determine how widespread the cusp phenomenon is, a major study of the core structure of globular clusters is being carried out by obtaining UBVR surface photometry for a total sample of 28 northern clusters using the CCD direct camera at the #1 0.9 m telescope at KPNO. Data for 11 of these clusters were obtained during a 21-24 May 1984 observing run. These clusters are NGC 4147, NGC 5024 (M53), NGC 5272 (M3), NGC 5904 (M5), NGC 6229, NGC 6254 (M10), NGC 6341 (M92), NGC 6517, NGC 6624, NGC 6760, and NGC 7078 (M15). Additional data will be obtained at KPNO and also at the Cerro Tololo Interamerican Observatory (CTIO) in a related study of a large sample of southern globular clusters. A scheduled 4 night run on the CTIO 4 m PFCCD in July 1984 was unfortunately snowed out.

Data from the northern and southern samples will be used (1) to test for the presence of central cusps (i.e., luminosity excesses over normal King core surface brightness profiles), and (2) to investigate the dependence of core structure on galactic position and cluster core metallicity. The results of these studies will be of fundamental importance to our developing understanding of cluster dynamical evolution.

3. PRELIMINARY RESULTS

The U-band CCD frames obtained at KPNO for M15 and NGC 6624 have been analyzed using the SURFBRT routines on the KPNO Interactive Picture Processing System (IPPS). These routines were developed by Karen and Steve Strom for analyzing surface brightness profiles of elliptical galaxies. The core structure of globular clusters can best be studied in the U-band where the effects of individual red giants are minimized. As can be seen in Figures 1a and 1b, both M15 and NGC 6624 exhibit cusp structure. In both cases the surface brightness profiles rise as a power law into the seeing disk. The indicated seeing disk radii are the measured HWHM of stellar images from the same CCD frames as the cluster cores.

While no detailed fits to the cluster profiles have as yet been carried out, the power law portions of the profiles for $r_{seeing} < r < 10''$ have slopes $d \ln \sigma / d \ln r \approx -0.75$. As discussed by Cohn elsewhere in this volume, core collapse in a cluster in which the heaviest stars are nonluminous remnants can result in a slope such as that estimated from these data, which is flatter than the slope predicted for a single mass component cluster. Surface brightness profiles will be obtained for all 11 clusters using software developed by Hertz and Grindlay (1984). These profiles will be used to place limits on the seeing deconvolved central slopes of the profiles.

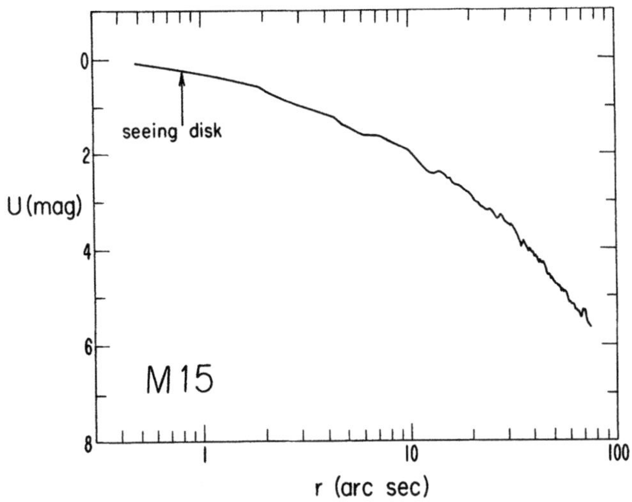

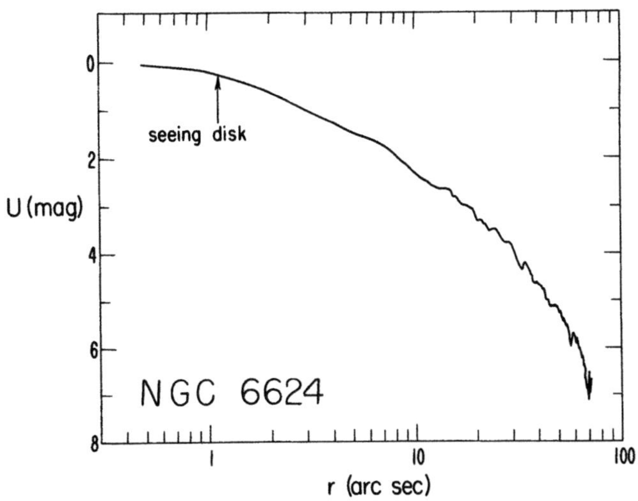

Figures 1a and 1b. U-band CCD surface brightness profiles of M15 and NGC 6624. The curves shown represent data points connected by straight line segments. Arrows indicate HWHM of the seeing profile. The zero point of the magnitude scale corresponds to the central value of the surface brightness.

The data obtained in this study will be used to investigate the dependence of core structure on galactic position (and hence tidal shocking rate) as well as on metallicity (and hence initial mass function). The question of whether the bulge clusters (which contain 7 of 9 of the high luminosity cluster X-ray sources) have evolved any differently than clusters farther out in the halo will be addressed. These data will also facilitate the construction of detailed, realistic models for evolving clusters using the Fokker-Planck methods developed by Cohn (1979, 1980).

REFERENCES

Cohn, H. 1979, Ap.J., **234**, 1036.

_________. 1980, Ap.J., **242**, 765.

Cohn, H. and Hut, P. 1984, Ap.J. (Letters) **277**, L45.

Djorgovski, S. and King, I. R. 1984 Ap.J. (Letters), **277**, L49.

Hertz, P. and Grindlay, J. 1984, Submitted to Ap.J.

Inagaki, S. 1984, M.N.R.A.S., **206**, 149.

Newell, B. and O'Neil, E. J. 1978, Ap.J. (Suppl.), **37**, 27.

ROTATIONAL FIELD AND VELOCITY DISPERSION IN GLOBULAR CLUSTERS: ω CEN AND 47 TUC

G. Meylan and M. Mayor
Geneva Observatory
CH-1290 Sauverny
Switzerland

ABSTRACT. Thanks to numerous and accurate radial velocity measurements obtained with CORAVEL, the rotational velocity fields, the spatial velocity dispersions from the centre to the edge and the ordered to random motions ratios are deduced for two globular clusters : ω Cen and 47 Tuc. Masses through virial theorem and M/L ratios are also given. A more comprehensive study will be published in Astron. & Astrophys.

1. OBSERVATIONS

The observations of accurate radial velocities have been collected with the CORAVEL photoelectric spectrometer (Baranne et al. 1979) mounted on the 1.5 m Danish telescope at Cerro La Silla, Chile. The number of observations amounts to 390 measurements of 302 stars in ω Cen and 265 measurements of 194 stars in 47 Tuc. The kinematical description given below has been obtained through the mean radial velocities of 298 member stars in ω Cen and 192 member stars in 47 Tuc. Typical uncertainties are 0.9 km/s per measurement in ω Cen and 0.6 km/s in 47 Tuc.

These observations were obtained through collaboration with astronomers of Geneva, Marseille, Copenhagen and ESO. New measurements in the nucleus of 47 Tuc have been obtained since the publication of 47 Tuc radial velocity catalogue (Mayor et al. 1983). Radial velocities in ω Cen are in course of publication.

A preliminary discussion of the kinematics of 47 Tuc has been published, including a few results concerning stellar duplicity (Mayor et al. 1984).

2. ROTATION V(r,z)

The rotational velocity fields V(r,z) result from a least-squares fit between the observed radial velocities and the projection onto the line of sight of a parametric form of the rotation curve. These functions

J. Goodman and P. Hut (eds.), Dynamics of Star Clusters, 93–96.

allow to estimate the main characteristics of the differential rotation in these two clusters :

- solid body rotation in the nucleus

  ω Cen : $\Omega_c = 4.6 \text{ km s}^{-1} r_c^{-1} = 1.3 \cdot 10^{-6} \text{ y}^{-1}$

  47 Tuc: $\Omega_c = 2.7 \text{ km s}^{-1} r_c^{-1} = 4.9 \cdot 10^{-6} \text{ y}^{-1}$
- the maximum velocity V_{max}
- the position of the maximum velocity
- the differential rotation in the outer parts
- the decrease of rotation towards the poles (non cylindrical rotation)

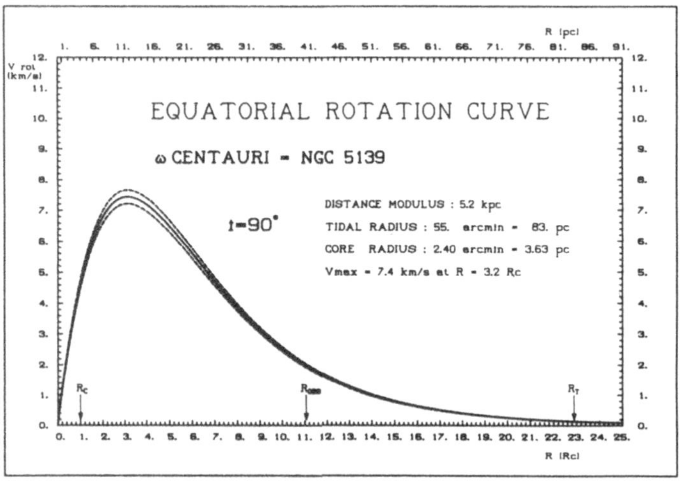

Figure 1.

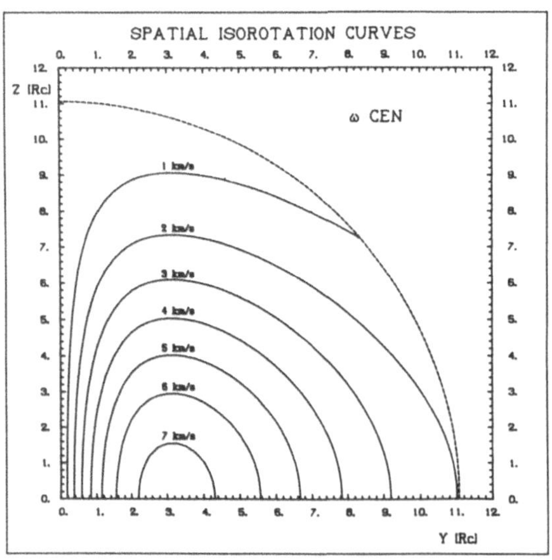

Figure 2. Spatial (non integrated) isorotation curves deduced from the rotational velocity law.

Figure 1 displays the equatorial rotation curve of ω Cen deduced from the fit. The non cylindrical rotation appears in figure 2. These results were obtained under the hypothesis of $i = 90$ degrees (equator-on).

3. VELOCITY DISPERSION

Contributions of rotation and integration along the line of sight having been eliminated, the spatial velocity dispersion component due only to random motions has been obtained in a few concentric shells : 8 for ω Cen (figure 3) and 7 for 47 Tuc (figure 4). These results show the same tendency in both clusters : slow and regular increase of the velocity dispersion from the edge of the clusters towards the centre with a strong value in the cores.

Figures 3 and 4. True velocity dispersion in the 8 and 7 concentric shells of ω Cen and 47 Tuc.

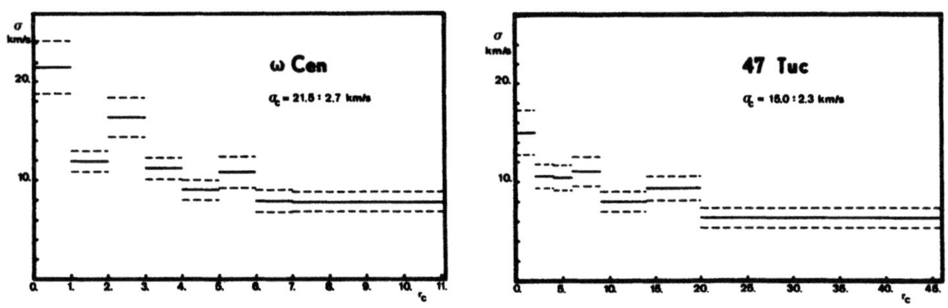

Deduction of the mean velocity of the cluster having been made, the distributions of radial velocities do not exhibit any difference from normal laws. Thus the large values of the velocity dispersions in the cores do not result from a few stars with exceptional velocities. These velocity dispersions are outstandingly larger than those obtained by Peterson & King (1975) (table 1) from the central surface brightness, integrated magnitude and models.

Table 1.

	Central dispersion from central surface brightness Peterson & King (1975)	Central dispersion from $c=\log r_t/r_c$ and integrated magnitude Peterson & King (1975)	Observed central dispersion from CORAVEL radial velocities
ω Cen	10.96 km/s	10.22 km/s	21.5 ± 2.7 km/s
47 Tuc	10.52 km/s	8.42 km/s	15.0 ± 2.3 km/s

This excess of dispersion in the cores of both these clusters corresponds to the presence of some concentrated and non luminous mass.

4. ORDERED / RANDOM MOTIONS RATIOS

The relative importances of ordered and random motions are given by the ratio v_o/σ_o , where v_o^2 is the mass-weighted mean square rotation speed and σ_o^2 is the mass-weighted mean square random velocity along the line of sight.

The results visible in figure 5 show that ω Cen and 47 Tuc are in close agreement with models of oblate axisymmetric systems with small ellipticity and slight anisotropy.

Figure 5. The dashed line shows the trajectory followed by $[v_o(i)/\sigma_o, \varepsilon_a(i)]$ when i decreases from 90°, while ε_{true} and δ are held constant. For both clusters results are given for i = 90° and 60°.

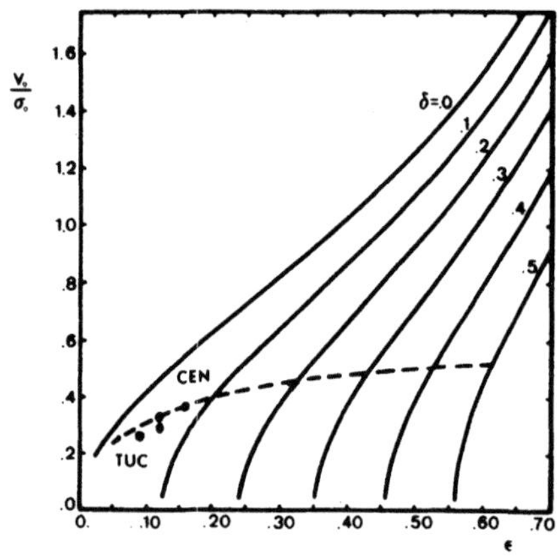

$i=90^o$	v_o/σ_o	v_{max}/σ_c
ω Cen	0.32±0.05	0.34±0.06
47 Tuc	0.26±0.05	0.31±0.06

5. MASSES AND MASS-LUMINOSITY RATIOS

Using the virial theorem, we obtain for ω Cen a mass of $3.8\cdot10^6$ $M_\odot$ and for 47 Tuc a mass of $0.91\cdot10^6$ $M_\odot$. With the absolute visual magnitude given by Harris and Racine (1979) the M/L_V ratio is equal to 3.5 ± 1.4 for ω Cen and 1.8 ± 0.8 for 47 Tuc, results in good agreement with Illingworth (1976).

BIBLIOGRAPHY

Baranne, A., Mayor, M., Poncet, J.L. : 1979, Vistas in Astronomy 23,279

Harris, W.E., Racine, R. : 1979, Ann. Rev. Astron. Astrophys. 17, 241

Illingworth, G. : 1976, Ap.J. 204, 73

Mayor, M., Imbert, M., Andersen, J., Ardeberg, A., Baranne, A., Benz, W., Ischi, E., Lindgren, H., Martin, N., Maurice, E., Nordstroem, B., Prevot, L. : 1983, Astron. Astrophys. Suppl. 54, 495

Mayor, M., Imbert, M., Andersen, J., Ardeberg, A., Benz, W., Lindgren, H., Martin, N., Maurice, E., Nordstroem, B., Prevot, L. : 1984, Astron. Astrophys. 134, 118

Peterson, C.J., King, I.R. : 1975, Astron. J. 80, 427

BV CONCENTRIC APERTURE PHOTOMETRY OF GLOBULAR CLUSTERS

Charles J. Peterson
Department of Physics & Astronomy
University of Missouri
Columbia, Missouri 65211

Globular clusters are among the most difficult types of objects to observe with standard photoelectric techniques. Among factors which may contribute to measurement error are the centering of apertures in the oftentimes diffuse cores of clusters, the random distribution of bright stars near the cluster centers, and the determination of an adequate sky correction in fields badly crowded by foreground stars. These problems have been well-discussed by a number of authors (see, e.g., Hanes and Brodie 1984, and the references they cite).

The most comprehensive published tabulation of photoelectric data has been the study by Kron and Mayall (1960). Other observers have observed more limited samples of clusters for determination of total magnitudes, for determination of colors for chemical abundance or interstellar absorption studies, and for comparison of observed luminosity growth curves with the predictions of dynamical models based on the spatial distribution of the individual stars within the clusters.

As many clusters still had not been studied adequately for proper comparison with dynamical models, we began in 1976 a program of observations at Cerro Tololo Inter-American Observatory of as many of the globular clusters of our Galaxy as possible. Multi-aperture photometric observations were obtained during 28 complete or partial nights during the years 1976 to 1980. Approximately 4000 V and B measurements were made with the 0.4, 0.9, 1.0, and 1.5 meter telescopes with the use of various series of aperture sizes; 101 clusters in the Milky Way Galaxy and 1 cluster which is a member of the Small Magellanic Cloud system were observed. These data together with data from all other sources are being used for a reassessment of cluster central surface brightnesses, cluster total magnitudes, and cluster structural parameters as defined in the self-consistent dynamical models of King (1966; see also Peterson and King 1975 and Peterson 1976).

As discussed elsewhere in this symposium, high resolution surface photometry shows that surface brightness profiles in the central regions of only a small number of the globular clusters of our Galaxy

J. Goodman and P. Hut (eds.), Dynamics of Star Clusters, 97–98.

deviate from the predicted profiles of the simple King models; hence, for most clusters the original concept of the core radius is still valid. Seitzer (this symposium) has argued that the tidal radii of the King models are valid limiting radii as set by the gravitational tidal field of the Galaxy; tidal or limiting radii may be determined observationally only by detailed star counts in the outer regions of clusters. Concentric aperture photometry converted to the radial variation of surface brightness ties together the high resolution photographic and CCD studies in the cluster centers with the deep star count studies in the outer regions to produce overall structural profiles of the clusters which can be compared to the models of King or others. Of importance to note is that the present observational work does show the existence of systematic errors in some of the older photometric data which in turn may have led to a few errors in our original estimations of cluster structural parameters.

A preliminary assessment of cluster color (B - V) as a function of radius is also now underway. The data suggest, in agreement with the conclusion of Hanes and Brodie (1984), that, in general, radial color gradients in globular clusters are extremely difficult to detect if they are real; repeated observations especially with very small aperture sizes show that single measurements may be subject to very large errors probably due to the statistical distribution of a few very bright stars near the cluster centers.

This work has been supported in part by the National Science Foundation.

REFERENCES

Hanes, D.A., and Brodie, J.P. 1984, M.N.R.A.S., pre-print.
King, I.R. 1966, Astron. J. **71**, 64
Kron, G.E., and Mayall, N.U. 1960, Astron. J. **65**, 581.
Peterson, C.J. 1976, Astron. J. **81**, 617.
Peterson, C.J., and King, I.R. 1975, Astron. J. **80**, 427.

AN EXTENSION OF THE SEARCH FOR SPECTROSCOPIC BINARIES IN M3[1]

Carlton P. Pryor
Department of Physics, University of Victoria
Victoria, British Columbia V8W 2Y2, Canada

David W. Latham and Martha L. Hazen-Liller
Harvard-Smithsonian Center for Astrophysics
60 Garden Street
Cambridge, Massachusetts 02138, USA

ABSTRACT. We have obtained 295 new radial velocities for the 112 giants in the globular cluster M3 previously observed by Gunn and Griffin. Our velocities have a typical accuracy of 0.8 km/s per measurement and have been combined with the Gunn and Griffin data in order to search for radial velocity variations over a time span of ten years. We find no convincing evidence that any of the giants observed are spectroscopic binaries with one notable exception, von Zeipel 164, which we believe is the first spectroscopic binary to be found in a globular cluster. Modelling of the velocity variations that would be expected in our data for a variety of binary populations confirms Gunn and Griffin's conclusion that binaries with separations of less than 10 AU must occur much less frequently among the giants of M3 than among the population I field stars.

1. INTRODUCTION

In 1980 we began programs to measure accurate radial velocities of faint stars in order to study the dynamics of clusters and the frequency of binaries in a variety of environments. Our basic approach was to obtain digital spectra with photon-counting Reticon detectors and echelle spectrographs on the telescopes at Mt. Hopkins. These spectra were then cross-correlated against a set of standard template spectra using modified versions of the analysis software originally developed for the CfA Redshift Survey (Tonry and Davis 1979).

In 1979 Gunn and Griffin (hereafter GG) published their excellent velocities for 111 giants in the globular cluster M3, and we undertook to remeasure several of their stars with the 1.5-m Tillinghast Reflector. Our initial results proved to have the same accuracies as the GG data. Moreover, no obvious velocity variables emerged when the two data sets were combined, further supporting GG's claim that globular clusters are deficient in binaries. Subsequently we undertook a major program to reobserve all of GG's objects

[1]Observations reported here were obtained in part with the Multiple Mirror Telescope, a joint facility of the Smithsonian and the University of Arizona.

J. Goodman and P. Hut (eds.), Dynamics of Star Clusters, 99–101.

one or more times, using the Multiple Mirror Telescope (MMT) as well as the 1.5-m telescope. In section 2 we summarize the radial velocity observations for the handful of giants in M3 which appear to have variable velocities.

Our data, when combined with that of GG, should allow the detection of most of the binaries with periods less than 10 years (their velocity amplitudes would exceed 15 km/s peak-to-peak for two 0.8 solar mass stars in circular orbits viewed edge-on) and should also provide some constraints on the frequency of binaries with longer periods. We have used a two-parameter model of the binary population (one parameter governs the overall fraction of the stars which occur in binaries while the other sets the lower cutoff to the allowed binary separations), and we have determined the parameter ranges that are not excluded by the data. The results of this modelling are summarized in section 3.

2. VELOCITY VARIABLES

We have found only one giant in M3, von Zeipel (vZ) 164, that shows good evidence for velocity variations due to orbital motion. The four observations covering ten years are listed in Table I. The total amplitude is at least 18 km/s and the period appears to be a few years. Note that vZ 164 is more than twenty core radii away from the cluster core, and we are fairly confident that the correct star was observed each time, because the field is not at all crowded. We believe that vZ 164 is a spectroscopic binary, the first to be found in a globular cluster.

TABLE I. Radial velocities for vZ 164, a giant in M3 that appears to be a spectroscopic binary.

Velocity	Error	Julian Day	Telescope
-141.21		2442199.70	Palomar
-159.78	0.7	2445477.74	MMT
-151.38	1.2	2445842.83	1.5 meter
-151.69	1.3	2445867.74	1.5 meter

There are two classes of mechanisms that may have removed primordial binaries from the sample of giant stars that we observed in M3. For long period, widely separated, and weakly bound systems, encounters with other stars tend to disrupt the binary. Binaries with orbital velocities larger than the cluster velocity dispersion (9 km/s at the center) have a good chance of surviving. For systems with circular orbits and one solar mass total, this corresponds to a separation of roughly 10 AU. On the other hand, primordial binaries that are too close will be modified or even disrupted by mechanisms such as mass loss or mass transfer during the evolution of the primary from a main sequence star to a giant. The giants in M3 have radii as large as 0.4 AU, but it is hard to see how mass loss or mass transfer could be effective at modifying binaries with separations larger than about 1 AU. Thus there should be a window of separations between roughly 1 and 10 AU where primordial binaries in M3 could survive, corresponding to a velocity window of roughly 30 to 9 km/s for systems with circular orbits and one solar mass total. It may be significant that the first spectroscopic binary to be found among the giants in M3 appears to be in this velocity window and is located in the outer region of the cluster, where stellar encounters are much rarer than in the cluster center.

GG found that two stars, vZ 764 and 911, fell off the cluster mean velocity of -146.9 km/s by +3.5 and -4.5 times the central velocity dispersion of the cluster (if the dispersion is calculated with the two high-velocity stars rejected). We observed these two stars several times, and we find no evidence for any velocity variation.

In addition to the two well-known pulsating variables vZ 318 and 803, our velocity data also provide evidence for small, seemingly random, velocity variations in some of the bightest cluster giants. The total range of variation can be as high as 5 km/s. We find six stars, each with at least five observations, for which the chi-square values of the differences have a probability of occurrence of less than 1 percent: vZ 490, 807, 858, 1053, 1397, and 1449. All six of these stars are among the brightest in the cluster, but only half the stars brighter than photovisual magnitude 12.8 appear to have this variability. Mayor et al. (1984) have recently reported the independent discovery of a similar phenomenon in 47 Tuc.

3. BINARY FREQUENCY

Harris and McClure (1983) have claimed that GG's failure to detect spectroscopic binaries among the giants in globular clusters is not inconsistent with the fairly high frequency of binaries that they found in a survey of field giants. They argued that the large diameters of the globular cluster giants observed by GG would remove short-period binaries with large orbital velocities from the sample. In addition, the smaller masses of the cluster giants would also reduce the velocity amplitudes for any binaries that might survive, compared to the younger field giants. We have taken advantage of the larger data set available to us (111 stars with multiple observations compared to GG's 33, and a time coverage of ten years compared to their four) and have modelled the residuals that would be expected in our data for a two-parameter family of binary populations. The details of this analysis will be reported elsewhere. In summary, an Abt and Levy (1976) distribution of binaries is rejected at the 96 percent confidence level unless all binaries closer than about 1 AU are removed from the sample, or alternatively the fraction of binaries is reduced from 0.54 to less than about 0.2.

REFERENCES

Abt, H. A., and Levy, S. G. (1976). Astrophys. J. Suppl. 30, 273.
Gunn, J. E., and Griffin, R. F. (1979). Astron. J. 84, 752.
Harris, H. C., and McClure, R. D. (1983). Astrophys. J. Lett. 265, L77.
Mayor, M., Imbert, M., Andersen, J., Ardeberg, A., Benz, W., Lindgren, H., Martin, N., Maurice, E., Nordström, B., and Prévot, L. (1984). Astron. Astrophys. 134, 118.
Tonry, J., and Davis, M. (1979). Astron. J. 84, 1511.

A SEARCH FOR CATACLYSMIC BINARIES IN THE GLOBULAR CLUSTER M3

M. M. Shara*
Space Telescope Science Institute, Baltimore
A. F. J. Moffat*
Dept. de Physique, Universite de Montreal
D. A. Hanes
Anglo-Australian Observatory

ABSTRACT. Deep Hβ narrowband and broadband images of M3 have been electronically blinked to search for cataclysmic binaries. Tests of the method on a known, faint cataclysmic enable us to set limits on the sensitivity of the technique. No bright ($M_B < 6$) emission-line (equivalent width > 12 Å) cataclysmic binaries exist in M3 between 4 and 30 core radii from the center. Low luminosity globular X-ray sources could still be weak-lined (E.W. < 12 Å) and bright ($M_B \simeq +5$ like some old novae) or strong-lined (E.W. ≃ 60 Å) and faint ($M_B > 7$ like dwarf novae).

INTRODUCTION

Binary stars (*except* for cataclysmic variables) seem to be at least 30 times rarer in globular clusters than in the Galactic disk (Webbink 1980, IAU Symp. No. 88). For example, a sensitive radial velocity search has failed to detect a single spectroscopic binary (Gunn and Griffin 1979, A.J. *84*, 752) amongst 111 giants in M3. No W Ursae Majoris stars have been found on deep plates of M55 (Trimble, 1976 BAAS *8*, 443) or NGC 6397 (Budding and Alexander 1978, NATO Globular Cluster Workshop, Cambridge). Not a single ecliping binary of any other type is known to be a definite globular cluster member.

There are tantalizing indications, however, that cataclysmic variable (CV) binaries (which comprise a red dwarf transferring matter via a luminous accretion disk onto a white dwarf companion) are very common in globulars. One possible and two near certain classical novae have been sighted in globular clusters in the last century (and more may have been missed). Novae recur every $\sim 10^{5-6}$ years (Ford, 1978, Ap. J. *219*, 595), thus implying that a) 10 to 100 old novae reside in an average globular cluster (with a few far enough from the cluster core to be resolvable) and b) novae are 10 to 100 times more common in globulars than in the Galactic disk. The recent

*Guest observers, Canada-France-Hawaii telescope.

J. Goodman and P. Hut (eds.), Dynamics of Star Clusters, 103–104.

spectroscopic confirmations of a dwarf nova in M5 (Margon et al. 1981 Ap. J. Lett. *247*, L89) also hint that a systematic deep search for CV binaries in globulars would be rewarded.

Practically all of the ~500 known field cataclysmics have been discovered on the basis of their optical brightening to m $\lesssim$ 13. So few CVs are known in globulars, we believe, because no one has surveyed for them down to faint enough magnitudes or with the right technique. The main objective of this program is to discover CVs in globulars to a limiting magnitude m $\simeq$ 21 systematically using the method described below.

OBSERVATIONAL METHOD

The distinguishing characteristics of CV spectra in quiescence are a blue continuum which is essentially free from absorption lines but dominated by Balmer emission lines. These emission lines make even faint CVs detectable in the following manner.

We have used the Canada-France-Hawaii 3.6 meter telescope +90 m. ITT image tube camera at prime focus to obtain Hβ narrowband (35 Å FWHM) and broadband (350 Å FWHM) plates of i) a 19th magnitude, moderate line-strength cataclysmic binary and ii) a 20' diameter field centered on M3. The plates have been digitized on a Space Telescope Science Institute PDS machine and displayed on a de Anza image display. An empirical density-to-density transformation is derived for each plate pair which nulls stars in the difference image Hβ narrow-Hβ wide. Stars with excess Hβ (i.e. Hβ in emission) then stand out from all others, even in crowded fields.

RESULTS

The cataclysmic binary used to test the observational technique, WY Sge (Nova 1783) is at $m_B \simeq 19$, and has equivalent width (Hβ) $\simeq$ 18 Å. It was immediately and easily re-discovered as the only Hβ-emission object on the Hβ narrow-Hβ wide image. From our experience with WY Sge we estimate that our technique works for stars $\gtrsim$1 magnitude brighter than the plate limit, and with E. W. (Hβ) $\gtrsim$12 Å.

No Hβ emission-line candidates (with M_B < 6 and E.W. (Hβ) > 12 Å) were found in M3 between four and thirty core radii from the cluster center.

CONCLUSIONS

Significant numbers of bright, strong emission-line cataclysmic binaries do not exist in M3 (except possibly near the cluster core). Weak-lined novalike objects and dwarf novae would have escaped detection in this first survey.

More rigorous data reduction is underway, as are deeper searches for very faint cataclysmics; both will be reported in detail elsewhere.

POSSIBLE CHARACTERISTICS OF SUPERNOVA INDUCED ENRICHMENT OF GLOBULAR CLUSTERS

Graeme H. Smith
Harvard Smithsonian Center for Astrophysics
60 Garden Street
Cambridge, MA 02138 U.S.A.

Within globular clusters, stars of similar temperatures and gravities commonly exhibit differences in the strength of the λ3883 CN band, the distribution of which is often bimodal. An archetype example of this phenomenon is NGC 6752 (Norris et al. 1981). As documented by Da Costa and Cottrell (1980), and Norris et al. (1981), the CN-rich giants in this cluster possess enhancements in nitrogen abundance by factors of 7-8 relative to the CN-poor giants. This may indicate that chemical enrichment of proto-cluster gas clouds took place as a result of the activity of massive stars, which manufactured nitrogen within their interiors and ejected it into the cluster gas while low mass stars were forming. The aim of this paper is to investigate whether the properties of NGC 6752 mentioned above might be consistent with supernova enrichment. Both the free and adiabatic expansion phases of supernova shells (Spitzer 1968) will be discussed in this context.

The Free Expansion Phase

A proto-cluster gas cloud is idealized to be spherical in shape, with an internal density distribution of the form $\rho(r) = \rho_o r^{-2}$, where r is the distance from the cloud center. Upon denoting the total cloud mass as M_6 (in units of $10^6 M_\odot$), and the cloud radius as R, the density distribution can be written as

$$\rho(r) = 5.42\times10^{-24} M_6 R_{100}^{-1} r_{100}^{-2} \quad \mathrm{gm\ cm^{-3}},$$

where a subscript of 100 denotes that a length is being expressed in units of 100 parsecs.

Suppose that a supernova explosion occurs at a distance of $r_{1,100}$ from the cloud center. Initially the ejecta will expand at constant velocity, until the mass of ambient material that has been swept up exceeds the mass of the stellar ejecta. The radius $s_{a,100}$ of the bubble at this time will be

$$s_{a,100} = 1.44\times10^{-2} (M_j/M_6) R_{100}^{1/3} r_{1,100}^{2/3},$$

J. Goodman and P. Hut (eds.), Dynamics of Star Clusters, 105–108.

where M_j is the mass (in units of $M_\odot$) of ambient material that has been swept up. The bubble will be contained within the cloud during this phase if the site of the explosion is within a distance

$$r_{A,100} \sim 579\ (M_6/M_j)^{1/2}\ R_{100}$$

of the cluster center. Taking M_j to be $\sim$ 10 times the mass of the stellar ejecta, which may in turn range from 1 to 10 $M_\odot$, the above relation indicates that for protoclouds of masses $M_6 > 10^{-3}$, $r_{A,100} > R_{100}$, and the supernova remnants will be contained at the conclusion of their free expansion phase.

The Adiabatic Expansion Phase

Once M_j has become sufficiently large, the supernova bubble will enter a phase of adiabatic expansion. The remnant will now become elongated, but for the purposes of the present calculations is nonetheless assumed to be spherical throughout the entire adiabatic phase. In addition, it is assumed to be small in size relative to $r_{1,100}$, so that the evolution can be treated as that of a remnant expanding into a uniform medium of density equal to that locally at $r_{1,100}$.

Throughout the adiabatic phase the total energy of the gas remains equal to the initial explosion energy E. The front of the remnant is a shock wave whose expansion for a gas of $\gamma = 5/3$ is governed by the equation $P = 2\lambda E/3V$, where P is the pressure immediately behind the shock wave, V is the volume enclosed within the remnant, and λ is defined as the ratio of the energy density behind the shock front to the average energy density within the remnant ($\sim$ 1.6; Spitzer 1968). In the strong shock approximation, the ratio of post-shock density to ambient gas density is 4. The post-shock temperature can then be calculated from the ambient density, the perfect gas relation, and the pressure. Following Spitzer (1968), it is assumed that radiative energy losses from the shock front become important when the post-shock temperature drops below 10^6 K. The radius $s_{b,100}$ of the remnant at this time is

$$s_{b,100} = \{3.04\times10^{-4}\ E_{50}\ M_6^{-1}\ R_{100}\ r_{1,100}^{2}\}^{1/3},$$

where E_{50} is the explosion energy in units of 10^{50} ergs.

The amount of ambient material that has been swept up by the remnant when it reaches this size is

$$M_B = 101\ E_{50}\ (M_\odot).$$

The mass M_B depends only on the supernova energy, and for an explosion of $E_{50} = 10$, $\sim 10^3\ M_\odot$ of ambient material will be swept up prior to the cessation of adiabaticity. Should the nitrogen produced by the precursor star become incorporated into this shell material, and if enriched stars can form from it, then such stars will possess abundance enhancements that reflect only the amount of nucleosynthesised material ejected by the supernova. Should stars of only a small mass range constitute the supernova precursors, the chemical enrichment induced by each would be similar, and a bimodal abundance distribution could be established within a cluster. The calculations of Renzini and Voli (1981) indicate that $\sim 0.1\ M_{\odot}$ of primary nitrogen could typically be ejected by intermediate mass (5-8 $M_{\odot}$) stars. Consider for example, a protocloud with an initial abundance of [A/H] = -1.5, [N/A] = 0. The mass of ambient nitrogen swept up would be 0.03 $M_{\odot}$. Upon distributing the ejected nitrogen throughout the $10^3\ M_{\odot}$ shell, enriched stars could be formed with a nitrogen enhancement of $\Delta[N/A] = [N/A]_{new} - [N/A]_{old} = 0.64$ relative to the CN-poor stars. Such a nitrogen difference is similar to that observed between the CN-rich and CN-poor giants in NGC 6752 (Da Costa and Cottrell 1980, Norris et al. 1981). Smith and Norris (1982) estimate that the excess mass of nitrogen present in the NGC 6752 CN-rich stars, relative to that in the CN-poor, is 10 $M_{\odot}$. To produce this amount, 100 supernova precursors are required. The combined explosions of these stars would sweep up $10^5\ M_{\odot}$ of gas by the end of adiabaticity, which is consistent with the total mass of the CN-rich population within NGC 6752.

It appears that supernova enrichment, providing that it occurs near the end of the adiabatic expansion phase, can explain some of the properties of the CN-rich stars in clusters such as NGC 6752. However no physical explanation can be given as to why enriched stars should form out of the remnant material just when its expansion ceases to be adiabatic. In addition, there are other observations and constraints which a supernova enrichment scenario must be able to accommodate, and two of the most important of these are as follows.

(i) In view of the apparently large number of intermediate mass stars needed in clusters such as NGC 6752, confinement of the supernova ejecta within the protocloud poses a serious problem. It may be that the stars which produce primary nitrogen do not explode as supernovae, but instead distribute their products by means of a stellar wind (e.g. Kennicutt [1984] finds a lower mass limit to Type II supernova precursors of 8 $M_{\odot}$).

(ii) The CN-rich giants in NGC 6752 are depleted in carbon by a factor of $\Delta[C/A] \sim 0.3$ relative to the CN-poor giants (Da Costa and Cottrell 1980, Norris et al. 1981). This phenomenon is common to other clusters such as M3 and M13 (Suntzeff 1981), and as discussed by Smith and Norris (1982), is difficult to account for in terms of a primordial enrichment framework. It suggests instead that the CN anomalies originate within the CN hydrogen burning regions of the cluster giants themselves. A recent discussion of this possibility is given by Langer and Kraft (1984).

REFERENCES

Da Costa, G. S. and Cottrell, P. L. 1980, Astrophys. J. (Letters), 236, L83.

Kennicutt, R. C. 1984, Astrophys. J., 277, 361.

Langer, G. E. and Kraft, R. P. 1984, Pub. Astron. Soc. Pac., 96, 339.

Norris, J. E., Cottrell, P. L., Freeman, K.C. and Da Costa, G. S. 1981, Astrophys. J., 244, 205.
Renzini, A. and Voli, M. 1981, Astron. Astrophys., 94, 175.
Smith, G. H. and Norris, J. 1982, Astrophys. J., 254, 594.
Spitzer, L. Jr. 1968, Diffuse Matter in Space, Interscience Publishers, John Wiley and Sons.

PRECOLLAPSE EVOLUTION OF GLOBULAR CLUSTERS

Lyman Spitzer, Jr.
Princeton University Observatory
Princeton, New Jersey 08544 USA

1. INTRODUCTION

The dynamical evolution of a cluster in which N, the number of stars, is large, results primarily from encounters between pairs of stars; these tend to modify the system so that it approaches a more probable configuration. In statistical equilibrium, the distribution of stars among states of different energy, E_i, per unit mass would be determined to a first approximation by the probability P_i, where

$$P_i \propto g_i \exp(-BE_i) \quad ; \qquad (1)$$

here g_i is the statistical weight of each state, proportional to the amount of phase space available, and B is a constant. Evidently dynamical evolution tends a) towards stellar states which are more tightly bound by their mutual gravitational attraction and have a correspondingly lower E_i, and b) towards the free state, where the volume of available phase space is very large. The only relatively stable state for an isolated cluster is one in which a central, non-evolving massive configuration (a compact object, or tight binary, or a hierarchical multiple system) has all the initial binding energy of the cluster and all other stars have escaped.

The general tendency of large-N clusters towards a combination of collapse and disruption has been known for some years. The different mechanisms by which such a cluster evolves and the possible evolutionary sequences resulting were reviewed ten years ago, at IAU Symposium No. 69 at Besançon (Spitzer, 1975). The present paper discusses the progress made in this field during the past decade; the topics considered are essentially those related to the precollapse phase until n, the number of stars per unit volume, approaches its peak value. Work on the postcollapse phase is reviewed in the following paper by D. C. Heggie. The extensive research in these fields prior to 1974 is mostly omitted from the present review, apart from a few passing references. Some of the topics discussed here have been treated in greater detail in the review by Lightman and Shapiro (1978).

The three sections which follow this Introduction discuss certain

J. Goodman and P. Hut (eds.), Dynamics of Star Clusters, 109–137.

physical processes which have been the subject of considerable research during the last decade. Section 2 treats the gravothermal instability, which appears to be a dominant process in a late phase of cluster evolution. The development of mass stratification, which is more important in the early phases, and which is associated with its own instabilities, is discussed in Section 3. Effects produced by the hardening of binaries are reviewed in the fourth Section. Discussion of the various additional effects which appear during the collapse phase, and presumably bring to a halt the contraction of the central core, are presented in the final Section 5.

It should be emphasized that these different physical processes are generally defined for idealized clusters. Analysis of each effect separately helps us to understand the evolution of actual clusters, and of the numerical models which approximate real clusters. The discussion here attempts to explain the phenomena observed in these models in terms of the separate idealized physical processes. However, it must be remembered that when all these processes are occurring at the same time, it may not be possible even in principle to isolate precisely the effects produced by each process.

This review is not in any sense complete, and necessarily reflects the author's special interests and background. Thus the effect of a massive black hole on cluster evolution is not discussed, despite the elegant and interesting theory of this subject developed by various authors. Similarly, several papers on the origin and early evolution of globular clusters are not included.

This survey deals essentially with isolated clusters, since there has been little work during this last decade on the effects produced by the tidal field of the Galaxy. However, one should keep in mind that evaporation of stars from many of the globular clusters is much increased by a tidal cut-off at a finite escape radius, r_t, and may be a dominant effect on the evolution of such clusters. The importance of this process is apparent in the results presented at this Symposium by Stodółkiewicz; according to his detailed model, enhanced evaporation leads roughly to the linear decrease of bound mass with time found by Hénon (1961) in his pioneering work. For some clusters, strong tidal shocks will increase the evaporation rate further. Fall and Rees (1977) have suggested that the surviving clusters we observe are those for which these processes do not lead to complete disruption in 10^{10} years.

One prominent characteristic of the research reviewed here is the wide variety of techniques which have been employed. Theoretical cluster models have been constructed using the fluid equations, several types of Monte Carlo programs, numerical integrations of the Fokker-Planck equations, N-body calculation of individual orbits, and hybrid programs making use of several of these techniques in the same models. The methods used in some of these approaches have been greatly extended and improved during the last ten years. As the following sections indicate, all this activity has substantially increased our theoretical understanding of how a globular cluster must evolve with time, though there are still major uncertainties concerning the processes that occur late in the core collapse phase.

The contact between this impressive structure of theory and astronomical observation is admittedly rather incomplete. Even during the precollapse phase, where theoretical problems seem minimal, the observational data cannot be said to verify the theory in any detail. My training under Sir Arthur Eddington, together with my own inclinations, lead me to believe that the theory, based on simple dynamical assumptions of unquestioned validity, has an independent credibility.

2. GRAVOTHERMAL INSTABILITY

2.1. Instability criteria

The numerous analyses of the gravothermal instability all stem from the classic paper by Antonov (1962), who investigated the stability of gaseous isothermal spheres bounded by a rigid, insulating outer spherical shell, with a radius denoted here as r_b. Antonov's work is based on the entropy, S, of the system, defined as

$$S = - \int f(\underline{r}, \underline{v}, t) \ln f(\underline{r}, \underline{v}, t) \underline{dr}\underline{dv} \quad , \tag{2}$$

where f is the usual phase-space distribution function (to which we shall usually refer as the velocity distribution function). Equation (2) is the so-called Boltzmann entropy, consistent with the one-particle distribution function considered in equation (1). We assume here, and throughtout all of Section 2, that all stars in the system have the same mass, m, and that no binary or multiple stars are present.

We now consider what changes in S are possible when the total mass, M, energy, E, and volume, V, of the system are held fixed. At all times the velocity distribution function, f, is assumed Maxwellian and the system is taken to be spherically symmetrical, since these two assumptions maximize S; changes in the system are produced by changes in δf, corresponding to changes in particle density and local mean square velocity. In an equilibrium system δS must vanish to first order in δf, but is proportional to $(\delta f)^2$. Antonov demonstrated that if the density contrast, D, between the center and the radius $r = r_b$ is less than 709, then δS is negative for all perturbations; the equilibrium state is the most probable one, and is thermodynamically stable. For a density contrast exceeding 709 the system will be unstable for some spherical perturbations, provided that the perturbed velocity distribution remains Maxwellian.

The physical principles underlying this effect were discussed some time ago by Lynden-Bell and Wood (1968), who named it the "gravothermal instability". They pointed out the importance of negative specific heat in driving such an instability in self-gravitating systems. In a bounded isothermal sphere with a sufficient central condensation, the central core can contract and become steadily hotter as it loses heat to the outer regions.

A series of investigators have analyzed the gravothermal instability criterion from several viewpoints. Hachisu and Sugimoto (1978) reconsidered Antonov's problem, computing δS for changes from an equilibrium state, with M, E and V kept constant, and assuming both spherical symmetry and a locally Maxwellian velocity distribution. Their mathematical method is quite different, however, and yields the critical D values for modes with various numbers of nodes. For the fundamental mode, a value of 709 is again found for D.

A more direct demonstration of instability is provided by a time-dependent analysis of perturbations, demonstrating that these grow exponentially under some conditions. Such an analysis has been carried out by Nakada (1978) for the same isothermal sphere considered by Antonov, with a short mean free path assumed for the self-gravitating particles. As D increases above a critical value, essentially the same as found by Antonov, the growth rate increases above zero, and is proportional to the thermal conductivity.

Yet another technique for demonstrating the appearance of the gravothermal instability is use of the classical "linear series" method. This method, developed originally for static systems in dynamical equilibrium, is based on the nature of a series of equilibrium configurations in which some parameter, μ, is assumed to vary. A static system is characterized by a potential energy W, which will depend on μ. If with increasing μ, W increases to a maximum and then decreases, the point of maximum W is called a "turning point"; in general the system will be dynamically stable on one side of a turning point and unstable on the other (Jeans, 1929).

This same technique has been applied (Lynden-Bell and Wood, 1968) to systems in thermodynamic equilibrium, especially to the bounded isothermal sphere considered by Antonov. In this case, -S, the negative of the total entropy, replaces W. The gas is in equilibrium if S is a maximum under any redistribution of mass and energy, again keeping mass, energy and volume constant. For the parameter μ we take the density contrast, D, between the center and the gas just inside the bounding sphere, and we consider a series of equilibria all of the same mass and volume, but of different temperature and energy. For D only slightly exceeding unity, gravothermal binding is unimportant and the system is stable. At D = 709, the entropy has a minimum value, and for greater D the system is thermodynamically unstable, in agreement with Antonov's results. The mathematical basis for this method as applied to clusters and the various conditions required for its application have been discussed by various authors--see Horwitz and Katz (1978) and Katz (1978).

Since M and V are constant along the series of equilibria considered here, one can visualize a given mass moving along the series if heat is slowly added or subtracted. Since dS = TdE under these conditions, it is clear that when S is an extremum so is E; for the critical configuration at which S is a minimum, E also has its minimum value. For application of the linear series method to clusters it is generally simpler to compute E rather than S along the sequence. A plot of E against 1/T is a spiral, with each extremum in E or 1/T corresponding to the onset of some instability (Katz, 1978).

Evidently there is no question about the existence of the gravothermal instability in a bounded isothermal gas sphere. However, actual stellar systems differ from these idealized spheres in at least two major respects. First, the mean free path is much longer than the dimensions of the system. Second, the velocity distribution is not Maxwellian to begin with, since in the absence of any confining surface, the velocity distribution function, $f(\underline{\underline{r}}, \underline{\underline{v}})$, must vanish for v greater than the escape velocity.

Because of the long mean free path (LP), $f(\underline{\underline{r}}, \underline{\underline{v}})$ will generally be anisotropic if the mean stellar kinetic energy varies systematically with distance from the cluster center. In such systems, $f(\underline{\underline{r}}, \underline{\underline{v}})$ must be essentially constant along a dynamical trajectory at any one time, if the relaxation time much exceeds the dynamical crossing time. Hence if the core becomes hotter than the outer regions, as one expects when the gravothermal instability develops, then far out in the system $f(\underline{\underline{r}}, \underline{\underline{v}})$ for radial orbits, which pass through the core, will tend to vary less steeply with E than will the corresponding distribution function for tangential orbits. Thus the perturbations possible for a LP system differ from the perturbations with local Maxwellian distributions which are possible in the case of short mean free path (SP). As a result the entropy increase possible in such perturbations is somewhat less in the LP case than in the SP case.

The instability expected in this LP case, when the initial equilibrium is again an isothermal sphere bounded by a reflecting shell, has been studied by Ipser and Kandrup (1980) and by Inagaki (1980). The former showed that the criterion for instability was the same as for the SP case analyzed by Antonov (1962). The latter analyzed the time-dependent growth of these perturbations in the LP case, though with an isotropic velocity distribution assumed in the Fokker-Planck equation. As expected, the growth rate varied inversely as the interparticle relaxation time. This latter study, like the parallel SP analysis by Nakada (1978) discussed above, gave eigenfunctions for the instabilities, though with non-Maxwellian values for the perturbed velocity distribution function. The physical differences between the SP and LP systems do not seem to have much qualitative effect on the initial appearance of the gravothermal instability, though of course the growth rates in these two cases depend very differently on the density and the velocity dispersion.

In systems which are not confined by a bounding surface, $f(\underline{\underline{r}}, \underline{\underline{v}})$ must be truncated at the escape energy. In such a configuration the entropy is not a local maximum, and the system is thermodynamically unstable to begin with. It is this instability that leads to evaporation of stars and resultant contraction of the system, effects fully studied more than a decade ago. However, it is not implausible that additional contraction, related to the gravothermal instability, will appear at the turning points computed without regard to the initial non-Maxwellian character of $f(\underline{\underline{r}}, \underline{\underline{v}})$. Analyses of such instabilities have been carried out by Katz (1980) for various energy-truncated models, in which f vanishes above some cut-off energy. The gravothermal instability is found to set in when the escape velocity $v_\infty(0)$ from the center is about 3.8 times the central velocity dispersion $v_m(0)$. As shown in Figure 1, a comparison with the Monte Carlo computations of Spitzer and Thuan (1972) shows that the core collapse seems to change its character

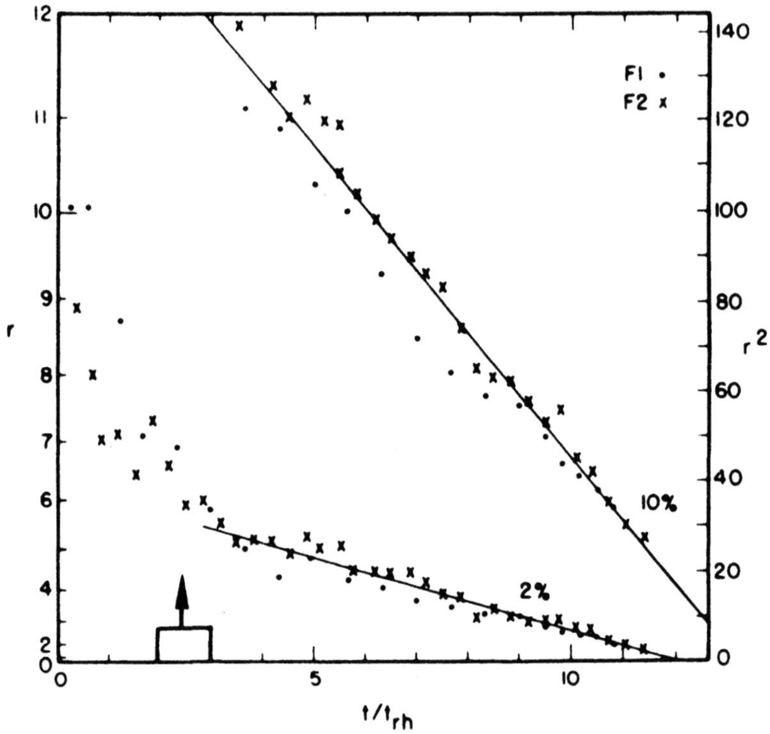

Figure 1. Comparison of Cluster Evolution with the Predicted Onset of the Gravothermal Instability. The plotted points show the radii containing 2 and 10 percent, respectively of the cluster mass for model F of Spitzer and Thuan (1972), plotted as a function of time since the cluster was born as a uniform sphere. For comparison, the arrow shows the time at which the cluster should become subject to the gravothermal instability, as computed by Katz (1980).

at about this time, with r_1^2 subsequently decreasing linearly with time; r_1 is the radius enclosing two percent of the mass. These results seem to provide some confirmation of the gravothermal instability in realistic systems.

2.2 Self-similar collapse model

To make further progress in this field, one must analyze the detailed consequences of the gravothermal instability. Hachisu et al. (1978) have carried out a time-dependent calculation for a confined isothermal gaseous sphere, with an initial density contrast, D, much exceeding the critical value of 709. The core can either expand and cool or contract and heat up. If it expands, the sphere gradually approaches another equilibrium configuration with $D < 709$. If the core starts to contract, it continues to shrink without limit, developing a structure in which ρ is constant at small r and varies as $r^{-\alpha}$ for larger r. The value of α depends on the functional form of the thermal conductivity; if the formula for a fully ionized gas is used, α is 2.4, the same value found by Larson (1970) in his analysis based on the moment equations.

A more realistic analysis, based on an improved formula for the heat flow in a globular cluster, has been given by Lynden-Bell and Eggleton (1980). They considered a solution of the self-similar type, as in the familiar Sedov solution for a supernova shell. Thus $\rho(r)$ is a function of a scaled radius r_*, equal to $r/r_c(t)$, where $r_c(t)$ may be taken as the core radius. Hence

$$\rho = \rho_c(t)\rho_*(r_*) \quad , \tag{3}$$

where $\rho_c(t)$ is the density at $r = 0$. Analogous equations apply for $M(r)$, the mass interior to r, for v_m, the rms random velocity and for F_c, the conductive energy flux per unit area at the radius r.

Their equation for the conductive heat flux, F_c, modifies the standard formula for a perfect gas to allow roughly for the long mean free path in a cluster. In the limit of short mean free path (SP), we have the usual result

$$F_c = C_1 \rho v_m \lambda \frac{d}{dx}\left(\frac{3kT}{m}\right) \quad , \tag{4}$$

where v_m is again the rms velocity, λ is the mean free path, and C_1 is a constant of order unity. All particles are again assumed to have the same mass, m. Here $\lambda dT/dx$ is the excess temperature between the starting point of the particles and the stopping point one collision later. In the LP limit, this excess temperature becomes HdT/dx, where H is the scale height of the particle orbits. In addition, in this case the number of orbits over a distance H required for one collision is t_r/t_d, the ratio of the collision time, or relaxation time, to t_d, the dynamical crossing time; this ratio is about equal to H/v_m. Hence F_c in equation (4) must be multiplied by the factor $(H/\lambda) \times (t_d/t_r)$, yielding the result

$$F_c = \frac{C_1 \rho H^2}{t_r} \frac{d}{dx} \left(\frac{3kT}{m}\right) . \tag{5}$$

If we set H proportional to the distance $v_m/(G\rho)^{1/2}$, take for t_r the expression (Spitzer and Hart, 1971a, equation (30), with $\Psi = 1$ for only one mass component present)

$$t_r = \frac{v_m{}^3}{15.4 G^2 \rho m \ell n \Lambda} , \tag{6}$$

and replace 3kT/m by $v_m{}^2$, we obtain the final equation adopted by Lynden-Bell and Eggleton (1980) for the conductive flux,

$$F_c = \frac{C_2 \rho G m \; \ell n(0.4N)}{v_m} \frac{dv_m{}^2}{dx} , \tag{7}$$

where C_2 is another constant.

A second assumption is that all perturbations vanish as r_* becomes large. Thanks to this assumption the outer regions of the cluster become irrelevant, and the cluster as a whole can be an isothermal sphere, either bounded or infinite, or a more complex configuration with an inner isothermal region surrounded by a halo of stars moving in more nearly radial orbits. For the actual LP case the velocity distribution must clearly be anisotropic, as we have seen in subsection 2.1. While this anisotropy is ignored in the self-similar solution, the analysis provides a first approximation for the gravothermal collapse of the inner regions of realistic clusters.

The time dependent functions $r_c(t)$, $M_c(t)$, etc. are identical with those obtained in self-similar solutions for simple model clusters with evaporation of stars (Gurevich and Levin, 1950a, King, 1958). In particular, for $r_c(t)$ and $M_c(t)$ we have

$$r_c(t) \propto (1 - t/t_0)^{(4-2\zeta)/(7-3\zeta)} , \tag{8}$$

$$M_c(t) \propto (1 - t/t_0)^{2/(7-3\zeta)} . \tag{9}$$

Evidently t_0 is the time at which the collapse becomes singular, with r_c and M_c approaching zero together, while ρ_c becomes infinite. The parameter ζ is the ratio of the relative change of energy to the relative change of mass for the cluster core, giving

$$\frac{1}{E_c} \frac{dE_c}{dt} = \frac{\zeta}{M_c} \frac{dM_c}{dt} . \tag{10}$$

For evaporation of stars from an isolated cluster as a result of distant two-body encounters, the escaping stars have a very small kinetic energy at infinity and ζ is a small negative number, approaching zero with increasing N. Another relation for self-similar collapse is

$$v_c{}^2 \propto \rho_c{}^{(1-\zeta)/(5-3\zeta)} . \tag{11}$$

For a gravothermal collapse, $\rho_*(r_*)$ has been determined numerically (Lynden-Bell and Eggleton, 1980) and is shown in Figure 2. This

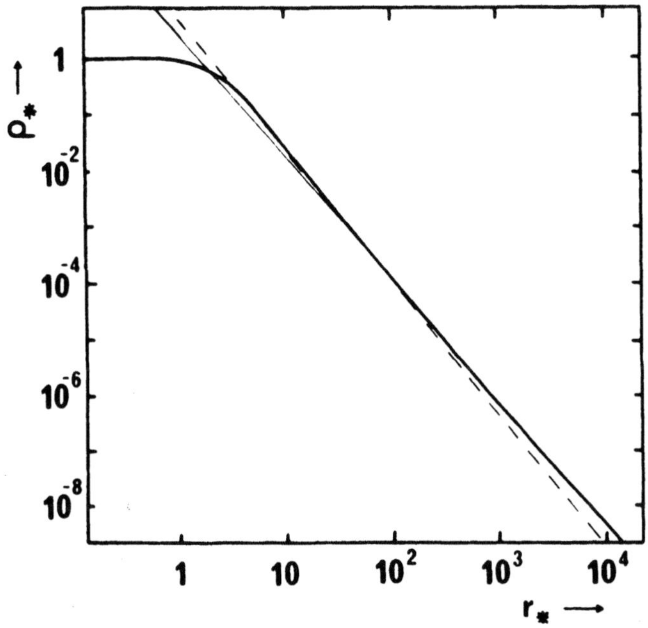

Figure 2. Density Profile of Similarity Solution for Gravothermal Collapse. The solid curve shows the density profile for the self-similar solution found by Lynden-Bell and Eggleton (1980); $\rho_* \equiv \rho(r,t)/\rho(0,t)$ is plotted against $r_* \equiv r/r_c(t)$, where $r_c(t)$ is the radius of the contracting core. The dashed line shows ρ_* varying as $r_*^{-2.4}$, while the thin straight line shows the asymptotic variation of ρ_* as $r_*^{-2.21}$.

determination is an eigenvalue problem for the parameter, ζ, whose value is found to be 0.737. The asymptotic logarithmic slope, α, of the $\ell n\rho_* - \ell n r_*$ plot is a simple function of ζ and equals 2.208. As a result of the inwards temperature gradient, this density distribution is slightly steeper than for the isothermal sphere, in which ρ varies asymptotically as r^{-2}. The systematic material velocity, w_r, is directed inwards for small r_* and outwards for large r_*, vanishing at the radius where $\rho_* = 0.0071$, corresponding to $r_* = 16$ in Figure 2.

2.3 Comparison with realistic cluster models

As a test for the importance of the gravothermal instability we may compare this idealized first approximation with detailed cluster models. In the early Monte Carlo models by Spitzer and Thuan (1972) the scale height, κ, of the core varies with time as $(1 - t/t_0)^{0.65}$, in comparison with the exponent 0.53 found in equation (8) for $\zeta = 0.737$. Also, the macroscopic velocity, w_r, was found to reverse sign when the local density was less than $\rho(0)$ by some two orders of magnitude, in rough agreement with the theory.

A more conclusive confirmation of self-similar gravothermal collapse is obtained in later realistic cluster models, which achieve greater precision and extend to later times. The central density in these later models increases by some 3 to 4 orders of magnitude as compared with an increase of only 2 orders of magnitude in the earlier work. The Fokker-Planck equation, with f taken to depend both on energy, E, and angular momentum, J, was used by Cohn (1979) in a detailed numerical computation of cluster evolution. A new Monte Carlo method, which follows the diffusion of stellar orbits in E, J space, but with two important differences from the Hénon (1971a,b) method, was developed and applied by Marchant and Shapiro (1980). Variations in the time step used give the correct mean changes of velocity during each orbital period, and radial variations in the number of shells (Hénon's superstars) per unit mass give the necessary additional data needed to follow the evolution of a core whose mass continually decreases. These two modifications much increase the precision of the calculations, especially for late stages of core collapse.

An important result obtained with these models is the variation of velocity dispersion, v_m, in the core with changing core density. Figure 3 shows the relationship obtained in these two types of models. The value of v_m at $r = 0$ is designated here as V_0. The slope of the dashed line, which is about the same as for the plotted data, corresponds to ρ_c varying as V_0^{20}, or to an exponent 0.10 in equation (11), as compared to the value 0.094 found for $\zeta = 0.737$. For the evaporative model of core collapse, $\zeta = 0$, and the predicted exponent in equation (11) is 0.20, twice the "observed" value obtained from Figure 3. These results provide rather strong confirmation that the gravothermal instability rather than evaporation is the dominant mechanism in the core collapse of isolated clusters.

The self-similar solution of Lynden-Bell and Eggleton received further confirmation from a much more extended Fokker-Planck computation by Cohn (1980). In this work f was averaged over J, thus ignoring

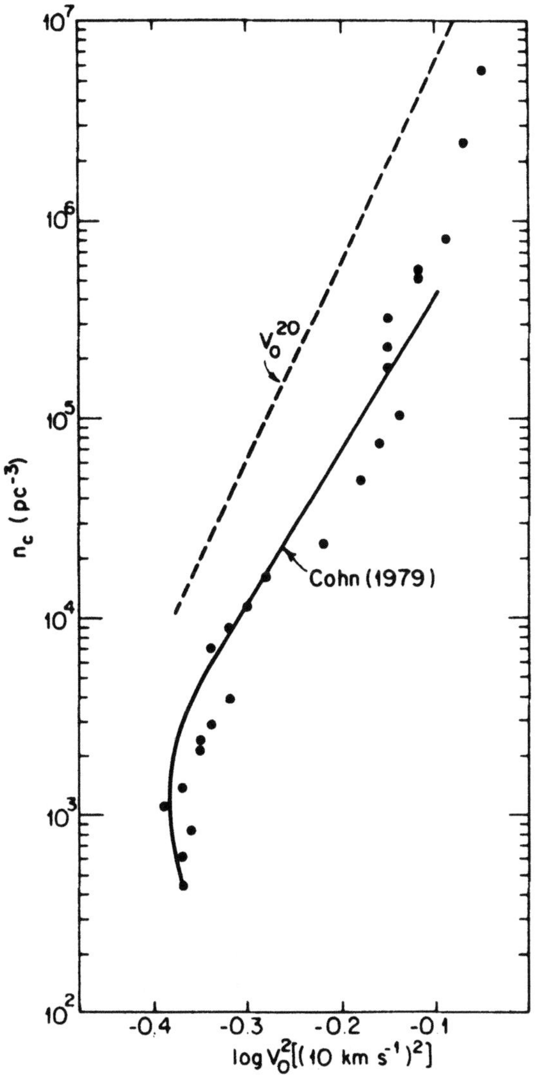

Figure 3. Variation of Central Density with Velocity Dispersion. The plotted points show n_c, the central density in the cluster core, as a function of V_0, the rms velocity dispersion at the center, computed by Marchant and Shapiro (1980) with a Monte Carlo program. The solid line shows these same quantities computed by Cohn (1979) from the time-dependent Fokker-Planck equation. For comparison the dashed line represents n_c varying as V_0^{20} during the evolution.

the anisotropy of the velocity distribution. With the simplification so achieved, the computations were extended to a central density increasing by almost 20 orders of magnitude, a somewhat formal extension, since the core mass at the end is much less than a solar mass! An interesting feature of the results is that at late times the density ρ outside the innermost core varies asymptotically as $\rho^{-2.23}$, in very close agreement with the result $\rho^{-2.21}$ obtained from the similarity solution.

Another noteworthy result of Cohn's work is that as the cluster collapses, the collapse rate, measured by $t_r(0)d\ell n\rho(0)/dt$, approaches a constant value equal to about 6×10^{-3} in the first Fokker-Planck computation, with non-isotropic velocities considered, and 3.6×10^{-3} in the second one, with f(E, J) averaged over J. This difference may result in part from the increased precision in the later computations, though one would expect some difference to result between these two types of velocity distribution. In any case, the collapse is slow in that hundreds of relaxation times are required for appreciable evolution.

The anisotropy of the velocity distribution associated with the gravothermal collapse is evidenced by the Monte Carlo models of Duncan and Shapiro (1982), who used the same techniques as did Marchant and Shapiro (1980). Their result, confirming earlier less complete computations by Spitzer and Shull (1975), showed the development in late collapse stages of marked velocity anisotropies outside the core but well within the half-mass radius r_h. At a radius enclosing one fourth the total mass the mean square radial velocity exceeded half the mean square tangential velocity by 20%. It is clear physically that when the velocity dispersion of the core increases the mean square radial velocity of all stars on radial orbits will tend to increase, since f(E, J) is constant along a dynamical trajectory. One of the ways in which kinetic energy is transmitted to the outer regions is by collisional deflections of stars in radial orbits, which increase the mean square tangential velocity of stars in the outer regions of the cluster. Not unexpectedly, for r within $3r_c$ the velocity distribution remains nearly isotropic. Evidently this complex process is not considered in the similarity solution obtained by Lynden-Bell and Eggleton.

Despite this appearance of velocity anisotropies in their detailed models, Duncan and Shapiro (1982) find that in all other respects their collapse solution is very similar to that obtained with isotropic velocities assumed. In particular, the density varies about as $\rho^{-2.2}$ for radii between r_c and r_h, in close agreement with the results noted above for isotropic velocity distributions, both in the self-similar solution and in Cohn's isotropic model. We have already seen that the variation of ρ_c with v_c in the anisotropic Monte Carlo models is in relatively close agreement with the self-similar solution, in which the velocity distribution is isotropic. Apparently within the half-mass radius r_h, the approximation of isotropic velocities makes very little difference in overall behaviour of a cluster during the collapse phase.

The effect of close encounters during the collapse phase has been studied by Goodman (1983), who finds that large-angle deflections do not make any important difference, even when the number of stars in the core is small. Similar conclusions for the evaporation of stars from clusters at earlier stages were reached by Spitzer and Thuan (1972), who compared their Monte Carlo results with Henon's analysis (1960) of close encounters, and also by Retterer (1979).

3. MASS STRATIFICATION

3.1 Instability criteria

An idealized theory showed some time ago (Spitzer, 1969) that an isolated cluster with two mass components, m_1 and m_2, cannot achieve kinetic equilibrium in the core if the total mass, M_2, of the heavier stars, is more than a certain critical fraction of the total mass, M_1, of the lighter stars. The physical basis for this result is that when $\rho_2(0)/\rho_1(0)$, the ratio of densities of the two components at the cluster center, much exceeds unity, the self-attraction of the heavier stars tends to give them an appreciable mean square random velocity v_{m2} which is greater than the equipartition value $(m_1/m_2)^{1/2}v_{m1}$. An approximate theory gave as a condition of equilibrium the following limit on M_2/M_1

$$\mathscr{S} \equiv \frac{M_2}{M_1}\left(\frac{m_2}{m_1}\right)^{3/2} < 0.16 \quad . \tag{12}$$

Vishniac (1978) has found that with a continuous mass spectrum a relation similar to equation (12) can be derived, with m_1 equal to the mass of the lightest stars present, provided that the total fraction of the cluster mass in stars of mass exceeding m_2 replaces M_2/M_1. The critical value of $\mathscr{S}$ so redefined is about 2; this order of magnitude increase over the corresponding value in equation (12) results in part from the fact that m_1 is substantially less than the mean mass of the stars lighter than m_2.

The equilibrium of two-component systems has received considerable attention in the last decade, with some half dozen papers published in this area. This work generally indicates that when equipartition is in fact present in a system confined by its own gravitational field, the computed equilibrium configurations become unstable if the mass fraction in the heavier stars exceeds some critical value, roughly comparable with that found in equation (12). Stable equilibria seem possible with greater mass fractions of the relatively heavy stars, but such equilibria are either not self-confined or are not characterized by equipartition.

Analyses of two-component isothermal systems confined by a rigid bounding shell of radius r_b have been carried through by Saito and Yoshizawa (1976), Lightman (1977) and Yoshizawa et al. (1978). These analyses determine the minimum energy, E_c, for different values of m_2/m_1 and M_2/M_1. If M_2/M_1 is either very large or very small this minimum energy, measured in terms of GM^2/r_b, has the same value -0.335 as for a one-component bounded isothermal sphere, with a density contrast, D, equal to 709. For intermediate values of M_2/M_1, E_c is positive and no equilibria of negative energy exist. For M_2/M_1 small, the condition that $E_c = 0$ corresponds to $\mathscr{S}$ about equal to 0.6, some four times greater than the upper limit in equation (12). For $\mathscr{S}$ equal to 0.16 and M_2/M_1 again small, $E_c = -0.25$. As E_c increases above this value, the confinement of the system by the pressure on the outer rigid shell presumably becomes increasingly important, and the system becomes less and less like an isolated cluster.

To analyze isolated systems a non-Maxwellian velocity distribution must be assumed, with f(E) vanishing for E positive, corresponding to stars which escape. In the well known models by King (1966), f(E) is truncated according to the law

$$f(E) = f_0(e^{-BE} - 1) \text{ for } E > 0, \quad f = \text{otherwise} \quad . \tag{13}$$

As in equation (1), E is the energy per unit mass. To give a finite system, the gravitational potential is assumed to vanish at a finite cut-off radius, r_t. Two-component systems based on this distribution have been analyzed by Merritt (1981) and by Kondrat'ev and Ozernoy (1982). They applied equation (13) separately for each of the two components, with masses m_1 and m_2. The latter authors assume that

$$B_1/m_1 = B_2/m_2 \quad . \tag{14}$$

If the central potential $\phi(0)$ has a large negative value, so that $-B\phi(0)$ is a large positive number, the cut-off is in the far tail of the Maxwellian distribution, and equation (14) gives equipartition of energy, with $m_1 v_{m1}^2 = m_2 v_{m2}^2$. For shallow potential wells equipartition of energy no longer follows from equations (13) and (14); for sufficiently small $|B\phi(0)|$, f(E) is proportional to BE for $v < v_\infty$, and the mean square velocities are equal for the two mass components. For example, if $m_2/m_1 = 3$, the ratio of kinetic energies of heavy to light stars increases to about 1.5 if $|B_1\phi(0)| = 3$ (Kondrat'ev and Ozernoy, 1982), approaching m_2/m_1 as $|B_1\phi(0)|$ decreases towards zero.

In the models by Merritt (1981) the ratio B_2/B_1 is adjusted to preserve equality of kinetic energies between the two mass components at the cluster center. Models based on this assumption are obtained for arbitrary values of M_2/M_1. However, the models with $\mathscr{S}$ much exceeding 0.16 seems quite unrealistic; for example, for a model with $\mathscr{S} = 1$ the fraction of heavy stars within the tiny central core, to which equipartition is limited, is only about 10^{-8} ! The stability of these various two-component King models has not been investigated.

Katz and Taff (1983) have considered two-component models in which the velocity distribution function has the following form, based on one proposed by Wilson (1975),

$$f(E) = f_0(e^{-BE} - 1 + BE) \quad . \tag{15}$$

Equation (14) was assumed, relating the values of B for the two components. The detailed one-component numerical models (as, for example, the Fokker-Planck model by Cohn, 1979) show a distribution function in the central regions which is intermediate between equations (13) and

(15). In the models by Katz and Taff (1983) the gravitational potential $\phi(r)$ was assumed zero at a boundary radius, r_t. The linear series method (§2.1) determined the critical condition at which instability sets in.

The results are shown in Figure 4, which plots the critical value of $k \equiv -B_1\phi(0)$, the dimensionless depth of the potential well, as a function of $\mathscr{A}$, defined in equation (12). Each critical value corresponds to a turning point in a plot of total energy against B_1. The different curves in Figure 4 correspond to different values of m_2/m_1; the arrow is drawn to indicate the critical value of $\mathscr{A}$ given in equation (12). Evidently if $\mathscr{A}$ increases much above 0.16, the maximum depth of the potential well consistent with the theoretical criterion for stability decreases rather steeply. As pointed out above, with decreasing $|B_1\phi(0)|$, equipartition disappears, and in the limit v_{m1} approaches v_{m2}. When $m_2/m_1 = 4$ and $\mathscr{A} = 0.54$, the deviations from equipartition at the cluster center are about 50%, and are even greater in the outer regions.

These results on the stability of two-component clusters are not entirely conclusive, in view of theoretical uncertainties concerning the linear series method as applied to systems with non-Maxwellian f(E),--see § 2.1. However, the general evidence certainly points to the conclusion that in realistic clusters kinetic equilibrium cannot be established in the core if the mass fraction of heavy stars exceeds a small value, and the system of heavy stars at the cluster center must continue to shrink compared to the light stars.

3.2 Collapse of realistic clusters

Ever since the early Monte Carlo computations (Spitzer and Hart, 1971b) it has been known that the heavier stars tend to move into the central core, where there is an evident tendency toward equipartition, and that the core of heavier stars then proceeds to collapse in much the same way as do the cores in single-component clusters. Subsequent computations (Spitzer and Shull, 1975; Saito and Yoshizawa, 1976; Angeletti and Giannone, 1977; Inagaki and Wiyanto, 1984) confirm this general result. There are no simple time-dependent theoretical solutions, such as the self-similar solution for the sphere subject to gravothermal instability, with which multi-component models can be compared. As a result general conclusions are somewhat more difficult to extract from these models.

One result which is of theoretical significance, though not surprising, was obtained by Saito and Yoshizawa (1976), who used Larson's (1970) fluid dynamic technique, based on moments of the Boltzmann and Fokker-Planck equations, to follow the evolution of an isothermal sphere with two mass components confined by a rigid shell. They assumed equal velocities for the two components initially. If the system energy was well above the minimum possible, corresponding to stability against either gravothermal or mass stratification collapse, they found that equipartition developed at the expected rate and the system approached a stable equilibrium. Lightman and Fall (1978) compared early Monte Carlo results with a simple analytical theory for the overall

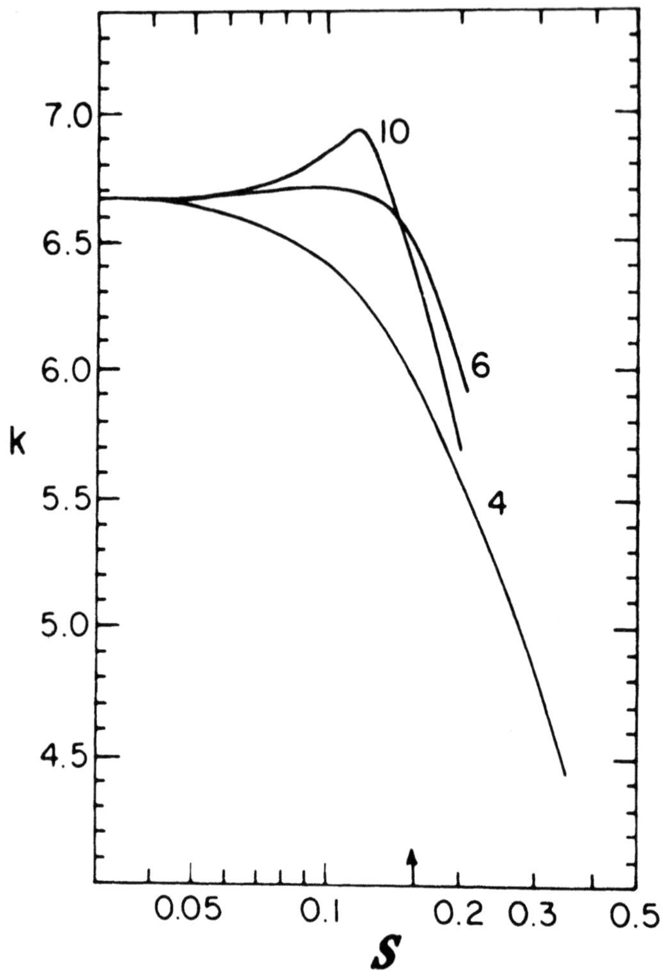

Figure 4. Theoretical Onset of Mass Stratification Instability. The curves show the maximum depth of the potential well (in dimensionless units) for which a cluster with stars of the indicated mass ratio is stable, according to the theory by Katz and Taff (1983). The quantity $\mathcal{S}$, defined in equation (16), was found to be less than 0.16 for stable equilibrium (see arrow) in an early theory by Spitzer (1969).

properties of the central core, and obtained reasonable fits. They considered two effects,--the mass stratification instability and evaporation from the core, on the assumption that the latter phenomenon is responsible for the collapse of one-component models; as we have seen, more recent computations indicate that this assumption seems invalid.

An extensive set of model two-component cluster computations was obtained by Inagaki and Wiyanto (1984), who used the time-dependent Fokker-Planck equation, with an isotropic velocity distribution, following Cohn (1980). With m_2/m_1 fixed at 2, they constructed models with M_2/M_1 equal to 0.01, 0.05, 0.11 and 1.0; the corresponding $\mathscr{S}$ values are 0.028, 0.14, 0.31 and 2.8. The results show that for the lowest $\mathscr{S}$ the central core soon approaches equipartition (with $(T_2 - T_1)/T_1$ equal to about 2%) and stays there until just before the final collapse. As $\mathscr{S}$ increases above unity, the minimum relative difference of central temperatures between the two components (i.e., the minimum relative difference of kinetic energies) increases, with $(T_2 - T_1)/T_1$ equal to about 17 percent for $\mathscr{S} = 2.8$. During the final collapse phase the core is mostly composed of heavy stars even for $M_2/M_1 = 0.01$. As shown in earlier studies, many of the heavy stars remain outside the core, and do not approach equipartition.

Inagaki and Wiyanto (1984) follow Lightman and Fall (1978) in explaining their data for core evolution by the predictions of mass stratification and the gravothermal instability, using for ζ in equation (10) the value 0.74 obtained both from the self-similar solution and the more realistic calculations for one-component clusters. In all cases, the core collapse rate, measured by $t_{r2} d\ell n\rho_2/dt$, since the core becomes composed largely of the heavier stars, is equal to the corresponding value which they find for a system composed entirely of heavy stars. This value equals about 3.5×10^{-3} (when the relaxation time they used is multiplied by 3/4 to agree with the different $t_r(0)$ used by Cohn (1980)), a rate which is essentially identical with that found also by Cohn (1980) for one-component systems with an isotropic velocity distribution. This general result strongly suggests that the final core collapse in the two-component clusters is due to the gravothermal instability. The chief effect of mass stratification in realistic clusters is to produce rather rapidly a concentration of heavy stars within the core and to shorten the time of evolution until the gravothermal stability becomes dominant. Also, since t_r varies as $1/\langle m\rangle$ for constant ρ, the dominance of heavy stars increases the final rate of collapse, though the time required for this collapse process, once it begins, is already very short.

The effect of stellar mass loss on cluster evolution has been considered by Angeletti, Dolcetta and Giannone (1980), who employed the fluid-dynamical approach developed by Larson (1970). Multi-component models, with 5 or 8 stellar groups, were considered, and stellar evolution during the cluster life was considered. The purpose of this ambitious work was to obtain a fit with the observed detailed properties of clusters. A certain qualitative agreement was obtained with M3, but the uncertainties in the initial conditions and the neglect of shock heating make the results somewhat tentative.

4. ENERGY EXCHANGE WITH BINARIES

4.1 Basic processes

The exchange of energy between single stars and binaries and its possible importance in globular clusters has been known since the pioneering work of Gurevich and Levin (1950b). These authors also pointed out the difference between "soft" and "hard" binaries. The former have net binding energies less than the translational energy of single stars and tend to gain energy and become softer as a result of gravitational interactions with passing stars. In contrast, "hard" binaries tend to lose energy and become harder. At the 1974 Besançon symposium, two important papers were in press giving much more extensive information on the interactions between single stars and binaries,--extensive numerical computations by Hills (1975a) and a very thorough analytical discussion by Heggie (1975).

The most complete numerical information on the complex interactions between single stars and binaries has since been obtained by Hut (1983), based on several hundred thousand orbital integrations. The results give definitive values for the mean exchange of energy in such encounters. Figure 5 shows the rate coefficient $\langle\sigma\Delta\rangle$ for this process, computed separately for Δ positive and Δ negative; σ is the cross section, while Δ is the relative change in binary binding energy. All stars have the same mass, m. The horizontal scale shows v_r, the velocity of the single star relative to the mass center of the binary, expressed in units of v_c, the incoming velocity for which the total energy of all three stars (regarded as mass points) vanishes. For three stars of identical mass m, $v_c^2 = 3Gm/2a$, where a is the semi-major axis of the binary orbit. Figure 5 shows clearly the "watershed effect", such that Δ tends to be negative for large v_r/v_c, and positive when v_r/v_c is small.

An important feature of such data is that they resolve a disagreement between Heggie and Hills in dE_b/dt, the average rate of increase of E_b, the binary's binding energy (defined as positive), in interactions with single stars. Heggie (1975) expressed his results in the form

$$\frac{dE_b}{dt} = n_s\langle v_r\sigma\Delta\rangle E_b = \frac{4(3)^{1/2}A}{35} \times \frac{n_s G^2 m_s{}^3}{v_{sm}}\ , \qquad (16)$$

where n_s is the particle density of stars and v_{sm} is the root mean square random velocity of the single stars. For the constant A, Heggie (1975) obtained a value about 45; the data by Hills (1975a) give a value of 17 for A. In the results by Hut (1983), plotted in Figure 5, A is not exactly constant with v_r/v_c (as indicated by the systematic change in separation between the open circles and the dashed line), but equals about 30 for a stellar kinetic energy about 10 percent of E_b.

Encounters between two binary systems can be important within an evolved cluster core, where under some conditions binaries can outnumber single stars. An asymptotic theory of this process has been developed by Spitzer and Mathieu (1980), who assume that the harder, more tightly bound binary behaves as a single star, and displaces one of the

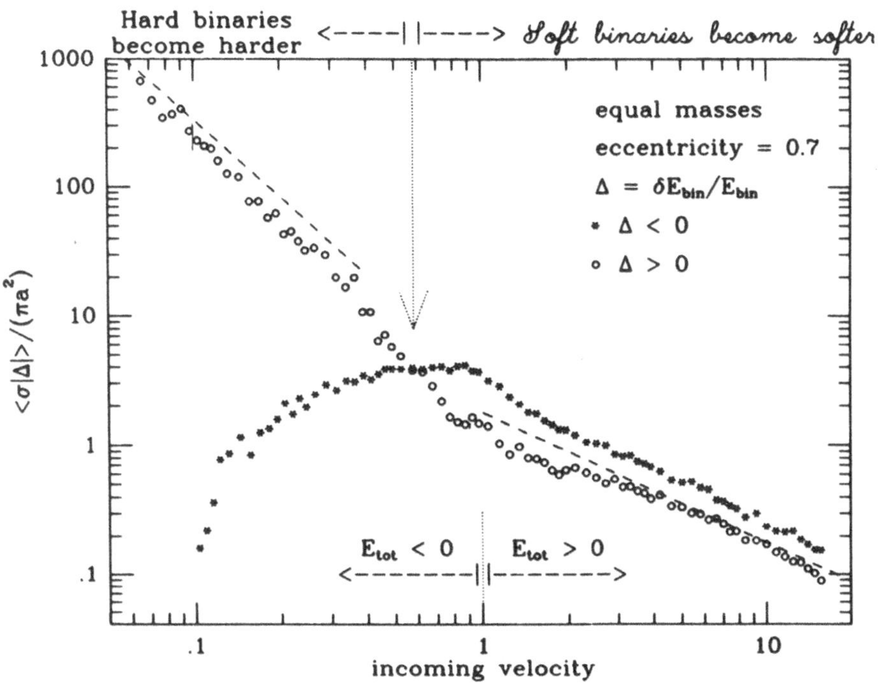

Figure 5. Energy Change in Collisions between Binaries and Single Stars. The plotted points represent average values of $\sigma\Delta$ in collisions between single stars and binaries, as a function of their relative velocity at infinity, computed by Hut (1983). The quantity σ is the cross-section, and Δ is the energy gained by the single stars. Averages over impact parameter and several angles characterizing each encounter are taken separately for Δ positive (open circles) and negative (solid circles); a is the semi-major axis of the binary. The dashed lines show theoretical results by Heggie (1975).

stars in the binary which it encounters. The resultant triple system then ejects one of the three stars, perhaps after further encounters have increased the orbital eccentricity of the loosely bound component so that it comes close to its accompanying binary, and acquires enough energy to escape. According to Heggie (1975) and Hills (1975a), the eccentricity of a binary is modified more rapidly by gravitational encounters than is the binding energy.

Direct computations of binary-binary encounters have been reported by Mikkola (1983a, 1984), who carried out roughly 5000 orbital integrations, assuming stars all with the same mass, and by Hoffer (1983), who reported some 42,000 orbits, mostly for soft binaries, with several mass groups. While these computations are not yet sufficiently detailed to yield all the cross-sections of interest, they should provide a check on various simplified models for such interactions. They agree with each other (and with Spitzer and Mathieu, 1980) in finding a high probability that in a strong interaction between two identical hard binaries, one is disrupted into two single stars; on the average, disruption results (Mikkola, 1983a) in some 60% of such collisions, with 30% leaving a three-body system. Both authors give preliminary fitting formulae for certain interaction cross-sections.

The formation of binaries by three-body encounters is, of course, a basic physical process of importance for globular clusters. It was shown early (Spitzer and Hart, 1971a) and confirmed by Heggie (1975) that for typical conditions in clusters this process is normally much too slow to have any significance. However, in a collapsing core, where the density becomes very high, but the number of stars involved is so low that the velocity dispersion remains small, this process can become significant as can the formation of binaries through tidal capture of one star by another. These two processes are discussed in Section 5.

4.2 Effect of primordial binaries on core collapse

The energy released by a hard binary, as it hardens further in encounters with passing stars, is transmitted chiefly to these single stars and increases the cluster energy (computed with each binary system replaced by a mass point). If an appreciable fraction of hard binaries is assumed to be present, produced when the stars themselves were being formed, they can provide an important source of heat to the cluster, tending to offset the collapse of the central regions. While this assumption does not seem to be supported by the limited observational data available (Trimble, 1980), we shall here explore its consequences.

A preliminary analysis of cluster evolution when primordial binaries are present was carried out by Hills (1975b), who treated the cluster core as a homogeneous sphere, and assumed that the energy released from binaries went entirely into heating the core. The chief driving force for core contraction was assumed to be evaporation. On these assumptions he found that for a mass fraction of 40% in binaries, and 10^5 stars present in the core, the density in the core initially rises slightly, but then decreases as the core expands. For somewhat lower mass fractions in binaries, the initial contraction is greater, but binary hardening still wins out over evaporation eventually, and the core re-expands. Somewhat similar computations have been made by Alexander and Budding (1979).

Computations for a realistic cluster containing primordial binaries have been made by Spitzer and Mathieu (1980), using a Monte Carlo method. All stars, both single and in pairs, were assumed to have the same mass. The energy released by star-binary encounters was computed from the results of Hills (1975a) and Heggie (1975), while for binary-binary encounters the asymptotic theory described in the previous section was adopted. The initial mass fraction in binaries was taken to be 20 or 50 percent in different models. Since each binary has twice the mass of a single star, mass stratification increased the fraction of binaries in the inner core to at least 90% by mass towards the end of the evolution; evidently binary-binary encounters predominated at this time.

Figure 6 shows the radius containing 2% of the system mass plotted against the time, for clusters with 20% of the initial mass in binaries. The plotted curves are for different values of A, the numerical constant in equation (16); as we have seen, the correct value is about 30. For the two models with A = 25, the total binary energy released somewhat exceeds the initial binding energy of the cluster, computed as though each binary were a mass point. Nevertheless, the collapse of the cluster core is apparently not averted, but only postponed, even though the core is composed almost entirely of binaries at the later times. This effect has two causes. First, about 60% of the energy released is carried entirely away from the cluster by escaping reaction products--mostly single stars. Second, of the energy retained in the cluster, only a fraction goes into heating the inner two percent of the mass,--the inner core where most of the binary energy is released, and where the collapse is mostly concentrated. The reaction products which are retained in the cluster have sufficient energies to give them apocenters well outside the core, and the angular momentum gained from gravitational encounters with passing stars hinders them from returning to the central core. Hence the energy of these products is gradually absorbed in an extended region of the cluster.

These models are, of course, approximate and neglect a variety of effects. The most important simplification is probably the neglect of stars with different masses. As is well known, heavy stars tend to displace lighter ones from binary systems, an effect shown clearly in computations by Hills (1975a) and by Hills and Fullerton (1980). As shown in earlier N-body computations (Aarseth and Lecar, 1975) for systems with a spectrum of masses, it is usually the heaviest stars which end up in a central binary, with other stars in elongated orbits or escaping. Evidently the rate of mass stratification would be increased by the concentration of massive stars in binaries. The net result would be a significant enhancement of core heating, since the detailed computations by Mikkola (1983b) show that the heating rate resulting from binary-binary collisions grows rapidly with increasing stellar mass.

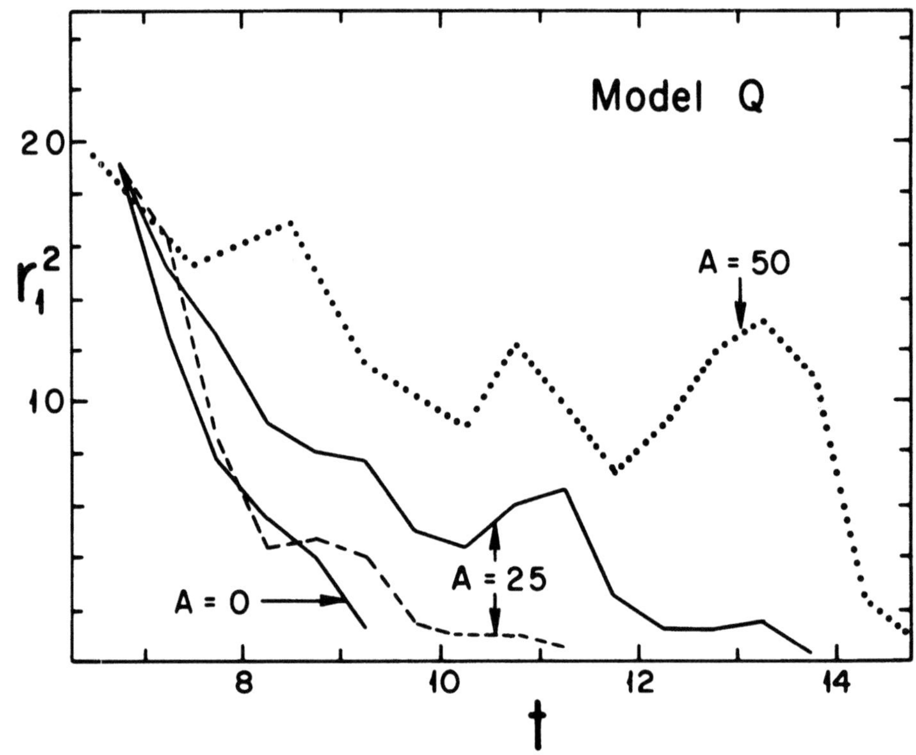

Figure 6. Evolution of a Cluster with 20 Percent of Mass in Binaries. The radius, r_1, containing the inner 2% of the cluster mass is plotted against time; both r_1 and t are in arbitrary units. Different values of A correspond to different coefficients assumed for the rate of energy gain by single stars from binaries. These computations, carried out by Spitzer and Mathieu (1980), show that binaries postpone the collapse by at most a factor two in time since the origin of the cluster.

5. PHENOMENA LATE IN THE COLLAPSE

5.1 Binary formation in three-body encounters.

As the core density increases, three-body encounters become significant and produce new binary stars. Some of these will be soft and will soon be disrupted, with a certain fraction giving up energy and becoming hard binaries. Other newly formed binaries will be hard to begin with. The formation rate, considered early by Agekyan and Anosova (1971), has been re-evaluated by Aarseth and Heggie (1976), who find that if N_c is the number of stars in a relatively uniform core, the rate of formation of hard binaries per unit time is given by

$$\frac{dN_b}{dt} = \frac{K}{N_c \ln(0.4N) t_r(0)} , \tag{17}$$

where $t_r(0)$ is the central relaxation time and K is a constant of order 10. Since we have seen that ρ_c changes on a time scale of several hundred $t_r(0)$, it is clear that a large fraction of stars will form binaries if N_c is as low as 10^2. This tendency is evident in direct dynamical computations of cluster evolution for systems with as few as 100 stars. Binaries formed in this way (which are sometimes referred to as "three-body binaries") are not, in general, very hard, and interact strongly with single stars and other binaries, becoming harder and heating the system.

The evolution of a collapsing core, subject to the formation of three-body binaries, has been considered in detail by Stodółkiewicz 1983), using a Monte Carlo method of Henon's type with extensive modifications; in particular the time step is varied with distance from the center, as in the work by Marchant and Shapiro (1980), and interactions between binaries and single stars are followed in numerical three-body integrations, with a number of input parameters chosen at random. Binary-binary encounters are treated more crudely; following Spitzer and Mathieu (1980), such encounters are assumed to disrupt the binary of lower binding energy. A distribution of stellar masses is assumed, following the familiar Salpeter formula. As in the models by Spitzer and Mathieu (1980), almost all the interactions involving energy generation from binaries occur in the very central regions, where the density is highest.

The models computed in this way show that the core collapse ceases when the central density has increased by a factor of 10^5 to 10^6 over its value in the initial state, taken to be an n = 5 polytrope (Plummer's model). The system then re-expands. It is not obvious why this result differs qualitatively from that found by Spitzer and Mathieu (1980), who found that core collapse was postponed but not averted. Possibly the three-body binaries were sufficiently softer in Stodolkiewicz's model so that a greater function of the energy released was retained in the central core. The presence of relatively more massive binaries, as compared to the individual stars, might conceivably tend in this same direction. Further computations of such models, perhaps using different computer codes as an overall check, would be important.

In this connection, the results obtained by Bettwieser and Sugimoto (1984) are of interest. Their calculations are somewhat approximate, based on moment equations, following Larson (1970), and assume the thermal conductivity derived by Lynden-Bell and Eggleton (1980)--see equation (7). The energy generation rate per unit mass is assumed proportional to $C\rho^k v_m^{-1}$, where v_m is the velocity dispersion, ρ is the local density, and k is taken to be 1 or 2. Since the energy released is assumed to be absorbed locally, it is not surprising that the collapse is halted, again when the density has increased by about 10^6 above its initial value. More unexpected is that after the subsequent expansion, additional collapses and subsequent re-expansions are observed at intervals of a few times the half-mass relaxation time, t_{rh}. This phenomenon occurs only if the constant C in the assumed energy generation formula is less than some critical value. Consideration of the actual cluster population, including the distribution of binaries of different mass at different times, would be required to verify the reality of these non-linear oscillations, with repeated fluctuations of central density by factors as large as 10^6.

A much more realistic analysis, again giving oscillations but of much smaller amplitude, has been carried through by McMillan and Lightman (1984a,b), who used a direct N-body dynamical integration for the core ($N_c \leqslant 100$), combined with a statistical approach (equivalent to the Fokker-Planck equation) for the outer regions. All stars were assumed to have the same mass, and no binaries were assumed present initially. This pioneering but complex program indicated that core collapse followed the similarity solution of Lynden-Bell and Eggleton (1980) until the number, N_c, of stars in the core had fallen to about 25, at which point the formation and hardening of a central binary liberated energy which reversed the core collapse. When the density had decreased to about a fourth of its peak value, the central binary was ejected from the core, recoiling from a strong interaction with a single star. The actual mechanism for this process was the formation of a triple star system which subsequently disrupted; as shown by Heggie (1975), input of energy from a binary to a single star usually occurs in this way. Subsequently the core collapsed again. These successive oscillations had relatively little effect on the outer regions of the cluster.

When a realistic mass spectrum is considered, the central core may be predominantly composed of compact, relatively heavy but almost invisible objects such as white dwarfs, neutron stars and black holes. If a few heavy black holes are present, they are the most likely components of the central, most tightly bound binary. The gravitational potential resulting from a core of faint objects surrounding a central binary could produce a rather minor central peak in the observed visual surface brightness of the cluster, an effect pointed out by Illingworth and King (1977) for a core of neutron stars and by Larson (1984) for a core of black holes.

While there are uncertainties and approximations associated with these various models, there seems very little doubt that binaries formed by three-body encounters can terminate and even reverse the process of core collapse when N is very small.

5.2 Tidal captures and collisions

Because of the finite size of a star, other complex processes become important when the high density of the collapsing core increases the rate of all two-body encounters. Fabian, Pringle and Rees (1975) pointed out that tidal dissipation during a relatively close encounter between two stars could lead to their mutual capture, producing a binary system. A binary formed in this way is sometimes called a "tidal-capture binary". Such a binary is initially in a very eccentric orbit, with a periastron separation equal to the initial distance of closest approach, while the apastron separation is determined by the excess of the energy dissipated over the initial kinetic energy. Successive periastron passages will dissipate additional energy, tending to produce a nearly circular orbit, with a radius equal to twice the initial periastron separation (provided that stellar rotation makes no significant contribution to the total angular momentum).

The least distance of closest approach, R_m, for capture has been computed in detail by Press and Teukolsky (1977) for main-sequence stars each of mass M_* and radius R_*. They find that R_m/R_* decreases from 5 to 2 as the relative velocity increases from 1 to 100 $[(M_*/M_\odot) \times (R_\odot/R_*)]^{1/2}$ km/s; for globular clusters 3 is a typical value. The corresponding impact parameter is much greater, and the cross-section varies as R_m. It follows that the cross-section for an actual collision ($R_m \leqslant 2R_*$) is comparable with that for tidal capture (R_m/R_* between 2 and 3); the effects of such stellar collisions are treated below.

A detailed review and analysis of tidal capture and of energy losses associated with somewhat more distant collisions has been given by Ozernoy and Dokuchaev (1982). Because of their close separation, tidal-capture binaries are relatively very hard and will produce relatively little heating in interaction with other stars, except under rather extreme conditions. Their formation represents a loss of energy initially, tending to accelerate core collapse; their subsequent concentration in the core, as a result of mass stratification, releases gravitational energy which provides a source of heating for other components of the cluster.

Such tidal-capture binaries will evolve into harder systems as a result of gravitational radiation. Mass flow between the close components will also affect the evolution of these hard binaries. Occasional encounters with field stars will accelerate the hardening; also, actual collisions will much enhance mass transfer to any compact component, increasing the X-ray flux (Krolik, Meiksin and Joss, 1984).

The overall effect of tidal dissipation and tidal capture on cluster evolution has not yet been explored with detailed realistic models. Several papers have analyzed this problem, treating the core with a simple homology model. In the paper by Inagaki (1984), the gravothermal instability is taken into account; the calculations indicate that the dominant effect produced by tidal-capture binaries is the mass stratification instability, resulting from their twofold mass increase over single stars. Three-body binaries are found to dominate in only a small fraction of clusters, but this result is uncertain. It is not obvious what terminates the collapse of a core composed of hard tidal-capture binaries. Strong encounters between such binaries are likely to eject all the reaction products from the cluster. Weak encounters between three hard binaries might be expected to form complex multiple-star systems, whose subsequent evolution may provide a substantial source of energy for the cluster as a whole.

A more precise theory of these effects involving tidal capture binaries would evidently be important, but would clearly involve a number of complications. Actual collisions between field stars should also be considered since, as noted above, the rate of such collisions is about the same, or even somewhat greater, than the rate of tidal capture. Physical collisions will produce a variety of effects, starting either with a pair of stars in a common envelope or a single star formed by coalescence. Subsequent evolutionary processes could lead, perhaps, to very massive stars, to supernovae or to black holes. Such effects are discussed in the review by Lightman and Shapiro (1978). While not much research has been carried out in this area during more recent years, actual collisions are clearly of potential importance in the late evolution of a globular cluster.

These various processes will clearly be affected by the nature of the stars which have concentrated in the core. For example, if the core is made up predominantly of compact objects, as suggested above (§5.1), the rate of tidal capture will be greatly reduced. As McMillan and Lightman suggest (1984b), it is possible that because of this effect tidal-capture binaries may have no significant evolutionary importance. Regardless of their effect in stellar evolution, the formation of binaries by tidal capture seems the most likely explanation for the observed presence of strong X-ray sources in a few clusters with relatively dense cores (Lightman and Grindlay, 1982, Krolik, 1984). Evidently the collapse of globular cluster cores is still a field with many challenging problems.

Helpful discussions with J. Goodman, P. Hut, S. Inagaki, D. C. Heggie, D. Lynden-Bell, J. P. Ostriker, S. L. Shapiro and S. D. Tremaine are gratefully acknowledged.

REFERENCES

Aarseth, S. J. and Heggie, D. C. 1976, Astron. and Astroph., 53, 259.

Aarseth, S. J. and Lecar, M. 1975, Ann. Rev. Astron. and Astroph., 13, 1.

Agekyan, T. A. and Anosova, Zh. P. 1971, Soviet Astron. AJ., 15, 411.

Alexander, M. E. and Budding, E. 1979, Astron. and Astroph., 73, 227.

Angeletti, L. and Giannone, P. 1977, Astroph. and Space Sci., 50, 311.

Angeletti, L., Dolcetta, R. and Giannone, P. 1980, Astroph. and Space Sci., 69, 45.

Antonov, V. A. 1962, Vestnik Leningrad Univ. (Series Math. Mech. and Astron.), 7, 135; English translation reprinted at end of this volume.

Bettwieser, E. and Sugimoto, D. 1984, Monthly Notices, Roy Astron. Soc., 208, 493.

Cohn, H. 1979, Ap. J., 234, 1036.
_______ 1980, Ap. J., 242, 765.
Duncan, M. J. and Shapiro, S. L. 1982, Ap. J., 253, 921.
Fabian, A. C., Pringle, J. E. and Rees, M. J. 1975, Monthly Notices, Roy. Astron. Soc., 172, 15P.
Fall, S. M. and Rees, M. J. 1977, Monthly Notices, Roy. Astron. Soc., 181, 37P.
Goodman, J. 1983, Ap. J., 270, 700.
Gurevich, L. E. and Levin, B. Ya. 1950a, Doklady Akad. Nauk. USSR, 70, 781.
_______ 1950b, Astron. Zh. (USSR), 27, 273; NASA translation TT F-11, 541, 1968.
Hachisu, I. and Sugimoto, D. 1978, Prog. Theor. Physics, 60, 123.
Hachisu, I., Nakada, Y., Nomoto, K. and Sugimoto, D. 1978, Prog. Theor. Physics, 60, 393.
Heggie, D. C. 1975, Monthly Notices, Roy. Astron. Soc., 173, 729.
Hénon, M. 1960, Ann. d'Ap., 23, 668.
_______ 1961, Ann. d'Ap., 24, 369.
_______ 1971a, Astroph. and Space Sci., 13, 284.
_______ 1971b, Astroph. and Space Sci., 14, 151.
Hills, J. G. 1975a, A. J., 80, 809.
_______ 1975b, A. J., 80, 1075.
Hills, J. G. and Fullerton, L. W. 1980, A. J., 85, 1281.
Hoffer, J. B. 1983, A. J., 88, 1420.
Horwitz, G. and Katz, J. 1978, Ap. J., 222, 941.
Hut, P. 1983, Ap. J. (Lett.), 272, L29.
Illingworth, G. and King, I. R. 1977, Ap. J. (Lett.), 218, L109.
Inagaki, S. 1980, Publ. Astr. Soc. Japan, 32, 213.
_______ 1984, Monthly Notices. Roy. Astron. Soc., 206, 149.
Inagaki, S. and Wiyanto, P. 1984, Publ. Astr. Soc. Japan, in press.
Ipser, J. R. and Kandrup, H. E., 1980, Ap. J., 241, 1141.
Jeans, J. H. 1929, Astronomy and Cosmogony, (Cambridge Univ. Press), 2nd edition, Chapt. VII.
Katz, J. 1978, Monthly Notices, Roy. Astron. Soc., 183, 765.
_______ 1980, Monthly Notices, Roy. Astron. Soc., 190, 497.
Katz, J. and Taff, L. G. 1983, Ap. J., 264, 476.
King, I. R. 1958, A. J., 63, 114.
_______ 1966, A. J., 71, 64.
Kondrat'ev, B. P. and Ozernoy, L. M. 1982, Astroph. and Space Sci., 84, 431.
Krolik, J. H. 1984, Ap. J., 282, 452.
Krolik, J. H., Meiksin, A. and Joss, P. C. 1984, Ap. J., 282, 466.
Larson, R. B. 1970, Monthly Notices, Roy. Astron. Soc., 150, 93.
_______ 1984, Monthly Notices, Roy. Astron. Soc., in press.
Lightman, A. P. 1977, Ap. J., 215, 914.
Lightman, A. P. and Fall, S. M. 1978, Ap. J., 221, 567.
Lightman, A. P. and Grindlay, J. E. 1982, Ap. J., 262, 145.
Lightman, A. P. and Shapiro, S. L. 1978, Rev. Mod. Phys., 50, 437.
Lynden-Bell, D. and Eggleton, P. P. 1980, Monthly Notices, Roy. Astron. Soc., 191, 483.

Lynden-Bell, D. and Wood, R. 1968, Monthly Notices, Roy. Astron. Soc., 138, 495.

Marchant, A. B. and Shapiro, S. L. 1980, Ap. J., 239, 685.

McMillan, S. L. W. and Lightman, A. P. 1984a, Ap. J., 283, in press.

________ 1984b, Ap. J., 283, in press.

Merritt, D. 1981, A. J., 86, 318.

Mikkola, S. 1983a, Monthly Notices, Roy. Astron. Soc., 203, 1107.

________ 1983b, Monthly Notices, Roy. Astron. Soc., 205, 733.

________ 1984, Monthly Notices, Roy. Astron. Soc., 207, 115.

Nakada, Y. 1978, Publ. Astr. Soc. Japan, 30, 57.

Ozernoy, L. M. and Dokuchaev, V. I. 1982, Astron. and Astroph., 111, 1.

Press, W. H. and Teukolsky, S. A. 1977, Ap. J., 213, 183.

Retterer, J. M. 1979, A. J., 84, 370.

Saito, M. and Yoshizawa, M. 1976, Astroph. and Space Sci., 41, 63.

Spitzer, L. 1969, Ap. J. (Lett.), 158, L139.

________ 1975, Dynamics of Stellar Systems, IAU Symp. No. 69, Ed: A. Hayli (Reidel: Dordrecht), p. 3.

Spitzer, L. and Hart, M. H. 1971a, Ap. J., 164, 399.

________ 1971b, Ap. J., 166, 483.

Spitzer, L. and Mathieu, R. D. 1980, Ap. J., 241, 618.

Spitzer, L. and Shull, J. M. 1975, Ap. J., 201, 773.

Spitzer, L. and Thuan, T. X. 1972, Ap. J., 175, 31.

Stodółkiewicz, J. W. 1983, Publ. Astron. Inst. Czechoslovak Acad. Sci., 56, 103.

Trimble, V. L. 1980, Star Clusters, IAU Symp. No. 85, Ed: J. Hesser (Reidel: Dordrecht), p. 259.

Vishniac, E. T. 1978, Ap. J., 223, 986.

Wilson, C. P. 1975, Dynamics of Stellar Systems, IAU Symp. No. 69, Ed: A. Hayli (Reidel: Dordrecht), p. 207.

Yoshizawa, M., Inagaki, S., Nishida, M. T., Kato, S., Tanaka, Y. and Watanabe, Y. 1978, Publ. Astr. Soc. Japan, 30, 279.

DISCUSSION

COHN: Is there no question on the basis of the calculations you reviewed that binaries must reverse core collapse?

SPITZER: There seems very little question, from the calculation I reviewed, that the formation of binaries by three-body processes, and the subsequent hardening of these binaries, can reverse core collapse.

GOODMAN: Do you see a conflict between the theoretical prediction that cluster cores should collapse rapidly and the observation that very few cores appear to have collapsed?

SPITZER: The number of cluster cores that should have collapsed, according to the computed values of t_{rh}, is not entirely certain, but could be appreciably less than half. The appearance of a cluster after collapse is also uncertain. A dense central core could still remain, but if this core is composed of neutron stars and black holes, it might produce only small effects on the distribution of visible stars. It is also possible that post-collapse expansion eliminates the relatively dense central core.

SUGIMOTO: Bettwieser and I think that there exist gravothermal oscillations of the core. When the central density is high, i.e., when the core has a cusp, the timescale of oscillation is much shorter than at the stage of less concentrated core. Therefore, in a probabilistic sense, comparable or less members of globular clusters are in the collapsed phase. This is conclusive as physics. In order to give a numerical value, however, in astronomical situations we have to take account of other effects such as escape of binaries from the system, segregation of binaries etc. Thus the detailed value in astronomical situations is an open question.

GRINDLAY: Can you elaborate further on your statement that tidal-capture binaries will not be a significant source of cluster heating and contribute to halting core collapse? I would think that the process of formation of tidal-capture binaries would be enhanced during core collapse and would remove significant binding energy from the core.

SPITZER: My statement was meant to imply that binaries formed by tidal capture would not, after their formation, be a significant source of heating for the cluster. This is because such binaries are very hard as a result of their close separation; hence, the cross-section for their close encounter with a single star is rather small, and, in addition, if a close encounter does occur, so much energy is likely to be transferred to the star that this will be ejected from the cluster and probably the binary also. On the other hand, the process of tidal capture is a dissipative one, reducing the kinetic energy of the cluster. Additional dissipation arises from close encounters between two stars which lose some of their kinetic energy but not enough to be captured. These dissipative processes tend to accelerate core collapse.

BETTWIESER: Can thermodynamic concepts be applied to N-body systems? I remember the calculations of R. Miller in 1972 showing that energy exchange between N-body systems and a foreign body can be quite the opposite you expect from thermodynamics.

SPITZER: While the conventional application of thermodynamics to N-body systems certainly involves some approximations, I believe that in most situations useful results can be obtained with this technique.

DYNAMICAL EVOLUTION OF GLOBULAR CLUSTERS AFTER CORE COLLAPSE

Douglas C. Heggie
University of Edinburgh
Department of Mathematics
King's Buildings
Edinburgh EH9 3JZ, U.K.

ABSTRACT. This review describes work on the evolution of a stellar system during the phase which starts at the end of core collapse. It begins with an account of the models of Hénon, Goodman, and Inagaki and Lynden-Bell, as well as evaporative models, and modifications to these models which are needed in the core. Next, these models are related to more detailed numerical calculations of gaseous models, Fokker-Planck models, N-body calculations, etc., and some problems for further work in these directions are outlined. The review concludes with a discussion of the relation between theoretical models and observations of the surface density profiles and statistics of actual globular clusters.

1. INTRODUCTION

At the last I.A.U. conference which dealt at length with the dynamics of clusters, Hénon (1975) briefly described two new Monte Carlo models. The evolution with time of the central density in his equal-mass model (published here for the first time, with Dr. Hénon's kind permission) is shown in Fig. 1. The rise in central density during the first half of the calculation reflects the phenomenon of core collapse, described in this volume in the paper by Prof. Spitzer. This paper is concerned with the evolution of models of stellar systems during the phase following the maximum central density.

That the problem is a significant one is suggested by two arguments. First, during the collapse phase, when ρ is increasing, the time until maximum ('complete core collapse') is of the order of a hundred times the current central relaxation time (Spitzer & Shull 1975b). Since the inferred central relaxation times in a few clusters are as short as 10^7 yr (Peterson & King 1975), such clusters are predicted to reach complete collapse in the next 10^9 yr or so. Hence it is reasonable to suppose that some clusters have already done so (cf. Lightman 1982). Second, there is increasing observational evidence that the central regions of some clusters do not fit the single-component, lowered-Maxwellian models which are often used,

J. Goodman and P. Hut (eds.), Dynamics of Star Clusters, 139–160.

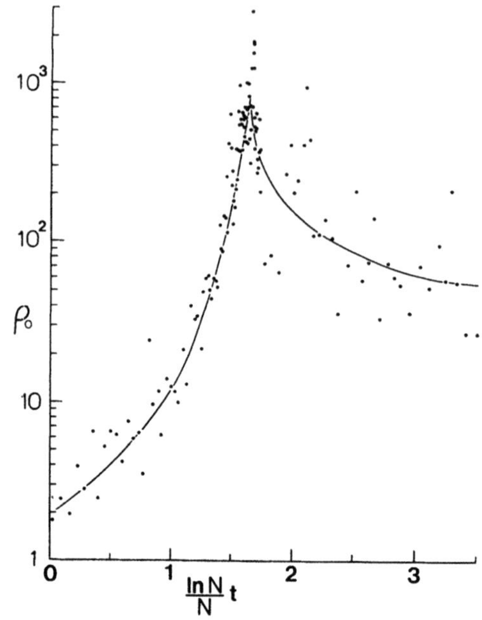

Figure 1. Evolution of central density, ρ_0, versus time, t, for Hénon's equal-mass model (Hénon 1975). Units are such that G = 1, M = 1 and E = -1/4, where G is the constant of gravitation, and M and E are total mass and energy, respectively. N is the number of stars in the system. The continuous curve was drawn by eye.

and it has been suggested that such objects may already have collapsed (Heggie 1980, Djorgovski & King 1984).

This review is mainly concerned with describing theoretical models for post-collapse evolution, and attempting to show how they fit together and how they can be related to more detailed simulations. The models are generally highly idealised and in the concluding sections we discuss avenues for future research in the direction of greater realism, and also the still tenuous link with observations.

2. MODELS

2.1 Hénon's models

The study of post-collapse evolution, as we would now call it, began with the long paper by Hénon (1961). He made several simplifications in the Fokker-Planck equation: (i) isotropic velocity distribution, (ii) self-similar evolution and (iii) a single mass component (though in other parts of his paper several mass components were considered). In searching for a solution he found it necessary to relax any assumption that the central density was finite, and finally obtained a solution behaving near the centre like the singular isothermal model, with velocity dispersion varying with time, and with a flux of energy from the central singularity. Hénon thought, with some support from

N-body simulations (von Hoerner 1960), that the energy in a real system might be supplied by the formation and evolution of binary stars.

Hénon devised two models, one tidally limited and the other (Hénon 1965) isolated. For the latter the density distribution is given in Fig. 2. The *form* of the time-evolution can be found by

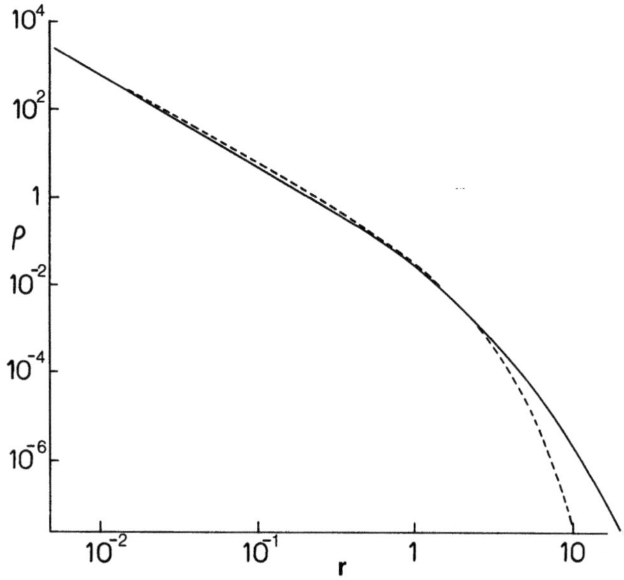

Figure 2. Density distribution (log ρ versus log r, where r is distance from the centre) for Hénon's isolated model (solid line) and Goodman's model (dashed line). Units are as in Fig. 1 for Hénon's model, (though the energy of the system changes with time), and Goodman's model has been scaled to the same total mass and the same velocity dispersion near the centre.

setting the elapsed time t equal to a typical relaxation time, and supposing that the total mass M is constant. Numerical values are given by Hénon, and yield, for example, the result

$$v_o^2(t) = v_o^2(0)\left(1 + 4.15\,\frac{v_o^3(0)}{GM}\,\frac{\ln N}{N}\,t\right)^{-2/3}, \tag{1}$$

where v_o^2 is the central one-dimensional velocity dispersion.

2.2 Goodman's model

In the past few years it has become quite common to approximate a star cluster by a self-gravitating gas (Hachisu *et al* 1978) with a coefficient of heat conduction designed to mimic the transport of energy by relaxation (Lynden-Bell & Eggleton 1980). This was used by Goodman (1984) for further development along the lines of Hénon's models. In particular he investigated the effect of the loss of mass which must accompany the emission of energy if binaries are responsible for this (Ozernoy & Dokuchaev 1982, Dokuchaev & Ozernoy 1982).

In many respects the results change only slightly. For instance, close to the centre the solution again deviates very little from the singular isothermal model. On the other hand, Fig. 2 shows that the outer parts differ substantially from the corresponding result of Hénon's model, although Goodman points out that the physical assumptions on which the two models are constructed may be expected to break down in different ways in the envelope of the cluster. For the dependence of the central velocity on time Goodman finds

$$v_o^2(t) = v_o^2(0)\left(1 + 11.9A\,\frac{v_o^3(0)}{GM(0)}\,\frac{1}{N(0)}\,t\right)^{-\frac{2}{3}(1+2\nu)} \qquad (2)$$

where A,ν are constants. In fact ν measures the rate of mass-loss, the favoured value being of the order of .0125. The constant A appears in the coefficient of thermal conductivity, and Goodman chooses A = 0.31 ℓn (0.4N) to ensure that the gaseous model evolves at the same rate as the isotropised Fokker-Planck model (Cohn 1980) in the homologous part of the *collapse* phase. (This is equivalent to the choice C = .104 of a corresponding constant in the formulation of Lynden-Bell & Eggleton (1980); the value C = .072 found by Inagaki & Lynden-Bell (1983), again by comparison with Cohn's result, is based on an incorrect expression for the relaxation time used by him.) Using Goodman's value for A, and comparing (1) and (2), we see that, in the limit of large N, the gaseous model evolves at a quite similar rate to the Fokker-Planck model.

2.3 The model of Inagaki & Lynden-Bell

The models of Hénon and Goodman have finite mass and energy, but exhibit two singularities. One, already referred to, is the central singularity in density and luminosity. But (1) and (2) also reveal a singularity at some time in the past, when the central velocity dispersion becomes infinite, and the scale radius of the models vanishes. What one would like, by contrast, is a model which joins on smoothly in the past to the density and velocity profile left behind by the core collapse. This is what is accomplished by the model of Inagaki and Lynden-Bell (1983). It evolves forward from an instant at which the density profile everywhere is the power-law profile $\rho = Ar^{-\alpha}$, where A is a constant and $\alpha \simeq 2.208$, left behind after core collapse in gaseous models (Lynden-Bell & Eggleton 1980).

The model still has an isothermal singularity at the centre (Fig. 3). It also has infinite mass and energy, and can therefore be applicable only to the inner parts of a system. The variation with time of the central velocity dispersion is given by

$$v_o^2(t) = v_o^2(0)\left(1 + 1.2\,\frac{G^2 m\rho_*(0)\ell nN}{v_*^3(0)}\,t\right)^{-2\,\frac{\alpha-2}{6-\alpha}}, \qquad (3)$$

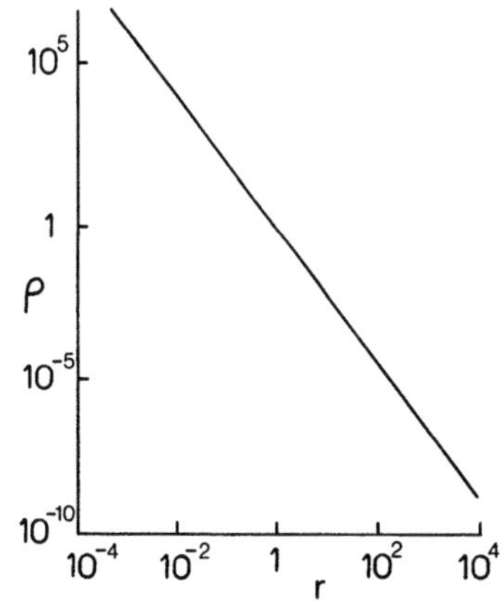

Figure 3. Density distribution in the model of Inagaki & Lynden-Bell. The variables used are the authors' homology variables ρ^*, r^*, which is equivalent (if G = 1) to scaling so that $v_o^2 = 2\pi$ and $L_o = 12\sqrt{2\pi^3} m C \ln N \times (\alpha - 2)$, where L_o is the central flux of energy, m is stellar mass, C is discussed in § 2.2, and $\alpha \approx 2.21$. (For Hénon's model, in the scaling of Fig. 2, the corresponding values are $v_o^2 = 0.400$, $L_o = .184\, m \ln N$.) The asymptotic slope is -2 (small r), and -2.21 (large r).

where the suffix * indicates quantities evaluated at the edge of the isothermal region, defined as the radius where $v_* = v_o \exp(-0.1)$. The scale radius increases with time and the scale density decreases, in such a way that the outer part of the system, where $\rho \propto r^{-\alpha}$, does not change. Thus one can think of the isothermal region, where $\rho \propto r^{-2}$, eating outward through the halo.

2.4 The central density

Maintenance of the strict self-similarity of any of the models discussed so far is dependent on the emission of energy at a precisely determined rate (see caption to Fig. 3). Besides the lack of realism in supposing that all the energy is emitted at one point, it is inconceivable that any given mechanism would emit precisely the amount of energy required. But it is reasonable (up to a point; cf. § 4.1 below) to suppose that energy is emitted at different places at a rate depending on local values of ρ and v. Under suitable conditions it is then found that the central parts of the isothermal region may be non-singular, the central density adjusting itself so that the core (i.e. the region where the density is not far from its central value) altogether emits energy at the rate that is required for the self-similar evolution of the above models. Further details on how the central density may be determined have been given elsewhere (Heggie 1984).

Two general consequences of this argument should be mentioned. In the first place the evolution is no longer self-similar, because in general the radius, r_*, of the edge of the isothermal region will evolve at a different rate from the core radius, r_c, defined by

$$r_c^2 = \frac{9v_o^2}{4\pi G\rho_o} \tag{4}$$

(which agrees with the definition in King 1966 when the concentration parameter is large). Second, for a wide range of idealised energy production mechanisms, the rate at which the central density decreases in post-collapse evolution is much smaller than the rate of increase during collapse; while the density (ρ_*) at the edge of the isothermal region in post-collapse evolves at much the same rate as the central density (ρ_o) in the collapse phase (Inagaki & Lynden-Bell 1983); *after* collapse the central density ρ_o lags further and further behind ρ_*. Similar remarks may be made about the reexpansion of the core radius and the core mass, but the central velocity dispersion *does* evolve on much the same time-scale in both collapse and post-collapse phases.

2.5 Evaporative models

The evaporative model of core evolution, which assumes that the core can be described by two time-dependent parameters such as mass and radius (see, e.g.,Lightman & Shapiro 1978), has proved a useful tool for the study of core collapse. The core evolution is implicitly assumed to be entirely homologous, the structure varying from one time to another by variations in the two parameters. Before it was realised that post-collapse evolution was essentially non-homologous (unless discussion is restricted to regions well *outside* the core), a number of simple generalisations of the model had been proposed which dealt with post-collapse evolution. Some discussions (Hills 1975, Alexander & Budding 1979, Dokuchaev & Ozernoi 1982) were concerned with binaries, and those by Shapiro (1977) and Duncan & Shapiro (1982) with a cluster containing a central massive black hole. The latter problem will be discussed below (§ 3.2), and the further remarks here will be confined to the discussions involving binaries.

In Hills' work it was assumed that the fraction of binaries, f_b, was constant and uniform, and the rate at which they emitted energy varied essentially as

$$\dot{E} \sim \frac{f_b E}{T_R \ell n N} \tag{5}$$

where E is the energy of the cluster and T_R is a relaxation time. Now in the evaporative model, and in models where the assumption of self-similarity after collapse is *not* made (Heggie 1984), core collapse ends when $\dot{E}$ becomes comparable with the rate at which energy is exchanged by two-body relaxation, which varies as E/T_R. Thus in Hills' model the arrest of core collapse depends crucially on the variation of ℓn N in (5). In fact it seems likely that variations in f_b would be far more important.

Perhaps the strongest indication that the simplest evaporative models are inappropriate when significant heating by binaries is included is the equation

$$\dot{N} = - \frac{\alpha N}{T_R}$$

for N, interpreted as the number of stars in the core, where α is a constant. This forces the number of stars in the core to decrease, contrary to the findings of more detailed models and simulations.

It is convenient to mention here some models by Retterer (1984), though they are much more elaborate than evaporative ones, especially in taking account of the detailed distribution of the binding energies of the binaries. Even the cluster is described by an additional parameter (which can be thought of as the central potential). Unfortunately, the method fails when the central potential becomes too large, and only small systems could be followed past core collapse.

2.6 Synthesis of the models

Something has been said already about the regimes, in space and time, to which the validity of the various models may be restricted. Now we shall attempt to show how in principle they may be put together to provide a relatively complete semi-quantitative description of post-collapse evolution, in much the same way that the evaporative model covers the collapse phase.

We suppose that, at the end of core collapse, the outer parts of the cluster (well beyond the *initial* core radius) have not changed very much. At the very centre is an isothermal region with a small but finite core radius, and between here and the outer regions is the power law profile $\rho \propto r^{-\alpha}$. (Numerical experiments indicate that the constant of proportionality can be fixed to about a factor of two by supposing that the profile joins smoothly onto the initial profile at the point with the same logarithmic slope.)

During the first phase of the post-collapse evolution, the edge of the isothermal region gradually expands through the power-law profile, and the central density and core radius adjust themselves as described in § 2.4. The central velocity dispersion varies as in eq. (3), and, after a time comparable to the duration of the collapse phase, reduces to its initial value; however the central density is now much greater than it was originally. Between the current core radius and the original core radius is a large region with approximately the r^{-2} isothermal profile, and well beyond the original core radius the density profile has still not greatly altered from its original appearance. (Numerical simulations, described in § 3.1 below, indicate that densities may be higher by a factor of up to 10).

Next, at the beginning of the second phase of post-collapse evolution, the structure of the system approaches that of the Hénon model described in § 2.1 (or Goodman's, § 2.2), though the central density adjusts according to the discussion of § 2.4. Thus, outside the current core radius is a large region with the isothermal r^{-2} density profile, which at larger radii gives way to one of the limiting forms

$\ln\rho \sim -\frac{9}{4}\ln(r/\alpha_1) - (r/\alpha_1)^{1/2}$ (for Hénon's isolated model) or

$\ln\rho \sim -2\ln(r/\alpha_2) - r/\alpha_2$ (for Goodman's model), where α_1, α_2 are radius scales (Goodman 1983). The central velocity dispersion evolves as in eq. (1) or (2). If, as we have supposed, the system is isolated, then it expands indefinitely at an ever diminishing rate.

3. SIMULATIONS

So far we have reviewed a number of models for the post-collapse evolution of star clusters. The solution of the corresponding equations generally requires some numerical work, but at least the models have a kind of universal significance independent of special initial conditions. We now turn to simulations, which often permit the inclusion of more realistic detail than the idealized models, and also require the specification of initial conditions and, generally, much more numerical effort. We discuss these simulations roughly in order of increasing elaboration.

3.1 Gaseous Models

These are the most idealized types of formulation, and also the easiest. However, there are serious disagreements between those of Bettwieser *et al* (Bettwieser & Sugimoto 1984, Bettwieser & Fritze 1984) and the present author. The former authors find that under quite a wide range of circumstances the post-collapse evolution is oscillatory, rather resembling the possibility of repeated core collapse previously considered by Fall & Malkan (1978). The present author, on the other hand, observes only a steady expansion in the post-collapse phase, even though models have been computed with the same equations, initial conditions, boundary conditions and energy generation as those of Bettwieser *et al*. There are differences in the scaling, the variables, and the numerical methods, and it is probably in one of these that the disagreement originates. Since the oscillating models are described by Bettwieser and by Sugimoto elsewhere in this volume, the following remarks will be confined to a recent calculation by the author. They must be regarded as tentative as long as the source of the disagreement remains undiscovered.

The initial conditions are a Plummer model, and units are those of Fig. 1. The system is surrounded by an impermeable enclosure of radius 20 times the half-mass radius. Energy is generated at a rate

$$\varepsilon \propto \rho^2 v^{-1} \tag{6}$$

per unit mass, a choice which was made for the purposes of comparison with the results of Bettwieser & Sugimoto. Details of calculations with a different choice of initial conditions and other assumptions on the energy generation are given in Heggie 1984. Our purpose here is to illustrate some of the statements in § 2.6.

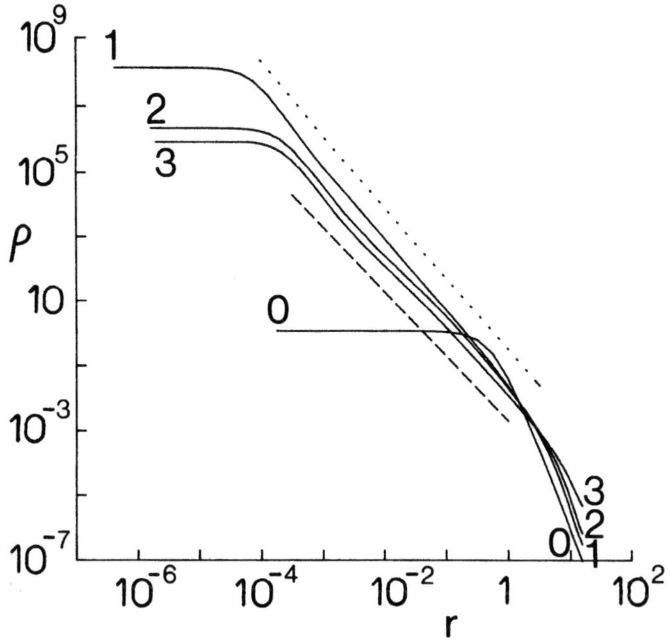

Figure 4. Density profiles log ρ against log r at different times in the evolution of a gaseous model. Initial and boundary conditions and units are stated in the text. The times of the profiles 0, 1, 2, 3 are given by $C \frac{\ln N}{N} t = 0, \; .167, \; .305, \; 1.111$. The dotted and dashed straight lines have logarithmic slope -2.21, -2 respectively.

Some density profiles are shown in Fig. 4, where profile 0 is the initial Plummer model. Close to the time of maximum central density (profile 1) the core contains some 10^{-4} of the total mass (a value governed by the choice of the constant of proportionality in (6)). It is surrounded by an extensive halo with logarithmic slope approximately -2.21, and beyond that is a steeper envelope still reflecting initial conditions. (One difference between gaseous models and Monte-Carlo simulations is that in the former there appears to be no sign of the formation of an envelope with logarithmic slope -3.5, contrary to what was found by Spitzer & Shull 1975a.) After a comparable time interval (profile 2) the central velocity dispersion has reduced to slightly below its initial value, but the central density is still very high, and the logarithmic slope of the bulk of the profile has changed from -2.21 to approximately -2. The slight differences between this profile and profile 3, taken much later, show how slow the evolution of the system becomes.

The quantitative rate of evolution of the system can be judged to some extent from the dependence of the central velocity dispersion on time (Fig. 5). After the maximum value of v_o is attained the dependence soon settles down to a form like (3), except that the coefficient must be reduced from 1.2 to about 0.6, and for α we have taken the value 2.22, instead of the value 2.208 of Lynden-Bell & Eggleton.

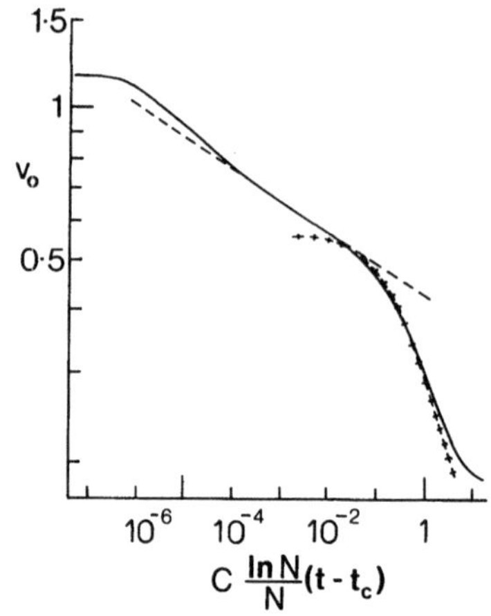

Figure 5. Dependence of root mean square central velocity v_o against time, given as $C \frac{\ell n\ N}{N}(t - t_c)$, where t_c is the time of maximum value of v_o. The scaling is as in Fig. 1, and the two theoretical curves are explained in the text.

(The larger value of α is consistent with the evolution of the model during core collapse, and the difference may be due to the relatively coarse time-step used in this calculation, but the reason for the discrepancy in the coefficient is unknown. Eq. (3), modified in this way, is plotted as the straight dashed line in Fig. 5. Note, however, that the origin of time is different in Fig. 5 and eq. (3).) The simulations begin to deviate again from this theoretical relation when v_o reduces to approximately the value it had at the beginning of core collapse, and thereafter for a period it follows quite closely eq. (2), which is the curved theoretical relation in Fig. 5. (Here we have set $\nu = 0$, since there is no mass-loss in the simulation, and also the origin of t has been chosen to optimise the fit; this was done by plotting v^{-3} against t. Such optimization is justifiable, since there is no reason to suppose that the time at which Goodman's solution becomes singular will equal the time of maximum v_o in the simulation.) Thereafter, the rate of evolution slows down as the entire system within the enclosing wall becomes more and more nearly isothermal. By this time it is evident that the boundary conditions are playing an important role in the structure of the system.

3.2 Fokker-Planck Models

It was Hénon (1975) who conjectured that the details of the mechanism of energy-generation did not matter, within wide limits, as far as the evolution of the rest of the system is concerned, though it is now realized that this can only apply to regions of the system outside the core. Hénon's conjecture was also confirmed by Stodółkiewicz (1982) in several Monte-Carlo solutions of the Fokker-Planck equation, modified in several different (but artificial) ways so as to avoid the central singularity which would otherwise develop. While gross features of the density profile and its evolution (at and beyond the innermost 10% of the mass) are virtually unaffected by the nature of the singularity, the escape of stars may be influenced by it.

Hénon (1975) also gave some results of two Monte-Carlo simulations with an artificial source of energy, and two density profiles, respectively early and late in the post-collapse phase, are shown in Fig. 6. These make the qualitative point that post-collapse evolution in this model is relatively very slow.

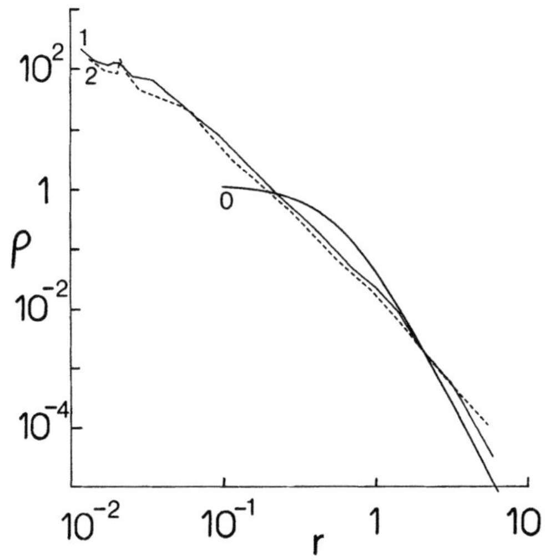

Figure 6. Density profiles, log ρ against log r, in Hénon's equal-mass model. The times of the Monte-Carlo profiles (1 and 2) are given by $\frac{\ln N}{N}$ t = 1.73 and 2.66, respectively, and the units used are those of Fig. 1. (Unpublished data presented here by kind permission of Dr. Hénon.) Large parts of the earlier profile can be fitted by a power law of the form $\rho \propto r^{-2.44}$. Also shown (0) is the analytic profile corresponding to the initial conditions.

We now come to solutions of the Fokker-Planck equations in which the artificial energy sources discussed so far are replaced by more realistic processes, in particular the formation and evolution of binary stars (Stodółkiewicz 1983; Cohn, this volume), or the presence of a central black hole (Marchant & Shapiro 1980, Duncan & Shapiro 1982). This work of Stodółkiewicz contains many important clues to the complicated behaviour of binaries, such as: the spatial distribution and numbers of binaries, the importance of binary-binary reactions when the total number is not small, the very restricted size of the region within which energetic interactions take place, the importance of the loss of mass by escaping binaries (as in Goodman's model) and so on. Cohn's calculations are more idealized, but show very beautifully the homological form of post-collapse expansion of the Hénon model.

The emission of energy by a central black hole is, in some respects, a much simpler process, in that the energy flux can be given fairly accurately as a relatively simple function of parameters in the core. This permits some fairly detailed comparison (Heggie 1984) between the simple models outlined in §§ 2.3-4 and the Monte-Carlo simulations, thanks also to the significant amount of quantitative information published by Duncan & Shapiro.

These Monte-Carlo results have been interpreted also in terms of a simple extension of the evaporative model for core evolution, in which the core is characterized by its radius and central density. But now there is an additional equation for the mass of the black hole, and additional terms for the fluxes of mass and energy which it produces (Shapiro 1977) and for mass encompassed by the core as it reexpands (Duncan & Shapiro 1982; see also McMillan, Lightman & Cohn 1981). With a suitable (and reasonable) choice of a free parameter which this last effect requires, good agreement with the growth of the black hole is obtained. Consistency with the evolution of the core parameters is less satisfactory, probably because of the homological assumption implicit in the model. The considerations of §2.4 account for these aspects of the simulations more successfully.

Density profiles are given by Marchant & Shapiro (1980), but the post-collapse phase was followed for only about 10% of the total collapse time. During this early post-collapse phase, the cusp around the black hole, with logarithmic slope -1.75, eats out through the steeper profile left behind by core collapse. This would be difficult to distinguish from the isothermal (slope -2) profile of the models discussed in § 2.

3.3 Hybrid Simulations

Very simple considerations (e.g. Lightman & Shapiro 1978) suggest that a single binary can emit energy at a rate comparable to the energy flux in core collapse, if the core contains around 100 stars. Furthermore, numerical simulations confirm that such considerations do correctly predict the conditions under which collapse ends and reexpansion begins (Duncan & Shapiro 1982, Heggie 1984). Therefore, the conditions for validity of the Fokker-Planck equation (especially, the assumption that N is large) are violated at the beginning of the post-collapse phase. Furthermore, since reexpansion of the core is much slower than its collapse (if the mechanisms of energy generation that have been studied are typical in this regard), this difficulty persists for some time. For this and other reasons (e.g. correct treatment of the behaviour of binary stars) it is desirable to simulate the central parts of the system by direct N-body integration, though one of the 'simpler' techniques based on the Fokker-Planck equation may be used further out. Precisely such a hybrid scheme has recently been described (McMillan & Lightman 1984a,b), and it represents an exciting and notable advance in the techniques of simulation.

Though still not fully implemented, a restricted version of the scheme has been applied to a very short interval of time around maximum core collapse. Several interesting aspects of the evolution of binary stars were observed, indluding oscillation of the core. In one cycle of the oscillation, core heating by a binary first causes expansion of the core, until a sufficiently energetic interaction ejects it from the core, whereupon a modest contraction occurs until a new energetic binary forms. (Such oscillations are causally unrelated to those observed by Bettwieser *et al.*) In an interesting discussion of the behaviour of binaries in post-collapse evolution, the authors predict that such oscillations will die out, even though the fraction of time during which an active binary is present diminishes steadily.

3.4 N-body simulations

Whether the results of N-body simulations, with at most about 1000 stars, are applicable to globular star clusters is not the point at issue here (but see the paper by Goodman in this volume). Rather, now that there is a much better theoretical understanding of post-collapse evolution, albeit in an idealized form, it would be worth reexamining the evidence of these simulations in order to learn how comprehensive these theories are, since N-body simulations, even more than hybrid ones, remove many of the simplifying assumptions on which the theory is based. (By post-collapse evolution, in this context, we mean the phase following the appearance of very energetic binaries in a dense core, as this is usually taken as the manifestation of core collapse.)

Hénon himself (Hénon 1965) compared his homological model with the early 25-body models of von Hoerner (1963). The fit is extraordinarily good, over 7 orders of magnitude in density. The fact that such a result could be obtained, in a simulation so far removed from the assumption of large N on which Hénon's model was based, suggests that much could be learned by reexamining the much larger systems that have been studied since von Hoerner's time.

4. UNSOLVED PROBLEMS

The purpose of this section is to indicate a selection of problems and avenues for further research. Some of these are specific questions which could be answered using existing techniques after a few days' work; others require a prolonged program of research of which even the outline may be quite unclear.

4.1 Gaseous Models

The use of gaseous models in this problem has proved so effective that further research along this highly idealized direction is worthwhile. Among the problems for solution may be mentioned the following:

(i) The value of C: while the formal dependence of the thermal conductivity on density and velocity dispersion is not in doubt, the constant of proportionality should be determined; perhaps this could be done by comparing the rate of evolution of a slightly perturbed isothermal system in the gaseous and Fokker-Planck models.

(ii) Gravothermal oscillations: the reason for the discrepancy between the author's results and those of Bettwieser *et al* should be settled; a third independent integration would be a useful contribution.

(iii) Stochastic energy sources: so far the energy-sources used have been time-independent functions of density, velocity dispersion and location. A time-dependent source should be tried, perhaps along lines suggested by the discussion of McMillan & Lightman (1984b).

(iv) Tidal effects: study of tidally truncated systems would be useful to check whether the Inagaki-Lynden-Bell solution (modified centrally) gives way to the gaseous analogue of the original Hénon model (Hénon 1961), just as isolated systems follow the Goodman model (Fig. 5). The self-similar gaseous analogue of the tidally truncated Hénon model should be found.

(v) Anisotropy: though Larson's model for the evolution of stellar systems (Larson 1970 a,b) does not deal with energy transport in the same way as the gaseous model, it is formally very similar, and looks almost as simple. It was also designed to deal with an anisotropic distribution of velocities, and post-collapse evolution should be studied with such a method. To begin with, searches for self-similar solutions analogous to those of Inagaki & Lynden-Bell and of Goodman would be desirable. (It is perhaps worth stating that no anisotropic self-similar solution for core collapse has yet been produced either.)

(vi) Rotation: to drop the assumption of spherical symmetry and discuss axisymmetric rotating clusters with two spatial dimensions would greatly complicate the gaseous model. But clusters do rotate, and it might at least be worth treating rotation as a perturbation by including small $P_2 (\cos \theta)$ terms in the equation, provided (as in (i)) a rational method could be found for determining the coefficient of angular-momentum transport.

(vii) Mass spectrum: work has begun (Bettwieser & Inagaki 1984) on the inclusion of stars of different mass, but comparison with Fokker-Planck simulations indicates that further work is needed. The most delicate problem is the rational estimation of coefficients for heat conduction and for energy exchange between different populations.

(viii) Dissipative effects: it would be desirable, and easily possible in principle, to incorporate these into the gaseous model, by adding a dissipative term in the same manner as the energy generation term (Bettwieser & Fritze 1984). Given a satisfactory solution to (vii) above, it would also be possible to treat the formation and segregation of two-body binaries.

4.2 Fokker-Planck models

While simulations based on the Fokker-Planck equation are much more time-consuming than those using the gaseous model, there is no doubt that it is a superior model of the essential relaxation processes which drive much of the evolution. The following problems appear to be desirable or feasible, and sometimes both:

(i) Self-similar models: in unpublished work the author obtained a self-similar isotropic model for core collapse which agrees quantitatively with the results of Cohn (1980), and Hénon's models presumably relate to late post-collapse evolution. It would be desirable to find the Fokker-Planck analogue of the Inagaki-Lynden-Bell solution.

(ii) Simulations with simple energy-sources: these have been helpful in developing the theory of gaseous models, and it would be desirable to study and interpret a comparable suite of Fokker-Planck simulations, along the lines of those described by Cohn in this volume. (The difference between this and the artificial energy sources used by Hénon and Stodółkiewicz is that the energy generation would depend in a known way on the local density and distribution of velocities.)

(iii) Realistic effects: Fokker-Planck models have generally tended to be less idealized than gaseous models, and some models of Stodółkiewicz which go well beyond core collapse (see, for example, his paper in this volume) also include such effects as anisotropy, tidal truncation, stellar mass spectrum, mass-loss in stellar evolution, tidal shocking, and formation and evolution of binaries. The ultimate aim of such realism is, of course, comparison with observation. However, unless the history of all existing observable globular clusters has been identical, many different simulations would be needed. An alternative approach would be to introduce the various realistic effects one at a time, rather as is suggested above for gaseous models, so that a theoretical understanding of their effects could be developed.

4.3 Other remarks on simulations

(i) N-body integrations: it is worth repeating the suggestion made at the end of § 3 that renewed attempts should be made at the theoretical interpretation of these simulations, based on the picture that is now emerging from more highly idealized calculations. These and the hybrid simulations are invaluable for elucidating the detailed behaviour of binaries.

(ii) Presentation of results: no author is to be blamed because he did not publish those details of his models which, ten years later, turn out to be of interest. But it is worth stating that there is current interest in density- and velocity-profiles, and on the time-dependence of central values of velocity dispersion and density. Some of these are primarily of theoretical interest, but the importance of velocity and density profiles surely extends also into the domain of comparison with observations. Another topic to which thought should be given is scaling of results, since a lack of uniformity here makes comparison of published results by different authors time-consuming and sometimes impossible. The scaling often used in N-body simulations (see Fig. 1) has much to recommend it for most purposes. One further point on the presentation of results is the use of tables; those given by Hénon (1961, 1965) make his models much more useful than models given only in a graphical representation.

5. COMPARISON WITH OBSERVATIONS

5.1 Lowered Maxwellian models

It is customary to compare observations of the stellar distribution in globular clusters with lowered Maxwellian models (King 1966), which are tidally truncated, single-component isotropic models with a nearly isothermal core. Such models provide a satisfactory fit to the surface brightness profiles of most clusters, and it is usually only when the attempt to fit such a model fails that the possibility that the cluster has passed core collapse is entertained. What evidence do we have that a cluster with a collapsing core should resemble a lowered Maxwellian model, or that a cluster in post-collapse evolution should not?

The lowered Maxwellian model is an approximation to the solution of the equation

$$\frac{\partial f}{\partial t} = \left(\frac{\partial f}{\partial t}\right)_c ,$$

(King 1965), where $f(\varepsilon)$ is the stellar distribution function expressed as a function of energy ε per unit mass, and the right-hand side is the Fokker-Planck collision term, in which the diffusion terms are evaluated for a Maxwellian, and this equation is satisfied at the centre of the cluster. By contrast, the Fokker-Planck equation for a cluster with an isotropic distribution of velocities is

$$\overline{\frac{\partial f}{\partial t}} + \overline{\frac{\partial \phi}{\partial t}} \frac{\partial f}{\partial \varepsilon} = \left(\overline{\frac{\partial f}{\partial t}}\right)_c \qquad (7)$$

where ϕ is potential per unit mass, and the bar denotes a phase-space average for a star of energy ε (see, for example, Hénon 1961). Thus the lowered Maxwellian is an approximation to an equation which differs from the Fokker-Planck equation, and so its inherent dynamical justification is doubtful. Therefore we should avoid drawing any dynamical inferences from the success, or lack of success, of attempts to fit lowered Maxwellians to observational data. Rather, if the attempt is to be made at all to do this with single-component, isotropic models (and more will be said about this below), then the models used should be those obtained by integration of equations such as (7), and not an approximation to a solution of an equation which is not used to study dynamical evolution.

It is likely that deviations of solutions of (7) from lowered Maxwellian models would be most pronounced in regions far from both the core and the tidal boundary. The core of the lowered Maxwellian model is nearly isothermal, and after a few central relaxation times it would be a satisfactory fit to the core of a solution of (7). On the other hand simple isothermal models might be equally satisfactory for this purpose.

5.2 Interpretation of surface brightness profiles

Even if we abandon lowered Maxwellian models and use solutions of (7), it is probable that the results of comparisons between observations and such simple models would be misleading, because the models differ from more realistic simulations in ways which may be essential. From the point of view of comparison with observations, it can be argued that the idealisation which is most suspect is the lack of a spectrum of masses, and the reason for this is observational. In their discussion of M15, Illingworth & King (1977) proposed that the distribution of visible stars was partly controlled by a centrally concentrated population of heavier, invisible neutron stars. There is some attraction in interpreting this population as a collapsed core (Heggie 1980), as only the heavier stars will undergo the process of core collapse. If so, an understanding of the observable properties of post-collapse clusters depends essentially on the presence of different masses. Indeed it has been shown by Larson (1984) that approximate two-component models provide a relatively natural explanation of the observed velocity dispersion in the cores of several globular clusters, as well as being consistent with the central parts of the surface brightness profiles.

Besides M15, a second cluster in which there is much interest is NGC 6624 (Djorgovski & King 1984). A large part of the surface brightness profile, starting just outside the seeing disc, has a logarithmic slope of -1, which would be given by projection of an isothermal. Thus the central parts of this cluster could be fitted by one of the single-component post-collapse models of § 2. If this interpretation is correct, it should be noted that the interpretations of M15 and NGC 6624 as post-collapse objects are essentially different. On the other hand, Grindlay's surface brightness profile of NGC 6624 (see his paper in this volume) resembles that of M15, and an isothermal profile is, he asserts, a poor fit.

5.3 Statistical comparisons

Another approach to the problem of seeking observational evidence on post-collapse evolution was pioneered by Lightman, Press & Odenwald (1978), who studied how a population of clusters (characterised by two core parameters) would change as the core evolved. While there is no theoretical support for the kind of evolution they considered (i.e. prompt dissipation of a cluster very soon after core collapse), their approach is well worth following up. An indication of what changes are needed if current ideas of post-collapse evolution are adopted is furnished by Cohn & Hut (1984). There are two reasons why their results are illustrative rather than definitive, however. First, they suppose that the *central* relaxation time is a constant multiple of the time until, or since, maximum collapse, whereas in post-collapse evolution it is the relaxation time at the *edge* of the isothermal region which behaves in this way. Secondly, if at least two mass-components are needed for an understanding of surface-brightness profiles, and it is only the less massive, brighter, uncollapsed component that we observe, then an understanding of the evolution of such systems will surely be needed also for the adequate interpretation of the statistics of the cores.

6. CONCLUDING REMARKS

The process of idealization has been quite successful in leading to an understanding of some essential ideas in the theory of post-collapse evolution. Indeed we can have some confidence that the idealized problem of the evolution of an isolated system containing a large number of point masses can even be described quite simply in a semi-quantitative manner. But it is sobering to realize that the essence of these ideas (and much of the detail) was already contained in Hénon's paper of 1961. He made the assumptions of isotropy and self-similarity (which are both wrong), obtained a solution with singularities in both space and time (which might have impelled one to reject it), and devised a *post hoc* explanation in terms of binary stars. And yet the ideas have stood the test of time, and it may fairly be claimed on Hénon's behalf that post-collapse evolution was essentially understood several years before the collapse phase was!

7. REFERENCES

Alexander, M.E. & Budding, E.: 1979, *Astron. Astrophys.* **73**, 227.
Bettwieser, E. & Fritze, U.: 1984, submitted to *P.A.S.J.*
Bettwieser, E. & Inagaki, S.: 1984, 'A Gas Model for Multi-Component Star Clusters: 1. Method and Comparison with Fokker-Planck,' in preparation.
Bettwieser, E. & Sugimoto, D.: 1984, *M.N.R.A.S.*, in press.
Cohn, H.: 1980, *Ap. J.* **242**, 765.
Cohn, H. & Hut, P.: 1984, *Ap. J. Lett.* **277**, L45.
Djorgovski, S. & King, I.R.: 1984, *Ap. J. Lett.* **277**, L49.
Dokuchaev, V.I. & Ozernoy, L.M.: 1982, *Astron. Astrophys.* **111**, 16.
Duncan, M.J. & Shapiro, S.L.: 1982, *Ap. J.* **253**, 921.
Fall, S.M. & Malkan, M.A.: 1978, *M.N.R.A.S.* **185**, 899.
Goodman, J.: 1984, *Ap. J.* **280**, 298.
Hachisu, I., Nakada, Y., Nomoto, L. & Sugimoto, D.: 1978, *Prog. Theor. Phys.* **60**, 393.
Heggie, D.C.: 1980, in J.E. Hesser (ed.), *Star Clusters*, Reidel, Dordrecht, p. 401.
Heggie, D.C.: 1984, *M.N.R.A.S.* **206**, 179.
Hénon, M.: 1961, *Ann. d'Astrophys.* **24**, 369.
Hénon, M.: 1965, *Ann. d'Astrophys.* **28**, 62.
Hénon, M.: 1975, in A. Hayli (ed.), *Dynamics of Stellar Systems*, Reidel, Dordrecht, p. 133.
Hills, J.G.: 1975, *Astron. J.* **80**, 1075.

Illingworth, G. & King, I.R.: 1977, *Ap. J. Lett.* 218, L109-12.
Inagaki, S. & Lynden-Bell, D.: 1983, *M.N.R.A.S.* 205, 913.
King, I.R.: 1965, *Astron. J.* 70, 376.
King, I.R.: 1966, *Astron. J.* 71, 64.
Larson, R.B.: 1970a, *M.N.R.A.S.* 147, 323.
Larson, R.B.: 1970b, *M.N.R.A.S.* 150, 93.
Larson, R.B.: 1984, 'Black-Hole Remnants in Globular Clusters', preprint.
Lightman, A.P.: 1982, *Ap. J. Lett.* 263, L19.
Lightman, A.P., Press, W.H. & Odenwald, S.F.: 1978, *Ap. J.* 219, 629.
Lightman, A.P. & Shapiro, S.L.: 1978, *Rev. Mod. Phys.* 50, 437.
Lynden-Bell, D. & Eggleton, P.P.: 1980, *M.N.R.A.S.* 191, 483.
McMillan, S.L.W. & Lightman, A.P.: 1984a, *Ap. J.*, in press.
McMillan, S.L.W. & Lightman, A.P.: 1984b, *Ap. J.*, in press.
McMillan, S.L.W., Lightman, A.P. & Cohn, H.: 1981, *Ap. J.* 251, 436.
Marchant, A.B. & Shapiro, S.L.: 1980, *Ap. J.* 239, 685.
Ozernoy, L.M. & Dokuchaev, V.I.: 1982, *Astron. Astrophys.* 111, 1.
Peaterson, C.J. & King, I.R.: 1975, *Astron. J.* 80, 427.
Retterer, J.M.: 1984, 'A Dynamical Model for Small Stellar Systems,' preprint.
Shapiro, S.L.: 1977, *Ap. J.* 217, 281.
Spitzer, L., Jr. & Shull, J.M.: 1975a, *Ap. J.* 200, 339.
Spitzer, L., Jr. & Shull, J.M.: 1975b, *Ap. J.* 201, 773
Stodółkiewicz, J.S.: 1982, *Acta Astron.* 32, 63.
Stodółkiewicz, J.S.: 1983, in J. Ruprecht & J. Palouš (eds.), *Star Clusters and Associations and their Relation to the Evolution of the Galaxy*, Astron. Inst. Czech. Acad. Sci., Prague, p.103.
von Hoerner, S.: 1960, *Z. Astrophys.* 50, 184.
von Hoerner, S.: 1963, *Z. Astrophys.* 57, 47.

DISCUSSION

OSTRIKER: What determines the core radius in your simplified post collapse models?

HEGGIE: I assume that the rate of energy generation (per unit mass) depends in a known way on the local density and velocity dispersion, v. If the dependence on density is sufficiently steep, energy generation is nearly confined to the core, and the total rate, E, can be expressed in terms of v and the core radius, r_c. In the models of Hénon, Goodman, or Inagaki & Lynden-Bell, at any time v is determined and also the luminosity L of the central source. Hence r_c is determined by supposing the $L = \dot{E}$.

WHITE: You mentioned that Hénon's model had a precollapse density exponent of about -2.4. Could you clarify whether this is a result of the fact that his calculations allowed anisotropic velocity distributions and thus might be more "realistic" than purely isoltropic models which give -2.21 or -2.23?

HEGGIE: Actually I found -2.44 for Hénon's profile. Larson, with his fluid dynamical model, found -2.4 in one case. However Haldan Cohn tells me he found -2.23 (if I recall correctly), and this was also obtained by Duncan & Shapiro. All these models allow for anisotropy, and so there is some disagreement to sort out here.

COHN: Do I understand correctly that there is a difference in the form of the energy input rate used in your calculations and that adopted by Drs. Sugimoto and Bettwieser?

HEGGIE: There is no difference in *form*. There is a difference in the constant of proportionality, which I chose in such a way as to produce a density increase during collapse similar to that in one of their calculations which produced oscillations. The fact that I had to choose a different constant to achieve the same density increase is due in part to an error in their specification of the problem they solved. Dr. Bettwieser tells me that their cluster had a total mass of $10^5 M_\odot$, not $10^6 M_\odot$ as stated in their paper.

BETTWIESER: A comment to the question of Dr. Cohn. Recently we made simulations with very different energy sources, extended as well as point sources. We find core oscillations provided energy input is sufficiently small. Also I want again to stress the point, that in Dr. Heggie's calculations a large energy flux directed inwards is not allowed for, since his dependent variables are singular at $L = -1$. But the latter is necessary in order to have gravothermal expansion and not only a simple thermal expansion.

HEGGIE: I use the variable $\ln(1 + L)$, where L is the luminosity, and this does not permit $L \leq -1$. But I always observe L to be positive. If a gravothermal expansion were to begin, I would first see L become slightly negative, before the singularity at $L = -1$ is reached. However, L does not become negative.

SUGIMOTO: One comment and one question: The comment is concerned with Inagaki-Lynden-Bell's self-similar solution. As you criticized it correctly, the time dependence of the energy generation is determined by the overall expansion. However, the energy generation is controlled by a completely different mechanism, i.e., by binary hardening. Thus it will deviate from the required value. Since the collapsed core solution is gravothermally unstable, the deviation will grow and, therefore, it is meaningless to consult with the similarity solution. This criticism is given in detail in Bettwieser and Sugimoto (1984, M.N.R.A.S. in press).

The question is concerned with your recent solution for general expansion in the post collapse phase. If we compare it with the collapse phase at the same central density, the expansion is slower than the contraction by a factor of say 20 when the central density has fallen appreciably and thus the energy generation rate has fallen appreciably. Since the temperature is decreasing outward everywhere, the system should be gravothermally collapsing if there is no energy source at the same timescale as the collapse at the same age. You imply that it is stopped and reversed. However, the power of reversing is ≈ 20 times weaker and seems to be able to be neglected as compared with the ponderamotive force to contraction. How can you interpret this expansion physically? Is it related with the fact that you used the variable $\ln(1 + L)$ for the flux? It does not all $L < -1$ but the negative large value of L is essential for the gravothermal expansion. Or does your calculation correspond to a case with high energy generation where the system became a thermal system and makes a general expansion for the energy input? Though you said that you computed the same case as ours (Bettwieser and Sugimoto 1984), such could be the case if our paper should contain any important typographical error in numerical values.

HEGGIE: I agree that the Inagaki & Lynden-Bell solution, with its singular density profile, does not occur in my simulation. I argue that the central singularity is replaced by a non-singular core which produces the same luminosity as the singular core would, and has the same velocity dispersion, v. Thus it is meaningful to compare v with the behaviour predicted by Inagaki & Lynden-Bell, and yet the question whether or not a certain singular density profile is stable is not relevant, since such a profile is not present.

The reason why expansion is observed, and not collapse, even with temperature decreasing outward and a low rate of energy generation, is that a larger part of the system is isothermal than at the corresponding time in the collapse phase. Thus the heat flux across the edge of the isothermal part is smaller than at the corresponding time during collapse, and a smaller rate of energy generation is sufficient not only to prevent collapse but also to sustain expansion.

The question on the rate of energy generation I used, and on typographical errors, are dealt with in my reply to Dr. Cohn. The use of the variable $\ln(1 + L)$ is defended in my reply to Dr. Bettwieser.

GOODMAN: I'd like to point out that binaries do not have to supply kinetic energy *directly* to the cusp: by ejecting mass, they decrease the binding energy of the cusp and hence drive its expansion even if the "heat" evolved by binary encounters is lost from the cluster. Also, to balance the ejection, stars have to flow inward, yielding up energy as they do so and heating the cusp. The latter mechanism is qualitatively the same as that by which a black hole heats a cluster.

KING: I share your awe of the accuracy with which Hénon predicted twenty years ago what we would be talking about today. You pointed out that many of his assumptions were wrong; I suggest to you that one of the characteristics of genius is to get the right answer from the wrong assumptions!

DIRECT FOKKER-PLANCK CALCULATIONS

Haldan Cohn
Department of Astronomy
Indiana University
Swain Hall West 319
Bloomington, IN 47405
USA

ABSTRACT. The past decade has seen the development of powerful numerical methods for studying star cluster evolution by direct integration of the Fokker-Planck equation. Cohn's basic algorithm for spherical systems of identical point masses and its application to the study of core collapse is reviewed. Merritt's extension of this method to treat systems containing a mass spectrum, and Goodman's extensions to include strong scattering and rotation are discussed. Results from direct Fokker-Planck computations of core collapse in single mass and multi-mass isotropic clusters and single mass anisotropic clusters are presented to illustrate the method. Calculations of pre- and post-collapse evolution with a central black hole and with heating by hard binaries are reported. Prospects for future development of the direct Fokker-Planck method will be discussed, with emphasis on the central goal of developing physically realistic models for interpreting Hubble Space Telescope observations of globular cluster structure.

1. INTRODUCTION

During the past decade I've been involved in the development of a robust numerical method for studying star cluster evolution, based on direct integration of the Fokker-Planck equation (Cohn 1979a,b, 1980). Unlike Monte Carlo simulation techniques that produce particular stochastic realizations of the Fokker-Planck equation, the direct method reviewed here employs a finite difference scheme to obtain a time-dependent solution. The direct method is particularly well suited to investigating core collapse in star clusters and subsequent post-collapse evolutionary phases. The application of the direct Fokker-Planck method to this challenging astrophysical problem is the focus of this review.

The general redistribution of stars in phase space produced by two-body gravitational scattering was early-on recognized as the physical process that drives star cluster evolution (Jeans 1929; Ambartsumian 1938; Spitzer 1940; Chandrasekhar 1942). Following this evolution numerically presents a significant computational challenge,

J. Goodman and P. Hut (eds.), Dynamics of Star Clusters, 161–178.

since the gravitational potential in which cluster stars orbit itself evolves in time along with the evolving phase space distribution. This joint evolution was decoupled in early numerical treatments of the Fokker-Planck equation by the adoption of a square well approximation to the cluster potential. This approach allowed a calculation of the rate of escape of stars from clusters due to small angle two-body scattering (Spitzer and Harm 1958), and also permitted the development of quasi-equilibrium models for star clusters (Mitchie 1963a,b; King 1966).

Henon's (1961, 1965) path breaking similarity solutions to the self consistent evolution problem led the way to subsequent detailed numerical treatments of the Fokker-Planck equation. Larson (1970) produced the first time dependent integration of the self consistent Fokker-Planck equation using a moment equation reduction. Powerful Monte Carlo simulation methods were developed and refined in the following decade (Henon 1973, Spitzer 1975).

During the two decades following the work of Spitzer and Harm (1958), much progress was made by plasma physicists in developing direct numerical methods for solving the Fokker-Planck equation. A considerable amount of this work has been directed towards the study of particle orbit diffusion in magnetic confinement devices (Killeen, Mirin, and Rensink 1976). There is a great similarity between this orbit diffusion process and that which occurs in star clusters. Motivated by this analogy, Russell Kulsrud and I undertook the modification of a plasma kinetic Fokker-Planck code to attack the problem of determining the quasi-equilibrium stellar distribution about a massive black hole at the center of a star cluster (Cohn and Kulsrud 1978; cf. Shapiro, this volume). Buoyed by the success of this venture, Jerry Ostriker and I took on the self-consistent problem (Cohn 1979a, b). Section 2 of this review presents an overview of the direct Fokker-Planck method that resulted from this work. Extensions by Merritt (1981) to treat a mass spectrum and Goodman (1983) to include strong scattering and cluster rotation are discussed. Section 3 reviews the application of this method to the study of core collapse and cluster life thereafter. Prospects for future work are discussed in Section 4.

2. THE DIRECT FOKKER-PLANCK METHOD

The direct Fokker-Planck method is so named because it directly attacks the orbit-averaged Fokker-Planck kinetic equation for the stellar distribution function, resulting in a time-dependent solution. As discussed by Cohn and Kulsrud (1978), this kinetic equation is derived from the full Boltzmann equation under the assumption that the dynamical time is much shorter than the relaxation time. To lowest order in t_d/t_r we find that the distribution function must satisfy,

$$f = f(E,J^2,t) \tag{2.1}$$

Here E and J^2 are the specific energy and square of the specific angular momentum of a star. The time dependence of f, which is of <u>first</u> order in t_d/t_r, is given by,

$$\frac{\partial f}{\partial t} = \frac{1}{P} \int \frac{dr}{v_r} \left(\frac{\partial f}{\partial t}\right)_c \qquad (2.2)$$

where $(\partial f/\partial t)_c$ is the local velocity space Fokker-Planck operator (Rosenbluth, MacDonald, and Judd 1957), v_r is the radial velocity, $P = \int dr/v_r$ is the orbital period, and the orbit-averaging integral extends between the two radial turning points of an orbit.

The Fokker-Planck equation (2.2) describes the diffusion of stars in (E,J^2)- space which is represented schematically in Figure 1.

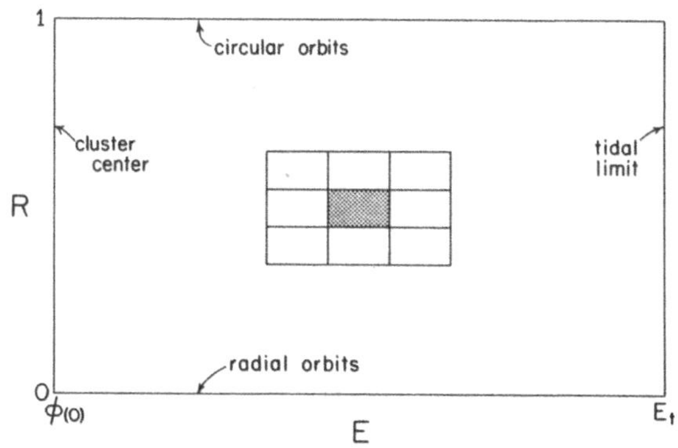

Figure 1. The domain of the orbit-averaged Fokker-Planck equation: (E,R)-space. The shaded region is a typical computational cell surrounded by eight neighbors.

It is convenient here to introduce a scaled angular momentum $R = J^2/J_{max}{}^2(E)$ where J_{max} is the angular momentum of a star of energy E. Each point in the rectangular region of (E,R)-space shown in Figure 1 represents a possible stellar orbit in a given spherically symmetric gravitational potential well $\phi(r)$. For the purpose of developing a finite difference form of the Fokker-Planck equation, (E,R)-space is partitioned into cells; such a cell and its eight immediate neighbors are shown in Figure 1. The occupation number for each cell is given by,

$$N(E,R) = 4\pi^2 \, P\,(E,R)\, f(E,R). \qquad (2.3)$$

In terms of N, the Fokker-Planck equation can be written as an equation of continuity in (E,R)-space,

$$\frac{\partial N}{\partial t} = - \nabla \cdot \mathbf{F} \qquad (2.4)$$

where $\mathbf{F}$ is the (E,R)-space particle flux vector and ∇ is the (E,R)-space divergence operator. The form of equation (2.4) indicates that only the diffusion of stars between neighboring cells (orbits) is considered in the Fokker-Planck approach.

The Fokker-Planck equation can be written in component form as,

$$\frac{\partial f}{\partial t} = -\frac{1}{A}\frac{\partial}{\partial E}\Big(D_{EE}\frac{\partial f}{\partial E} + D_{ER}\frac{\partial f}{\partial R} + D_E f\Big) - \frac{1}{A}\frac{\partial}{\partial R}\Big(D_{RE}\frac{\partial f}{\partial E} + D_{RR}\frac{\partial f}{\partial R} + D_R f\Big) \quad (2.5)$$

where the coefficients A, D_{EE}, D_{ER}, etc. depend on E, R, and $\phi(r)$. The "D" coefficients depend additionally on f(E,R); this dependence can be written schematically as,

$$D = \int\frac{dr}{v_r}\int dE'\ G(E,E',r)\ \bar{f}(E',r) \quad (2.6)$$

where G is a particular function of the indicated arguments that can be derived from the results of Rosenbluth, MacDonald, and Judd (1957) and $\bar{f}$ is the isotropized (R-averaged) distribution function. The approximation of an isotropized background has commonly been made in stellar dynamical Fokker-Planck studies (c.f. Spitzer and Hart 1971; Marchant and Shapiro 1980), in order to simplify the computation of diffusion coefficients. However it would be possible to consider the full two-dimensional dependence of the background distribution, at the cost of some additional computational complexity, as been done in plasma kinetic studies.

Computing the evolution of a star cluster involves a self-consistent potential problem since the potential and distribution functions are related by Poisson's equation,

$$\nabla^2\phi(r) = 4\pi\ G\ \rho(r),\ \text{where} \quad (2.7a)$$

$$\rho = \int d^3v\ f. \quad (2.7b)$$

Thus as f evolves in time according to equation (2.5), ϕ must also evolve. In practice, the evolution of ϕ is handled implicitly by use of adiabatic invarients. These are quantities that remain constant, to all orders of t_d/t_r, as ϕ evolves in time. Two convenient choices of adiabatic invarient are J^2 and the radial action $Q = \int dr\ v_r$. The evolution of a star cluster may be thought of as the smooth trajectory in a (f,ϕ) function space shown schematically in Figure 2. This evolution is tracked computationally by an alternating direction method indicated by the stair-step curve. The distribution function is first advanced, with ϕ held fixed, by use of a discrete form of equation (2.5). The potential ϕ is then advanced, with f held fixed as a function of adiabatic invarients, by solving equation (2.7a). This procedure is exact to second order in the time step (Cohn 1979a,b).

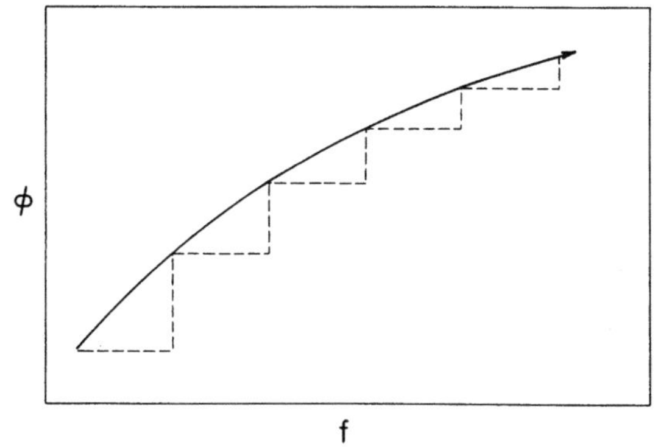

Figure 2. Schematic representation of cluster evolution in (f,ϕ)-space. The solid curve is the true trajectory while the dashed curve is the computational approximation.

The discussion to this point has focused on the two-dimensional (E,R)-space Fokker-Planck equation. As I discovered after some years in two-space, one-space is yet simpler. The two-dimensional Fokker-Planck equation (2.5) can be reduced to a one-dimensional E-space equation by averaging over R. Geometrically this amounts to projecting along the R dimension in Figure 1. Each cell in E-space now has only two neighbors. When equation (2.5) is averaged over R and zero flux boundary conditions are applied at R=0 and R=1, the energy-space Fokker-Planck equation of Henon (1961) and Cohn (1980) results. Besides significantly reduced CPU time and core storage requirements, another important advantage of the one-dimensional Fokker-Planck equation is the existence of the Chang and Cooper (1970) differencing scheme. As discussed in Section 3.3, no two-dimensional generalization of this one-dimensional scheme has been found which preserves all of its beneficial properties.

In order to treat a system with more than one mass group, an additional distribution function is added for each additional component. There is a separate Fokker-Planck kinetic equation corresponding to equation (2.5) for each distribution function, with coefficients that can be derived from the results of Rosenbluth, MacDonald and Judd (1957). Cohn (1984) and Inagaki and Wiyanto (1984) have carried out calculations of the evolution of two-component clusters, while Merritt (1981) has extended the direct Fokker-Planck method to treat systems with a continuous mass spectrum. In these three studies, the isotropized Fokker-Planck equation was used. When a large number of mass components are used, as in Merritt's (1981) study, the computation is effectively two-dimensional. Thus inclusion of the angular momentum dependence of the distribution functions would increase the dimensionality to three. This increase of computational complexity has not yet been attempted by practitioners of the direct Fokker-Planck method.

As demonstrated by Cohn and Kulsrud (1978) and by McMillan, Lightman, and Cohn (1981), the direct Fokker-Planck method can be extended to include additional processes besides two-body scattering by adding terms to the Fokker-Planck equation that describe loss of stars and loss or gain of energy. Such source and sink terms can often be calculated by orbit-averaging local (position dependent) rates following equation (2.2). An example of such a local rate is given by equation (3.4) below for the binary heating rate. The orbit-averaging process in this case results in an additional contribution to $\langle \Delta E \rangle$, the first order rate of charge of stellar energy.

Goodman (1983) has developed two additional extensions of the direct Fokker-Planck method: the inclusion of strong scattering in the isotropic case, and a modification of the anisotropic approach to treat axisymmetric, rotating clusters.

3. RESULTS

3.1. Core Collapse in Single Mass Systems

A particular advantage of the direct Fokker-Planck method, for studying core collapse, is the flexibility possible in zoning the orbital phase space. As discussed by Cohn (1979a,b) it is useful to adopt a uniform zoning in an auxillary function of energy which is chosen so as to allocate as many zones as desired to any interesting region of orbital phase space. As core collapse proceeds, the fractional phase space volume occupied by the cluster core becomes quite small. It is straightforward, however, to keep the number of energy zones in the core approximately constant.

The power of the direct Fokker-Planck method for studying core collapse is illustrated by Figure 3 which shows the evolution of the density profile of a single mass star cluster (Cohn 1980). This calculation used the isotropic form of the direct Fokker-Planck method discussed in Section II. The development of a power law central cusp is clearly evident in Figure 3. The slope of the cusp produced by the direct Fokker-Planck method, $d \ln \rho / d \ln r = -2.23$, is in good agreement with the value of -2.21 that was obtained by Lynden-Bell and Eggleton (1980), on the basis of the conducting gas sphere model. While the core of a globular cluster runs out of stars long before the central density has increased by 20 orders of magnitude, there will be about 10^3 stars remaining in the core when the density has increased by six orders of magnitude, assuming a total of 10^6 stars in the cluster. Figure 3 shows that the self-similar solution is well established by this time.

The long-term fate of the density cusp that develops during core collapse is a issue of fundamental importance in the study of globular cluster dynamical evolution. If a central cusp is a long-lived structural feature, then it may offer a means of identifying post-collapse clusters (Cohn and Hut 1984). Sections 3.4 and 3.5 discuss two alternative scenarios for post-collapse cluster evolution that has been investigated with the direct Fokker-Planck method: black hole driven reexpansion and three-body binary driven reexpansion.

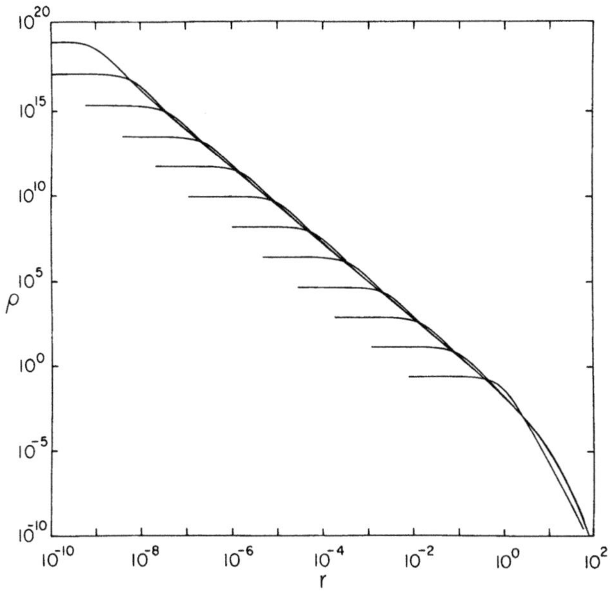

Figure 3. Evolution of the space density profile during core collapse in a single mass cluster. The central density increases in time as the core radius shrinks.

Direct Fokker-Planck calculations of core collapse have indicated an important difference between the cusp that develops as a result of core collapse and the cusp expected about a massive black hole. While the slopes of the density profiles for these two cases are similar ($d \ln \rho / d \ln r = -2.23$ and -1.75, respectively) there is a significant difference in the slopes of the velocity dispersion profiles ($d \ln v^2 / d \ln r = -0.23$ and -1.0, respectively). The much steeper cusp in velocity dispersion profile that results from the presence of a massive black hole should provide a means of distinguishing these situations observationally (cf. Bahcall and Wolf 1976).

3.2. Core Collapse in Multi-Mass Systems

Single mass models for globular clusters are clearly highly idealized, since observed mass spectra span at least a decade in mass. The isotropic direct Fokker-Planck method has been extended to treat both systems containing a small number of discrete mass components (Cohn 1984, Inagaki and Wiyanto 1984) and systems with a continuous mass spectrum (Merritt 1981). The latter study was concerned with the evolution of galaxy clusters under the action of two-body relaxation and tidal stripping and thus did not investigate core collapse in any detail. The former two studies were concerned with core collapse in a two-component star cluster and obtained similar results.

Figures 4a,b show the evolution of the space and surface density profiles for two components differing in mass by a factor of two (Cohn

1984). The fraction of the cluster mass in the light and heavy components is 90% and 10% respectively. On the basis of Spitzer's (1969) criterion, such a cluster is expected to be unstable to mass segregation.

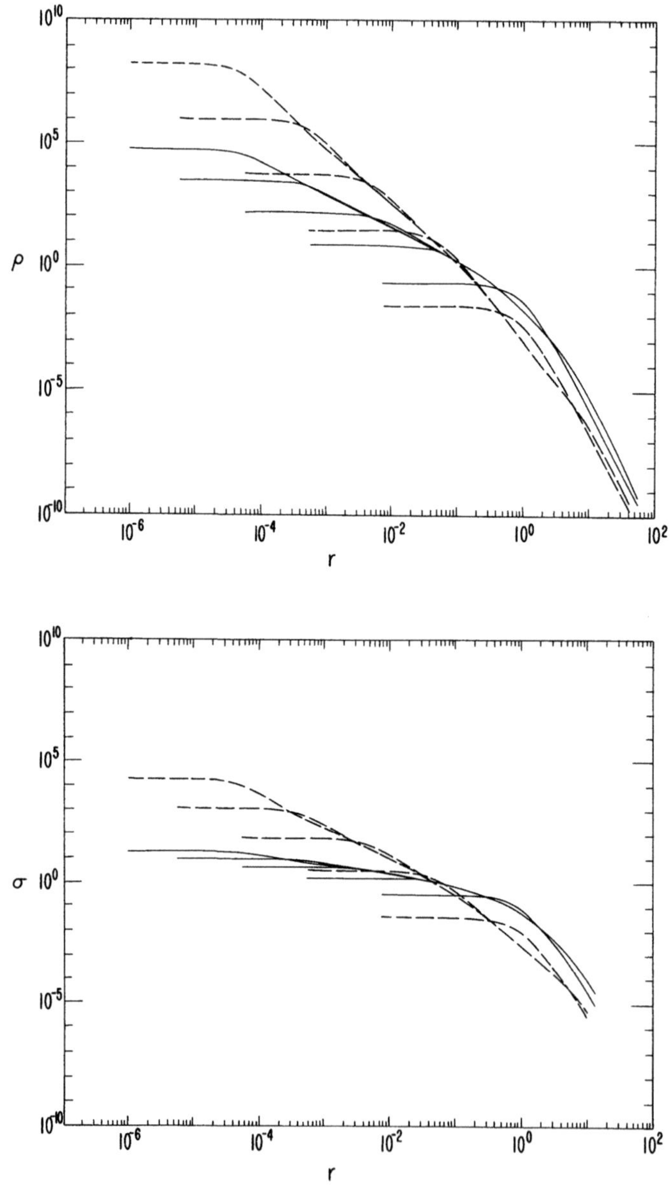

Figure 4. (a) Evolution of the space density profile during core collapse in a two-component cluster; (b) corresponding evolution of the surface density profile. Solid lines indicate the less massive component while dashed lines indicate the more massive component.

The development of mass segregation is clearly evident in Figures 4a,b. While the two components intitially have the same density distribution (Plummer's model), the heavy component soon dominates the central part of the cluster. By the end of the computation, the evolution of the distribution of heavy stars closely resembles a one-component core collapse. In particular the slope of the heavy star cusp is $d \ln \rho/d \ln r = -2.23$, projecting to a surface density cusp slope $d \ln \sigma/d \ln r = -1.23$. The light stars also exhibit a central power law cusp; however the slopes of the space density and surface density are much flatter, -1.3 and -0.3 respectively. The relation between the slopes of the density cusps for the heavy and light components can be computed on the basis of the results of Bahcall and Wolf (1977) and Cohn (1980). The former study showed that if $p = d \ln f/d \ln E$, then

$$\frac{p_1}{p_2} = \frac{m_1}{m_2} \tag{3.1}$$

where the subscripts refer to mass component. Since the heavy component is distributed as in a one-component core collapse, $p_2 = 8.2$ (Cohn 1980). Denoting the density cusp slope by $\beta = - d \ln \rho/d \ln r$, the gravitational potential in the central part of the cluster has slope $d \ln \phi/d \ln r = 2-\beta_2$, with $\beta_2 = 2.23$. Combining these results with the relation $d \ln \rho/d \ln \phi = p + 3/2$, it follows that,

$$\beta_1 = 1.89 \frac{m_1}{m_2} + 0.35 \tag{3.2}$$

This scaling relation agrees well with the results shown in Figure 4a.

The most massive stars in a present day globular cluster are likely to be nonluminous neutron stars and heavy white dwarfs (Gunn and Griffin 1979). The stars that dominate the luminosity function -- red giants in the visual and horizontal branch stars in the UV (King 1980) -- are less massive than these nonluminous remnants. In this case, it follows from equation (3.2) that the observable surface brightness cusp that develops as a result of core collapse can be rather flatter than for a one-component cluster. Thus, depending on the mass spectrum, core collapse might occur without producing a large deviation from a normal King-type surface brightness profile.

As reviewed by Heggie elsewhere in this volume, recent conducting gas sphere model treatments of post-collapse evolution suggest that the central cusp established during core collapse evolves to a singular isothermal, $\rho \propto r^{-2}$ (Inagaki and Lynden-Bell 1983, Heggie 1984, and Goodman 1984). These studies considered a one-component cluster; if the nonluminous heavy stars in a multi-component cluster are distributed as a singular isothermal then the less massive stars will have a density gradient,

$$\beta_1 = 2 \frac{m_1}{m_2} \tag{3.3}$$

(Cohn and Hut 1984). As discussed by Djorgovski and King (1984) the observed surface brightness cusp in NGC 6624 is consistent with a projected singular isothermal slope of -1, although a somewhat flatter slope is also consistent with their data.

Fokker-Planck calculations of core collapse in a cluster with a continuous mass spectrum are clearly needed. Based on the results of two-component models (Cohn 1984, Inagaki and Wiyanto 1984), it seems likely that the most massive component will segregate to the center of the cluster. However if the difference in mass between components is small it is possible there may be some qualitative changes in the results.

3.3. Development of Anisotropy in Core Collapse

Comparison of the full two-dimensional Fokker-Planck calculations of core collapse presented by Cohn (1979a,b) with the isotropized one-dimensional calculations of Cohn (1980) shows few differences in the evolution up to the end of the former calculation, which corresponds to a central density enhancement by a factor of 10^3. The self-similar solution is only beginning to set in at this point. This particular set of two-dimensional calculations was terminated due to the cumulative effect of a slow but secular numerical nonconservation of energy. As discussed by Cohn (1980), adoption of the Chang and Cooper (1970) differencing scheme in the one-dimensional code reduced the energy loss rate by over two orders of magnitude. This greatly improved energy conservation permitted the isotropized calculations to be carried much farther into the self-similar regime.

Generalization of the Chang and Cooper (1970) scheme to a two-dimensional phase space has not proved straightforward. No such two-dimensional scheme appears to be available in the literature. I've investigated several alternative generalizations as has Jeremy Goodman; no scheme we've tried performs quite as well as the one-dimensional version, although all improve energy conservation relative to the simple centered differencing used by Cohn (1979a,b).

Figure 5 shows some interesting -- though preliminary -- results obtained with a two-dimensional direct Fokker Planck code using a particular heuristic generalization of the Chang and Cooper (1970) scheme. This calculation was run to an density enhancement factor of 10^6, well into the self-similar regime. The figure shows the ratio of transverse velocity dispersion to radial velocity dispersion near the onset of the gravothermal instability and at a later time. As was suggested by the earlier results of Spitzer and Shull (1975), anisotropy extends well inside the half-mass radius at late times. The degree of anisotropy at late times is not large -- about 10% -- but it is definitely present. If this calculation were to be continued beyond core collapse it is plausible that the degree of anisotropy would continue to increase. This speculation will be tested in future work.

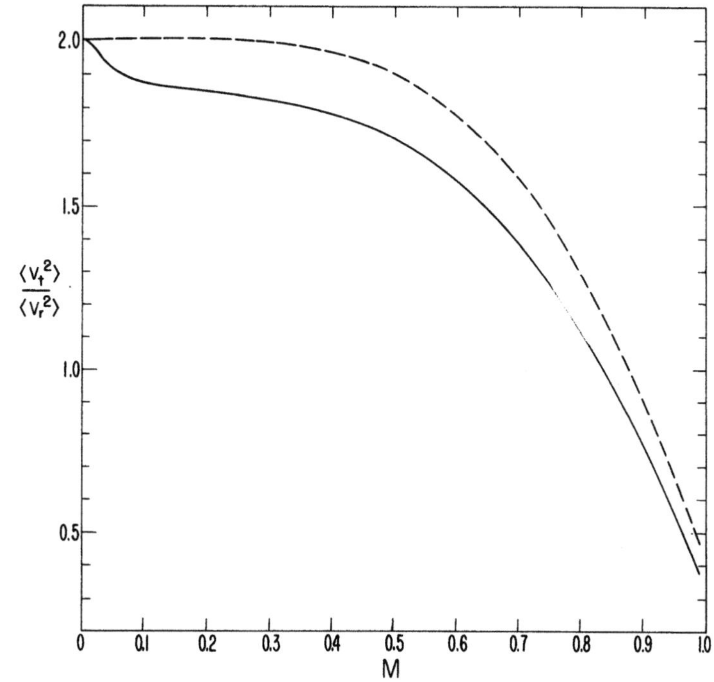

Figure 5. Evolution of anisotropy during core collapse. Abscissa is Lagrangian mass coordinate; ordinate value of 2.0 corresponds to isotropy. Dashed curve corresponds to t/t_{rh} = 12.7; solid curve to t/t_{rh} =15.8.

3.4. Post-Collapse Evolution with a Central Black Hole

As discussed by Shapiro elsewhere in this volume, the presence of a central black hole in a star cluster ultimately reverses core collapse and causes a general expansion of the cluster. It is straightforward to extend the direct Fokker-Planck method to include a central black hole by modifying the central boundary condition and adding the tidal destruction sink term given by Cohn and Kulsrud (1978). McMillan, Lightman, and Cohn (1981) have reported direct Fokker-Planck calculations of the evolution of black hole powered active galactic nuclei. Figure 6 shows previously unreported results I have obtained for the evolution of a globular cluster with a central black hole. The cluster begins as a n = 5 polytrope with a black hole mass of M_{bh} = 50 $M_\odot$ and a total cluster mass of 3 x 10^5 $M_\odot$. At the start of the run, the cusp surrounding this "seed" black hole only extends out to about 10^{-3} r_{core}. After an evolution of about 2 x 10^{10} yr, however, the combined effect of core collapse and black hole growth by accretion of tidal debris results in a cusp that extends out to about 0.1 r_{core}. (Since the profile no longer exhibits a flat core at this time, r_{core} is operationally defined as the outermost value of radius at which d ln ρ/d ln r = −1.5, which is the value at the edge of a King-type core).

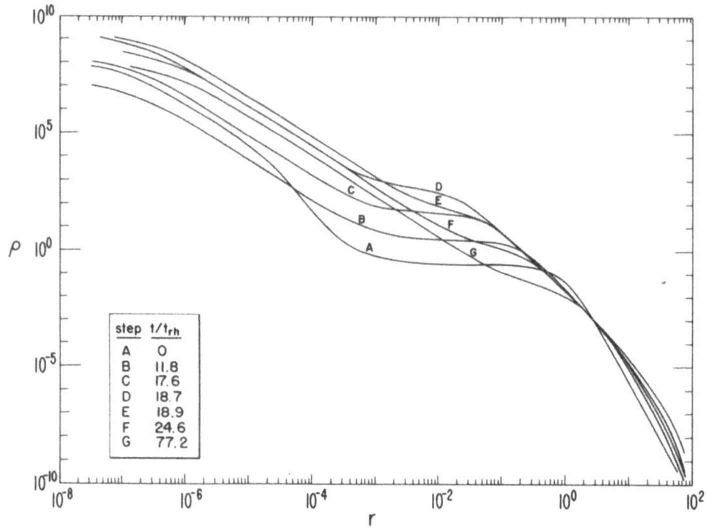

Figure 6. Evolution of density profile in a cluster containing a central black hole. For the adopted cluster structure parameters, $t_{rh} \approx 10^9$ yr.

The slope of the cusp closely agrees with the value of -1.75 first obtained by Bahcall and Wolf (1976). The collapse of the core is reversed at a time of about 2×10^{10} yr and the core radius (defined as above) gradually increases in time while the density at r_{core} decreases. Density profiles F and G in Figure 6, represent evolution times of 2.5 and 7.7×10^{10} yr respectively, indicating that the evolution rate slows markedly in the late post-collapse phase. At late times, the central black hole has grown to about 3×10^3 $M_\odot$. The density profile is characterized by a central cusp, a flatter transitional region, and an outer halo of steeper slope. This detailed structure is substantially different from that shown in Figure 3 which represents core collapse without a black hole. The cusp that develops in the presence of a central black hole also differs substantially from a normal core collapse cusp in that the velocity dispersion scales as $\langle v^2 \rangle \propto r^{-1}$ in the former case and $\langle v^2 \rangle \propto r^{-0.2}$ in the latter.

3.5. Post-Collapse Evolution with Three-Body Binaries

As Heggie has reviewed elsewhere in this volume, it now appears likely that post-collapse evolution of globular clusters is driven by energy release from hard binaries. Piet Hut and I have recently incorporated this additional physical effect into the direct Fokker-Planck method. Several alternative approaches to this extension are possible; we have investigated the addition of a heating term to the Fokker-Planck equation that is the product of the three-body binary formation rate multiplied by the total energy input per hard binary up to the time of ejection. In this approximation, hard binaries are "burned" instantaneously on the cluster evolution timescale. The adopted form of the heating term is

$$\dot{E} = 90\, n^2\, G^5\, m_*^5 v_1^{-7} \qquad (3.4)$$

where $\dot{E}$ is the energy input rate per unit mass, n is the local density, m_* is the stellar mass, and v_1 is the one-dimensional velocity dispersion. Other alternatives for treating hard binaries are discussed in Section 4.

Figure 7 shows the evolution of the density profile that results with the inclusion of three-body binaries. The binary heating rate normalization was chosen for this particular run so that the total number of stars in the cluster is 5×10^4. The time corresponding to each curve in Figure 7 is given in units of t_{rh}, which is of order 10^8 - 10^9 yr for reasonable choices of core structure parameter values. For this range of half-mass relaxation times, the evolution of the system extends beyond a Hubble time.

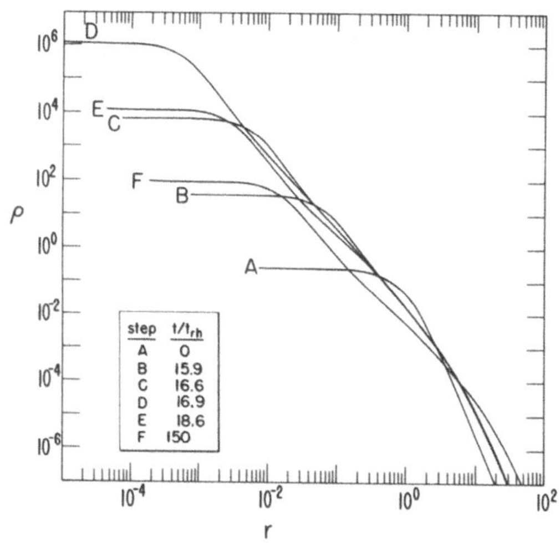

Figure 7. Evolution of density profile with inclusion of three-body binary heating.

The evolutionary sequence shown in Figure 7 begins with a core collapse phase which is halted when 27 stars remain in the core. A subsequent expansion then ensues on an ever increasing time scale; note that the time elapsed between profiles E and F is seven times longer than the time up to E.

Figure 8 shows the evolution of the density gradient profile. At the time of maximum central density (curve D) the profile is close to the $\rho \propto r^{-2.2}$ power law core collapse solution. During the subsequent expansion phase (curves E and F), no single power law characterizes the profile. Instead, d ln ρ/d ln r fluctuates between -1.9 and -2.5 for $1.5\, r_{core} \leq r \leq 300\, r_{core}$. While there is a very slight flattening of the profile between the initial drop off from the core and the halo it is nowhere near as pronounced as the transition region between a black hole cusp and the surrounding halo (Figure 6). The late-time cluster profiles E and F in Figures 7 and 8 resemble a $W_o = 17$ King model. While a cluster undergoing core collapse departs from the King sequence at $W_o = 8.5$ (Cohn 1980), the "ignition" of a binary heat source evidently returns the cluster to a truncated isothermal configuration of higher central concentration.

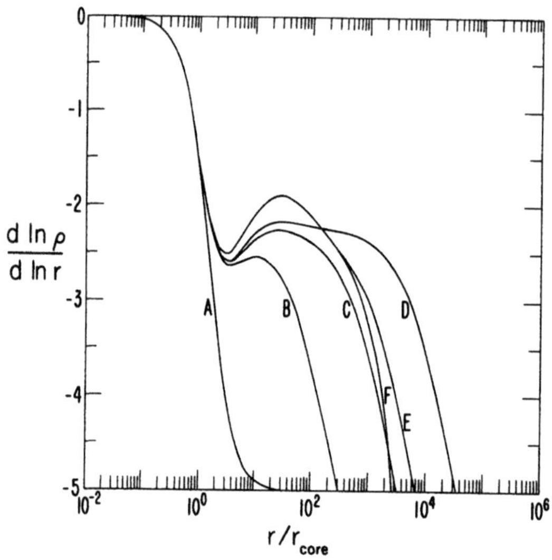

Figure 8. Evolution of density gradient profile with inclusion of three-body binary heating. The abscissa is radius scaled by instantaneous core radius. The times corresponding to each curve are the same as in Figure 7.

Figure 9 shows the time evolution of Lagrangian mass shells containing the indicated fractions of the initial cluster mass. While the core collapse process is nonhomologous, the post-collapse expansion becomes essentially homologous with the exception of the outer mass shells which are affected by the presence of an outer "absorbing" boundary. Within a few t_{rh} following core collapse, each radius expands as $\tau^{2/3}$ where τ is the time since core collapse. This scaling, which agrees nicely with predictions of Henon (1965), McMillan, Lightman, and Cohn (1981) and Goodman (1984), indicates that the time scale for energy input by binaries is equal to half mass relaxation time everywhere in the cluster. Thus, the late post-collapse phase is characterized by constant $\tau/t_{rh} = 6$ as well as constant $\tau/t_{rc} = 10^6$. During the late pre-collapse phase, only the latter ratio is constant.

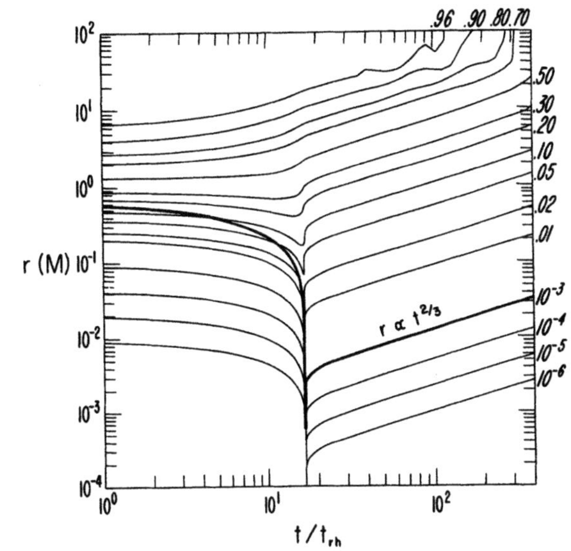

Figure 9. Evolution of mass distribution with inclusion of three-body binary heating. Light curves represent radii containing indicated mass fractions. Heavy curve represents core radius.

4. DISCUSSION

The results presented in Section 3 demonstrate the power of the direct Fokker-Planck method for investigating both pre- and post-collapse phases of globular cluster evolution. While significant progress has been made, much work remains toward the goal of constructing a evolutionary sequence of realistic cluster models. In this section I'll discuss some extensions of the work reviewed in Section 3 that are necessary for the achievement of this goal.

While two-mass direct Fokker-Planck calculations of the pre-collapse phase have been carried out (Section 3.2), no corresponding multi-mass post-collapse calculations exist. The latter calculations are crucial for predicting the surface brightness profile of a post-collapse cluster that contains nonluminous stellar remnants. A realistic mass spectrum should be adopted, with stellar mass loss that results from normal stellar evolution included (Applegate 1983). There do not appear to be any obstacles to the simultaneous treatment of these various refinements.

Most of the work reviewed in Section 3 is based on the one-dimensional (isotropized) Fokker-Planck equation. However, as described in Section 3.3, some degree of anisotropy does develop during core collapse. A two-dimensional, multi-mass calculation would provide an important test of the one-dimensional results. Such a calculation is probably best carried out using a supercomputer.

The theory of the evolution of a globular cluster with a massive central black hole now appears to be well understood. As discussed in

Section 3.4, the cluster structure that results in this case has several unique observational signatures. Hubble Space Telescope observations of globular cluster cores should improve the observational basis for confirming or denying the presence of massive black holes in globular clusters. Should there be any positive evidence, then detailed multi-mass calculations would be in order.

The work reported in Section 3.5 on post-collapse evolution with three-body binaries represents a first step towards a full understanding of binary-driven post-collapse evolution. As discussed by Ostriker and by Stodolkiewicz elsewhere in this volume, binaries formed by two-body tidal dissipative capture should also have an important influence on globular cluster evolution. Direct Fokker-Planck calculations of the effect of this process on cluster evolution are now being carried out.

The approach for incorporating three-body binaries using a heating term discussed in Section 3.5 represents only one of several alternative treatments. A second approach that Piet Hut and I plan to implement is to use a two-species (single stars plus binaries) Fokker-Planck equation; this allows the actual distribution of binaries to be tracked in time. The detailed single-binary interaction cross sections of Hut (1983) will be used together with heuristic binary-binary cross sections. A third approach for treating three-body binaries involves tracking a small number of individual binaries using a Monte Carlo method similar to that of Henon (1975), while using the direct Fokker-Planck method for single stars. By investigating several alternative approaches for treating binaries, it will be possible to assess the robustness of the results reported in Section 3.5.

The detailed evolving globular cluster models that will result from the research program just outlined should make direct contact with observations of cluster structure. In the next few years a wealth of new data is expected to result from both ground-based and Hubble Space Telescope studies. The direct Fokker-Planck method will be a powerful tool for providing the models with which to interpret this data.

REFERENCES

Ambartsumian, V. A. 1938, Ann. Leningrad State Univ., No. 22.
Applegate, J. 1983, B.A.A.S., **15**, 920.
Bahcall, J. N. and Wolf, R. A. 1976, Ap.J., **209**, 214.
__________. 1977, Ap.J., **216**, 883.
Chandrasekhar, S. 1942, Principles of Stellar Dynamics, (Chicago: University of Chicago Press).
Chang, J. S. and Cooper, G. 1970, J. Comp. Phys., **6**, 1.
Cohn, H. 1979a, Ph.D. Thesis, Princeton University.
__________. 1979b, Ap.J., **234**, 1036.
__________. 1980, Ap.J., **242**, 765.
__________. 1984, B.A.A.S., **16**, 495.
Cohn, H. and Hut, P. 1984, Ap.J. (Letters), **277**, L45.
Cohn, H. and Kulsrud, R. M. 1978, Ap.J., **226**, 1087.
Djorgovski, S. and King, I. R. 1984, Ap.J. (Letters), **277**, L49.
Goodman, J. 1983, Ph.D. Thesis, Princeton University.
Goodman, J. 1984, Ap.J., **280**, 298.

Gunn, J. E. and Griffin, R. F. 1979, A.J., **84**, 752.
Heggie, D. C. 1984, M.N.R.A.S., **206**, 179.
Henon, M. 1961, Ann. d'Ap., **24**, 369.
_________. 1965, Ann. d'Ap., **28**, 62.
_________. 1973, in Dynamical Structure and Evolution of Stellar Systems, ed. L. Martinet and M. Mayor (Geneva: Observatory).
Hut, P. 1983, Ap.J. (Letters), **272**, L29.
Inagaki, S. and Lynden-Bell, D. 1983, M.N.R.A.S., **204**, 913.
Inagaki, S. and Wiyanto, P. 1984, submitted to Pub. Astr. Soc. Japan.
Jeans, J. H. 1929, Astronomy and Cosmogony, (Cambridge: Cambridge Univ. Press).
Killeen, J., Mirin, A. A., and Rensink, M. E. 1976, in Methods in Computational Physics, Vol. 16, ed. B. Alder, S. Fernback, and M. Rotenberg (New York: Academic Press).
King, I. R. 1966, A.J., **71**, 64.
_________. 1980 in Globular Clusters, ed. D. Hanes and B. Madore, (Cambridge: Cambridge Univ. Press).
Larson, R. 1970, M.N.R.A.S., **147**, 323.
Marchant, A. B. and Shapiro, S. L. 1980, Ap.J., **239**, 685.
McMillan, S. L. W., Lightman, A. P., and Cohn, H. 1981, Ap.J., **251**, 436.
Merritt, D. 1981, Ph.D. Thesis, Princeton University.
Mitchie, R. W. 1963a, M.N.R.A.S., **125**, 127.
_________. 1963b, M.N.R.A.S., **126**, 499.
Rosenbluth, M. N., MacDonald, W. M., and Judd, D. L. 1957, Phys. Rev., **107**, 1.
Spitzer, L. 1940, M.N.R.A.S., **100**, 396.
_________. 1969, Ap.J.(Letters), **158**, L39.
_________. 1975, in IAU Symposium 69, Dynamics of Stellar Systems, ed. A. Hayli (Dordrecht: Reidel).
Spitzer, L. and Harm, R. 1958, Ap.J., **127**, 544.
Spitzer, L. and Hart, M. H. 1971, Ap.J., **164**, 399.
Spitzer, L. and Shull, J. M. 1975, Ap.J., **200**, 339.

DISCUSSION

SPITZER: Could you please summarize the differences in behavior (density profile, rate of collapse, etc.) between your isotropic one-component model and the corresponding anisotropic model?

COHN: Apart from the modest anisotropy that develops in the latter model, there are very few differences in the results. In particular, the $\rho \propto r^{-2.23}$ law holds for anisotropic core collapse and the collapse rate is the same as for the isotropic case.

INAGAKI: How is the distribution function in the post-collapse phase? Is the Maxwellian part expanding?

COHN: I haven't yet plotted the energy dependence of the distribution function for post-collapse clusters. My impression is that there is a depletion in the distribution function at energies corresponding to the center of the potential well, so that it may be non-Maxwellian at the most bound energies.

CLOSE ENCOUNTERS

Jeremy Goodman
130-33 Theoretical Astrophysics
California Institute of Technology
Pasadena, CA 91125
U.S.A.

ABSTRACT

The mathematical treatment of encounters is discussed briefly, and it is pointed out that in a spherical system with an isotropic distribution function and equal-mass stars, the relevant equations can be analytically orbit-averaged. Even in small systems, close encounters have little effect on core collapse. Near a black hole, however, close encounters may cause an accumulation of stars on large, very eccentric orbits.

1. INTRODUCTION

It is well known that two-body relaxation of large stellar systems is dominated by the effects of distant encounters (in three spatial dimensions--in two dimensions, close encounters dominate [Rybicki 1970]). The Fokker-Planck equation, which is based on the approximation that scatterings are weak and frequent, is particularly appropriate for inverse-square forces and has been very successfully applied to the dynamical evolution of star clusters. Why should one be interested in close gravitational encounters?

First of all, the relaxation effects of close encounters are smaller than those of weak ones by a factor which scales only logarithmically with N , the number of particles in the system. Therefore it is important to check that omission of close encounters from evolutionary calculations does not make for important quantitative errors. Secondly, we may sometimes be interested in processes that occur only by large energy transfers in a single encounter: for example, the ejection of stars with a finite velocity at infinity. Thirdly, the formalism that has been invented to treat close encounteres is of intrinsic interest. This formalism has a surprising simplicity, which must be traceable to the fact that the keplerian two-body problem can be solved exactly.

After briefly indicating the principal formulae, I will pass on to

J. Goodman and P. Hut (eds.), Dynamics of Star Clusters, 179–187.

some applications. This is strictly a review talk, everything here has appeared in the literature.

2. FORMALISM

The statistical treatment of close encounters starts from $R(\vec{v},\vec{v}')$, the probability per unit time that the test star suffers an encounter changing its velocity from $\vec{v}$ to $\vec{v}' = \vec{v} + \Delta\vec{v}$ per unit $d^3\Delta\vec{v}$. An explicit expression for R is

$$R(\vec{v},\vec{v}') = \frac{4G^2}{|\Delta v|^5} \int dm_F\, m_F^2 \int_{\vec{w}\cdot\Delta\vec{v}=0} d^2\vec{w}\, f_F\left(\vec{w} + \vec{v} + \Delta\vec{v}\,\frac{m_T+m_F}{2m_F},\, m_F\right), \tag{1}$$

where $f_F(\vec{v}_F,m_F)$ is the local distribution of field stars in velocity and mass, and is normalized to the total number density of field stars.

Equation (1) is exact in the sense that $\Delta\vec{v}$ is not assumed to be small compared to $\vec{v}$. Notice that the velocity integral is over a plane in velocity space perpendicular to $\Delta\vec{v}$ (but not, in general, passing through the origin). This is a consequence of energy and momentum conservation. It is, however, a remarkable peculiarity of the inverse-square law that the jacobian is independent of the integration variables.

Formulae equivalent to equation (1) have been given by Agekian (1959) and Hénon (1960a) for isotropic f_F and by Ipser and Semanzato (1983) and Goodman (1983) in the general case. Gryzinski (1959; 1965a,b) treated Coulomb scattering at great length, and his methods appear to be equivalent to equation (1).

The time evolution of a distribution $f_T(\vec{v},\vec{x},t;m_T)$ of test particles embedded in the field distribution f_F is governed by

$$\frac{Df_T(\vec{v},\vec{x},t;m_T)}{Dt} = \int d^3\vec{v}'\, f_T(\vec{v}',\vec{x},t;m_T)\, R(\vec{v}',\vec{v}) - f_T(\vec{v},\vec{x},t;m_T) \int d^3\vec{v}'\, R(\vec{v},\vec{v}'), \tag{2}$$

where

$$\frac{D}{Dt} \equiv \frac{\partial}{\partial t} + \vec{v}\cdot\frac{\partial}{\partial\vec{x}} - \frac{\partial\Phi}{\partial\vec{x}}\cdot\frac{\partial}{\partial\vec{v}} \tag{3}$$

is the convective derivative along the unperturbed test-star orbits in the smooth mean cluster potential Φ. I will refer to (2) as the "master equation", but it is really just a form of the Boltzmann equation. The Fokker-Planck equation (cf. Rosenbluth, MacDonald and Judd 1957) can be derived from the master equation by expanding R through second order in $\Delta\vec{v}$.

The master equation (2) is cumbersome for evolutionary calculations because the collisional term on the righthand side fluctuates along an eccentric orbit, while the average strength of the operator is so small that many orbits must be followed before f_T changes significantly. In the parlance of numerical analysis, the master equation is "stiff": it contains two very different timescales. The remedy is to average over the short timescale and obtain a reduced equation for the long-term evolution. The possibility of doing this is the most important numerical advantage that statistical methods, whether based on the Fokker-Planck equation or on the master equation, have over direct N-body schemes.

Hénon (1961) and more recently Cohn (1980) showed that the orbit average can be done analytically for the Fokker-Planck equation if f_F and f_T are assumed to be functions of time and binding energy $E = \phi - v^2/2$ only, where $\phi \equiv -\Phi$ is defined to be positive in the cluster. It has been shown (Goodman 1983) that the same can be done for the scattering kernel and for the master equation if all stars have the same mass, m. The orbit-averaged rate of scatterings from E to $E' = E + \Delta E$ per unit E'; denoted by $K(E,E')$, is

$$p(E)K(E,E') = \frac{8\pi^2 G^2 m^2}{|\Delta E|^3}\left\{[2q(E) - \Delta E p(E)]\int_{E_{min}}^{E'} dE_F\, f(E_F) + \int_{E'}^{\phi(0)+\Delta E} dE_F\, f(E_F)[2q(E_F - \Delta E) - \Delta E p(E_F - \Delta E)]\right\} \quad \text{if } E' < E,$$

$$= \frac{8\pi^2 G^2 m^2}{|\Delta E|^3}\left\{[2q(E') + \Delta E p(E')]\int_{E_{min}}^{E'} dE_F\, f(E_F) + \int_{E'}^{\phi(0)} dE_F\, f(E_F)[2q(E_F) + \Delta E p(E_F)]\right\} \quad \text{if } E' > E, \tag{4}$$

in terms of the functions $q(E)$ and $p(E)$, where $4\pi^2 q(E)$ is the volume of phase space at binding energies between E and $\phi(0)$, and $p(E) \equiv -dq(E)/dE$:

$$q(E) \equiv \frac{4}{3}\int_0^{\phi^{-1}(E)} dr\, r^2[2\phi(r) - 2E]^{3/2}, \tag{5}$$

and E_{min} is the minimum binding energy in the cluster (often 0). From now on it will be assumed that the test and field stars represent the same population, so the subscript "F" has been omitted from f. The orbit-averaged master equation is

$$\left[p(E)\frac{\partial}{\partial t} + \frac{\partial q}{\partial t}\frac{\partial}{\partial E}\right] f(E,t)$$
$$= \int dE'\ p(E')f(E')K(E,E') - p(E)f(E)\int dE'K(E,E') \ . \qquad (6)$$

The remaining convective term on the left represents the adiabatic adjustment of the distribution function to the changes in the mean cluster potential.

3. EVAPORATION FROM AN ISOLATED CLUSTER

Hénon (1969) computed the rate of escape of stars by close encounters from a Plummer model. For equal-mass stars, he found

$$N^{-1}\frac{dN}{dt} = -\frac{4.26 \times 10^{-3}}{N}\left[\frac{GM}{R_0^3}\right]^{1/2} ,$$

$$E^{-1}\frac{dE}{dt} = -\frac{1.34 \times 10^{-3}}{N}\left[\frac{GM}{R_0^3}\right]^{1/2} , \qquad (7)$$

where R_0 is the Plummer core radius, M the total mass, and E the total energy of the system. Most importantly, he found that the escape rate is considerably higher for unequal mass stars. For example, if the stars are divided into two equinumerous groups, the mass of every star in the first group being m_1, and in the second m_2, then the escape rate for $m_2/m_1 = 2$ is approximately 3 times greater than for $m_2/m_1 = 1$, and approximately 30 times greater for $m_2/m_1 = \infty$.

Hénon thought that close encounters were the only way that stars could escape from an isolated system, on the grounds that as stars diffuse to weaker and weaker binding energies, their orbit-averaged diffusion rates tend to zero (Hénon 1960b). Spitzer and Shapiro (1972) have shown, however, that escape does occur diffusively via an extended halo of radial orbits that continue to pass through the core. The diffusive mechanism is faster than ejection by close encounters because it enjoys the $\ln\Lambda$ factor.

4. EQUILIBRIUM DISTRIBUTION IN A SQUARE WELL

Petrovskaya (1970a, 1970b), Kaliberda (1971), and Retterer (1979) calculated "equilibrium" test-star distribution functions in a square-well potential with finite escape velocity. The field star distribution was taken to be either a maxwellian or a lowered maxwellian. Under these idealized conditions, the convective terms in the master equation vanish and the distribution function is independent of position, so that it is not necessary to perform the orbit-average.

I use the term "equilibrium" somewhat loosely. There can be no true equilibrium because the test stars gradually evaporate. These workers computed the most slowly-decaying eigenmode of the master equation: because the field-star distribution is fixed, the evolutionary equations are linear. The results for the test-star distribution agree well with Fokker-Planck calculations, although for $N \sim 10^2$-10^5 there is a slight ($\sim$10%) excess of stars in the tail relative to the Fokker-Planck solutions. Both the Fokker-Planck and the close-encounter solutions agree with a lowered maxwellian; the agreement is poorer for $m_T/m_F \gtrsim 0.5$ than for equal-mass stars.

5. CLOSE ENCOUNTERS IN CORE COLLAPSE

I have applied equation (6) to the evolution of a cluster of equal-mass stars (Goodman 1983, 1984), using a modified version of a computer program kindly lent to me by Haldan Cohn. In this code (Cohn 1980), the potential is recalculated at each time step to make it consistent with the distribution function.

Even with the simplifications made in deriving equation (6) from (2), evolutionary calculations with the master equation are expensive. I used a uniform grid of 100 cells in energy space, of which 10 cells corresponded to the core of this system. Since all energy changes computed were multiples of the cell width, the effective value of $p_{max}/p_{min} \equiv \Lambda$ for this system was 10 to 100; hence $\ln\Lambda = 2.3$-4.6 . This corresponds to a system with 25 particles in the core, and the relative importance of close encounters compared to distant ones is greatest for small N .

Nevertheless the results of this calculation agree very well with those of Cohn (1980), who used the orbit-averaged Fokker-Planck equation. Cohn found that the late stages of core collapse proceed homologously, as predicted by Lynden-Bell and Eggleton (1980). In this phase the radial density profile is a power law in the inner parts of the system:

$$\rho(r) \propto r^{-\beta} \quad \text{with} \quad \beta = 2.23 \ . \tag{8a}$$

He also found that the central three-dimensional velocity dispersion $v_m^2(0)$, central density $\rho(0)$, and central relaxation time $t_r(0)$ (as defined by Spitzer and Hart (1971)) obey

$$v_m^2(0) \propto \rho(0)^{\gamma} \quad \text{with} \quad \gamma = 0.10 \ , \tag{8b}$$

$$t_r(0) \frac{d \ln\rho(0)}{dt} = \xi \quad \text{with} \quad \xi = 3.6 \times 10^{-3} \ . \tag{8c}$$

My calculations confirm (8a) and (8b) to two figures. In (8c) I found $\xi = 2.5 \times 10^{-3}$ to 4.6×10^{-3} , depending on what I took for the effective value of $\ln\Lambda$, which of course enters the definition of $t_r(0)$.

These results constitute strong confirmation of the accuracy of the Fokker-Planck equation in describing strictly two-body relaxation in core collapse. Neither Cohn's calculation nor my own, however, takes any account of three-body effects, which must become very important once the number of particles in the core is less than 10^2.

6. CLOSE ENCOUNTERS NEAR A BLACK HOLE

Lin and Tremaine (1980) have considered the rate of ejection of stars from a stellar cusp surrounding a black hole.

The calculation is very easy if one makes the following assumptions: (i) The masses of the stars are equal. (ii) The smooth mean potential ϕ in the cusp is dominated by the hole, $\phi = GM_h/r$. (iii) The Bahcall and Wolf (1976) formula

$$f(E) \approx 2 \frac{n_0}{(2\pi\sigma_0^2)^{3/2}} \left[\frac{E}{\sigma_0^2}\right]^{1/4} \tag{9}$$

adequately describes the distribution function between $E = \sigma_0^2/2$ and a few times this value, where σ_0 is the one-dimensional velocity dispersion and n_0 the stellar number density in the surrounding cluster core.

Under these assumptions, the number of stars ejected from the cusp per unit time with positive energies at infinity per unit mass greater than E_0 is

$$\dot{N}(E_0) = \int_{E_0}^{\infty} dE \int_0^{\infty} dE' \; 4\pi^2 p(E')f(E')K(E',-E) \; . \tag{10}$$

In the keplerian potential of the black hole,

$$q(E') = \frac{\pi}{3\sqrt{2}} \frac{G^3M_h^3}{E'^{3/2}} \; . \tag{11}$$

Because p , q , and f are power laws, the integrals involved in computing K and S are elementary, and the result is (Lin and Tremaine 1980)

$$\dot{N}(E_0) = 110 \frac{G^5M_h^3m_*^2n_0^2}{\sigma_0^{11}} \left|\frac{E_0}{\sigma_0^2}\right|^{-1} . \tag{12}$$

It is convenient to re-express this result in dimensionless form in terms of the total number of stars in the cusp

$$N_c = 4\pi^2 \int_{\sigma_0^2/2}^{\infty} dE \; p(E)f(E) \tag{13}$$

and the reference relaxation time defined by Spitzer and Hart (1971) evaluated with the parameters appropriate to the core surrounding the cusp:

$$t_{rf} \frac{\dot{N}(E_0)}{N_c} = \frac{3.5}{\ln\Lambda} \left| \frac{E_0}{\sigma_0^2} \right|^{-1} . \tag{14}$$

For situations of interest, $\ln\Lambda$ is ≤ 18 ; thus at least a fifth of the stars in the cusp will be ejected with $|E_0| \geq \sigma_0^2$, every relaxation time. (Relaxation times in dense galactic nuclei may be as short as a few $\times 10^9$ years.) Furthermore, Lin and Tremaine show that if we put $E_0 = \sigma_0^2$, the ejection rate (12) is $\gtrsim 3$ times larger than the consumption rate of the hole for typical globular cluster parameters with $M = 10^3 M_\odot$. The same can be true of a galactic nucleus.

Why is the ejection rate so large? Principally because, under the $E^{1/4}$ law, at any given point in the cusp most of the stars are already close to the escape velocity: 84% have $v > 0.5 v_{esc}$ and 19% have $v > 0.9 v_{esc}$. From the details of the calculation it can be seen that the main contribution to $\dot{N}(E_0)$ comes from stars with binding energies $E' \approx |E_0|/12$. On the other hand, because the maximum energy that can be gained in a local close encounter is v_{esc}^2 , these stars suffer the encounters that eject them near pericenter where $v_{esc}^2/2 > E' + |E|$. Thus the potential well of the black hole serves as a catalyst for ejection by permitting the stars to gain large kinetic energies which are then occasionally exchanged in close encounters.

As stressed to me by Jerry Ostriker, the population of eccentric orbits by this mechanism tends to exaggerate the apparent mass of the black hole, because when one looks straight at the cusp, these stars are moving directly along the line of sight, and one sees their full space velocity. Indeed, Duncan and Wheeler (1980) and Binney and Mamon (1982) have argued that the cusp in M87 could be explained by velocity anisotropy alone without a black hole.

It should be pointed out, however, that for values of σ_0 that are typical of dense galactic nuclei, the 90° deflection distance is comparable to a stellar diameter; thus physical collisions will probably influence the ejection rate.

REFERENCES

Agekian, T.A. 1959, Sov. Astr.--A.J. 3, 46.
Bahcall, J.N. and Wolf, R.A. 1976, Ap. J. 209, 214.
Binney, J. and Mamon, G.A. 1982, M.N.R.A.S. 200, 361.
Cohn, H. 1980, Ap. J. 242, 765.
Duncan, M.J. and Wheeler, J.C. 1980, Ap. J. 237, L27.
Goodman, J. 1983, Ap. J. 270, 700; (errata) 1984, Ap. J. 278, 893.
Gryzinski, M. 1959, Phys. Rev. 115, 374.
Gryzinski, M. 1965a, Phys. Rev. 138, A305.
Gryzinski, M. 1965b, Phys. Rev. 138, A322.
Hénon, M. 1960a, Ann. d'Ap. 23, 467.

Hénon, M. 1960b, Ann. d'Ap. 23, 668.
Hénon, M. 1961, Ann. d'Ap. 24, 369.
Hénon, M. 1969, Astr. Ap. 2, 151.
Ipser, J. and Semanzato, R. 1983, Ap. J. 271, 294.
Kaliberda, V.S. 1971, Sov. Astr.--A.J. 14, 773.
Lin, D.N.C. and Tremaine, S. 1980, Ap. J. 242, 789.
Lynden-Bell, D. and Eggleton, P.P. 1980, M.N.R.A.S. 191, 483.
Petrovskaya, I.V. 1970a, Sov. Astr.--A.J. 13, 647.
Petrovskaya, I.V. 1970b, Sov. Astr.--A.J. 13, 957.
Retterer, J.M. 1979, A.J. 84, 370.
Rosenbluth, M.N., MacDonald, W.M. and Judd, D.L. 1957, Phys. Rev. 107,1.
Rybicki, G.B. 1970, in I.A.U. Coll. No. 10, Gravitational N-Body Problem, ed. M. Lecar, p. 22.
Spitzer, L. and Hart, M.H. 1971, Ap. J. 164, 399.
Spitzer, L. and Shapiro, S.L. 1972, Ap. J. 173, 529.

DISCUSSION

INAGAKI: You got a very good agreement with Haldan Cohn's results. Does the homologous evolution continue even though the number of particles in the core becomes less than 100?

GOODMAN: Yes, it does, in my numerical calculations. Many of the assumptions upon which the calculations are based break down, however, when $N_{core} < 100$. For example, the relaxation time becomes comparable to the dynamical time, so that the validity of orbit-averaged equations is questionable in this regime. More importantly, three-body interactions become important for $N_{core} < 100$, whereas both Cohn's calculations and my own neglect these effects. My calculation shows merely that the increasing importance of close two-body encounters as N_{core} decreases does not disturb the homological evolution.

SEVERNE: In comparing the master equation for close encounters and the Fokker-Planck equation, it would seem to me that the former has the sounder theoretical foundation and thus justifies the latter, rather than inversely. Indeed, both equations contain the basic assumption that velocity transfer is local and instantaneous. This assumption is however well satisfied only in the case of close encounters. Could you comment on this?

GOODMAN: The approximation of local and instantaneous encounters is very reasonable, even for the Fokker-Planck equation. In a self-gravitating system of N particles and half-mass radius R, the close-encounter distance is of order R/N, and the mean interparticle distance is of order $(R/N)^{1/3}$. Since the Chandrasekhar formula weights all logarithmic intervals in impact parameter equally, two thirds of the effects of encounters can therefore be attributed to impact parameters smaller than the mean interparticle distance; these encounters are certainly local. In fact, the inhomogeneity of the system is probably important only for impact parameters of order R/e or larger, and if these are incorrectly represented by the local impulse approximation, the relative error in the calculated relaxation rate is of order $(\ln N)^{-1}$.

DYNAMICAL EVOLUTION OF MULTI-COMPONENT CLUSTERS

Shogo Inagaki
Department of Astronomy
University of Kyoto
Kyoto 606
Japan

ABSTRACT. Presence of stars with disparate masses causes great differences in the dynamical evolution of star clusters from the evolution of single component clusters. One remarkable effect is acceleration of the evolution. Another effect is destabilization or stabilization. In two-component clusters equipartition at the cluster centre is nearly achieved if Spitzer's (1969) condition is satisfied. In multi-component clusters equipartition at the cluster centre is achieved if either the range of stellar mass is very narrow or the mass spectrum is very steep. Global equipartition is never achieved. Post-collapse evolution of multi-component clusters is discussed briefly and some remained problems are presented.

1. INTRODUCTION

Several pioneering works on multi-component clusters have been done. Michie (1963) calculated the escape rate from multi-component clusters. Spitzer (1969) obtained the condition for equipartition in two-component clusters. Hénon (1969) estimated the escape rate due to close encounters. Spitzer and Hart (1971) and Hénon (1971) made Monte Carlo simulations of multi-component clusters and revealed some basic properties of the dynamical evolution of multi-component clusters. Saslaw and De Young (1971) considered the condition for equipartition in clusters with continuous mass spectra. Aarseth (1974) made N-body simulations of isolated clusters and emphasized the importance of binaries. Angeletti and Giannone (1977b) found the importance of mass loss from stars. Recently Larson (1984) constructed a model of two-component clusters in post-collapse stage.

In this paper I do not mention every aspect of multi-component clusters discussed by the above authors but I would like to clarify the physical processes of gravitational encounters by focusing attention to the late evolution of cores of globular clusters. Therefore I do not mention on the effects of stellar mass loss or escape of stars from clusters, which are important in the early stages of dynamical evolution.

This paper comprises following subjects. Section 2 discusses

J. Goodman and P. Hut (eds.), Dynamics of Star Clusters, 189–205.

dynamical evolution of two-component clusters, which are the simplest multi-component clusters. Section 3 discusses the evolution of clusters with components more than two. Section 4 discusses briefly the post-collapse evolution of multi-component clusters. Finally several remained problems are presented in section 5.

2. TWO-COMPONENT CLUSTERS

2.1. Basic Properties of Two-Component Clusters

In single component clusters there is only one motive force of evolution, that is the gravothermal instability (see e.g. Spitzer 1985). In the gravothermal instability of single component clusters, the energy is transported from the cluster core to the halo. On the other hand, in multi-component clusters, in addition to the gravothermal instability, energy exchange between components plays an important role.

One remarkable effect of presence of disparate masses is the acceleration of evolution. It takes about fifteen half mass relaxation times to reach the infinite central density in an isolated single component cluster (Cohn 1980, Marchant and Shapiro 1980). Here the half-mass relaxation time, t_{rh}, is defined by

$$t_{rh} = \frac{0.06M^{1/2}R_h^{3/2}}{\langle m\rangle\, G^{1/2}\log\Lambda} \tag{1}$$

(Spitzer and Hart 1971), where R_h is the half mass radius, G is the gravitational constant and $\log\Lambda$ is the Coulomb logarithm.

The time needed for complete collapse becomes significantly shorter in multi-component clusters (e.g. Spitzer and Hart 1971). In a two-component cluster with $m_2/m_1 = 2$ and $M_2/M_1 = 0.11$, the collapse time is about one half of the single component cluster (see table 1a), where m_i is the mass of a star in the i-th component and M_i is the total mass of the i-th component. In two-component clusters with $m_2/m_1 = 5$, the collapse time is much shorter: In a cluster with $m_2/m_1 = 5$ and $M_2/M_1 = 0.072$, the collapse time is about one tenth of the single component cluster. For detail see tables Ia and Ib.

TABLE Ia. The time (in the unit of t_{rh}) required for the complete collapse in two-component clusters with $m_2/m_1 = 2$.

M_2/M_1...	0.0	0.001	0.01	0.05	0.11	1.0	9.0
t_{cc}...	15.4	15.3	13.6	10.2	8.5	9.6	13.6

TABLE Ib. The same as table 1a but for $m_2/m_1 = 5$.

M_2/M_1...	0.001	0.005	0.014	0.072	0.30
t_{cc}...	14.8	10.8	4.2	1.7	1.9

The acceleration of evolution in two-component clusters can be understood as follows. In order that a single component cluster evolves, the energy must be transferred from its core to its halo. Since the relaxation time at the halo is very long, it takes considerable time for the single-component cluster to evolve. On the other hand, a two-component cluster can evolve if the energy is transferred from the massive component to the less massive component. The relaxation time at the core is relatively short so that the two-component cluster can evolve much faster than the single-component cluster.

Another remarkable effect of presence of disparate masses is destabilization. An isothermal single component cluster is gravothermally unstable if the density contrast, ρ_c/ρ_b, is larger than 709 (Antonov 1962, Lynden-Bell and Wood 1968), where ρ_c is the central density and ρ_b is the density at the boundary. On the other hand, an isothermal two-component cluster with $m_2/m_1 = 10$ and $M_2/M = 0.3$ is gravothermally unstable with very small density contrast, i.e., $\rho_c/\rho_b > 19$ (Yoshizawa et at. 1978), where M is the total mass of the cluster.

The destabilization can be understood as follows. In order that a single component cluster causes the gravothermal instability, the energy must be transported from the core to the halo (Lynden-Bell and Wood 1968, Hachisu and Sugimoto 1978). In other words, the halo works as a heat reservoir. Therefore an extended halo is necessary for the single component cluster to cause the gravothermal instability. In a two-component cluster, the energy can be deposited in the less massive component so that no halo is necessary to cause the gravothermal instability.

The presence of disparate masses also causes stabilization (Katz and Taff 1983). The isothermal cluster with $m_2/m_1 = 10$ and $M_2/M = 0.003$ is stable if $\rho_c/\rho_b < 5012$ (Yoshizawa et al. 1978). This stabilization can be understood as follows. Figure A1 of Yoshizawa et al. shows that $\rho_2(0)/\rho_1(0)$ is constant (= about eight) along the marginally stable states near this model. This means that the development of the halo does not affect the stability or that the stability is determined by the state of the core. In other words, the instability is caused by the exchange of energy between components in the core. However, if the halo becomes too extended (for example, density contrast of the less massive component exceeds 709), the less massive component becomes unstable as a single component cluster. The density contrast at this stage is about $\rho_2(0)/\rho_1(0)$ times 709 so that it is quite larger than 709. In this sense, the stabilization in two-component clusters is deceptive.

2.2. Dynamical Evolution of Two-Component Clusters

It has been well known that mass stratification occurs in two-component star clusters (e.g. Spitzer and Hart 1971, Saito and Yoshizawa 1976, Angeletti and Giannone 1977a). These authors employed either a Monte Carlo technique or a fluid dynamical approach. Recently Cohn (1979, 1980, 1985) devised a new technique to integrate the orbit-averaged Fokker-Planck equation (e.g. Lightman and Shapiro 1978) numerically. Cohn's method solves the orbit-averaged Fokker-Planck equation very accurately and numerical noises produced are very small. Therefore Cohn's method is the most suitable for the detailed study of slow evolution of large-N (say 10^6 stars) clusters due to two-body encounters. Inagaki and Wiyanto (1984, hereafter referred to as IW) made several simulations, using Cohn's method with the assumption of isotropy in velocity space. Assumption of isotropy will be enough for the study of evolution of the core because anisotropy will be small there. Moreover Cohn (1985) confirmed that in a single component cluster there is no differences in the density profile and in the collapse rate even if anisotropy of the velocity distribution is taken into account.

IW adopted Plummer's model as initial models. The density distribution of Plummer's model is given by

$$\rho_i(r) = \frac{3M_i}{4\pi r_0^3} \frac{1}{[1 + (r/r_0)^2]^{5/2}} \tag{2}$$

and its velocity dispersion is given by

$$\langle v_i^2 \rangle = \frac{GM}{2r_0} \frac{1}{[1 + (r/r_0)^2]^{1/2}} , \tag{3}$$

where r_0 is a scale length and G is the gravitational constant. Equations (2) and (3) show that the density profiles of both components are similar and the velocity dispersions are the same. In other words, the kinetic temperature of a component is proportional to the mass of a constituent star of the component. Figure 1 shows the evolution of the central density and temperature for the model with $m_2/m_1 = 2$ and $M_2/M_1 = 0.01$. The central potential is adopted as the time axis. In this model $\rho_2(0) = 0.01\rho_1(0)$ initially so that the initial overall evolution is governed by the less massive component. In other words, the less massive component collapses like a single component system. The massive component is considered to be floating in the potential well made by the less massive component. In such a case temperature of the massive component drops swiftly and the temperatures of the both components become nearly equal at about 8 t_{rh}. It should be noted that the temperature difference remains large at the halo (see figure 3 of IW) since the relaxation time is large there. After this epoch until the density of the massive component dominates the density of the less massive component, the difference in central temperatures between components remains negligible while the less massive component is collapsing. During this stage the density of the

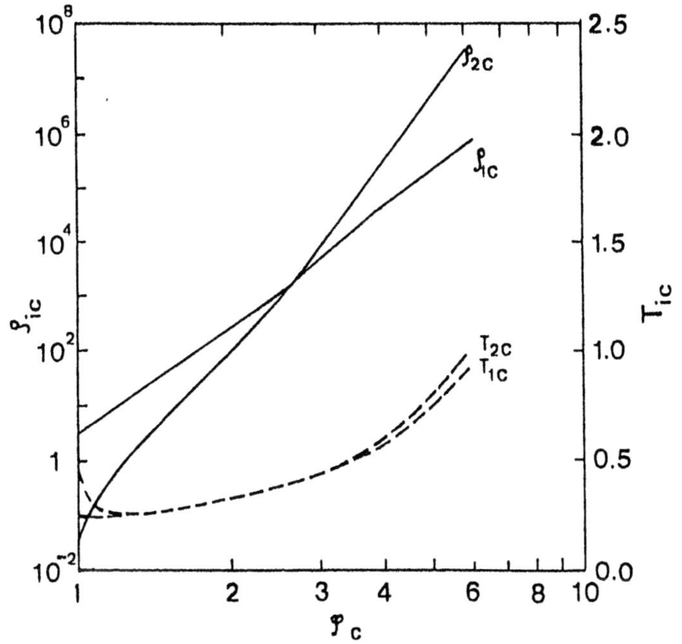

Figure 1. Time evolution of the central density and temperature for each component as a function of the central potential for the model with $m_2/m_1 = 2$ and $M_2/M_1 = 0.01$.

massive component increases more rapidly than that of the less massive component. The evolution at this stage is explained well by Lightman and Fall's (1978) model (see IW).

After the density of the massive component dominates that of the less massive component, the massive component causes the gravothermal instability and collapses independently from the less massive component. At this stage the temperature difference between components becomes large again. It should be noticed that even though the fraction of the massive stars is very small, mass stratification occurs finally and the temperature difference becomes large. In this sense equipartition holds only temporally.

If the fraction of the massive stars is large, for example, $M_2/M_1 = 1.0$, there is no initial stage such that the density of the less massive component dominates that of the massive component. Figure 2 shows the evolution of the central density and temperature of the model with $m_2/m_1 = 2$ and $M_2/M_1 = 1.0$. The temperature difference decreases initially and reaches the minimum value at about $t = 5.6t_{rh}$. At this epoch $(T_{2c} - T_{1c})/T_{1c} = 0.17$, where T_{ic} is the central temperature of the i-th component. The temperature difference grows monotonically after this epoch.

Figure 3a shows how the minimum of the relative temperature difference along the evolutional sequence, $[(T_{2c} - T_{1c})/T_{1c}]_{min}$, depends on M_2/M_1 for the models of $m_2/m_1 = 2$. It is seen that the temperature difference is larger as M_2/M_1 is larger. This value is larger for the

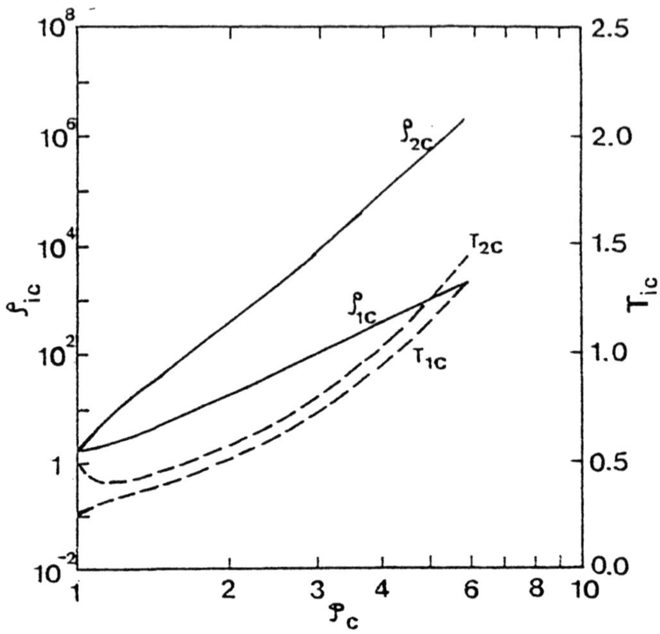

Figure 2. The same as figure 1 but for $M_2/M_1 = 1.0$.

models of $m_2/m_1 = 5$: $[(T_{2c} - T_{1c})/T_{1c}]_{min}$ is about 50% for the model of $M_2/M_1 = 0.1$ (figure 3b). Figures 3a and 3b show that the relative temperature difference is constant for both large M_2/M_1 and small M_2/M_1. The dependence of the relative temperature difference on M_2/M_1 is the

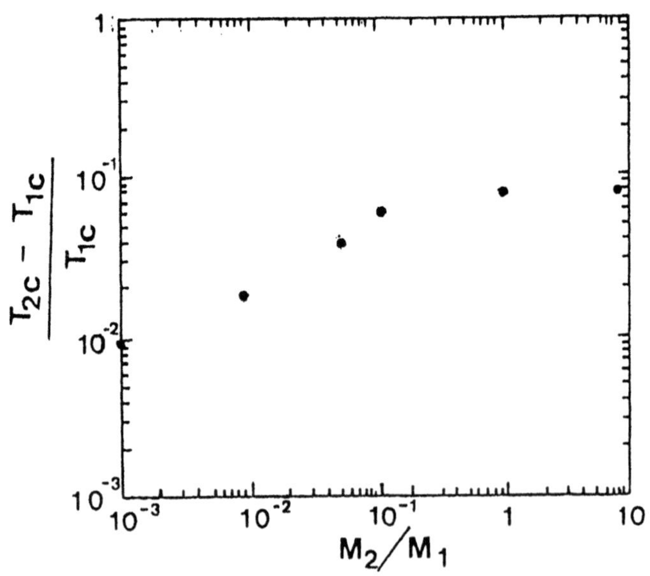

Figure 3a. The dependence of the minimum of the temperature difference on the fraction of the total mass of the massive component for the models of $m_2/m_1 = 2$.

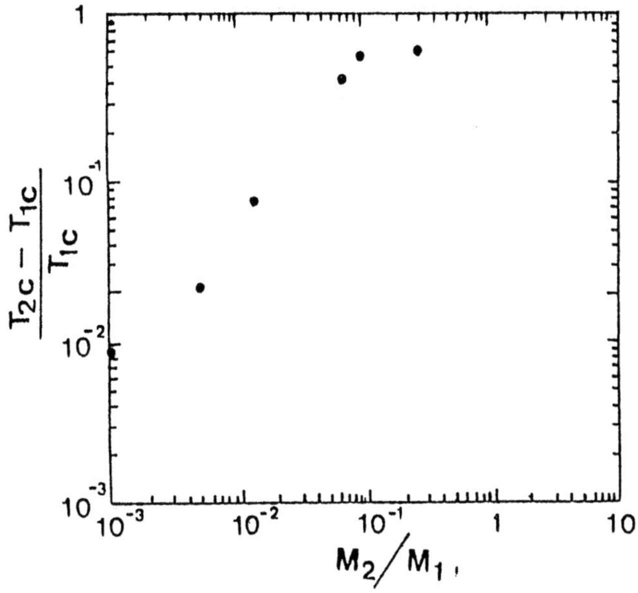

Figure 3b. The same as figure 3a but for $m_2/m_1 = 5$.

largest near the model with $M_2/M_1 = 0.16(m_2/m_1)^{-3/2}$, the value obtained by Spitzer (1969) as the critical value for existence of equipartition.

The evolution of the massive component after the central density of the massive component dominates the central density of the less massive component is nearly the same as that of the single component clusters. The collapse rate defined by $\xi_2 = t_{r2,c}\, d\ln \rho_{2c}/dt$ is the same as that of the single component clusters and $\langle v_{2c}^2\rangle \propto \rho_{2c}^{0.1}$, which is also the same relation as the single component clusters, where $t_{r2,c}$, ρ_{2c} and $\langle v_{2c}^2\rangle$ are the central relaxation time, density, and velocity dispersion, respectively, of the massive component. These relations holds even in the models of $m_2/m_1 = 5$ where the temperature difference is quite large.

IW started their simulations from Plummer's model, where equipartition does not hold initially. The evolution from models where the equipartition holds initially can be conjectured. Let us consider how clusters evolve if we start simulations from two-component Wilson's model, for example. The isotropized-orbit-averaged Fokker-Planck equation can be written in the form

$$\frac{\partial q}{\partial E}\frac{\partial f_i}{\partial t} - \frac{\partial q}{\partial t}\frac{\partial f_i}{\partial E} = 16\pi^2 G^2 \ln\Lambda \sum_j m_j^2 \frac{\partial}{\partial E}\{f_i [\int_{\varphi(0,t)}^{E} f_i f_j' q' (\frac{\partial \ln f_i}{\partial E} - \frac{m_i}{m_j}\frac{\partial \ln f_j'}{\partial E'}) dE' + q\int_E^0 f_i f_j' (\frac{\partial \ln f_i}{\partial E} - \frac{m_i}{m_j}\frac{\partial \ln f_j'}{\partial E'}) dE']\} \qquad (4)$$

(Inagaki 1980), where

$$q = \frac{1}{3} \int_0^{\varphi^{-1}(E)} (2E - 2\varphi)^{3/2} r^2 dr. \tag{5}$$

In Wilson's model ln $f_i(E)$ is convex upwards because of the cut-off of the distribution function. In this case, equation (4) predicts that the energy flows outwards in the core region. This means that the core contracts (Inagaki 1980). As the core contracts, keeping equipartition in the core, the density of the massive component grows more rapidly than does the density of the less massive component (see IW). It should be reminded that the distribution function keeps convex upwards because the evolution of the distribution function is characterized by the extention of isothermal part into low energy region (see figure 1 of Cohn 1980). If the density of the massive component becomes high enough to cause the gravothermal instability, the gravothermal instability of the massive component takes place and the temperature difference between components becomes larger. Thus even if we start from models with equipartition initially, the final state is the gravothermal instability of the massive component and the temperature difference emerges.

Global equipartition in the sense that the average temperatures throughout a cluster do not depend on components is not achieved in any case. If the initial state is far from equipartition, the temperature at the halo does not change so much because the relaxation time is very long there. Therefore equipartition is not achieved in this case (see figure 8 of IW). Even if global equipartition holds initially, the temperature difference becomes large finally because of the final collapse of the massive component.

3. MULTI-COMPONENT CLUSTERS

Spitzer and Hart (1971) simulated the evolution of three-component clusters and found that the most massive stars condense strongly in the central region and the least massive stars concentrate in the halo. Spitzer and Shull (1975) made more detailed study of the evolution of three-component clusters and found that the final collapse of the most massive component is in the same way as do the cores of single component clusters. Hénon (1971) simulated evolution of a five component cluster with a flat mass spectrum and showed that evolution proceeds in the direction away from equipartition. Recently Stodółkiewicz (1985) made Monte Carlo simulations of globular clusters, using a realistic mass function.

In spite of the effort of these authors, some of the basic properties of the evolution of multi-component clusters had been unclear until Inagaki and Saslaw (1985, hereafter referred to as IS) made systematic simulations of multi-component clusters. IS integrated the isotropized-orbit-averaged Fokker-Planck equation, using Cohn's (1980) scheme. They assumed simple power law mass spectra of the form,

$$dM \propto m^{-\alpha} dm \qquad \text{for} \quad m_{min} \leq m \leq m_{max}. \tag{6}$$

They made several simulations with different m_{max}/m_{min} and α (Table II).

TABLE II. The time (in the unit of t_{rh}) required for complete core collapse in multi-component clusters. In the table n is the number of components.

	$\frac{m_{max}}{m_{min}} = 10$ (n = 15)	$\frac{m_{max}}{m_{min}} = 10$ (n = 5)	$\frac{m_{max}}{m_{min}} = 4$ (n = 5)	$\frac{m_{max}}{m_{min}} = 2.8$ (n = 5)	$\frac{m_{max}}{m_{min}} = 2$ (n = 5)
$\alpha = 1.0$	2.6	4.0	---	7.5	10.4
$\alpha = 2.5$	1.9	2.0	4.3	6.6	10.0
$\alpha = 4.0$	4.3	4.9	5.7	7.3	----
$\alpha = 6.0$	9.0	10.4	9.6	9.7	11.1

The time for the complete core collapse in multi-component clusters is shown in Table II. One important conclusion drawn from tables I and II is that the collapse time is mainly determined by m_{max}/m_{min} and α not by the number of components. In multi-component clusters, the fastest collapse time in the unit of the half mass relaxation time is attained when α is about 2.5. The shortest collapse time is about 10 t_{rh} for five-component clusters with $m_{max}/m_{min} = 2$. This collapse time is comparable with that of the two-component cluster with $m_2/m_1 = 2$ and $M_2/M_1 = 0.05$. The collapse time of the fifteen-component cluster with $m_{max}/m_{min} = 10$ and $\alpha = 2.5$ is 1.9 t_{rh}, which is nearly the same as the collapse time of the two-component cluster with $m_2/m_1 = 5$ and $M_2/M_1 = 0.072$. Thus the shortest collapse time for given m_{max}/m_{min} can be obtained even with a two-component cluster with a certain value of M_2/M_1. The collapse time does not become shorter indefinitely even if very large m_{max}/m_{min} is adopted. The shortest collapse time is about the same for $m_{max}/m_{min} \gtrsim 5$.

Another important question is what is the condition for equipartition in multi-component clusters. Saslaw and De Young (1971) claimed that 'stable equipartition' could not be achieved for any reasonable mass spectrum. 'Stable equipartition' is the state such that equipartition is unaffected by small changes in the mass spectrum. Vishniac (1978), however, suggested that equipartition is possible if the mass spectrum of a cluster is sufficiently steep. Unfortunately, his analysis was based on the assumption that the density profiles of different components are homologous. This assumption is generally not satisfied because the density profiles of non-dominant components are affected by the gravitational attraction of dominant components.

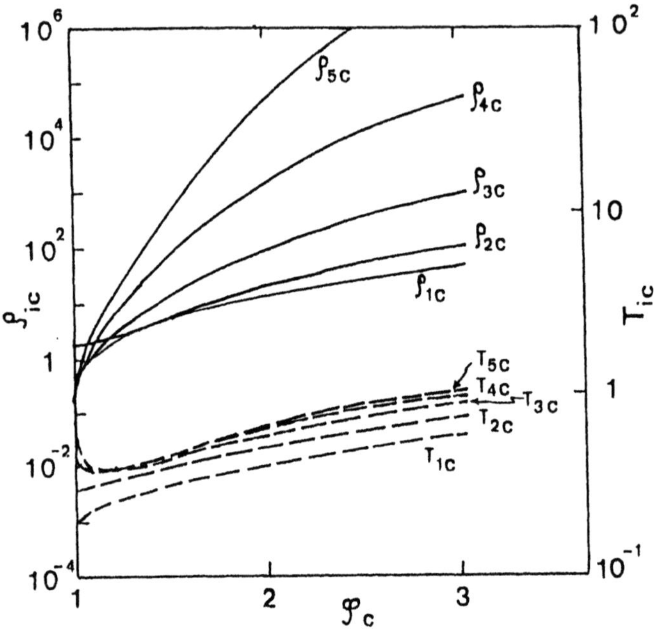

Figure 4. Time evolution of the central density and temperature for each component as a function of the central potential for the model with $m_{max}/m_{min} = 10$ and $\alpha = 2.5$.

Thus it was necessary to examine the condition for equipartition in evolving models of globular clusters. IS adopted Plummer's models characterized by equations (2) and (3) as initial conditions. Figure 4 shows time evolution of the central density and temperature of a five component cluster with $m_{max}/m_{min} = 10$ and $\alpha = 2.5$. It is seen that initially the temperatures of the massive components decrease rapidly and the forth and the fifth components become in equipartition. This equipartition, however, does not last long. When the density of the most massive component dominates significantly at later stages, the central temperature of the most massive component becomes higher than the central temperatures of the other components. Vishniac (1978) found that $\alpha \gtrsim 2.5$ is the necessary condition for equipartition. Figure 4, however, shows that $[(T_{5c} - T_{1c})/T_{1c}]_{min}$ is as large as 47%.

The steeper is the mass spectrum, the smaller is the temperature difference. For example, $[(T_{5c} - T_{1c})/T_{1c}]_{min}$ is about 9% for the model of $m_{max}/m_{min} = 10$ and $\alpha = 4$. In this model the central temperatures of the second to the fifth components become nearly equal at intermediate stage of the evolution. Figure 5 shows time evolution of the central density and temperature of the model with $m_{max}/m_{min} = 10$ and $\alpha = 6.0$. It is noticed that at this very steep mass spectrum equipartition at the cluster centre is nearly achieved. Thus it is seen that the mass spectrum

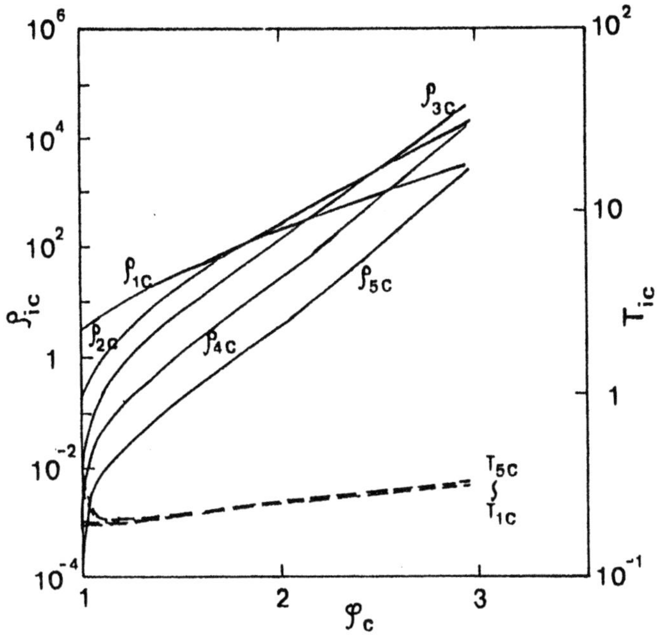

Figure 5. The same as figure 4 but for α = 6.0.

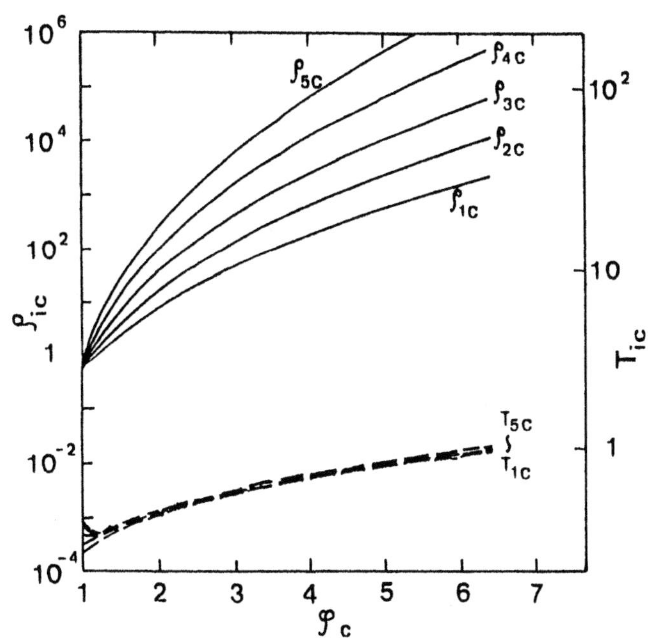

Figure 6. The same as figure 4 but for m_{max}/m_{min} = 2 and α = 1.0.

must be as steep as $\alpha \gtrsim 6$ in order that a cluster with $m_{max}/m_{min} = 10$ achieves equipartition at the centre. These tendencies are unchanged even though the number of components are as large as fifteen (see table 3 and IS for detail).

When the ratio of the maximum mass to the minimum mass of a star is 2, the results are quite different. Even for a very flat spectrum with $\alpha = 1$, $[(T_{5c} - T_{1c})/T_{1c}]_{min}$ is only 7% (figure 6). Table III shows the minimum of the relative central temperature difference between the most massive component and the least massive component, $[(T_{nc} - T_{1c})/T_{1c}]_{min}$, where n is the number of components. From table III we see that either $\alpha \gtrsim 6$ or $m_{max}/m_{min} \lesssim 2.8$ is necessary for the achievement of equipartition at the cluster centre.

TABLE III. The minimum of the relative central temperature difference, $[(T_{nc} - T_{1c})/T_{1c}]_{min}$, along the evolutional sequence.

	$\frac{m_{max}}{m_{min}} = 10$ (n = 15)	$\frac{m_{max}}{m_{min}} = 10$ (n = 5)	$\frac{m_{max}}{m_{min}} = 4$ (n = 5)	$\frac{m_{max}}{m_{min}} = 2.8$ (n = 5)	$\frac{m_{max}}{m_{min}} = 2$ (n = 5)
$\alpha = 1.0$	1.20	1.40	----	0.14	0.07
$\alpha = 2.5$	0.68	0.47	0.21	0.12	0.06
$\alpha = 4.0$	0.20	0.09	0.10	0.07	----
$\alpha = 6.0$	0.06	0.03	0.04	0.03	0.03

As the number of components increases, the fraction of the mass contained in a component decreases. Therefore it is rather difficult to see whether the final collapse of the core occurs in the same way as do the cores of single component clusters. However, if we define the rate of core collapse by $\xi \equiv t_{rc} d\ln\rho_c/dt$, we see that this quantity has the same value as the single component clusters. Here the central relaxation time is defined by

$$t_{rc} = \frac{(2\langle\bar{v}_c^2\rangle/3)^{3/2}}{2\pi G^2 \bar{m}\rho_c \ln \Lambda}, \qquad (7)$$

where $\langle\bar{v}_c^2\rangle$ and $\bar{m}$ are the average values of the central velocity dispersion and the mass, respectively. The average values are taken with the weights of the central number densities, i.e., e.g.,

$$\bar{m} = \frac{\sum_i m_i n_{ic}}{\sum_i n_{ic}} \qquad (8)$$

where n_{ic} is the central number density of the i-th component. The relation between the central velocity dispersion of the most massive component and its central density is also the same as in single-component clusters, i.e. $\langle v_{nc}^2 \rangle \propto \rho_{nc}^{0.1}$.

4. POST-COLLAPSE EVOLUTION OF MULTI-COMPONENT CLUSTERS

Recently several basic properties of post-collapse evolution have been revealed (see Heggie 1985). However, most studies were made under the assumption that the cluster consists of stars with the same mass. The only exception is the work by Stodółkiewicz (1982, 1985) who allows disparate masses.

As is mentioned in sections 2 and 3, the time required for the complete core collapse is shorter in multi-component clusters than in single component clusters. Therefore multi-component clusters set in the post-collapse phase earlier than single component clusters. Moreover, if we take into account that post-collapse evolution is powered by hard binaries, the post-collapse phase starts much earlier in multi-component clusters than in single component clusters. According to Heggie (1975), the formation rate of hard binaries is proportional to

$$Q(x) = \frac{14\sqrt{2}\pi^2 G^2}{3} \frac{(m_1 m_2)^4 m_3^{5/2}}{(m_1+m_2)^{1/2}(m_1+m_2+m_3)^{1/2}} \beta x^{-9/2} , \qquad (9)$$

where stars with the masses m_1 and m_2 form a binary and the star with the mass m_3 is recoiled from the binary. For the meaning of other symbols, see Heggie (1975). Equation (9) shows that if m_1 and m_2 are larger than m_3, the formation rate of hard binaries is larger than in the same mass case. Therefore in multi-component clusters massive stars form binaries more efficiently than in single component clusters.

TABLE IV. The time (in the unit of t_{rh}) when persistent hard binaries are formed in N-body simulations.

	N = 250	N = 1000
single component cluster	14.1	21.0
α = 2.5	3.7	3.1
α = 4.0	8.9	3.3
α = 6.0	3.8	5.9

Table IV shows the time of the onset of the post-collapse stage when persistent hard binaries are formed in N-body simulations. These simulations were performed with NBODY5 code developed by Aarseth (1984, 1985). The code can regularize not only strong two-body encounters but also violent three-body encounters. Therefore the code is suitable for the study of post-collapse evolution of small-N systems. From table IV we see that the onset of the post-collapse phase is much earlier in multi-component clusters than in single component clusters. Dependence on mass spectra is not clear from table IV. More detailed studies will be necessary.

Larson (1984) constructed a model of two-component clusters in post-collapse phase. He assumed that $\rho_2 \propto r^{-2.5}$, $m_2\langle v_2{}^2\rangle/m_1\langle v_1{}^2\rangle = 1.16$, and M_2/M is a few percent. From these assumptions he derived that $m_1 = 0.8$ M⊙ and $m_2 = 2.8$ M⊙. He thus considered that the massive stars are unseen black holes and the less massive stars are visible stars. He found that the calculated profiles of the light and velocity dispersion of visible stars are in agreement with observed profiles of the light and velocity dispersion, respectively, of globular clusters. It is interesting to examine whether his model is reproduced in an evolutional model of a two-component cluster.

We can make some predictions on the post-collapse evolution of two-component clusters. From the study of pre-collapse evolution by IW, we know that the density profile of the massive component is proportional to $r^{-2.2}$ at the time of complete core collapse. The density profile of the less massive component is much flatter (see figure 4 of IW). If we take account of the binary formation, hard binaries of massive stars will form when the central density of the massive stars becomes high enough. Thereafter expansion of the massive component begins. As the density of the massive component becomes smaller, the density of the less massive component also becomes smaller. In single component clusters, the post-collapse evolution is characterized by the expansion of isothermal region (Inagaki and Lynden-Bell 1983). From this analogy we expect that the region where equipartition is achieved expands. The radius of this region, r_*, is characterized by the equation, $t_{eq}(r_*) \simeq t - t_{cc}$, where t_{eq} is the time required for equipartition and t_{cc} is the time of the complete core collapse.

5. FUTURE PROBLEMS

(1) Construction of models of real globular clusters: In section 3 we see that $m_{max}/m_{min} \lesssim 2.8$ or $\alpha \gtrsim 6$ is necessary for the achievement of equipartition at the cluster centre. In real globular clusters the mass of the most massive stars will be larger than 1 M⊙ though they are unluminous. On the other hand, the mass of the least massive stars will be about 0.1 M⊙. The exponent in the mass spectrum is $1.3 < \alpha < 3$ for M3 (Gunn 1980). Therefore deviation from equipartition is expected in real globular clusters. However, most models of globular clusters (e.g. Da Costa and Freeman 1976, Illingworth and King 1977) were constructed under the assumption of equipartition at the cluster centre. Therefore

it is necessary to construct models of real globular clusters, taking account of deviation from equipartition.
(2) Post-collapse evolution of multi-component clusters: As is seen in section 4 we know only some qualitative features of post collapse evolution of multi-component clusters. It is, therefore, necessary to study post-collapse evolution of multi-component clusters quantitatively, using detailed evolutional models.
(3) Effects of tidal dissipational encounters: In single component clusters, two-body binaries formed by tidal dissipational encounters govern the evolution in late stages (Inagaki 1984, Ostriker 1985, Hut and Inagaki 1985). It is necessary to consider the role of two-body binaries in a cluster with realistic mass spectra.
(4) Construction of realistic evolutional models of globular clusters: Some basic properties of gravitational encounters in multi-component clusters are discussed in section 3. To understand the evolution of real globular clusters, it is necessary to construct more realistic evolutional models, taking account of stellar evolution and of tidal effects due to the Galaxy, etc.

I thank Professor W.C. Saslaw for stimulating discussions and Dr. S.J. Aarseth for providing his N-body program and for helpful advice.

REFERENCES

Aarseth, S.J.: 1974, Astron. Astrophys. **35**, 237.
Aarseth, S.J.: 1984, in Multiple Time Scales, eds. J.U. Brackbill and B.I. Cohen, Academic Press.
Aarseth, S.J.: 1985, in IAU Symposium 113 (this issue).
Angeletti, L. and Giannone, P.: 1977a, Astrophys. Space Sci. **50**, 311.
Angeletti, L. and Giannone, P.: 1977b, Astrophys. Space Sci. **58**, 363.
Antonov, V.A.: 1962, Vestnik Leningrad Univ. **7**, 135.
Cohn, H.: 1979, Astrophys. J. **234**, 1036.
Cohn, H.: 1980, Astrophys. J. **242**, 765.
Cohn, H.: 1985, in IAU Symposium 113 (this issue).
Da Costa, G.S. and Freeman, K.C.: 1976, Astrophys. J. **206**, 128.
Gunn, J.E.: 1980, in Globular Clusters, eds. D. Hanes and B. Madore, Cambridge University Press, p. 271.
Hachisu, I. and Sugimoto, D.: 1978, Prog. Theor. Phys. **60**, 123.
Heggie, D.C.: 1975, Monthly Notices Roy. Astron. Soc. **173**, 729.
Heggie, D.C.: 1985, in IAU Symposium 113 (this issue).
Hénon, M.: 1969, Astron. Astrophys. **2**, 151.
Hénon, M.: 1971, Astrophys. Space Sci. **13**, 284.
Hut, P. and Inagaki, S.: 1985, 'Globular Cluster Evolution: Timescales for Physical Processes', Astrophys. J., to be submitted.
Illingworth, G. and King, I.R.: 1977, Astrophys. J. Letters **218**, L109.
Inagaki, S.: 1980, Publ. Astron. Soc. Japan **32**, 213.
Inagaki, S.: 1984, Monthly Notices Roy. Astron Soc. **206**, 149.
Inagaki, S. and Lynden-Bell, D.: 1983, Monthly Notices Roy. Astron. Soc. **205**, 913.

Inagaki, S. and Saslaw, 'Equipartition in Multi-component Gravitational Systems', Astrophys. J., to be submitted (IS).
Inagaki, S. and Wiyanto, P.: 1984, 'On Equipartition of Kinetic Energies in Two-component Star Clusters', Publ. Astron. Soc. Japan **36**, in press (IW).
Katz, J. and Taff, L.G.: 1983, Astrophys. J. **264**, 476.
Larson, R.B.: 1984, 'Black Hole Remnants in Globular Clusters', preprint.
Lightman, A.P. and Fall, S.M.: 1978, Astrophys. J. **221**, 567.
Lightman, A.P. and Shapiro, S.L.: 1978, Rev. Mod. Phys. **50**, 437.
Lynden-Bell, D. and Wood, R.: 1968, Monthly Notices Roy. Astron. Soc. **138**, 495.
Marchant, A.B. and Shapiro, S.L.: 1980, Astrophys. J. **239**, 685.
Michie, R.W.: 1963, Monthly Notices Roy. Astron. Soc. **126**, 331.
Ostriker, J.P.: 1985, in IAU Symposium 113 (this issue).
Saito, M. and Yoshizawa, M.: 1976, Astrophys. Space Sci. **41**, 63.
Saslaw, W.C. and De Young, D.S.: 1976, Astrophys. J. **170**, 423.
Spitzer, L., Jr.: 1969, Astrophys. J. Letters **158**, L139.
Spitzer, L., Jr.: 1985, in IAU Symposium 113 (this issue).
Spitzer, L., Jr. and Hart, M.H.: 1971, Astrophys. J. **166**, 483.
Spitzer, L., Jr. and Shull, L.M.: 1975, Astrophys. J. **201**, 773.
Stodółkiewicz, J.S.: 1982, Acta, Astron. **32**, 63.
Stodółkiewicz, J.S.: 1985, in IAU Symposium 113 (this issue).
Vishniac, E.T.: 1978, Astrophys. J. **223**, 986.
Yoshizawa, M., Inagaki, S., Nishida, M.T., Kato, S., Tanaka, Y. and Watanabe, Y.: 1978, Publ. Astron. Soc. Japan **30**, 279.

DISCUSSION

KING: Can you give us some simple physical feeling for why a multi-component model evolves faster than a single-component model?

INAGAKI: In order that a single-component cluster evolves, heat must be transported from the core to the halo. It takes a few half-mass relaxation times for such a transportation. On the other hand, a multi-component cluster can evolve if energy is transported from the massive component to the less massive component *in the core*. About one relaxation time is enough for the exchange of energy between components.

APPLEGATE: I have calculated the evolution of young globular clusters, such as those found by Ken Freeman in the LMC, and find that the energy exchange between the different mass components is very important. One must, however, include stellar evolution in these systems because the massive stars evolve off the main sequence and loose mass on timescales comparable with the relaxation time.

INAGAKI: I agree with you.

GASEOUS MODELS

Daiichiro Sugimoto
Department of Earth Science and Astronomy,
College of Arts and Sciences, University of Tokyo

ABSTRACT

We can understand physics of self-gravitating system in terms of gaseous models in so far that their global natures and effects of self-gravity are concerned. Here summarized what are known in idealized gaseous models. They include gravothermal collapse/expansion in linear and non-linear regimes, and post-collapse evolution with gravothermal oscillation. Also discussed are their relations with discrete system and with treatment in statistical mechanics.

1. ROLE OF GASEOUS MODEL

Self-gravitating system shows some peculiar behaviors when it is seen from common sense of normal thermodynamics and statistical mechanics. They come originally from the infinite range of the gravitational force. If the gravitational energy is included, the internal energy is not extensive, i.e., not proportional to the mass of the system any more, but proportional to the square of the mass. It could bring about some phenomena out of common sense, because the common sense in thermodynamics tacitly assumes extensiveness of the internal energy. (Comments and warnings are, of course, given in standard text books. However, in an example of surface tension, the surface energy is proportional to $M^{2/3}$. It depends more weakly on mass than the extensive quantities, and in this sense it lies on the opposite side of the gravothermodynamics.)

The most prominent phenomenon is gravothermal instability or catastrophe, i.e., the instability of an isothermal system contained in an adiabatic wall, which was first pointed out by Antonov (1962) and later analyzed somewhat more closely by Lyndel-Bell and Wood (1968). They showed that the system is unstable if the density contrast between the center and a point just inside the adiabatic wall exceeds the critical value of 709, and they discussed that the instability develops into the collapse of the core.

J. Goodman and P. Hut (eds.), Dynamics of Star Clusters, 207–218.

In their approach only a single value of temperature was considered throghout the system. However, the gravothermal instability or collapse of the core is, in itself, the developement of spatial structure of the system. Therefore, a theory was desirable which treats the specific entropy and the temperature as functions of spatial coordinate.

Such work was done by Hachisu and Sugimoto (1978) by means of linearized stability theory. For the spatial structure and its evolution we have a wealth of knowledge from the theory of stellar structure and evolution. The gravothermal catastrophe is the same phenomenon as the gravitational contraction of the star except for the outer boundary conditions. Computer code for calculating steller evolution can easily be applied for calculating the gravothermal catastrophe of finite amplitude. Such computations were done by Hachisu et al. (1978) and also by Lynden-Bell and Eggleton (1980).

The aim of Hachisu and Sugimoto (1978) and of Hachisu et al. (1978) was to construct physics of the self-gravitating system which will be described in the next section. The aim of Lynden-Bell and Eggleton (1980) was somewhat different. In addition to correct Hachisu et al's (1978) expression for the heat conductivity, they found the existence of similarity solution and showed that the central density reaches infinity in a finite time. They seemed to be more interested in finding functional form to compare with astronomical observation than in exploring physics of self-gravitating system.

When we talk about gaseous models we should discuss how they simulate or represent real stellar systems which are almost collision-free systems. In gaseous models we can utilize many concepts which matured in the theory of stellar evolution, and we can establish physical concepts in well defined forms as will be discussed in the next section. In particle systems, on the other hand, physics has not been posed as well defined but talked about rather vaguely. It is true in particular when the final state of the system is talked about. Though, physics of gaseous model is an approximation to the particle systems, it can play a role of a guide to explore physics of the particle systems. Then we can ask what are same or different and what are limitations of the gaseous model. Here we should be reminded that physics of ideal gas was constructed before physics of interacting gas and that the concepts from the ideal gas proved a powerful tool to investigate the interacting gas.

2. CONCEPTS ESTABLISHED IN GASEOUS MODELS

For the sake of definitness we shall confine ourselves to physics of a gas system which is enclosed within a spherical adiabatic wall and which is described by equation of states for ideal gas. Once physics is established for such case, it is relatively easy to extend it to introduce isothermal wall, other equations of states, multi-components of gas particles etc., though they will not be discussed in the present paper. Here we shall summarize the results only as lemmas (L), and corollaries (C) in weak sense. Most of them are discussed in HS (Hachisu and Sugimoto 1978), HNNS (Hachisu, Nakada, Nomoto and Sugimoto 1978) and BS (Bettwieser and Sugimoto 1984; see also Sugimoto and Bettwieser 1983).

2-1. Linear regime (HS)

It is assumed that the gas is isothermal and in hydrostatic equilibrium with the density contrast,

$$D \equiv \rho_c/\rho_b, \tag{1}$$

where ρ_c and ρ_b are the densities at the center and just inside the wall. For a redistribution of heat or specific entropy hydrostatic equilibrium is recovered and we consider change in thermodynamical quantities after this hydrostatic readjustment. We shall not talk the hydrostatic readjustmet explicitly. Therefore, every quantity of response is an "apparent" quantity in the sense that the effect of gravitational interaction has been transferred into it.

If we describe changes in temperature and pressure distributions as

$$\delta \ln T(q) = \int_0^1 F(q,q')\delta\sigma(q')dq' \quad , \tag{2}$$

$$\delta \ln P(q) = \int_0^1 G(q,q')\delta\sigma(q')\, dq' \quad , \tag{3}$$

where $\underline{q}$ is the mass fraction coordinate and $\delta\sigma$ is non-dimensional specific entropy. The inverse of the tensor specific heat F(q,q') is related to the response function of the pressure G(q,q') by

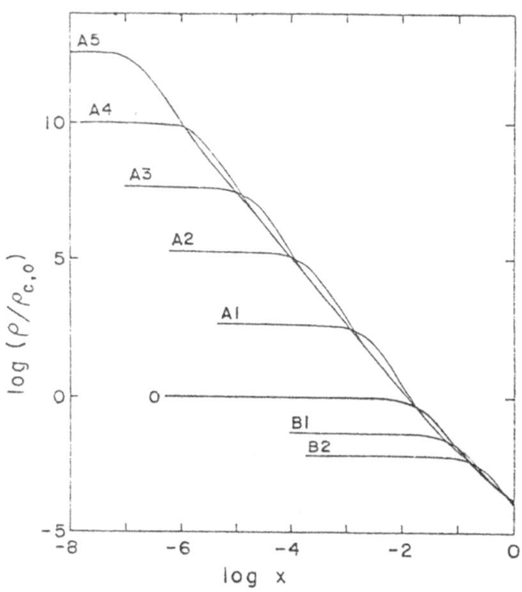

Fig. 1. Gravothermal collapse (sequence A) and expansion (B). The initial state is indicated with Stage 0. The density distribution is plotted against the radial coordinate $\underline{x}$. The outer adiabatic wall corresponds to log x = 0. Stage B2 which is the final state of the gravothermal expansion is now in a state of thermal system. Taken from HNNS.

$$F(q,q') = (2/5)[G(q,q') + \delta(q-q')]. \quad (4)$$

Here the delta function in the right hand side describes local specific heat and the function G describes the gravothermal effect.

L1: G(q,q') is negative in the central region and positive in the outer region. If its effect overrides the effect of the delta-function, the system is regarded to have "negative specific heat" as a whole.

L2: The system is thermally unstable when $D > D_{cr} = 709$. This is "gravothermal instability".
C2-1: When $D < D_{cr}$, we call it a thermal system to which the usual concepts of thermodynamics can be applied. In fact the gravo-thermodynamics tends to the usual thermodynamics in the limit of $D \to 1$.
C2-2: When $D > D_{cr}$, we call it a gravothermal system and its behavior can be different from one imagined in the usual thermodynamics.
C2-3: As is expected from the formalism of linearized theory, the gravothermal instability proceeds in both ways (Fig. 1), i.e., to contraction as well as expansion of the core according as the initial perturbation.

The entropy of the system is defined as

$$S = M\int_0^1 s(q)dq, \quad (5)$$

where M is the total mass of the system and s is the specific entropy.

L3: The equilibrium isothermal state lies at a local maximum of entropy for $D < D_{cr}$ and at a local minimum of entropy for $D > D_{cr}$.
C3-1: Both in the cases of contraction and expansion the gravothermal instability proceeds in the direction to increase S as is described by

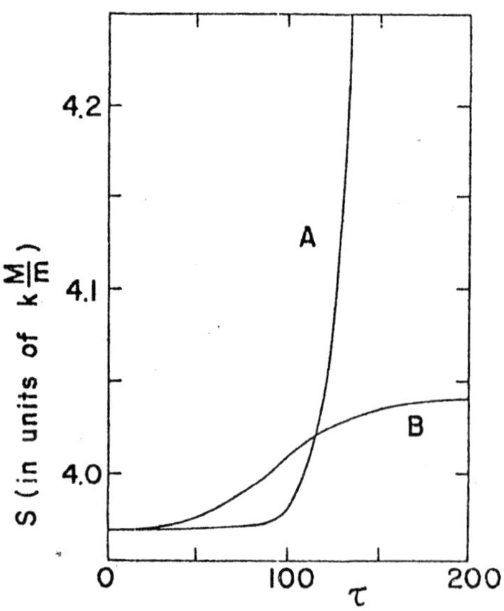

Fig. 2. Total entropy of the system is plotted against the non-dimensional time for the gravothermal contraction (A) and expansion (B) of Figure 1. In the gravothermal contraction the total entropy increases indefinitely within the non-general relativistic regime. In the gravothermal expansion it levels off to a value which corresponds to a local maximum for the system having the same energy, and the system tends to a stable thermal system. Taken from HNNS.

the second law of thermodynamics (Fig. 2). It is an irreversible process; entropy is produced as a result of heat transport.

2-2. Non-linear regime (HNNS)

When the amplitude becomes finite, the gravothermal instability is regarded to be in non-linear regime.

L4: As the instablity proceeds, the entropy of the system increases by heat conduction as an irreversible process, which is described by

$$\frac{ds}{dt} = \int_0^1 L_r \frac{\partial}{\partial q} \left(\frac{1}{T}\right) dq, \tag{6}$$

where L_r is the heat flux through a spherical shell at the radial distance r.

L5: For gravothermal expansion S increases but levels off to the value that corresponds to the state of stable equilibrium for the thermal system with the same energy (Fig. 2).

L6: For gravothermal contraction S increases indefinitely (Fig. 2) within a finite time.

L7: For the gravothermal contraction the central region contracts while the outer shells expand somewhat. The system results in a core-halo structure.

L8: The mass of the contracting core decreases in time (HNNS; Lynden-Bell and Eggleton 1980).
C8-1: If the total mass of the system is not large enough as compared with the mass of a gas-particle, the mass of the core could become as small as the mass of the gas-particle.
C8-2: If it is large enough, the radius of the core could become smaller than the gravitational radius of the core. The result is the collapse of the core into a black hole.

Similar things can be discussed also in the context of cosmology where the thermodynamics of the black hole and the radiation field plays a role. Sugimoto et al. (1981) discussed that the development of spatial structure or, non-uniformity if we state more exactly, is formed in the universe consistently with the second law of thermodynamics. Concerning the initial state of the universe Sugimoto et al. (1981) and later but independently Frautschi (1982) showed that at the Planck time its state was at the global maximum of entropy if we consider only the causally connected region or, in other words, the region within a particle horizon. As the universe expands, such region expands relative to a co-moving volume and the state falls out of the change in the state with the maximum entropy. In other words, the state of thermal equilibrium, i.e., the state with the maximum entropy, changes as the boundary condition of the system changes by the expansion of the universe, and the real system is left behind and now lies out of thermal equilibrium. Therefore, the entropy of the comoving volume can increase thereafter, and it is associated with generation of non-uniformites through the gravothermal catastrophe.

L9: The core collapse is described by similarity solution of almost isothermal core surrounded by an envelope of the density distribution $\rho \sim r^{-\alpha}$ with $\alpha = 2.2$. Its central density becomes infinitely high in finite time (Lynden-Bell and Eggleton 1980).

2-3. Post-collapse evolution (BS)

The gravothermal collapse is a result of the change in the mean field. When the central density becomes high enough, binaries of gas-particles (or stars) will be formed. This can be regarded as the development of two-body correlations. The binaries play a role to input energies into the mean field. To investigate the post-collapse evolution a gaseous model with energy source is considered. Since it will be discussed by Bettwieser in this Symposium, here we will summarize only their results.

Energy input by binaries is considered in analogy with the nuclear energy input in a main-sequence star. If the outermost shells of our system are regarded as the place where the entropy generated in the inner core can be dumped away, both the inner core of our system and the main-sequence star are similar dissipative open systems. The main-sequence star lies in a steady state.

L10: The core of our system makes gravothermal oscillation repeating gravothermal contraction and expansion.

L11: If the amount of the energy input into the mean field is neglected, it tends to a limit cycle encircling the collapsed singular isothermal solution. This is another state of the dissipative open system.

The difference between the main-sequence star and our system comes from the difference in the functional form of heat conductivity. In the main-sequence star the time scale of heat transport is shorter in the outer shells, while in our system it is shorter in the inner core. Therefore, the outer shells respond quickly to a change of the inner core in the case of the main-sequence star but they do not in our system. This is also the reason why the size of the core is relatively large in the interior of the star but is small in our system.

L12: Because of a net energy input to the mean field from the binaries, the energy of the mean field increases secularly. Finally, it becomes so high that the system can not be regarded to be the gravothermal system any more but a thermal system.
C12-1: When the system becomes a thermal system, the gravothermal oscillation does not occur any more, and the system makes an expansion on input of energy until an equilibrium thermal state is reached.
C12-2: If the rate of the energy input is too high (BS), the secular

change dominates over the limit cycle so that the core expands simply to a thermal state.

L13: The expansion phase of the gravothermal oscillation is driven by the gravothermal instability. Therefore, it requires no energy input after an initial small amount of expansion has taken place. Further expansion takes place with almost the same time scale as the contraction.

During the main phase of the gravothermal expansion the so-called gravitational energy release $\varepsilon_g \equiv -T\,ds/dt$ just balances the conduction loss $\varepsilon_{cond} \equiv dL_r/Mdq$, i.e., $\varepsilon_g = \varepsilon_{cond}$, in the core region. If one used inappropriately long time step Δt in numerical computation, ε_g could not be computed correctly but the conduction loss could balance the energy generation ε_{bin} from binaries, i.e., $\varepsilon_{cond} = \varepsilon_{bin}$, within a small but finite allowance of the assumed accuracy. Then one would obtain a slow general expansion. Heggie (1984 and in this Symposium) obtained such general expansion in his numerical computation with long time steps. This seems to be an interpretation of his results, but some more studies will be necessary in order to conclude that the gravothermal oscillation should take place in our idealized system.

Inagaki and Lynden-Bell (1983) showed that there exists a self-similar solution also for the post-collapse expansion. However, their solution determines the required strength and time change of the energy generation which has nothing to do with physical mechanism. In actual, however, the energy generation is determined by binaries, i.e., by local process which is independent from global mechanism controlling expansion of the system. There is no guarantee at all that the energy generation is tuned so as to match such prescribed requirement. If the energy generation is slightly deficient, for instance, the expansion is slowed down and deviates from the similarity solution, because the configuration and temperature distribution of the similarity solution correspond to those of a gravothermally contracting system.

3. RELATION WITH PARTICLE SYSTEM OR GRAVITATIONAL MANY-BODY PROBLEM

There are several important time scales in evolution of the many body system. The shortest is the time scale of violent relaxation t_{viol} which is of the order of the crossing time. The second is the time scale of heat transport t_{heat} which is of the order of the two-body close encounter. The third is the time scale of the growth of two-body correlations or the formation of binaries t_{bin} which is of the order of two-body tidal capture or three body collision. The fourth is the time scale of the growth of many-body correlations t_{corr}. Because many-body correlation is formed successively through collision of particles with a lower order correlation, its time scale t_{corr} is longer than but is of the order of t_{bin}. Finally, many-body correlations will prevail the whole system.

Let us consider a system consisting of N particles (stars) which are enclosed with a specularly reflecting wall. For such a system the orders of magnitude of these time scales are related by $t_{heat} \sim (N/\ln N)t_{viol}$, and $t_{bin} \sim n^{-2}$ where n is the particle density. In many numerical computations these time scales are not always well separated so that different physical phenomena appear in mixture. If we choose appropriate set of initial parameters, i.e., relatively large value of N and low density n, we can separate and visualize phenomena of different time scales.

Because the particle system is not the main topics of the present talk, I will show only one example. It was calculated recently by Bettwieser in contact with myself for a system with N = 100 by means of a computer code NBODY3 which was kindly supplied by Professor S. Aarseth. The computation covers through $200t_{viol}$. Figure 3 shows time evolution of logarithm of the density in the innermost region in which 10 particles are contained. Initially the particles were distributed uniformly within the radius R_0 and with a constant velocity dispersion. The total energy in units of $GM^2/4R_0$ is E = -581 upto t = $47t_{viol}$ and E = -170 after t = 50 t_{viol}. We see that the fluctuation is a factor of about $10^{1/2}$, that the gravothermal catastrophe takes place, and that the formation of hard binaries excites the gravothermal oscillation.

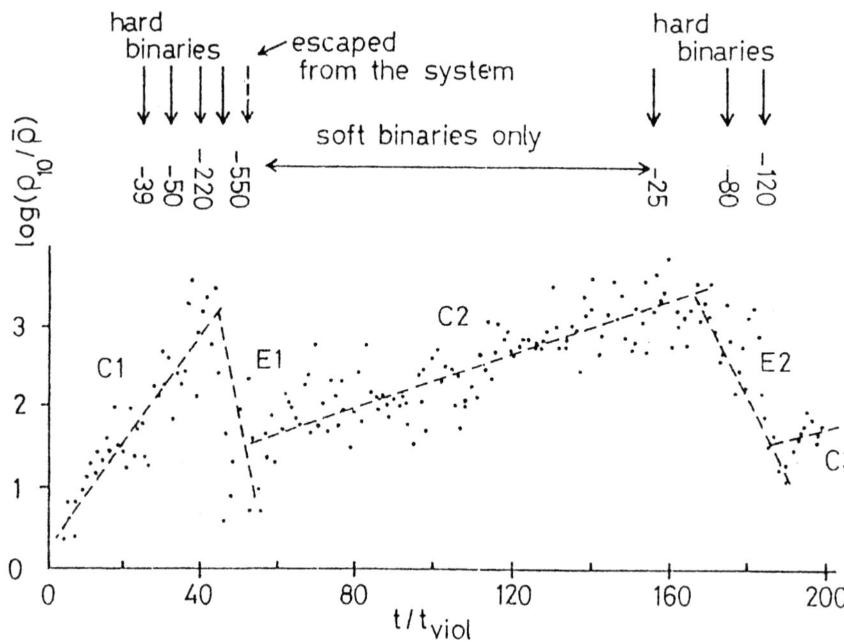

Fig. 3. Time evolution of the gravitational 100 body system. Phases of contraction (C) and expansion (E) are clearly seen. For each hard binary the time of its formation or escape is indicated by an arrow to which its energy is also attached. See the text for more details.

Consulting with such numerical computation and the lemmas given in the preceding section, we may conjecture a picture for the evolution of self-gravitating particle system as follows.

Phase 1: Violent relaxation. It takes place in the time scale t_{viol}. It may be described by collisionless Boltzmann equation in μ-space which includes only the mean fild. An equilibrium state will be reached for $t \to t_\infty$.

Phase 2: Gravothermal catastrophe. The infinite time cited in Phase 1 is not a real infinity but the time $t_{viol} \ll t_\infty \ll t_{heat}$. Therefore, two-body collisions proceed, and the equilibrium state in the sense of Phase 1 will be shaffled. Then a new equilibrium state will be settled which has a higher entropy. If the condition for the gravothermal system is met, this is the real progress of the gravothermal instability. This phase can still be described in μ-space.

Phase 3: Gravothermal oscillation. When the time becomes of the order of $t_{bin}(\gg t_{heat})$, binaries will be formed. Now the distribution function can not be described only by multiplication of a single particle distribution functions, but at least the two-particle distribution function in binary space is necessary. It can be divided into the part of independent particles and the part of two-body correlations, i.e.,

$$f(1,2) = f(1)f(2) + g(1,2). \tag{7}$$

Correspondingly, the energy can also be divided into the energy E_{mf} which is associated to the mean field and translations of single particles and multiplets, and the energy of correlations E_{corr} which corresponds to the binding energy of the multiplets, i.e.,

$$E = E_{mf} + E_{corr} \quad . \tag{8}$$

If the particle numbur $\underline{N}$ is large enough, there could be a situation where E_{corr} is still low compared with E_{mf} in their absolute values. Then the gravothermal oscillation can take place.

Phase 4: Prevalence of Correlations. If the system is contained in a specularly reflecting wall, $\underline{E}$ is kept constant. As the correlations devlop, E_{corr} becomes negative and large in its absolute value. Correspondingly E_{mf} becomes negative and small in its absolute value in the first place, and then can become even positive. This is a reaction to the mean field from the development of the correlations. Then the system becomes to be regarded as thermal system, if we see only the mean field, i.e., if the multiplets are considered to be constituents of the system and if E_{corr} is regarded as sub-constituent energy (such as binding energy of molecule). Then the more probable state becomes one for the thermal system, which is an isothermal state with weak density contrast. The gradual change to such state is the secular change during the gravothermal oscillation.

Phase 5: Final state. If we consider only the two-body correlation, the final state will be a uniform distribution of binaries because E could becomes much greater than E.

We have to notice that there is an important difference between the gaseous model and the usual statistical treatment of the problem. In the gaseous model we discuss the evolution of the system, while in the statistical mechanics we discuss only the final state in many cases. Therefore, the final state of Phase 5, which is obtaind from H-theorem in binary space for instance, does not necessarily imply that there is no evolution through Phases 2-3 of the gravothermal catastrophe and oscillation. In order to understand physics correctly for a system with possible strong correlation, we have to discuss their evolution by means of non-equilibrium statistical mechanics or, at least, by introducing different levels of infinite times t_∞ as done in the present section.

Nevertheless, we may ask what is the real final state as inferred from the statistical mechanics. After the formation of binaries the formation of more body correlations will proceed. Finally N' body correlation will develop where N' is of the order of N. Then we shall ask what is the spatial distribution of such N'-multiplets. However, it can not be answered by the statistical mechanics, because the number of N'-multiplets is too small to be treated in the statistical mechanics.

This work is supported in parts by Scientific Research Fund of the Ministry of Education, Science and Culture (57540120, 59540136).

REFERENCES

Antonov, V.A.: 1962, Vest. Leningr. Gos. Univ., 7, 135.
Bettwieser, E. and Sugimoto, D. (BS): 1984, Month. Notices Roy. Astron. Soc., 208, 493.
Frautschi, S.: 1982, Science, 217, 593.
Hachisu, I., Nakada, Y., Nomoto, K., and Sugimoto, D. (HNNS): 1978, Prog. Theoret. Phys., 60, 393.
Hachisu, I. and Sugimoto, D. (HS): 1978, Prog. Theoret. Phys., 60, 123.
Heggie, D.C.: 1984, Month. Notices Roy. Astron. Soc., 206, 179.
Inagaki, S. and Lynden-Bell, D.: 1983, Month. Notices Roy. Astron. Soc., 204, 913.
Lynden-Bell, D. and Eggleton, P.P.: 1980, Month. Notices Roy. Astron. Soc., 191, 483.
Lynden-Bell, D. and Wood, R.: 1968, Month. Notices Roy. Astron. Soc., 138, 495.
Sugimoto, D. and Bettwieser, E.: 1983, Month. Notices Roy. Astron. Soc., 204, 19P.
Sugimoto, D., Eriguchi, Y., and Hachisu, I.: 1981, Prog. Theoret. Phys. Suppl., 70, 154.

DISCUSSION

SHAPIRO: One must always be wary of "hydrodynamical" models of intrinsically collisionless or Fokker-Planck (secularly collisional) systems. Of course, one can always take *moments* of the collisional Boltzmann equation and get equations which, under suitable approximations, "look like" gas-dynamical equations. This is reliable in many instances and is exactly what R. Larson (1970) did in his pioneering work on the Fokker-Planck equation over a decade ago. How exactly do your equations compare with his? What is the analog of your conduction equation in his equations? Should one not be skeptical of the gas-dynamic approach to the extent that it differs from Larson's formalism?

SUGIMOTO: Our equations can be obtained from the moment equations when the third moment is truncated by means of the heat conduction equation. Larson (1970) took account of the higher moments upto the fourth moment. Therefore in his equations non-local heat flow and deviation from Maxwellian distribution function were taken into account. Here, we can ask what are described by each moment. As far as the mean field is concerned our equation describes the essential point. In other words we see that the gravothermal contraction is described very well by our model. Higher moments will introduce only quantitative difference because the deviation from the Maxwellian distribution makes mainly a local effect which can be transferred to a correction to the equation of state. Concerning the heat conduction, Bettwieser and Sugimoto (M.N.R.A.S., submitted) have recently computed a 1000-body problem, and calculated moments from numerical results and compared it with gas-dynamical concepts. They have found that the gas-dynamical concepts can be applied fairly well. Development of correlations are described by higher moments or in many-body space. Thus, it was not taken into account even in Larson's (1970) work. However, we can take account of its main effects phenomenologically by means of the energy generation rate by binaries, for instance. Anyhow, the gaseous model of an idealized system is appropriate to explore the fundamental behavior of self-gravitating systems. When this behaviour is understood exactly in idealized systems, there will be no difficulties to include astronomical details though they may be very complex. Thus, we can and have to follow two different approaches to the problem.

LARSON: I have some reservations about whether "gravothermal expansion" can occur in real star clusters. I can imagine that it could occur in a conducting gas sphere with an adiabatic boundary, where the presence of the boundary allows heat to be trapped in the outer part of the system; this trapped heat might then in some circumstances be conducted back into the core and drive expansion. Real clusters, however, do not have an adiabatic boundary but are surrounded by an infinite heat sink, namely the rest of the universe, so that heat can not accumulate in the outer part of the cluster. All of the observations that I know of show velocity dispersions decreasing monotonically outward, and this is also true for all of the models I know of for systems not surrounded by an adiabatic boundary. Thus, in real systems the sense of heat flow should always be outward.

SUGIMOTO: In the system the timescale of heat transport is much longer (even 10^6 times though depending on the central density) in outer shells. In this sense the relatively outer shells can be regarded as the "adiabatic wall" to a good approximation. This is an essential difference of our system from the stellar structure (In the stellar interior heat is transported by radiation and its timescale is much shorter in outer shells.) See my answer to Grindlay's question as for the relation with observations.

GRINDLAY: My question follows naturally from Dr. Larson's remark as well as from the viewgraph you now have on the screen. The question is: in a real globular cluster (of high mass), what is the maximum difference in "temperature" or velocity dispersion between the inner and outer core for a cluster in the expansion phase? As I mentioned in my talk yesterday, I suspect the X-ray globular NGC 6712 may be in such a post-collapse expansion phase and we are beginning a major program to measure V_{disp} vs. R. What is the maximum *increase* in V_{disp} v. radius R we might expect to measure if indeed the cluster is in an expansion phase?

SUGIMOTO: As can be seen in one of my figures (Figure 2 of Sugimoto and Bettwieser 1983 M.N.R.A.S. 204, 19P), the maximum possible temperature difference between its peak and the central value is $\log(T_{peak}/T_{center}) \simeq 0.05$. The peak lies at the edge of the core. The value given above is one at the stage when the mass fraction of the core is equal to 1×10^{-3}. Both in earlier and later stages of expansion, i.e., for smaller and larger core mass, respectively, the temperature difference is smaller.

KING: In a real cluster, where the density drops to zero at the boundary, what quantity corresponds to the magic density contrast of your physical idealization?

COHN (after Sugimoto referred the question to him): The best observational parameter to look at in order to determine whether the gravothermal instability is likely to have occurred is your W_0 parameter or equivalently the concentration parameter c. Clusters are expected to become unstable for $W_0 \geq 8.5$ that is $c \geq 2.0$. Indeed the clusters with possible cusps satisfy this criterion.

KING: That is very interesting. In the linear-sequence sense to which Spitzer referred yesterday, $W_0 = 8.5$ is very close to the turning point.

GRAVOTHERMAL OSCILLATIONS

Erich Bettwieser
University Observatory Göttingen
Geismarlandstr. 11
3400 Göttingen
Federal Republic of Germany

ABSTRACT. Stellardynamical systems are modelled by gaseous spheres. The properties of an isothermal equilibrium configuration with a central singularity are discussed in the context of post-collapse evolution. For the numerical calculations heating effects are taken into account. Evolution towards the equilibrium configuration is monotonically only if much energy is delivered from hardening processes of binary stars. Since the equilibrium configuration is gravothermally unstable a minute emission of energy is sufficient to turn gravothermal contraction into gravothermal expansion. In such case equilibrium is approached by oscillations of the core.

1. THE SINGULAR ISOTHERMAL SPHERE ('SIS')

The discussion of Hénon (1975) is a cornerstone of post-collapse evolution of globular clusters. He assumed a self-regulated, central energy source, which reverses core collapse leading to a general expansion of the cluster. Post-collapse evolution assuming strict self-similarity has been discussed by Inagaki and Lynden-Bell (1983) (henceforth ILB for short). The inner part of the ILB solution is isothermal with the density decreasing as $1/r^2$. The expansion of this isothermal region is triggered by the energy input. Similar post-collapse expansion has been found for finite systems too, e.g. by Lightman and McMillan (1985) using a N-body hybrid method, by Cohn (1985) within the scheme of Fokker-Planck theory and, in the framework of gaseous self-gravitating systems, by Heggie (1984) and Bettwieser and Sugimoto (1984, henceforth BS for short).

Self-similar solutions demand systems infinite in radius and mass. However, there is an equilibrium state of a self-gravitating sphere of mass M confined within an adiabatic wall of radius R, which is related to the ILB-theory.

J. Goodman and P. Hut (eds.), Dynamics of Star Clusters, 219–230.

It is the singular, isothermal solution (**SIS** for short), with the density ρ and temperature T given by

$$\rho = M/(4\pi R^3)\cdot(r/R)^{-2}; \quad T = GMm/(kR)\cdot 1/2 \qquad (1)$$

where k and m are Boltzmann's constant and the mass of the gas particles, respectively. The singular, isothermal equilibrium solution has the following properties: (i) The total entropy of **SIS** is rather low in comparison with intial models, hence a collapsed state cannot evolve towards **SIS** without eating negentropy. Energy has to be fed into a region with the highest temperature or must be removed from another region with a low temperature (e.g. by star evoporation from the boundary). (ii) In BS the gravothermal instability of **SIS** is demonstrated by the calculation of the inverse tensor of specific heat (cf. Fig.1) and the computation of eigenfunctions and corresponding eigenvalues, which extremize the second variation of the total entropy.

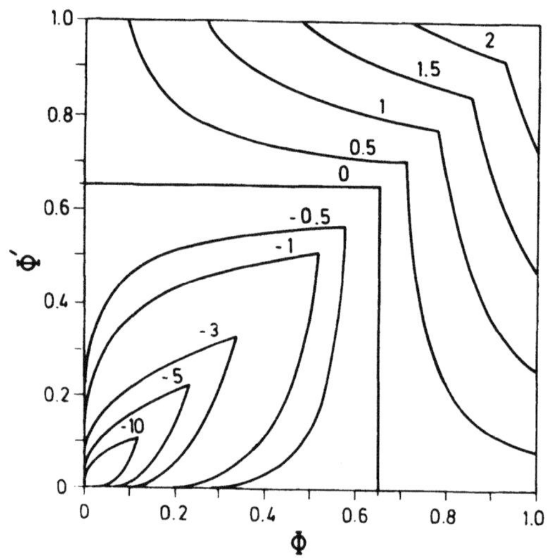

Fig.1. Tensor of the inverse specific heat $G(\phi,\phi')$ drawn as function of normalized mass coordinates. Note the large negative region and the singularity in the very centre. Functional form resembles those of regular isothermal solutions with very high density contrast.

Self-gravitating systems naturally have negative specific heats. SIS is one of the most extreme example for such case. (iii) SIS is unstable against spontaneous development of anisotropy of the velocity dispersion (Bettwieser et al., 1984). This is concluded from the functional form of the specific heats and from the numerical observation, that SIS if taken as an initial model in a pre-collapse calculation, does only marginally change its density distribution but becomes anisotropic on a secular timescale. In fact, we have constructed anisotropic and singular equilibrium configurations.

Let us turn to the general consequences of these properties of SIS. (i) Evolution triggered by heat added into the central core region will approach gradually to a structure nearby SIS. (ii) Provided effects of self-gravity dominate over the heating due to the energy fed into the core, gravothermal expansion or (and) gravothermal contraction can ensue. This will approximately be the case whenever the energy put into the core during one high density phase of evolution is small in comparison to the binding energy of the core. (iii) The post-collapse configuration becomes more and more anisotropic, and if gravothermal effects are more important than thermal effects (i.e. low energy input) even the region near the very center is expected to become anisotropic in the post-collapse phase. In the next chapter we discuss numerical results related to the topics (i) and (ii).

2. GRAVOTHERMAL OSCILLATIONS OF GASEOUS SPHERES

We investigated the evolution of gas models with a heat source designed to represent the kinetic energy added into the core by superscattering processes with hard binary stars. The model equations are given in BS for a system composed of single stars, and they are generalized to multi-mass systems in Bettwieser and Inagaki (1984). The energy generation rate per unit mass is assumed to be given by

$$\varepsilon = C/(1.95\ M_\odot L_\odot^{-1})\cdot[\rho/10^{10}\ M_\odot\ \mathrm{parsec}^{-3}]^2\cdot \\ \cdot[10\ \mathrm{km\ s}^{-1}/\sigma]^\ell \qquad (2)$$

σ is the three-dimensional velocity dispersion. In most models calculated the parameter ℓ was chosen to be unity (seven) for single-mass (multi-mass) systems, respectively (cf. BS and Heggie, 1984). For the sake of definiteness the outer boundary condition was treated to be a fixed adiabatic wall. At the centre the energy flux and the velocity of mean mass motion was set to zero. Hence the code can in principle calculate a configuration such as SIS being singular at the centre. For multi-mass systems the velocity dispersions are assumed to be equal initially. As a consequence evolution is also due to the tendency to equipartition of the kinetic energies. Starting from Plummer's model we computed its time evolution for different values of the energy input rate. The value of the constant C was changed by many orders of its magnitude. For extremely large (small) values of the constant C the core becomes a perfectly thermal (gravothermal) system, respectively.

For pre-collapse models and for the first case (C large) our results are in agreement with those obtained by Heggie (1984) or by Cohn (1980, 1985). The core suffers from gravothermal collapse in the first place, until the central density reaches a threshold beyond which the energy input becomes effective. The isothermal region expands and the decreasing density contrast can even fall below the critical value of 709. Then the system evolves eventually to a completely isothermal configuration which is also uniform in density.

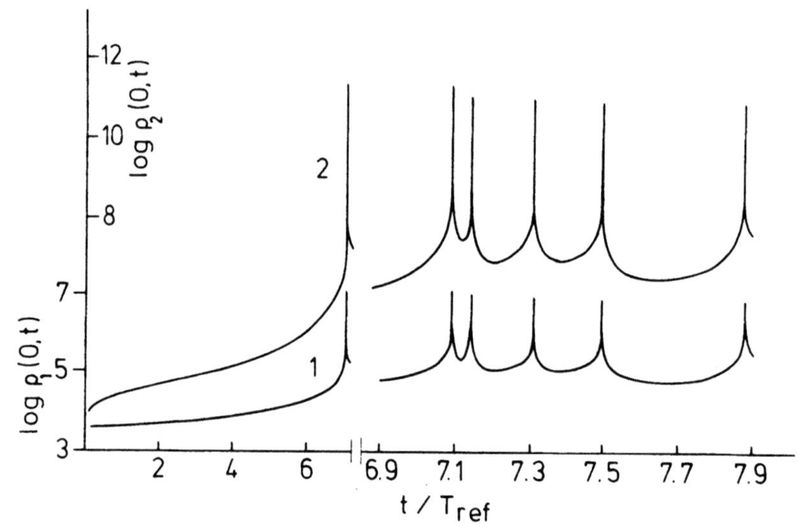

Fig.2. Time evolution of the central density ($M_\odot$/pc^3) for a two-component gas-system with a weak heat source ($C = 10^4$, $\ell = 7$, $M = 10^6 M_\odot$). Time in units of the Spitzer-Hart reference time.

For the second case (C small) the core contraction is halted and reversed at a higher value of the central density. Due to our numerical results, the energy released per mass within one local relaxation time is about constant (as a function of C). Hence the maximum of the density and the minimum of the core mass are proportional to C^{-1} and $C^{0.36}$, respectively. In fact, the collapsed state can have a very small core, with a mass fraction below 0.001. The core begins to expand and reaches a state of miminum density with a core mass of about 0.02 of the total mass for a broad range of values of C. Hence core heating is not the driving force for the phase of expansion. After this stage the core starts to contract again. As depicted in Fig.2 the system oscillates between phases of high and low central densities. In this case a two component system with a mass ratio of two was calculated for a mass fraction of the heavier component of 0.1. Energy was put into both components and Eq.(2) was used with the density replaced by the total density of the two components. The example shown is typical among our results; core oscillations have been found for one, two and three component systems; with different rates of the energy input; for extended energy sources as well as for a point source; for configurations isotropic or anisotropic in velocity space, and for different initial models and boundary conditions (Bettwieser and Fritze 1984). For these calculations three, in several aspects different numerical codes are used (cf. BS, Bettwieser and Inagaki 1984, Bettwieser et al 1984).

2.1. Physical Mechanism

The non-linear oscillations with such a great amplitude as seen in Fig. 2 is regarded to be a gravothermal flip-flop. The contracting phases are the well-known gravothermal catastrophe. Since the specific heat is negative in the core, contraction is driven by a self-sustaining heat flow directed outwards. With the accelerating increase of the central density energy input becomes suddenly appreciable. The core is supplied with a short energy impulse at a high central density. (It is worth to note here that for higher values of C energy is gradually put into the core at an earlier stage of the contraction, i.e. during a phase of slower evolution). Once the expansion commences the rate of emission of energy is quenched, since the central density decreases rapidly and the velocity dispersion changes only slightly. After the energy impulse is fed into the core there is a short phase of a thermal expansion of it. The inner core expands somewhat faster than the outer regions, since the energy input is concentrated towards the centre. As a result the slope of the temperature profile changes its sign and a heat flow directed inwards appears. Since the core is self-gravitating the heat flowing inwards enhances the bump in the temperature distribution as seen in stage 3 of Fig.3. Most of the time the expansion instability of self-gravitating systems is responsible for the evolution of the system. More and more heat flows into the central region accompanied by a decrease of its temperature. But there is a competing heat flow going outwards from the peak of the temperature profile to the halo. The total increase in entropy due to this latter process can be much larger than the entropy production invoked by perturbations, which lead to the gravothermal expansion (Hachisu and Sugimoto 1978). Therefore, the heat flowing outwards from the bump in temperature wins the game, and as a consequence the bump is smeared out. Finally, the core becomes almost isothermal and, since the density contrast is large, it enters again the phase of gravothermal contraction.

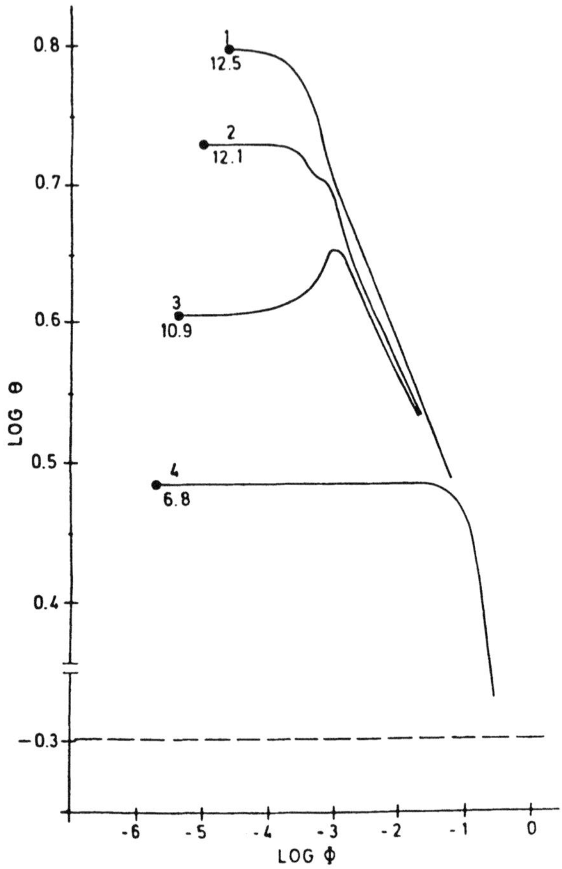

Fig.3. Temperature vs. mass (ϕ in units of the total mass of the cluster) for a one-component system, which suffers from gravothermal oscillations ($C = 10^4$, $\ell = 1$, $M = 10^5\ M_\odot$). At stage 3 the logarithm of density dropped from 12.5 to 10.9 as indicated in the figure, and the inversion in temperature is most pronounced.

It is worth to notice the following facts: (1) The energy generation is used mainly to turn the switch from unstable contraction to unstable expansion, provided the value of C is well below the threshold to excite the oscillations. Therefore the total amount of energy fed into the core during one cycle is virtually unaffected by even drastic changes of the energy generation rate (cf. table 1 in BS). But the value of C determines the maximum of the density and the typical time between successive peaks. (2) The evolution followed in the plane central temperature vs. central specific entropy looks as if the basic process works like a refrigeration cycle. The cyclic change in entropy and temperature comes from in- and out-flow of heat. The steady absorption of binding energy leads to a secular decrease of the average temperature. With each cycle the system as a whole is drifting more and more to SIS. The maximum size of the isothermal region gradually extends into the halo region. But even after many cycles the core is still regular in its density distribution. (The central singularity of SIS or of the self-similar ILB solution are mathematical artefacts). (3) The coefficient of heat conductivity is tailored to simulate the evolution of a stellar dynamical system. To a first approximation the dynamics of the core can be regarded to be independent from the outer shells, since the latter evolve on a much longer secular timescale. Therefore the appearance of the oscillations is such as if an isothermal core would be embedded in an adiabatic sphere, which follows the core's movement. (Here we disregard the secular drift during the cyclic changes of the core parameters). All the time any perturbation is renormalized to be in the linear regime. This is the reason why even a minute absorption of binding energy is sufficient to turn the contraction instability into the expansion instability. However, this remark does not mean, that the system can be divided into the weakly coupled parts of core and halo. Such system would fail to have gravothermal instabilities. (4) With increasing values of C the gravothermal effects and effects of core heating interfere. But the latter process is unimportant as long as C is below the threshold to excite oscillations (cf. table 1 in BS). When the energy supplied becomes too large the thermal perturbation is in the non-linear regime. Then an amount of energy is added into the core, which is comparable to the core's binding energy. As a result the reexpansion of an isolated cluster proceeds indefinitely. (5) In the case of multi-mass systems the kinetic energies of the core are near to equipartition during the whole post-collapse stage. The time between successive peaks in the density is shorter for a multi-mass system than for a single-mass system.

3. DISCUSSION AND COMPARISON WITH OBSERVATIONS

The role of the central binaries is often thought to reverse the collapse and serve as an efficient source of energy. If the cluster oscillates such role is played only in the long run of the evolution. The net flux of energy emitted will be reduced by inelastic two-body encounters, hence a small value of C may be more realistic. In a real star cluster the competition between heating, cooling, gravothermal and external effects, such as tidal heating, determines the net dynamical evolution and the final fate. The conditions to excite gravothermal oscillations are almost unaffected by the parameters of the model. In this sense the preliminary investigations are confronted with observations. If those clusters which have short relaxation times in their cores, follow an evolution prescribed by a small value of the heating efficiency C, it is comprehensible why no collapsed object is observed despite the core collapse time is shorter than the age of the cluster. The duration of the oscillation phase can be much longer than the time needed for its initial contraction, and even much longer than the age of the cluster. After a cluster passed through many cycles of core oscillations, its spatial structure is nearly that of SIS. The cluster's thermal relaxation is then well pronounced. The total amount of energy delivered by the central machine can add up to a significant fraction of the core's binding energy, either because of many energetically insignificant events (C small) or by one very efficient process (C large). As a result the observed brightness distribution of the inner halo oscillates around a 1/r profile. These oscillations around the average trend should be observable, and it seems that the theoretical prediction is met in NGC 6624 (cf. Djorgovski and King, 1983).

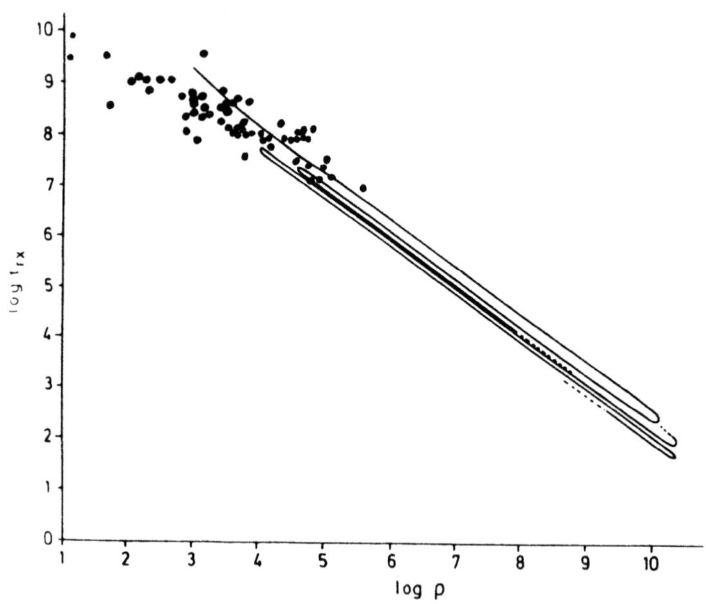

Fig.4. In the plane central relaxation time (years) vs. central density ($M_{\odot}/pc^3$) galactic globular clusters are represented by dots (data from Peterson and King, 1975). For the same model as in fig. 3 part of the evolution track is indicated.

Recently Djorgovski and Penner (1984) found several new post-collapse clusters with a brightness distribution near to those of the singular isothermal sphere. But the total fraction of such clusters is estimated to be only a few per cent, and hence in conflict with the timing estimate based on thermally expanding models. Therefore, such models (C large) seem to be ruled out in the sense that most of the clusters with short relaxation times approach the singular isothermal sphere much more slowly, and hence evolve very different from the model's prediction.

Let us look at this problem from the following point of view. In the cases of thermal expansion (C large), either triggered by accretion onto a central black hole (cf. Bettwieser et al. 1984) or by hardening of binary stars, evolution is much slower in the post-collapse stage as compared with cases where C is small. In fact, for C large the timescale of evolution greatly exceeds the central relaxation time (cf. section 3.1 in Heggie 1984). Therefore such a post-collapse object tends to stay a long time far outside the range of parameters observed. Since tidal forces can remove the outermost stars with increasing ease we encounter the problem that at present there should be remnants of partly or completely dissolved globular clusters.

For the evolution driven by the gravothermal instabilities the ordinate of Fig.4, i.e. the central relaxation time, is the typical timescale of evolution of the central parts of the cluster. Note that the high density phases are too short to have a good chance to be observed. In low density phases the cluster's structure is not so much different from its appearance before the initial collapse. The same is true for the outer halo region. The evolution track in Fig.4 is for a case with C being small. If many clusters make a similar evolution the core densities in the observed ensemble of clusters will vary over many orders of magnitude, even if initially the densities would be confined within narrow limits. At the stage of lowest density the observation of an oscillating core would be most probable. At such stage the core mass depends but weakly on the choice of the initial model or the parameters of the energy input. The core mass as well as the energy fed during one cycle is adjusted by the system as a whole (cf. BS 1984; Bettwieser and Fritze 1984). Within rather broad limits such quantitative and integral aspects of the evolution are almost independent from the details of the model calculated. Concerning the quality of evolution, however, there is the important difference between the slow thermal expansion and the more rapid gravothermal oscillations. Since the fundamental instabilities drive the latter and since the energy source is used mainly to turn a switch, it is not surprising that we rediscover Hénon's conjecture about self-controlled post-collapse evolution.

References

Bettwieser, E., Inagaki, S., 1984. to be submitted.
Bettwieser, E., Fritze, U., 1984. *Publ. Astron. Soc. Japan*, 36, 3024.
Bettwieser, E., Fricke, K.J., Spurzem, R., 1984. to be submitted.
Bettwieser, E., Sugimoto, D., 1984. *Monthly Not. Roy. Astr. Soc.* , 208, 439. ("BS")
Cohn, H., 1985. cf. this volume.
Cohn, H., 1980. *Astrophys. J.* 242, 765.
Djorgovski, S., King, I.R., 1983. preprint.
Djorgovski, S., Penner, H., 1984. preprint.
Heggie, D.C., 1984. *Monthly Not. Roy. Astr. Soc.*, 206, 179.
Hachisu, I., Sugimoto, D., 1978. *Prog. theor. Phys.* , 60,123.
Hénon, M., 1975. Dynamics of Stellar Systems. 69, p. 122, ed. Hayli, A., Reidel, Dordrecht/Holland.
Inagaki, S., Lynden-Bell, D., 1983.*Monthly Not. Roy. Astr. Soc.*, 205, 913. ("ILB")
Lightman, L.P., McMillian, S., 1985. cf. this volume.
Peterson, C.J., King, I.R., 1975. *Astron.J.*, 80, 427.

DISCUSSION

HUT: The Fokker-Planck post-core collapse calculations, which Haldan Cohn reported today did not show gravothermal oscillations. So you expect those to occur also in the Fokker-Planck approach?

BETTWIESER: The mechanism of gravothermal oscillations is not restricted to any specified model realization of a N-body system. Dr. Cohn discussed the class of those post-collapse solutions, which is generated by core heating rates above the upper limit to excite oscillations.

COHN: You said that with too large a normalization of the binary energy input rate the cluster simply expands rather than oscillates. How does your normalization compare with that used by Dr. Heggie?

BETTWIESER: Douglas Heggie uses normalized quantities, while our calculations are in physical terms.

HUT: Your N-body calculation using a few hundred particles in which you found oscillations must be usbject to large fluctuations in temperature, much larger than the temperature difference needed to drive gravathermal oscillations.

BETTWIESER: Concerning the 100-body calcaulation Daiichiro Sugimoto discussed briefly, I think there is a definite indication for core oscillations, on the other hand the level of fluctuations is too large to decide whether gravothermal effects play a role.

HEGGIE: (Addressed to Bettwieser & Sugimoto) I know you have studied the stability of the singular isothermal solution, but I gather from Dr. Sugimoto's question to me yesterday that you have also shown that the Inagaki & Lynden-Bell solution is unstable. Could you summarize your arguments and conclusion about the stability of the Inagaki & Lynden-Bell solution?

SUGIMOTO: I have not calculated stability of the I-L-B solution directly, because it is not an equilibrium solution to which the usual linearized stability analysis can be applied. Moreover, the I-L-B solution does not have any mechanism to control its energy generation rate. Therefore, I talked "stability" in the following senses: i) If the energy generation rate is not exactly equal in its intensity and time change to that required by the I-L-B solution, the expansion does not satisfy the I-L-B solution. ii) If the actual energy generation rate is appreciably lower than required, i.e., lower than heat conduction, the core will contract because the temperature is decreasing outwards and the state is in a stage of finite amplitude gravothermally contracting instability. These are self-evident from physics of the gravothermal catastrophe. We need not to perform any calculation for it. I-L-B's similarity solution is not a solution in equilibrium. In this sense it is difficult to apply concepts of the linearized stability theory. Discussion of stability in the frame of the expanding solution is not always impossible. However, in this particular case the expansion is as fast as the time scale of gravothermal expansion (instability). Therefore, this great amount of energy input controls the behavior of the system. Concerning this point, I have not developed any detailed calculation. We can ask in another way. Is the configuration gravothermally unstable or stable, if the energy input is turned off? This question is easily answered with certainty. Because the temperature is decreasing outward everywhere, the state corresponds to a gravothermal contraction of finite amplitude. It will therefore collapse. In I-L-B's similarity solution no physical mechanism is involved. Therefore, it is impossible to develop a stability analysis due to the associated changes in the energy generation rate.

APPLEGATE: It would be very useful if calculations presented in non-dimensional form would be related to physical units. This would help to determine if gravothermal oscillations last long enough to be observable and in comparing Dr. Bettwieser and Sugimoto's energy generation rate with Dr. Heggie's and Dr. Cohn's.

BETTWIESER: The fact, that there are several galactic globular clusters with rather short relaxation times in their cores is consistent with an interpretation, that these globulars are in post-collapse stages in between collapsed phases. The expanded phases of gravothermal oscillations can be described by standard King models.

SUGIMOTO: Concerning observational points, it is important to distinguish gravothermal expansion from thermal expansion. The gravothermal expansion takes place on the same time scale as the gravothermal contraction which is controlled by the relaxation time in the core. On the other hand the thermal expansion is controlled by the rate of energy input. The gravothermal expansion is much faster than the thermal expansion. Therefore, in the case of the gravothermal expansion and oscillation, most of the globular clusters would be observed in the expanded state without the density cusp. In the case of the thermal expansion, most of the globular clusters are in the slowly expanding state where the density cusp still exists in the central region. This difference may be used to decide observationally which expansion is the case in the real systems. The last figure shown by Bettwieser is in favor of the gravothermal oscillation.

INAGAKI: Inyour calculation if C is small, you have oscillations and if C is large, the core expands for ever. However, in both cases the collapsing phase is self-similar. I do not understand why the difference in energy generation rate causes the qualitative difference in evolution in self-similar stages.

SUGIMOTO: After the first beat of energy input the gravothermal expansion continues without any energy input from binaries. If C is too large, the energy input is so large that the system gets too much energy in the first expansion phase. Then the system will have so much energy that it is not a gravothermal system any more but a thermal system. We can see it clearly in the final phase of such expansion: As more energy is fed, the temperature finally begins to increase as in the case of a normal thermal system.

BINARY FORMATION AND INTERACTIONS WITH FIELD STARS

Piet Hut
Institute for Advanced Study
Princeton, New Jersey 08540, USA

ABSTRACT. Binaries provide an energy source in dense stellar systems. Exothermic gravitational interactions in star clusters can play a role similar to that of nuclear reactions in single stars. These gravitational interactions can be modeled in a laboratory setting, in the form of numerical binary-single star and binary-binary scattering experiments. Gravitational cross sections obtained this way can be applied to model star cluster evolution, just as nuclear cross sections are used as input data in stellar evolution calculations. References are given to detailed descriptions of gravitational cross sections, and a useful new example of an application is given: the rate at which hard binaries form in a homogeneous stellar background, as the solution of an integral equation describing the combined effects of creation, destruction, hardening and softening of binaries.

1. THE ROLE OF BINARIES IN STELLAR DYNAMICS

An interesting analogy can be made between binary reactions in star cluster evolution and nuclear reactions in stellar evolution. In both cases negative binding energy is accumulated in internal degrees of freedom, since exothermic reactions dominate statistically over endothermic reactions. Energy conservation produces positive energy in external degrees of freedom, in the form of excess kinetic energy in the particles remaining as the reaction products. After a few scatterings this excess energy is distributed over the thermal heat reservoir of the star or star cluster. In this way nuclear or binary reactions provide the balance in the energy budget, by replacing the energy lost at the surface by escaping photons or stars, for a star or star cluster, respectively. The analogy is only approximate, since the stars which manage to completely escape from an isolated globular cluster carry only a very small amount of excess energy to infinity. Therefore, the heat loss of a globular cluster core might be described as a local escape of stars from core to halo, or in more physical terms as an outwardly directed heat flow which fuels the expansion of the outer layers.

J. Goodman and P. Hut (eds.), Dynamics of Star Clusters, 231–249.

This analogy is helpful in understanding the overall evolution of star clusters, although there are other important qualitative differences as well. For example, nuclear reaction rates increase at higher density, and the same is true for gravitational binary-single star and binary-binary reactions. However, nuclear reaction rates increase steeply with increasing temperature, whereas binary reactions show the opposite behavior, releasing less energy with higher dispersion velocities in the star system. Even without considering binaries, there are important differences between stellar-dynamical systems and the gaseous interior of a single non-degenerate star. For example, increasing the density at constant temperature will make a star more opaque, thus lowering the conductivity. Increasing the density in a stellar system, while holding the velocity dispersion fixed, will increase the rate of two-body relaxation effects, and thus increase the effective conduction of energy through the system (*cf.* Lynden-Bell and Eggleton, 1980).

With these cautions in mind, I will review our knowledge of binary dynamics on three levels. The first level is a microscopic description of gravitational scattering processes between binaries and single stars in Sect. 2, and between binaries and binaries in Sect. 3. On the second level we can determine local statistical quantities to describe the effects of binaries on their immediate surroundings, in a thermodynamic rather than statistical-mechanics approach, as detailed in Sect. 4. The third level concerns the behavior of binaries in a global and realistic, dynamically changing environment, and the influence of the binaries on the evolution of that environment, as discussed in Sect. 5. Many recent developments on all three levels will be mentioned below. For reviews of earlier work and general background, see Heggie (1975b, 1980), and for a one-page summary of the effects of binaries in star clusters see the allegoric description by Lynden-Bell (1975).

2. BINARY-SINGLE STAR SCATTERING

Let us start with an intuitive definition of a differential cross section $\frac{d\sigma}{dy}$ for gravitational binary-single star scattering in which the binary binding energy changes by a net amount y. This definition can be introduced by closely following the language of laboratory experiments in atomic (nuclear, particle,...) physics, even though our astrophysical laboratory is a purely numerical one modeled by a computer. For simplicity let us consider gravitational interactions between stars in the approximation of point particles, postponing a discussion of internal degrees of freedom and finite-size effects of the stars to Sect. 5.

First we have to set up a target "plate" filled with binary stars, each well separated from the others so that each binary can be treated as an unperturbed two-body system. Next we prepare a bundle of single stars which all move parallel to each other. Again we only consider low-intensity beams with interstellar distances large enough to guarantee sufficiently undisturbed rectilinear motions.

We can now perform an experiment by aiming a bundle of single stars at right angles to a target plate of binaries, and subsequently observing the characteristics of the debris of the reactions. Many different kind of experiments are possible, with different prescriptions for the parameters describing the incoming bundle and the target plate. Let us start with a simple case. Take a monochromatic bundle of stars, all of equal mass m_3 (*i.e.* all stars have the same energy and hence the same velocity). Prepare a target plate of binaries which all consist of two stars with masses m_1 and m_2. These two masses do not have to be equal but are the same for all binaries in the target. Furthermore, give all binaries the same binding energy x, or equivalently the same semi-major axis.

When performing this experiment, we can measure the energy with which the debris leaves the target. This debris will consist of single stars and binaries. Occasional triple systems might emerge (only if the energy of an incoming star is lower than the binding energy of a binary), but they are all unstable, and will sooner or later decay in a single star and a binary. The reason is the absence of dissipation or simply the time-reversibility of Newton's equations: stable hierarchical systems are stable in the future *and* in the past. Therefore a capture will lead to an escape in the generic case, with the exceptions occupying a lower-dimensional subspace in the space of initial conditions (Chazy, 1929) which implies a vanishing probability of occurrence for any finite number of experiments.

Let us first study those scattering events where the binaries are left intact, and the same incoming star escapes again (possibly after a temporary capture). In this case we define the differential cross section $\frac{d\sigma}{dy}(x,y;V)$ as follows: the expression

$$\frac{d\sigma}{dy}(x,y;V)dy$$

gives the total cross section for all scattering processes which start with a binary of energy x and an incoming star with velocity V (with respect to the center of mass of the binary), and end with the same binary having an energy between y and y + dy.

As in the familiar case of atomic physics, this total cross section is defined as follows: Take a target plate with a large number of replicas of the same binary, but with random orientation, phase, and eccentricity (according to a thermal distribution, *i.e* $f(e) \propto e$, *cf*. Heggie, 1975a). Now aim a beam of single stars at this target, and measure the fraction of stars causing a transition into the right energy range y,y + dy. Multiply the beam area by this fraction and divide by the number of target binaries, and *voilà*: this is the (effective) cross section.

Of course, rather than taking up cosmic billiards, it is simpler to perform computer experiments to determine three-body scattering cross sections. Instead of a dilute target plate of binaries, consider now a large number of replicas of the same binary, again with random phase, etc. Now shoot (*i.e.*, calculate numerically the orbit of) one single star at each of these binaries, with impact vector $\vec{p}$ chosen randomly within a fixed area A perpendicular to the incoming direction ($\vec{p}$ measures the amount and direction of the off-center aim at infinity, with respect to the center of mass of the binary; $p = |\vec{p}|$ is the impact parameter).

In the limit of an ensemble consisting of an infinite number of replica's of such a system, the *fraction* of the experiments resulting in a new binary binding energy between y and y + dy is by definition

$$\frac{1}{\|A\|}\frac{d\sigma}{dy}\,(x,V;y)dy$$

where $\| A \|$ is the total surface area of A, and with the *caveat* that only those area's A may be used which are large enough to intersect *all* orbits of incoming stars which can possibly cause a transition to a new binary binding energy in the range y,y + dy. This is the precise definition of a cross section as an "effective surface area." Of course, for indificual experiments the above expression describes only the *probability* for the stated outcome.

Here we have arrived at the central problem of numerical scattering experiments: how does one determine the maximum impact parameter? An underestimate will result in too small a measured value of the cross section, since some of the relevant scattering processes are left out. An overestimate, however, will waste computer time and can easily make the number of "hits" too small to be statistically significant. Unfortunately, there is an extra complication. It would be extremely inefficient to perform a whole new series of numerical experiments for any additional question one wants to ask. Instead, the only reasonable way to get simultaneous information about many different parameters of interest is to mimic a particle accelerator experiment: first run a series of continuous experiments, the outcome of each of which is stored on tape; then analyze the one set of data later in a number of different ways. And there is the problem: different questions asked afterwards require different values for the maximum impact parameter...

This problem can only be solved satisfactorily by a combination of two approaches. First, the density of numerical experiments can be made to fall off with increasing impact parameter, so that one has a better coverage of near-central collisions than of rather wide encounters. Secondly, an extended (and, alas, time-consuming) series of test runs can be carried out in *every* new and previously unexplored parameter domain (*e.g.* for different incoming velocity of the third star, different eccentricity of the binary, etc.).

Given these formidable difficulties, a systematic and unbiased survey of three-body scattering became possible only recently with decreasing cost of computer time. The pioneer in the field of gravitational three-body scattering is Hills, who was the first to accumulate scattering statistics, starting in the sixties. In the light of the central problem outlined above, Hills chose to perform the majority of his experiments for binary orbits with zero initial eccentricity and at zero impact parameter, *i.e.* most of his collisions were exactly head-on. In this way he obtained a wealth of information, and obtained the first numerical three-body scattering cross sections (Hills, 1975) by making estimates about the dependence of energy exchange as a function of impact parameter based on a smaller amount of off-axis scattering experiments. However, it was not possible to accurately estimate the systematic errors introduced by the zero-impact-parameter approach. Other early papers reporting three-body experiments (*e.g.* Saslaw *et al.*, 1974, followed by Valtonen, 1975) did not express the results in a form which could be translated into cross sections.

In the same period Heggie (1975a) published an extremely detailed and comprehensive paper which is still *the* standard reference on analytical approximations for the determination of three-body scattering cross sections. He used a large arsenal of different approximations in many different parameter regimes, such as the use of impulsive approximations, of adiabatic invariants, and of phase-space-volume methods where the memory of initial states was assumed to be lost. Monaghan (1976a,b; 1977) applied a slightly different technique to the analysis of stochastic decay of (temporarily) bound three-body systems.

An extensive series of numerical experiments without any *a priori* limitations on the initial conditions was carried out much later, when large amounts of computer time were available. Here several months of C.P.U. time on a VAX 11/780 computer enabled literally millions of scattering experiments to be performed (Hut and Bahcall, 1983). To the extent that the results of these experiments had been predicted by Heggie (1975a), they agreed remarkably well, generally within a factor two and often significantly better. Of course, only the numerical results could accurately determine the behavior in the transition regions between the domains of validity of Heggie's analytic estimates, but the details of the asymptotic behavior had been correctly predicted. Some additional analytic work was needed to describe and interpret some other details such as the dependence of differential cross sections on eccentricity of the original binary orbit, and the behavior of exchange scattering, and was published by Hut (1983a).

Another result of the more recent calculations answered a controversy which had existed in the literature for many years, concerning the average amount of energy increase of hard binaries in a field of single stars. Hills' (1975) numerical results fell below Heggie's (1975a) analytical prediction, yielding a value only 40% of that of Heggie's estimate, and it was not clear how this difference arose (*cf.* Spitzer and Mathieu, 1980, who decided to use a value 60% as large as Heggie's estimate in their model calculations of globular cluster evolution). Was Heggie's analytical assumption of phase space mixing for strong scattering not accurate enough or did Hills' choice of zero impact parameter for most of his numerical experiments introduce systematic effects? The answer turned out to depend somewhat on the degree of hardness of the binary. For only moderately hard binaries ($E_{bin} \sim 10kT$) an average binding energy increase of 60% of Heggie's estimate was measured, in closer agreement with Hills' results. For very hard binaries ($E_{bin} \sim 500kT$), however, the measured value rose to 80% of Heggie's estimate (Hut, 1983b, 1985). Indeed, Spitzer and Mathieu's (1980) inspired guess turned out to have been the most accurate overall, especially since they modeled the effects of moderately hard binaries.

As a guide to the literature of gravitational three-body scattering, the following list names those papers which describe more than a thousand numerical experiments each: Saslaw *et al.*, 1974; Hills, 1975, 1983a,b; Heggie, 1975a; Valtonen, 1975; Valtonen and Aarseth, 1977; Valtonen and Heggie, 1979; Hills and Fullerton, 1980; Fullerton and Hills, 1982; Hut and Bahcall, 1983; Hut, 1983b,c, 1984a,b,c; Hut and Paczynski, 1984; Bahcall *et al.*, 1985.

The most detailed information available at present is an atlas of equal-mass differential scattering cross sections, with nearly a hundred different graphs which show separately the contributions of all kind of different processes to the final differential cross sections for energy exchange between binary and single star in a three-body scattering experiment (Hut, 1984a).

All experiments mentioned above involved a Monte Carlo sampling of initial conditions, in order to obtain a physicist's description of gravitational three-body scattering in terms of cross sections. From the point of view of a mathematician interested in the three-body system as a dynamical system (*cf.* Alekseev, 1981), additional insight into the extremely rich microscopic structure of the space of orbits has been obtained from a series of experiments for a grid of initial conditions, determined by stepwise varying several parameters independently while keeping the other parameters fixed (Hut, 1983c).

Specific astrophysical applications of three-body scattering experiments have been discussed recently in a variety of papers. The effects of three-body scattering on the formation and evolution of X-ray sources in globular clusters are discussed by Hut and Verbunt (1983a,b); Krolik(1983, 1984); Krolik *et al.* (1984); Hut and Paczynski (1984), and earlier references which can be found in these papers. The effects of perturbations of passing field stars on wide binaries are discussed by Hut (1984b) and Bahcall *et al.* (1985).

The analogy between gravitational scattering and atomic physics can be extended to a classification of the intermediate and final states. For example: ionization, charge exchange and resonant scattering all have their gravitational counterpart (Heggie, 1975a; Hut and Bahcall, 1983). Also, detailed-balance relations between each scattering process and its reverse can be derived in similar fashion (Heggie, 1975a). An important and interesting practical difference is that in gravitational scattering classical mechanics causes computational difficulties where quantum mechanics makes life easier. In classical mechanics there do not exist simple highly symmetric low-lying energy levels (let alone a spherically symmetric ground state). However, also in atomic physics there are cases where classical approximations are called for, when studying scattering processes with atoms where the outer electron occupies a very highly excited state. The techniques used in some of these approximations (*cf.* Shakeshaft and Spruch, 1979 and references therein) have been adopted successfully for the gravitational scattering case (Hut, 1983a; Heggie and Hut, 1975).

3. BINARY-BINARY SCATTERING

The number of different types of outcome is much larger in the case of four-body scattering than for three-body scattering. Not only are there more dynamical degrees of freedom and therefore more possibilities to exchange energy (and to exchange stars themselves), there is also a qualitatively new type of outcome: the formation of hierarchical triples. In binary-single star scattering, a temporary triple system can be formed but will "always" decay (in the physical, not in the mathematical sense; see the previous section). A heuristic way to realize this is to consider time reversal: a hierarchical stable triple will generally be stable in both future and past, and can therefore not have been formed in a single binary-single star scattering event. However, it is possible to shoot a single star at a hierarchical triple in such a way that two binaries result which subsequently move away from each other. Because this process is not an exceptional one requiring fine-tuning, also the reverse will occur regularly: the formation of stable hierarchical triples in binary-binary scattering (Mikkola, 1983, finds a relative rate of 20% for this process among strong interactions between hard binaries).

The first four-body scattering experiments were reported by Harrington (1974) and Saslaw *et al.* (1974), who were mainly interested in the details of the decay modes and did not present their results in a form suitable to distill scattering cross sections. Only very recently has progress been made in this field, where the numerical difficulties are so much more formidable than in the three-body scattering case. Hoffer (1983) has reported the results of 40,000 binary-binary scattering experiments. His results mainly concern soft binary encounters, limited to binary orbits of zero eccentricity; his hard binary encounters were not followed long enough to determine the final outcome. Following Hills (1975), Hoffer carried out part of his scattering experiments at zero impact parameter, which introduces another uncertain element in the derivation of cross sections. His main results are a determination of the average amount of energy exchange between the binding energy of the binaries and the kinetic energy of the single stars and center-of-mass motion of the binaries emerging from the scattering process, for a variety of different mass combinations.

Mikkola (1983a,b; 1984a,b,c) has reported the results of ca. 20,000 binary-binary scattering experiments. He has limited himself to considering only stars of equal mass, but at the same time treating the scattering process to a very high degree of accuracy, using sophisticated regularization techniques, some of which of a type which has not been applied to N-body experiments before. Most interesting is the simultaneous four-body regularization proposed by Heggie (1974), where additional dimensions are introduced to simultaneously eliminate *all* possible two-body collision singularities, as a generalization of the Kustaanheimo-Stiefel regularization (*cf.* Stiefel and Scheifele, 1971; Stiefel, 1973). This is the best type of coordinate transformation and dimensional extension possible, since it has been shown that three-body collision singularities can not be regularized (*cf.* Siegel and Moser,

1971). However, the equations of motions are considerably more complicated in this formalism, where the relative motion of the four bodies are described by 49 first-order equations (!), as compared to 18 in the standard Newtonian formalism (where the three relative positions and velocities have three Cartesian components each).

Mikkola has obtained cross sections for a variety of processes, for soft as well as hard binaries. These processes can be divided in four categories, in which (1) two binaries emerge from the scattering experiments (containing either the original or exchanged binary members); (2) one star escapes and a hierarchical triple is formed; (3) two single stars escape leaving a binary behind; (4) all four stars escape in different directions. The final errors in his cross sections are a combination of statistical and systematic effects. Statistical limitations follow from the limited number of experiments for each choice of parameters, while systematic uncertainties are introduced because several percent of the experiments were halted before the outcome was determined because of limitations on computer time. Nonetheless, the resulting cross sections represent a major improvement in our knowledge of binary-binary scattering.

4. FORMATION AND EVOLUTION RATES OF BINARIES IN A STATIC STELLAR BACKGROUND

The scattering cross sections reviewed in the previous two sections contain an overwhelming amount of information which can be applied to the modeling of dynamically evolving star clusters, as is discussed in the next section. But before applying these cross sections to a complicated realistic system, it is illuminating to first consider a simplified application, by studying the behavior of binaries in a homogeneous, static background of field stars. This approach, although informative, is formally not really correct since in stellar dynamics as well as in gas dynamics no consistent solutions exist which allow a static homogeneous distribution of matter. Therefore this assumption has been called the "Jeans swindle," after Jeans (1902) who took up the same assumption of a static background to derive linear stability criteria against gravitational collapse of gas clouds.

Before applying the three-body reaction rates to a static backdrop of single stars, let us quickly classify the main types of binaries and their average behavior. The most important feature of binary--single star scattering is energy exchange between external and internal degrees of freedom. Hard binaries, with an orbital velocity much larger than typical field star velocities, behave differently from soft binaries, for which the orbital velocity is much lower than that of the field stars. A fundamental law of three-body stellar dynamics is: *hard binaries tend to become harder while soft binaries tend to become softer*. This can be described heuristically by the following equipartition argument (Gurevich and Levin, 1950; Heggie, 1975a).

A fast star moving past a slowly revolving binary will on average lose some energy to the binary. However, trying to speed up the binary members will put them in wider orbits with an actually *lower* velocity (loosely speaking a Kepler orbit seems to have a negative 'specific heat', a general phenomenon for gravitational interactions; *cf.* Lynden-Bell, 1973). Hard binaries, on the other hand, can capture a slowly incoming field star under formation of a bound triple system. After some time, generally orders of magnitude longer than the initial binary period, one of the stars is ejected more or less stochastically. The velocity with which it reaches infinity is typically of order of the internal binary velocities, and therefore much larger than the initial field star velocity. The binary has to increase its binding energy in order to give off this energy, thereby shrinking and increasing its orbital velocity. Of course, not all hard binaries harden during each encounter with a field star, nor do soft binaries loosen up monotonically; both processes take place for both types of binaries, but the net energy balance has a different sign.

We are now in a position to predict the formation and subsequent evolution of a binary population against a background of single stars. Starting without any binaries, very soft binaries will soon be formed in large numbers, simply because the phase space available increases rapidly with decreasing binding energy. The easiest way to see this is to start with the quantum mechanical description of the hydrogen atom, and to approach the classical limit by increasing the size of the atom. In this way, by taking the limit of very highly excited states, the correspondence principle tells us that we will arrive at the proper classical description (*cf.* Heggie, 1972). Another way of seeing this is by recalling the way in which quantization was first attempted by Bohr and Sommerfeld who basically counted volumes in phase space in units of Planck's constant h, using the classical action integrals

$$\oint p_i \, dq_i = n_i h$$

over a full period of the generalized coordinates q_i, with the quantization conditions of integer values for the n_i. Thereby they computed the degeneracy of each energy level as being proportional to the *classical* amount of phase space available at that energy. Seen in this light, mentioning the correspondence principle is a sophisticated way of expressing that Bohr and Sommerfeld did our classical homework already.

The argument runs as follows. The distribution function $f(E)$ which describes the population of energy levels of the hydrogen atom in the classical limit can be written as

$$f(E) = w(E)\rho(E)e^{-E/kT}$$

where the three factors at the right are the statistical weight function, which measures the amount of degeneracy within one energy value, the density of levels of different discrete energy values, and the thermal Boltzmann factor, respectively.

The density of energy levels can be found simply from the relation between the energy E and the principal quantum number $n \gg 1$:

$$E \propto \frac{1}{n^2}$$

which implies

$$\rho(E) = \frac{dn}{dE} \propto E^{-3/2} \quad .$$

The degeneracy per energy level can be counted directly from the range of the quantum numbers $l = 0,1....,n-1$ and $m = -l,....+l$ as

$$w(n) = \sum_{0}^{n-1} (2l + 1) = n^2$$

and therefore

$$w(E) \propto E^{-1} \; .$$

Combining the equations above gives the correct classical distribution function for binary stars describing a state of thermal equilibrium:

$$f(E) \propto E^{-5/2} e^{-E/kT} \; .$$

It is immediately obvious that the equilibrium distribution found above is rather pathological: it diverges *both* for very hard and for very soft binaries! For soft binaries the divergence is not a serious problem. As mentioned above, we expect many very soft binaries to form in very wide orbits since so much phase space is available for them. These binaries form and break up at such high rates that most of them only traverse a tiny part of one orbital revolution before breaking up again, and for all practical purposes these short, formally momentarily bound encounters between stars play no role of any significance.

The divergence at the hard side, for binaries with $E >> kT$ is much more serious. In the limit of point masses, a binary can absorb an unlimited amount of (negative) binding energy, which automatically translates via the Boltzmann factor into an unlimited population of these low-lying energy levels. Here we have a clear indication that in practice a self-gravitating system can never attain true thermodynamic equilibrium, even though it perpetually strives to reach that state, as we will see below ("the great tragedy of self-gravitating systems," as Jeremy Goodman expressed it).

What will happen instead? Starting from a background of single stars only, at first many soft binaries will be created, most of which will be destroyed soon afterwards. After a while, some binaries will harden against the odds, and an occasional binary might be formed at an already somewhat higher binding energy. Those two effects will provide a slow but steady increase in the number density of hard binaries, in an effort to approach the equilibrium distribution. For hard binaries, at any given value for the binding energy $E_{bin} >> kT$ a balance will be reached between the incoming flux of binaries which were softer and are hardening to reach E_{bin} and the outgoing flux of binaries which already had an energy around E_{bin} and are presently moving towards even larger values of binding energy. The flux in the opposite direction is much smaller, since hard binaries predominantly tend to harden.

Heggie (1975) showed that for hard binaries the average energy increase per strong scattering encounter varies inversely proportional to the cross section for such an encounter. Therefore, even though the cross sections for strong encounters decrease with time, on average hard binaries harden at a constant rate. This implies a flat distribution of binaries, all traveling at the same speed towards larger values of binding energy, as modeled by Retterer (1980ab, 1984) who used the strong-scattering limit of Heggie's analytic expressions for the binary hardening rates. Goodman and Hut (1985) have followed up on Retterer's work using the results of detailed numerical experiments. These results were directly derived from the experimental data and were used in tabulated form; for practical reasons they are published in graphical form (Hut, 1983b, 1984a, 1985). Whereas Retterer calculated the time-dependent progressive hardening of binaries, Goodman and Hut solved an integral equation for the time-independent steady-state solution. This integral equation forces an exact balance at every value of the binary binding energy between the binaries arriving at that energy value (via creation, hardening or softening) and the binaries leaving that energy value (via ionization, hardening or softening). Numerically, this equation was solved by discretization and matrix inversion.

Fig. 1 shows both the equilibrium distribution (short-dashed line) and the steady-state solution (full line) as derived by Goodman and Hut (1985). Indeed, the steady-state solution approaches a constant value for large $E_{bin} \gg kT$ and thereby falls far below the exponentially rising equilibrium distribution. Fig. 1 agrees qualitatively with Fig. 3 of Retterer (1980a), with a quantitative difference of about a factor of two in the near-constant value of the distribution function for very hard binaries with $E_{bin} = 100$ kT: Retterer found a value of 0.7 whereas Goodman and Hut found a value of 0.3 (in Retterer's units).

The long-dashed line in Fig. 1 indicates the effects of introducing a finite escape velocity. For very hard binaries ($E_{bin} > 100$ kT) there is a significant chance that a binary will escape from the cluster after a strong encounter because of the large recoil velocity, which explains the rapid fall-off at large x-values. Introducing a finite escape velocity in a homogeneous background amounts to adopting a model of a star cluster in a finite-depth square-well potential. This computation (Goodman and Hut, 1985) forms the logical next step following Spitzer and Härm's (1958) derivation of the rate of escape of single stars due to two-body encounters in a finite-depth square-well potential: it gives the rate of escape of binaries due to three-body encounters, as discussed below. Figs. 2 and 3 show that the formation of full-grown hard binaries is not unlike that of most plants and simpler animals: many are created but while they grow up they are decimated at every stage. A quantitative measure is given in the figure captions of Figs. 2 and 3, showing that most hard binaries are born and have spent their youth in the dangerous areas where $E_{bin} \ll kT$. Only those few which manage to grow harder than 2.9 kT finally face favorable odds for survival.

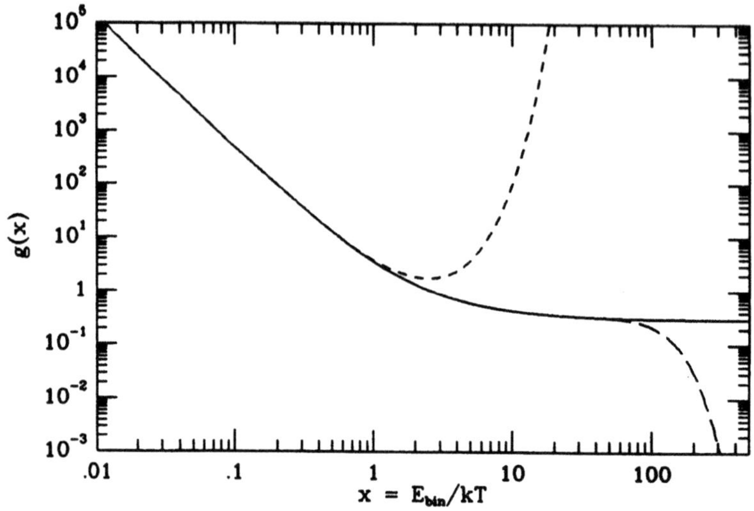

Fig. 1. The steady-state distribution of binaries g(x) in a homogeneous static stellar background. The binary binding energy x is given in units of $kT = m\sigma^2$, where σ is the one-dimensional velocity dispersion of the stars all of which have mass m. The short dashes follow the equilibrium distribution. The long dashes indicate the effects of a finite square-well potential, of depth $\phi = 6kT/m$.

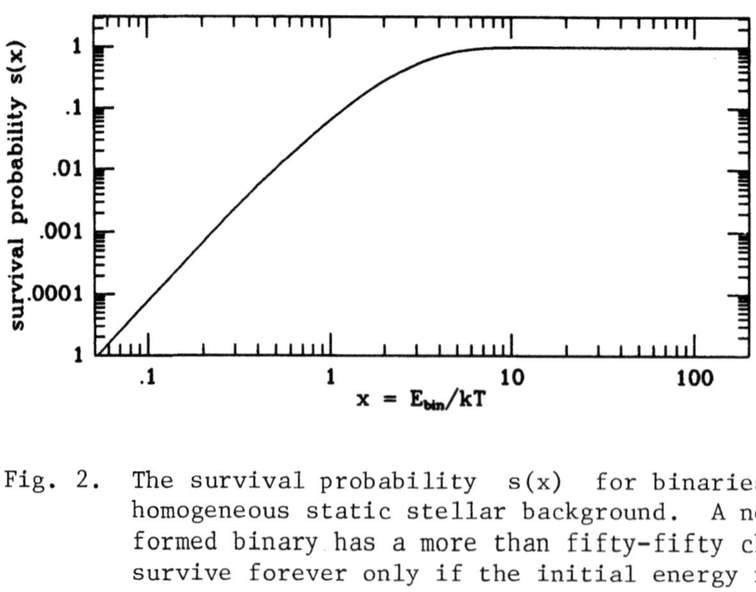

Fig. 2. The survival probability s(x) for binaries in a homogeneous static stellar background. A newly formed binary has a more than fifty-fifty chance to survive forever only if the initial energy $x > x_s = 2.9$.

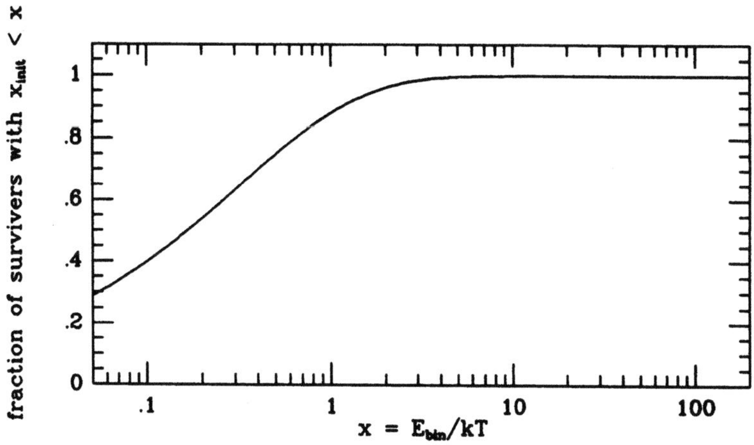

Fig. 3. The ordinate shows the fraction of permanent binaries which are born with an energy x_{init} less than x. The value $x = x_c = 0.17$ marks the median energy for the creation of binaries which survive forever.

For the modeling of the energy generation in star clusters, the most important result of these new calculations presented in the figures is an accurate determination of the formation rate of permanent hard binaries. The value found by Goodman and Hut (1985) for the creation rate of these "immortal" binaries is

$$C = (0.90 \pm 0.05) \frac{n^3 G^5 m^5}{\sigma^9}$$

per unit volume and per unit time, where n denotes the density of field stars, each of which has a mass m, σ the one-dimensional velocity dispersion of the field stars, and G the gravitational constant. The fact that the coefficient is so close to unity is somewhat fortuitous; if we had chosen to express our result in terms of the three-dimensional velocity dispersion instead, the coefficient would have been two orders of magnitude larger. In astrophysical units the creation rate of permanent binaries reads

$$C = (1.36 \pm 0.08) \times 10^{-12} \left(\frac{n}{10^5 pc^{-3}}\right)^3 \left(\frac{m}{M_\odot}\right)^5 \left(\frac{10 km/s}{\sigma}\right)^9 pc^{-3} yr^{-1}.$$

Heggie guessed a value of 1.3×10^{-12} for the coefficient in this expression (Heggie, 1980, eq. 19), adding that "the numerical coefficient was not well established." The fact that his guess turned out to lie within the tight error bars of the experimental value is especially remarkable because of the extreme sensitivity to the choice of natural units, given the high powers occurring in the equations above. Clearly, only at the last stages of core collapse can densities be reached which are high enough to dynamically form hard binaries, and then only in a limited volume with dimensions much less than a parsec (*cf.* Heggie's review article in these proceedings).

5. FORMATION AND EVOLUTION RATES OF BINARIES IN A DYNAMIC STELLAR BACKGROUND

In a realistic self-gravitating background of single stars, binary formation and subsequent hardening will be much more complicated than the simple picture sketched in the previous section for a static homogeneous background. Two major complications arise. Within the approximation of point particles the first problem of calculating the self-consistent interplay of binaries and single stars in the combined evolving potential is already rather formidable. However, a realistic description has to take into account the fact that stars can tidally capture each other, and can physically collide and sometimes even merge in the process. This second problem leads us outside stellar dynamics and introduces many uncertainties concerning mass loss, tidal effects, stellar evolution in (detached, semi-detached or contact) binaries, etc. Both problems are addressed in a variety of different approaches in a number of different contributions to the present symposium, and no attempt will be made here to review all these approaches, most of which are very recent and are still under development. Spitzer's review of pre-core-collapse models and Heggie's review of post-core-collapse models in the present proceedings give many references to the literature.

Nearly a quarter century ago Hénon (1961) published his Ph.D. thesis, in which he not only predicted the phenomenon of core collapse, but even went so far as to suggest the solution, at least on the level of point-particle approximations: the formation of binaries. It is encouraging that we now have begun to finally fill in the quantitative details of Hénon's suggestion, thereby solving a long-standing mathematical physics problem, so simple in its formulation: what is the long-term evolution of a system of self-gravitating point masses?

The non-gravitational effects which are expected to be crucially important in the later stages of the evolution of globular clusters are only beginning to be explored now (*cf.* Ostriker's review in the present proceedings). Possible complications such as *e.g.* repeated merging of stars leading to the formation of a relatively massive star are essentially unexplored as yet. Such a star might evolve on a time scale of order a million years before undergoing a supernova explosion which will cause a significant amount of mass to escape from the globular cluster core. These effects cannot yet be accurately estimated but they might turn out to have an importance comparable to that of stellar ejection by three-body encounters.

Now that our understanding of core collapse of globular clusters has improved dramatically over the last few years, the stage is set for further investigations as to the precise character of the central energy source. We know that stars can shine mainly because they utilize nuclear energy and in some cases energy from gravitational contraction. What maintains the central oven which enables globular clusters to boil off stars continuously at their surface: gravitational 'binary burning,' or stellar collisions and explosions? The answer is likely to contain elements from either type, and theoretical studies should explore these and other possibilities. The prospect of observational data with unprecedented resolution from Space Telescope (*cf.* Bahcall, these proceedings) forms an extra encouragement to study the complicated dynamics of the long-term evolution of globular clusters.

ACKNOWLEDGEMENTS

I acknowledge many lively discussions with members and visitors of the I.A.S., especially Haldan Cohn, Jeremy Goodman, Douglas Heggie, Shogo Inagaki and Seppo Mikkola, and I thank Jeremy Goodman for comments on the manuscript. This work was supported in part by the National Science Foundation through grant PHY-8440263

REFERENCES

Alekseev, V. M. 1981, *Amer. Math. Soc. Transl. (2)* 116, 97. (Translation of Ninth Math. Summer School (Kaciveli, 1971), Izdanie Inst. Mat. Akad. Nauk Ukrain. SSR, Kiev, 1972, pp 212-341)

Bahcall, J.N., Hut, P. and Tremaine, S.D. 1985, to be published in *Astrophys. J.*

Chazy, J. 1929, *J. Math. Pures Appl.* 8, 353.

Fullerton, L.W., and Hills, J.G. 1982, *Astron. J.* 87, 175.

Goodman, J. and Hut, P. 1984, to be submitted to *Astrophys. J.* (paper V).

Gurevich, L.E. and Levin, B.Yu., 1950, *Astr. Zh.* 27, 273; transl. in *NASA TT F-11*, 541.

Harrington, R.S. 1974, *Celest. Mech.* 9, 465.

Heggie, D.C. 1972, Ph.D thesis, Cambridge, U.K.

Heggie, D.C. 1974, *Celest. Mech.* 10, 217.

Heggie, D.C. 1975a, *Mon. Not. R. astr. Soc.* 173, 729.

Heggie, D.C. 1975b, in *Dynamics of Stellar Systems*, IAU Symposium No. 69, ed. A. Hayli, p. 73 (Dordrecht: Reidel).

Heggie, D.C. 1980, in *Globular Clusters*, ed. D. Hanes and B. Madore, p. 281 (Cambridge, U.K.: University Press).

Heggie. D.C. and Hut, P. 1985, to be submitted to *Astrophys. J.* (paper IV).

Hénon, M. 1961, *Ann. Astrophys.* 24, 369.

Hills, J.G. 1975, *Astron. J.* 80, 809.

Hills, J.G. 1983a, *Astron. J.* 88, 1269.

Hills, J.G. 1983b, *Astron. J.* 88, 1857.

Hills, J.G., and Fullerton, L.W. 1980, *Astron. J.* 85, 1281.

Hoffer, J.B. 1983, *Astron. J.* 88, 1420.

Hut, P. 1983a, *Astrophys. J.* 268, 342 (paper II).

Hut, P. 1983b, *Astrophys. J. Lett.* 272, L29.

Hut, P. 1983c, *Astron. J.* 88, 1549.

Hut, P. 1984a, *Astrophys. J. Suppl.* 55, 301.

Hut. P. 1984b, scheduled for publication in *Nature*.

Hut, P. 1984c, in *International Symposium G. Lemaître*, Louvain-La-Neuve, Belgium, October 1983 (Dordrecht: Reidel).

Hut, P. 1985, to be submitted to *Ap. J.* (paper III).

Hut, P. and Bahcall, J.N 1983, *Astrophys. J.* 268, 319 (paper I).

Hut, P and Paczyński, B. 1984, *Astrophys. J.* 284,

Hut, P. and Verbunt, F. 1983a, *Nature* 301, 587.

Hut, P. and Verbunt, F. 1983b, in *Cataclysmic Variables and Low-Mass X-Ray Binaries*, Seventh North American Workshop, ed. D.Q. Lamb and J. Patterson (Dordrecht: Reidel)

Jeans, J. 1902, *Phil. Trans. Roy. Soc.* 199A, 49.

Krolik, J.H. 1983, *Nature* 305, 506.

Krolik, J.H. 1984, *Astrophys. J.* 282, 452.

Krolik, J.H., Meiksin, A, and Joss, P.C. 1984, *Astrophys. J.* 282, 466.

Lynden-Bell, D. 1973 in *Dynamical Structure and Evolution of Stellar Systems*, ed. L. Martinet and M. Mayor, Third Advanced Course of the Swiss Society of Astronomy and Astrophysics at Saas Fee (Geneva Observatory, Sauverny, Switzerland).

Lynden-Bell, D. 1975, in *IAU Symposium 69, Dynamics of Stellar Systems*, ed. A. Hayli, (Dordrecht: Reidel), p. 443.
Lynden-Bell, D. and Eggleton, P.P. 1980, *Mon. Not. R. astr. Soc.* 191, 483.
Mikkola, S. 1983a, *Mon. Not. R. astr. Soc.* 203, 1107.
Mikkola, S. 1983b, *Mon. Not. R. astr. Soc.* 205, 733.
Mikkola, S. 1984a, *Mon. Not. R. astr. Soc.* 207, 115.
Mikkola, S. 1984b, *Mon. Not. R. astr. Soc.* 208, 75.
Mikkola, S. 1984c. Ph.D. thesis, Turku, Finland.
Monaghan, J.J. 1976a, *Mon. Not. R. astr. Soc.* 176, 63.
Monaghan, J.J. 1976b, *Mon. Not. R. astr. Soc.* 177, 583.
Monaghan, J.J. 1977, *Mon. Not. R. astr. Soc.* 179, 31.
Retterer, J.M. 1980a, *Astron. J.* 85, 249.
Retterer, J.M. 1980b, Ph.D. thesis, Univ. of California, Berkeley.
Retterer, J.M. 1984, preprint.
Saslaw, W.C., Valtonen, M.J., and Aarseth, S.J. 1974, *Astrophys. J.* 190, 253.
Shakeshaft, R., and Spruch, L. 1979, *Rev. Mod. Phys.* 51, 369.
Siegel, C.L, and Moser, J.K. 1971, *Lectures on Celestial Mechanics* (Springer: Berlin, Heidelberg, New York).
Spitzer, L., and Härm, R. 1958, *Ap. J.*, 127, 544.
Spitzer, L., and Mathieu, R.D. 1980, *Astrophys. J.* 241, 618.
Stiefel, E.L. 1973, in *Recent Advances in Dynamical Astronomy*, ed. B.D. Tapley and V. Szebehely (Reidel: Dordrecht), p. 3.
Stiefel, E.L. and Scheifele, G. 1971, *Linear and Regular Celestial Mechanics* (Springer: Berlin, Heidelberg, New York).
Valtonen, M.J. 1975, *Mem. R. astr. Soc.* 80, 61; 77.
Valtonen, M.J., and Aarseth, S.J. 1977, *Rev. Mex. Astron. Astrophys.* 3, 163.
Valtonen, M.J., and Heggie, D.C. 1979, *Celest. Mech.* 19, 53.

DISCUSSION

KING: With Space Telescope we should be able to resolve, in nearby globular clusters, soft binaries with E/kt less than a few tenths. Is it reasonable to look for these binaries?

HUT: Unfortunately, this will not be a very interesting study from a dynamical point of view because the equilibrium distribution of wide binaries is nearly the same as that of a random distribution of uncorrelated single stars. The excess correlation is given by the Boltzmann factor exp $(E_{bin}/kT) << 1$ for soft binaries.

TERLEVICH: In order to apply your calculations to open clusters and very young globular clusters; have you thought about including a mass function and therefore mass loss from stellar evolution?

HUT: Yes, I have thought about that, but I have not yet found the time to include these effects.

SPITZER: For the similar problem of H-atom ionization, the infinite number of high-n atoms is eliminated by consideration of the perturbation experienced by atoms with very extended electron orbits. Have you considered similar effects which would eliminate the infinity in the computed population of soft binaries?

HUT: Yes, and I can quote a result recently derived by Scott Tremaine (private communication): In every self-gravitating system the number of binaries which survive long enough for the two stars to complete at least one revolution around each other, is of order unity ($\sim$6) independent of N, the total number of stars in the system. The much larger number of very soft binaries expected theoretically does exist only temporarily and is formed by random short-living encounters.

SHAPIRO: How is your steady-state solution modified if, still in the context of the Newtonian point-mass approximation, you included the dissipative (hardening) effect due to gravitational radiation in binaries? One might imagine systems which, though nonrelativistic ($v/c << 1$), decay due to gravitational radiation on timescales *shorter* than other relevant timescales (i.e., relaxation) in the problem. Have you considered this effect?

HUT: I have not yet included this effect. If one were to include tidal formation of very close binaries, this effect could be important, together with other effects such as magnetic braking. Very hard binaries in globular clusters are also in this respect analogous to cataclysmic variables (*cf*. Hut & Verbunt, 1983, *Nature* 301, 587).

LARSON: This is a remark related to Shapiro's question about possible effects of gravitational radiation. In my poster presentation I speculate that the cores of globular clusters are dominated by heavy remnants with a spectrum of masses possibly extending up to several thousands of solar masses. The two most massive objects are likely to form a central binary that absorbs a substantial fraction of the binding energy of the system. It was recently pointed out to me by Kip Thorne that such a binary consisting of two moderately massive black holes would decay by gravitational radiation in a time shorter than the Hubble time. So maybe gravitational radiation would have significant effects in a condensed core of heavy remnants.

GOODMAN: The question of energy loss by gravitational radiation was set as an exercise in Jerry Ostriker's dynamics class, and I performed it. For point masses, the amount of gravitational radiation lost per relaxation time is comparable to the binding energy of the cluster. However, the amount of radiation energy released has such a steep dependence on impact parameter that the effect goes away when one puts in a finite size (e.g., 10km) for the stars.

JERNIGAN: Soft binaries are not clearly observable. How many hard binaries will be formed in a real cluster? How does one define a hard binary in light of the range of field parameters?

HUT: In a real cluster at most a few dynamically formed hard binaries will be present in the core at any given time. A hard binary is defined as having a binding energy $E_{bin} \gg kT = m\sigma^2$ where the one-dimensional velocity dispersion σ is a *local* average.

DIRECT N-BODY CALCULATIONS

Sverre J. Aarseth
Institute of Astronomy
Madingley Road
Cambridge CB3 OHA
England

ABSTRACT. The main principles for direct integration of large point-mass systems are outlined. Most particles are advanced by the Ahmad-Cohen neighbour scheme, using fourth-order force polynomials and individual time-steps. Close encounters and persistent binaries are handled by two-body regularization, whereas extreme triple and quadruple configurations are treated as unperturbed systems by special regularization techniques. As an illustration of these methods, we discuss some recent results of an isolated system containing 1000 particles of unequal mass, with special emphasis on the post-collapse phase.

1. INTRODUCTION

The simple formulation of the classical N-body problem permits individual orbits to be obtained by direct numerical methods. This approach allows the evolution of small stellar systems to be followed over significant times without introducing simplifying assumptions. However, the need to study such systems during the critical stages of core collapse for as many particles as possible poses a severe challenge for numerical techniques. Although globular clusters cannot be studied in this way, the direct approach is still very useful for a wide range of investigations.

The interest in systematic N-body studies has been rather sporadic (cf. Wielen, 1967; van Albada, 1968; Aarseth, 1974; Giannone and Molteni, 1984). In addition, open star clusters have received some attention; e.g. Wielen (1971); Hayli (1971); Aarseth (1973), and most recently through the extensive study of Terlevich (1984). Further work is desirable along several lines. Firstly, more approximate methods, among which are the Monte-Carlo, Fokker-Planck and moment method, need to be compared in some detail with direct calculations. Moreover, there is much to be learnt from numerical experiments relating to particular aspects, such as the effect of close encounters, core heating by binaries and the influence of a mass spectrum on the evolution. Realistic cluster simulations should also be explored further, especially in response to improved observational data. Hence it appears that numerical experiments

J. Goodman and P. Hut (eds.), Dynamics of Star Clusters, 251–259.

constitute an important tool for increasing our understanding of dynamics and the hope is that such work will also inspire improved theoretical developments.

2. INTEGRATION METHODS

The implementation of an efficient algorithm for large N-body systems requires careful considerations of several key aspects. In order of increasing complexity the main features can be categorized under four headings: (i) Force polynomials; (ii) Individual time-steps; (iii) Neighbour scheme; (iv) Close encounters. These central topics will be discussed briefly in the following since a complete description can be found elsewhere (Aarseth, 1985).

Since the number of interactions increases as N(N-1)/2, it is desirable to preserve past force values by using a polynomial representation. Practical considerations such as initialization, restarts, machine accuracy and increasing storage suggest that remembering four previous force evaluations is a good compromise. This is equivalent to representing the force polynomial for each particle by a set of three divided force differences which are held in store. In addition, contributions from the fourth difference are added as a corrector at the end of each time-step. This procedure permits the equations of motion to be advanced by significantly larger time-steps for a given accuracy, whereas low-order predictions can be made at intermediate times.

The mean particle separation tends to increase significantly with central distance. Since the time-scale for interactions is essentially controlled by the local density, it is advantageous to introduce individual time-steps for each particle. In this way, consistent solutions are obtained for a wide range of time-scales while maintaining the relative accuracy of the orbit. This scheme requires additional coordinate prediction at intermediate times but the net gain is still significant, bearing in mind the large range of density contrasts. Each time-step is selected from a relative accuracy criterion, involving the force as well as three derivatives. Consequently the time-step for a given particle is reduced appropriately during periods of strong interactions, followed by an increase if the motion is no longer subject to large force fluctuations. Moreover, the orbital information contained in the high-order force derivatives allows a more proper control of any errors.

The two features discussed so far combine into an efficient treatment for small systems. As the particle number increases, however, it is more cost-effective to separate the total force into two components, where the so-called irregular force is due to a relatively small number of neighbours. The neighbour force is updated on a time-scale determined by the local fluctuations, whereas the smoother regular component due to all other members need only be recalculated on a longer time-scale, with intermediate values obtained by prediction. The Ahmad-Cohen (1973) scheme has proved itself in a variety of calculations; it appears to be more efficient than a single force polynomial method for systems exceeding about 50 particles. In this way, a factor approaching 10 can be gained since it is possible to have up to ten irregular force summations for each total force evaluation, and the latter effort is by far the dominant one for large N. This scheme requires additional procedures, particularly relating to a consistent neighbour treatment, where the effect due to membership changes inside a given sphere is updated during each regular force calculation.

Close encounters play a dominant role in the systems studied by direct methods, which are of necessity limited to a few thosand interacting mass-points. The formation of energetic binaries add to the numerical problems, since special treatments are needed for encounters between single particles and binaries as well as binary collisions. The classical two-body regularization of Kustaanheimo and Stiefel (1965, hereafter KS) transforms the equations of motion for two colliding particles into a non-singular form, and the external perturbation is also included. This representation is especially useful for dealing with persistent binaries of high energy, since in addition to the regularized form only the tidal perturbation is required, thereby reducing considerably the numerical effort. Moreover, unperturbed centre of mass motion may be adopted in the case of relatively distant perturbers, or equivalently, small semi-major axis. The KS method has proved to be extremely powerful in cluster simulations and has made it possible to study advanced stages of evolution which would otherwise not be feasible. More recently, the treatment of critical triple encounters has been included using three-body regularization (Aarseth and Zare, 1974). Only the most compact triple configurations are selected, since for simplicity such sub-systems are assumed to be isolated but such cases are also the most difficult for the KS treatment. Following the success of this new development, an analogous four-body regularization has now been introduced. This employs the global N-body regularization of Heggie (1974), using the numerical formulation of Mikkola (1984). The implementation for colliding binaries again ignores the external perturbation, but provisional tests show that suitable cases do occur in simulated systems, especially those with an initial binary population.

It is evident from the above discussion that several powerful techniques can be successfully combined when using direct integration. Needless to say, the programming complexities are considerable but the results which can be obtained justify the exercise. As an illustration of a large-scale calculation, we present some new results which emphasize the post-collapse cluster evolution.

3. EVOLUTION BEYOND CORE COLLAPSE

The implementation of triple regularization has enabled a large N-body system to be followed through the critical stage of core collapse and for a considerable time afterwards. We study an isolated cluster with N = 1000 mass-points selected from a power-law mass function with index -2.75 and a modest maximum mass ratio of 5:1. The system is initially spherical, having isotropic velocities with a virial ratio 0.25 half-way between rest and approximate equilibrium. The cluster evolution has been followed over more than 155 initial crossing times, requiring at least 500 hours of VAX machine time.

The behaviour near core collapse may be studied in several different ways. Figure 1 shows the average central potential, based on the five most strongly bound single and centre of mass particles but excluding the internal contribution from hard binary components. There is a gradual trend towards a maximum value, reached after about 42 crossing times, with a phase of large fluctuations followed by a gradual decline. This behaviour is also reflected in the central density, determined from the maximum density within an individual neighbour sphere which usually involves about 30 core particles. Note that either of these representations do not depend on the density centre itself. In addition, an alternative description based on the innermost 1% and 4% of the total mass with respect to the density centre shows a similar behaviour. From this evidence we conclude that core collapse is essentially halted after about 42 crossing times, equivalent to less than six half-mass relaxation times, and is followed by a prolonged phase of expansion. However, there are considerable fluctuations near maximum collapse, some of which are associated with processes involving binaries.

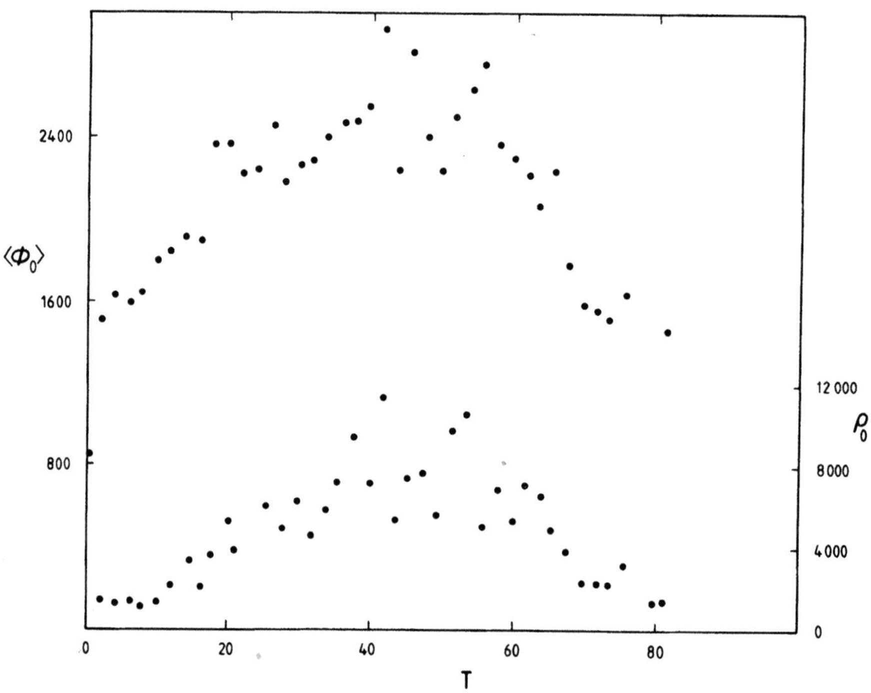

Figure 1. Illustration of core collapse. The upper data points show the average central potential based on the five most strongly bound single and c.m. particles. The lower data points show the central density as determined from the maximum neighbour density (vertical scale on the right). Time is in units of the standard crossing time, $2\langle R\rangle/\langle V\rangle$, where $\langle R\rangle$ is the half-mass radius and $\langle V\rangle$ is the equilibrium rms velocity.

An attempt to determine the core radius by generalizing the expression for an equal-mass system does not give a meaningful result. Instead Figure 2 displays the cumulative mass as a function of radius, with the cluster divided equally into heavy and light particles, at two epochs (about 28 and 76 crossing times). Mass segregation is already quite pronounced at the early stage, in qualitative agreement with theoretical expectations. The curves for the light particles are fairly straight in the inner cluster region, indicating an approximate isothermal structure. Lack of resolution does not permit conclusions about a possible density cusp or core radius. However, the main feature of substantial core expansion indicated by Figure 1 can be seen in more detail in Figure 2. It is also significant that the compact sub-system of predominantly heavy particles, formed during the pre-collapse phase, does not show much evidence of subsequent shrinkage. This behaviour has observational implications if the most massive particles are associated with non-luminous neutron stars, since the luminous component would then only undergo a modest core collapse.

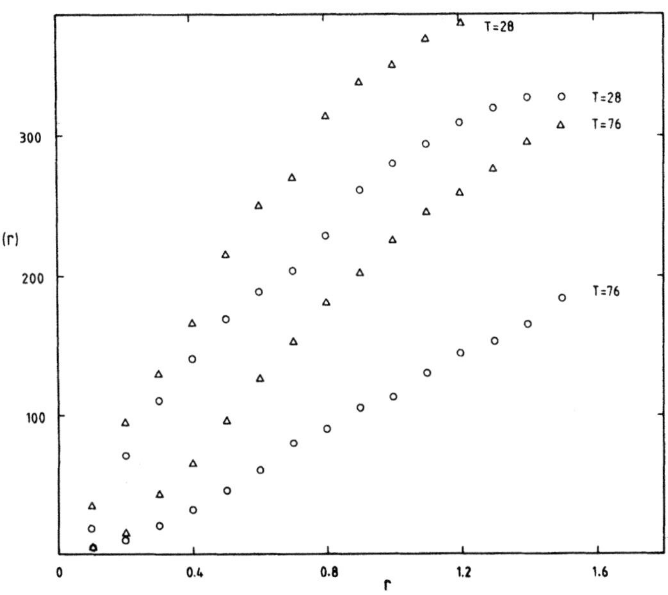

Figure 2. Cumulative mass as function of central distance. Results are shown separately for heavy and light particles (open triangles and circles) of equal total mass at early and late times (28 and 76 crossing times).

Binaries also control the later stages of core collapse in this simulation. The energy sink behaviour of central binaries in 250-body systems has been discussed in some detail before (Aarseth, 1974). Because of the smaller maximum mass ratio and larger N employed here, the first persistent hard binary does not occur until about 28 crossing times, but from then onwards an increasing amount of energy is absorbed by a few hard binaries. It is remarkable that the number of such binaries rarely exceeds two. The total energy stored in hard binaries is displayed in Figure 3. At the time of maximum core collapse, as determined from Figure 1, the dominant binary has acquired 7% of the total cluster energy. Detailed examination shows that energetic binaries, which tend to have massive components, are strongly concentrated towards the centre by mass segregation effects. This greatly increases the probability of binary collision and destruction of the least bound pair. On the other hand, close encounters with single particles may produce sufficient recoil to take a binary orbit outside the core, from which it subsequently spirals inwards again. With a typical semi-major axis of 0.0001 length units (initial half-mass radius) at this stage, such encounters are relatively infrequent but the increased centre of mass motion provides significant core heating via the two-body mechanism.

Extreme interactions may even result in escape from the cluster itself. This process is illustrated in Figure 3 which shows two such events occurring after about 49 and 69 crossing times and resulting from extremely energetic triple interactions, with negligible external perturbations. These binaries account for 35% and 42% of the initial total cluster energy, respectively. In each case, the terminal centre of mass velocity, evaluated at infinity, exceeds the overall rms velocity. Consequently, the corresponding velocity of single particles escaping after such interactions can be quite substantial, with the current examples producing two terminal velocities of about eight times the rms value. Distant particles exceeding the parabolic escape velocity are removed from the calculations, leaving the cluster more bound, whereas an escaping binary also carries away negative binding

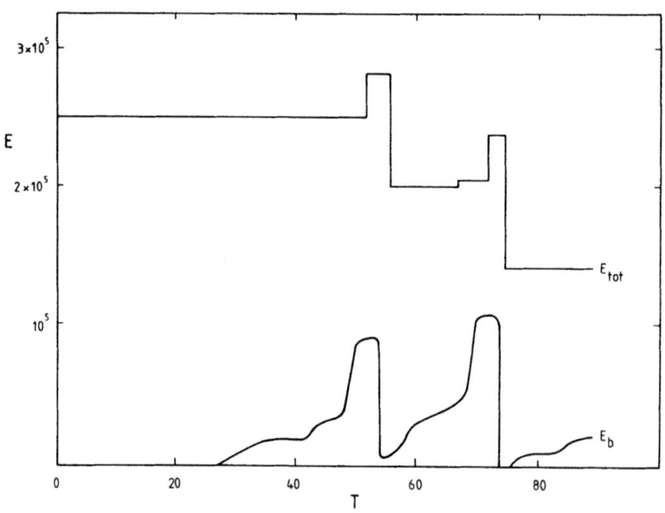

Figure 3. Binary evolution in the cluster. The lower curve shows the total energy contained in hard binaries, while the upper curve represents the total energy of all the bound particles in the cluster. Time is in units of the crossing time and energies are positive here.

energy. The upper curve in Figure 3 shows both effects, with the extremely energetic single escaper preceding the removal of the binary. The difference between the upper and lower curve represents the energy binding the cluster itself; at the end of the phase displayed here this energy is only half the initial value. Hence the corresponding cluster expansion implies a significantly increased time-scale for further evolution.

Binaries continue to play an important role in the subsequent evolution, not included in Figure 3. Eventually, after 130 crossing times, a third energetic binary is ejected from the cluster after a strong and almost isolated triple interaction. In this case, the binding energy corresponds to 58% of the initial total energy of the cluster and the terminal escape velocity exceeds four times the rms velocity. The recoil velocity of the corresponding single escaping particle is even more extreme, exceeding ten times the rms value. In all there are 18 energetic escapers, with terminal velocity above the rms value, out of a total of about 100 escapers. Hierarchical configurations containing energetic binaries provide the most important mechanism for large energy transitions, sometimes resulting in exchange of companions and immediate escape with high velocity. The combined kinetic energy of the fast escapers represents some 70% of the initial cluster energy, with the three extreme escapers associated with the largest binary recoils accounting for over 50%. On the other hand, the three escaping binaries carry away a binding energy equivalent to 135% of the initial energy of the whole cluster. Thus the final cluster is left with a greatly reduced negative binding energy, amounting to no more than about 35% of the initial value when allowance is made for one remaining binary with nearly 2%. We may therefore conclude that binaries form and absorb energy at sufficient rates to maintain a relatively low central density and velocity dispersion, thereby effectively controlling the whole cluster evolution.

At this stage the calculations were terminated, the main objective of investigating a significant phase of post-collapse core evolution having been achieved. Only some of the more spectacular features have been discussed here, whereas a detailed analysis will be presented in due course.

4. REFERENCES

Aarseth, S.J. (1973). In Vistas of Astronomy, ed. A. Beer. Pergamon Press, Oxford, P.13.

Aarseth, S.J. (1974). Astron. and Astrophys. 35, 237.

Aarseth, S.J. (1985). In Multiple Time Scales, eds. J.U. Brackbill and B.I. Cohen. Academic Press, New York, to be published.

Aarseth, S.J. and Zare, K. (1974). Celest. Mech. 10, 185.

Ahmad, A. and Cohen, L. (1973). J. Comp. Phys. 12, 389.

Hayli, A. (1971). Astrophys. and Space Sci. 13, 309.

Giannone, G. and Molteni, D. (1984). Astron. and Astrophys., in press.

Heggie, D.C. (1974). Celest. Mech. 10, 217.

Kustaanheimo, P. and Stiefel, E.L.(1965). J. reine angew. Math. 218, 204.

Mikkola, S. (1984). Ph.D. Dissertation, University of Turku, Finland.
Terlevich, E. (1984). Ph.D. Dissertation, University of Cambridge.
van Albada, T.S. (1968). Bull. Astron. Inst. Netherlands, **19**, 479.
Wielen, R. (1967). Veroeff. Astron. Rechen Inst. Heidelberg, No. 19, 1.
Wielen, R. (1971). Astrophys. and Space Sci. **13**, 300.

DISCUSSION

SEVERNE: To what extent is your code reversible (reproduction of the initial state upon velocity inversion)? Has it been applied to the actual study of distribution functions for the systems investigated?

AARSETH: It is possible to make a satisfactory reversal and obtain the initial conditions after an escaping particle has moved about five to ten half-mass radii outside a system. Simple distribution functions in one variable (i.e., energy, velocity) can easily be obtained at different times.

KING: These are extremely valuable calculations; but I regret that while you spend your time producing them, there is not someone else working with you to analyze them. This involves a lot of careful work and thought, and you need a collaborator who will give it the attention that it deserves.

A UNIFIED N-BODY AND STATISTICAL TREATMENT OF STELLAR DYNAMICS

Alan P. Lightman
Harvard-Smithsonian Center for Astrophysics
and
Stephen L. W. McMillan
University of Illinois, Champaign-Urbana

ABSTRACT

We summarize the methods of a new "hybrid" computer code for stellar dynamics. All particles in the inner spatial region are followed exactly via a direct N-body code and all particles in the outer spatial region are treated statistically via a distribution function and Fokker-Planck type methods. An intermediate region, with features of both, allows exchange of particles and energy between the outer and inner regions. We apply our code to the period just before core collapse and just after and summarize the results.

1. INTRODUCTION

From theoretical work in stellar dynamics, we have known for forty years that the core of a self-gravitating system of point masses will collapse, with an ever-decreasing number of particles and an ever-increasing density. (See, e.g. Spitzer, 1985 for a review.) Such a collapse must eventually be halted by such processes as stellar collisions and binary activity. Recent comparisons of the collapse theory with observations of actual globular cluster cores suggest that some globular clusters have already collapsed and are now embarked upon post-collapse evolution (Lightman, 1982; Cohn and Hut, 1984). In the same recent period, theoretical investigations have focused on post-collapse evolution, using statistical methods appropriate to the "large-N" approximation and an assumed central heat source to drive the expansion (Stodolkiewicz, 1982; Goodman, 1984; Inagaki and Lynden-Bell, 1983; Heggie, 1984; Sugimoto and Bettwieser, 1984; see Heggie, 1985 for a review). Most of these studies indicate the post-collapse core expands, with an ever-increasing number of particles.

For an understanding of the evolution shortly before core collapse and shortly thereafter, the purely statistical methods are inadequate. This is because, for binaries the agent of collapse reversal and no initial binaries, the core will always collapse down to about 100 particles before reversal, independent of any other initial conditions (Lightman and Fall, 1978; Heggie, 1984). Statistical methods fail for such a small number of particles. On the other hand, the core is typically surrounded by a vastly larger number of particles, also evolving on the time scale of interest and requiring a statistical treatment. To allow study of the core during the critical period mentioned above, where both "small-N" effects and "large-N" effects are important, we have developed a hybrid computer code, in which particles are followed exactly in the inner spatial regions via a direct N-body code and statistically in the outer spatial regions via a Fokker-Planck type code. In effect, our method provides a bridge between the purely statistical epochs of the pre- and post-collapse phases of evolution.

2. SUMMARY OF THE HYBRID CODE

2.1 Physical Approximations

The principal physical approximations are that all stars are treated as point masses, of initially equal mass, and binaries are initially absent. These approximations reduce the problem to one of pure stellar dynamics, with no essential free parameters.

J. Goodman and P. Hut (eds.), Dynamics of Star Clusters, 261–274.

2.2 Radial Zones and Methodology

The computer code that unifies the "large-N" and "small-N" regimes, treating stars statistically in some places but following orbits directly in others, is a "hybrid", combining many of the features of previous codes. The system is first divided into three *fixed* radial zones:

(*i*) an "N-body region", referred to below as "NB", defined by $r \leq r_N$,
(*ii*) an "inner Fokker-Planck region" ("FP1"), having $r_N < r \leq Kr_N$, and
(*iii*) an "outer Fokker-Planck region" ("FP2"), with $r > Kr_N$.

The radius r_N is arbitrary, constant in time, and chosen as the unit of length in all that follows. K is a constant and N is the number of stars within r_N at any given time. The choices of K and an initial value for N are governed mainly by economic considerations; typical values are $K \sim 3$, $N \sim 100$ - 150. Progressing outward from the center of the system, the three zones represent increasing degrees of approximation, as described below.

Given the three radial zones, a number of regions in energy and angular momentum space (E-J) space may be defined, according to the pericenter (r_-) and apocenter (r_+) of the corresponding orbits:

$$\begin{aligned} \text{I}: &\quad r_+(E,J) \leq r_N, \\ \text{II}: &\quad r_+(E,J) > r_N,\ r_-(E,J) < r_N, \\ \text{III}: &\quad r_-(E,J) > r_N. \end{aligned}$$

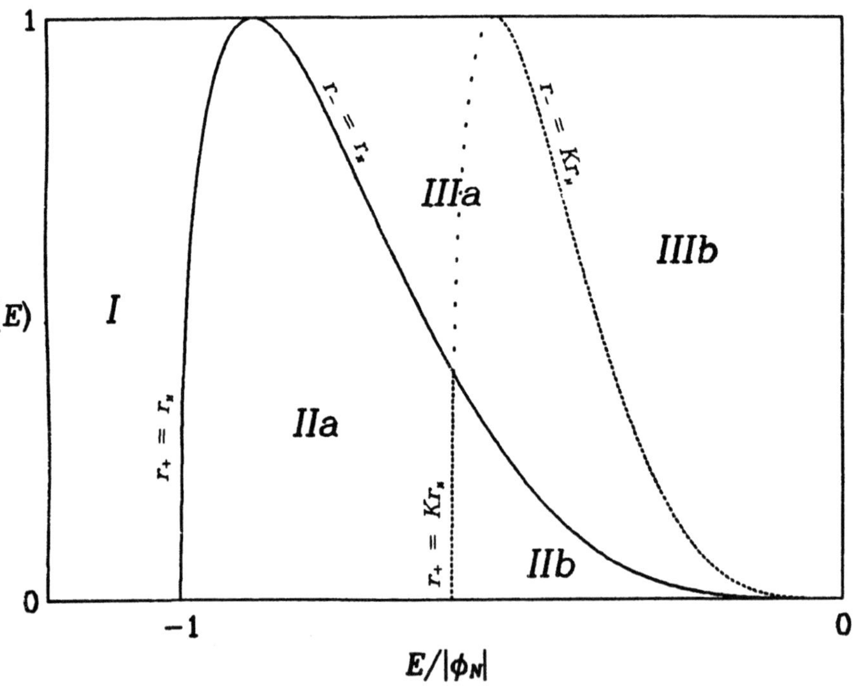

Figure 1: Regions of E-J space corresponding to different types of orbits.

Region II may be subdivided further into IIa, having $r_+ \leq Kr_N$, and IIb, with $r_+ > Kr_N$. Similarly, it will be convenient to denote that part of region III with $r_- \leq Kr_N$ as IIIa and the remainder of that region as IIIb. All these regions are shown schematically in figure 1. The methods used to treat a star's orbit depend on both the radial zone and the phase-space region under consideration.

(a) Stars inside r_N.

Within NB, *all* stars are followed exactly using a three-dimensional N-body code. Since the presence of hard binaries is of such critical importance to the late evolution of isolated small-N systems, it is anticipated that it might be similarly important here also. For this reason, the force law is taken to be a simple inverse square — no softening of the potential is introduced, so hard binaries, should they occur, are correctly followed.

The program actually used is a somewhat modified fifth-order predictor-corrector procedure of the type developed by Aarseth (1968, 1974), employing individual timesteps and including a two-body Kustaanheimo-Stiefel (1965) regularization procedure, enabling simultaneous "close" encounters between up to twenty pairs of stars to be treated efficiently. The most major modification is, of course, that only stars within a fixed radius are considered. When a star crosses r_N moving outward, it is removed from the N-body integration, as follows. First, all stars within r_N are integrated up to the current time. The escaper's contribution to the force and first three force derivatives felt by all other bodies in NB may then be calculated and subtracted out; its contribution to the binding energy is similarly removed (enabling energy conservation to be conveniently checked at any later time). Once this has been done, the N-body system has "forgotten" about the existence of the star and the integration can proceed. When it becomes necessary to insert incoming stars, a similar procedure is followed; because of the pairwise nature of the expression for the force and higher derivatives, all of these quantities may be determined for the newcomer simultaneously with the application of the corrections to the rest of the system, with no extra computational effort. The treatment of stars in the outer two zones (see below) means that no additional special measures need be taken to account for their interaction with a star crossing r_N.

In the runs described below, the N-body part of the code ran on an FPS-120B Array Processor. This device, while considerably faster than the VAX 11/780 used for the remainder of the program, suffered from quite severe limitations on the amount of memory available.

(b) Stars with $r_N < r < Kr_N$.

In the inner Fokker-Planck region, the trajectories of all stars are integrated in one dimension only. As a star enters this region, it is replaced by a spherical shell with the same mass and velocity, and the motions of all such shells are followed using a simple second-order integration scheme. The radial force per unit mass experienced by a shell of mass m_i, radius r_i and angular momentum J_i per unit mass is

$$a_i = -\frac{M_{NB} + \frac{1}{2}m_i}{r_i^2} + \sum_{\substack{shells\ j:\\ r_j < r_i}} \frac{m_j}{r_j^2}\,, \tag{1a}$$

and the radial equation of motion is

$$\ddot{r}_i = a_i + \frac{J_i^2}{r_i^3}\,. \tag{1b}$$

Here, M_{NB} is the total mass of all the stars in the N-body region (equal to Nm_* in the present case), regarded as located at $r = 0$. The extra $\frac{1}{2}m_i$ in the first term takes into account the self-gravity of a shell.

Forcing spherical symmetry on the potential greatly simplifies the integration. However, information on the original star's trajectory is not discarded. Its angular coordinates are also known, since the plane of the orbit is known and the coordinate in that plane, θ, satisfies

$$\dot{\theta} = \frac{J}{r^2}, \tag{2}$$

enabling full three-dimensional position and velocity information to be retrieved at any time (in particular, when the shell must be inserted into NB). The term "shell" should not be taken to mean that the object under consideration here represents more than one star, as it does in the work of Spitzer and coworkers (e.g. Spitzer and Hart, 1971). Rather, it simply reflects the fact that, dynamically, it reacts and contributes only to a spherically smoothed mass distribution and thus, for the purposes of the one-dimensional integration, behaves as though it were indeed a shell of material, albeit one with "hidden" parameters, such as orbital orientation, etc.

The one-dimensional force calculation described above allows shells to react to changes in their bulk distribution. It does not, however, permit the two-body effects that drive the evolution over long timescales. The two-body perturbation spectrum experienced by a shell can be generated in a manner similar to that described by Shapiro and Marchant (1978). The expected values of the orbital perturbations are obtained by integrating the local energy and angular momentum diffusion coefficients along a segment of the unperturbed orbit,

$$\langle Q\rangle(r_1, r_2) = \int_{r_1}^{r_2} \frac{dr}{v_r(E, X; \phi(r))} \langle Q\rangle_t, \tag{3}$$

for $Q = \Delta E$, $(\Delta E)^2$, etc. Here we have made the substitution $X = J^2$; and v_r and $\phi(r)$ denote the radial component of velocity and gravitational potential at radius r, respectively. The actual orbital perturbation is then generated by randomly selecting $(\delta E, \delta X)$ from a two-dimensional Gaussian distribution having means $\langle \Delta E\rangle$ and $\langle \Delta X\rangle$ and second moments $\langle(\Delta E)^2\rangle$, $\langle(\Delta X)^2\rangle$ and $\langle \Delta E \Delta X\rangle$. The component integrals are accumulated as the shell's orbit is calculated and the random perturbation is applied when any of a number of criteria is met:

(*i*) one of the perturbations, $\langle \Delta E\rangle^2$, $\langle(\Delta E)^2\rangle$, $\langle \Delta X\rangle^2$, $\langle(\Delta X)^2\rangle$, becomes unacceptably large, or

(*ii*) the shell is about to leave FP1, either at r_N or at Kr_N, or

(*iii*) the shell passes apocenter in FP1. In this case, the shell is then relocated to the apocenter of its new orbit and its acceleration (and that of all other shells in the system) is corrected before the integration continues.

Because the determination of the phase-space distribution function and the subsequent integration over energy for every shell at every timestep would be prohibitively expensive, to compute the diffusion coefficients we assume that the distribution is Mawellian at low energy and power-law at high energies, as given by the "large-N" calculations.

Because of their potential dynamical importance, it is essential that hard binaries not be "lost" as they cross r_N moving outward. Thus, when the center of mass of such a pair leaves NB, it is replaced by a shell of mass $2m_*$. However, in addition to all the usual parameters (angular momentum, orbital orientation, label, etc.) associated with a shell, the pair's binding energy at removal is also stored. Perturbations to the shell's orbit are determined and applied as described above, with the following two changes. First, the additional mass of the binary is taken itno account when applying the diffusion coefficients. Second, as shown by Heggie (1975), the binding energy, B, of a hard binary moving through a field of stars with density n and velocity dispersion $\langle v^2\rangle$ increases at a rate

$$\dot{B} \approx 10\, G^2 m_*^2 n \langle v^2\rangle^{-1/2}, \tag{4}$$

so, in addition to the perturbations to the orbital parameters, the binding energy is increased by the (usually small) amount

$$\delta B = \int_{r_1}^{r_2} \frac{dr}{v_r} \dot{B} \tag{5}$$

whenever a perturbation is applied.

No attempt is made to follow either the orientation or the eccentricity of the relative orbit of the pair. The transit time of the center-of-mass shell through FP is likely to be many binary orbital periods, so it is simply assumed that both quantities are random at reinsertion, with the eccentricity e having a distribution $\Phi(e) = 2e$ (Jeans, 1929). Reinsertion of the pair is accomplished by choosing some point on the relative orbit, inserting the two components separately and allowing the N-body code to perform the necessary initialization and regularization. It should be emphasized that only hard binaries are treated in this way. Pairs emerging from NB with $B < \frac{1}{2}\langle v^2 \rangle$ are broken up and their components treated separately.

(c) Stars outside Kr_N

Following the general approach laid down earlier, stars in FP2 are followed in (E, X) space only. Furthermore, and in contrast to the treatment in FP1, the potential and background distribution function used to perturb the orbits is FP2 is held fixed. This approximation, which we hope to eliminate in future work, must lose its validity on a time of order the relaxation time at Kr_N.

Whether the potential in FP2 is held fixed or allowed to vary, the diffusion and perturbations of stars in FP2 is treated as follows: The IIIb stars, those that never come within Kr_N, may be treated using any of the large-N techniques mentioned earlier — the Cohn (1979) and Shapiro-Marchant (1978) methods are most appropriate for the configuration used here. Accordingly, region IIIb is covered with a grid or realized in the form of "superstars" and the evolution is followed by considering fluxes of stars between neighboring grid cells or random perturbations to the superstars' orbital parameters, generated much as described above. In general, one would expect the relaxation time at large radii to be much greater than that in the central regions, so very few FP2 timesteps need be taken during a typical run covering many core relaxation times.

This outermost portion of the calculation communicates with the shell integration via the IIb and IIIa stars that are "in transit" through FP2. Given the potential outside Kr_N, the orbits, and the perturbations applied to the orbits, of shells outside Kr_N can be calculated. When an outward-moving shell crosses Kr_N, it is perturbed by an amount calculated, as before, from the moments

$$\langle \Delta E \rangle_K = 2 \int_{Kr_N}^{r_+} \frac{dr}{v_r} \langle \Delta E \rangle_t ,$$

etc., and "absorbed" into the IIb or III distribution in FP2. As mentioned above, no orbital phase or orientation information is retained. Instead, region IIb/IIIa in phase-space is divided into 1500 bins (100 in E, 15 in X). Associated with each bin are three numbers: the number of stars it contains, $\mathcal{N}$, the mean rate, $\mathcal{R}$, at which stars from that bin are to be inserted into FP1 and the time of the next insertion. On insertion, the shells "hidden" orbital orientation is generated randomly. When a shell is added to or removed from a bin, the occupation number and insertion rate are adjusted accordingly. This procedure has the desirable property that it allows the bin occupation number to react rapidly to fluctuations in the rate at which shells arrive from FP1; the reinsertion rate changes on a timescale $\mathcal{R}^{-1}$ — less than the return time for an individual shell by a factor of order $\mathcal{N}^{-1}$ — thus tending to smooth the irregularities out. The (extremely) rare event of a binary shell leaving FP1 is treated by attaching a "flag" to the bin in question indicating the binding energy and the number of insertions which must occur before the binary returns.

2.3 Summary of Error Corrections

The method used to integrate the shell equations of motion, being only of second order, is inherently much less accurate than the N-body integration. However, the energy error can be calculated at each timestep at no extra computational cost and a correction can be applied. The underlying reason for the reduced accuracy is the fact that shells can cross and, when they do, the force experienced by each of the crossing pairs changes discontinuously. The neglect of shell crossings would cause the energy to increase secularly. By keeping track of the relative ranks of shells before and after each time step, the net number of crossings for each shell can be determined and an energy correction applied.

Another source of accumulated error stems from the fact that a finite number of perturbations drawn from a given distribution will generally not have the same mean as the distribution itself. To avoid undesirable spurious fluxes of energy and angular momentum from one part of phase space to another, care must be taken to ensure that, over a period of a few crossing times, the desired perturbation distribution is correctly generated. Accordingly, regions II and IIIa are divided into 100 cells on a 10×10 grid. As each perturbation is applied, the actual perturbations, $(\delta E, \delta X)$, the calculated mean perturbations, $(\langle \Delta E \rangle, \langle \Delta X \rangle)$, and the identity of the shell perturbed are recorded against the appropriate cell. Once a sufficient number (chosen to be 15) of perturbations has accumulated in a given cell to ensure adequate statistics, the energies and angular momenta of the stars in question are adjusted to force the averages of the actual perturbations to the desired mean values. It is assumed that no additional correction need be applied to force the second moments of the distribution in a similar way, although such a correction could be applied if necessary.

2.4 Tests of the Code

We performed a number of tests on our code, checking energy conservation and other such desireable properties. One test, described here, was the approach to thermal equilibrium under appropriate conditions. As shown by Antonov (1962) and by Lynden-Bell and Wood (1968), a collection of N point masses interacting only gravitationally and confined within a sphere will attain and maintain thermal equilibrium provided certain conditions are satisfied: (1) binary interactions must be ignored; (2) if E and M are the total energy and mass, respectively, of the confined particles and R_s is the radius of the spherical shell confining the particles, then E must exceed the value -0.335 GM^2/R_s; and (3) the central density must be less than 709 times the density at R_s.

Our test consisted of placing a reflecting shell at $R_s = K r_N$, with K=3, introducing an effective cutoff (or "softening parameter") in the short range interaction of particles to prevent hard binary formation, and beginning the system far from thermal equilibrium. In particular, we initialized the system with 300 particles in NB, at constant density and with flat velocity distribution, and no shells in FP1. The total energy satisfied E=-0.165 GM^2/R_s. Violent oscillations ocurred for the first ten crossing times, after which the system began to relax toward a thermal distribution. After about thirty central relaxation times, the mean velocity dispersion leveled off at its equipartition value and, by sixty central relaxation times, stochastically departed from this value by only about 4%. Moreover, the velocity dispersions of both the particles in NB and the shells in FP1 were essentially identical after thirty central relaxation times, at which time approximately 200 of the initial particles were in the form of shells. Both the density profile and the velocity distribution approached those of an isothermal distribution appropriate to the fixed parameters of the system, with the one exception that, at about sixty central relaxation times (the end of the run), the high velocity tail of the distribution (with squared velocities larger than 2.5 times the mean) had approximately 25%too much kinetic energy. Because of the steep increase of relaxation time with velocity, these particles and shells had not yet had time to relax.

This test proved extremely useful as a diagnostic tool. Errors in the algorithms used in the generation and correction of perturbations and the redistribution of the errors incurred at r_N invariably showed up as secular trends in the core parameters and/or the heating or cooling of some radial zone or zones. Furthermore, the similar and correct behavior of both shells and particles suggests that no important error is being introduced by the artificial boundary at r_N and the different methods of treatment inside and outside of this boundary.

3. APPLICATIONS AND RESULTS FOR CORE BOUNCE

3.1 Initialization

In the most recent calculations of core collapse, the Fokker-Planck treatment of Cohn (1980) and the fluid treatment of Lynden-Bell and Eggleton (1980), the late stages of collapse become homologous and have the solution

$$n(r,t) \propto r^{-\alpha}\,,\ r \gg R_c(t) \tag{6a}$$

$$R_c(t) \equiv \left(\frac{3\langle v^2\rangle_c}{4\pi G m_* n_c}\right)^{1/2} \propto \left(1 - \frac{t}{t_{coll}}\right)^{2/(6-\alpha)} \tag{6b}$$

$$n_c(t) \equiv n(o,t) \propto R_c^{-\alpha} \tag{6c}$$

$$\langle v^2\rangle_c \equiv \langle v^2\rangle(o,t) = 2Gm_* N_c R_c^{-1} \propto N_c^{(2-\alpha)/(3-\alpha)}\,. \tag{6d}$$

Here time t is measured from the beginning of the homologous phase, r is the radial distance from the center of the (spherically symmetric) system, n is the density, R_c the core radius, m_* the mass of a typical star in the core, $\langle v^2\rangle$ the squared velocity dispersion, $N_c = (2\pi/3)n_c R_c^3$ the number of stars in the core, $\alpha = 2.23$, and t_{coll}, the collapse time, is 200 - 300 initial central relaxation times. Subscripts c on all quantitites will denote central (core) values. The equality in Equation (6d) expresses the virial relationship. In the region $r \lesssim R_c$, the distribution is that of an isothermal sphere, with an approximately constant density. Henceforth, we shall refer to the above solution as the CL-BE distribution.

Ideally, we would like to initialize our system with a configuration corresponding to a snapshot of the CL-BE distribution, at a late stage of the collapse, but not so late that the statistical approximation has yet broken down. This would involve at least several hundred particles in the core and an N-body region capable of following a similar number of particles. For practical reasons, we have been limited to an initial value of the number of particles in the N-body region, N, equal to 100. (N changes in time; in particular, it increases as the collapse proceeds.) For an initial value of R_c/r_N we take 0.6. We then choose a random distribution of N-body particles and Fokker-Planck shells representing the CL-BE distribution with these values of N and R_c/r_N. Initially, the number of particles in the core, N_c, is 55 and the number of shells in FP1 is 220. All times will be measured in units of the time-independent quantity t_N, defined by

$$t_N \equiv \frac{r_N}{|\phi_N|^{1/2}} \quad \text{at} \quad t = 0, \tag{7}$$

where ϕ_N is the central potential. The time scale t_N is approximately a quarter of the initial N-body-region crossing time. With our assumed initial parameters and distribution, the core relaxation time

$$t_{Rc} \equiv \frac{\langle v^2\rangle_c^{3/2}}{15.4\, G^2 m_*^2 n_c \ln\Lambda} \tag{8}$$

intitially satisfies $t_{Rc} = 6t_N$. Here $\ln\Lambda \approx \ln 0.4\, N_c$ is the gravitational Coulomb logarithm.

Since stars are treated as point particles, there is no overall length or time scale imposed upon the problem, except by reference to observed globular clusters and the number of particles in the core. With the scaling laws given by the CL-BE distribution and characteristic values $N_c \sim 10^4$ and $R_c \sim 1$ pc for a globular cluster at the time it begins its homologous collapse, we may attach the following physical dimensions to the core at the time it has reached $N_c \approx 55$: $r_N \sim 2 \times 10^{-3}$ pc, $t_N \sim 9$ yrs., $t_{Rc} \sim 50$ yrs.

3.2 Overall Behavior

The overall behavior of the system can be seen in Figure 2. This shows the time-dependence of two measures of the central density. The solid line shows the mean density within $0.25r_N$ of the geometric center of the system (defined at large radii), and the dotted line shows the mean density within the core using Von Hoerner's (1963) definition of the core radius.

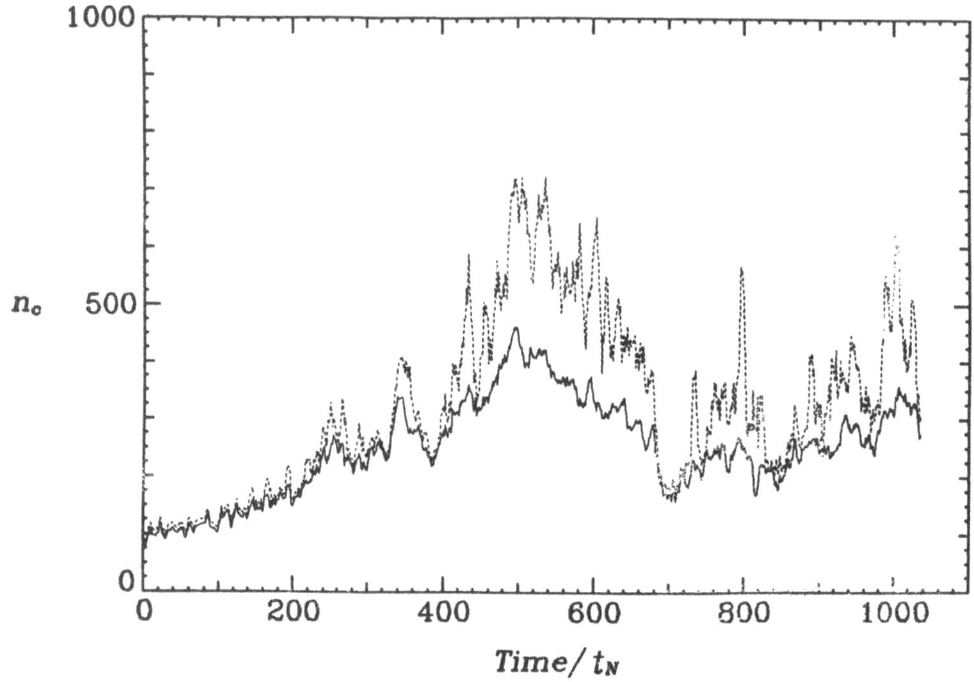

Figure 2: Central density as a function of time.

The evolution may be conveniently divided into a number of distinct phases:

(A) The initial collapse phase, from $t = 0$ to $t \approx 500\,t_N$. Here, the system continues to collapse along the "large-N" track, following the CL-BE solution, until the number of stars in the core becomes very small ($\sim$ 25) and the contraction is halted by the formation and hardening of a hard binary.

(B) The hard binary continues to harden, driving an expansion phase for $500\,t_N \lesssim t \lesssim 680\,t_N$. This phase continues until the binary is ejected from the core by a violent triple encounter.

(C) The recollapse phase, $t \gtrsim 680\,t_N$, in which, the internal heat source removed, the core begins to collapse again. Once the central density increases sufficiently, new binaries form and begin to harden and, by the end of the run, are injecting energy into the central regions at a rate almost sufficient to start a second expansion.

3.3 Homologous Initial Collapse Phase

The projected core collapse time is best obtained by fitting a straight line to the decreasing central relaxation time as a function of time (any homologous collapse should show such a linear behavior), giving a collapse time of 110 initial central relaxation timescales. See Figure 3. We suggest that this is somewhat shorter than the values of 200-300 found by Cohn (1979, 1980) because of difficulty in balancing the diffusion coefficients in the Fokker-Planck region with the actual encounters in the N-body region. In addition, deviations from isothermality will be higher in any stochastic realization with a small number of particles, and this speeds up the evolution.

The behavior of the "central" parameters R_c, n_c, $\langle v^2 \rangle_c$, again with the two definitions of the term "central" described above, are displayed in Figure 4, for the initial phase. Using the collapse time determined above, these parameters are fitted with the CL-BE curves and the fit is extremely good. Further evidence for the homology of the collapse is given in Figure 5, which shows that the density profile follows a one-parameter family throughout this phase. It is miraculous that the homology continues long after the statistical approximation has broken down.

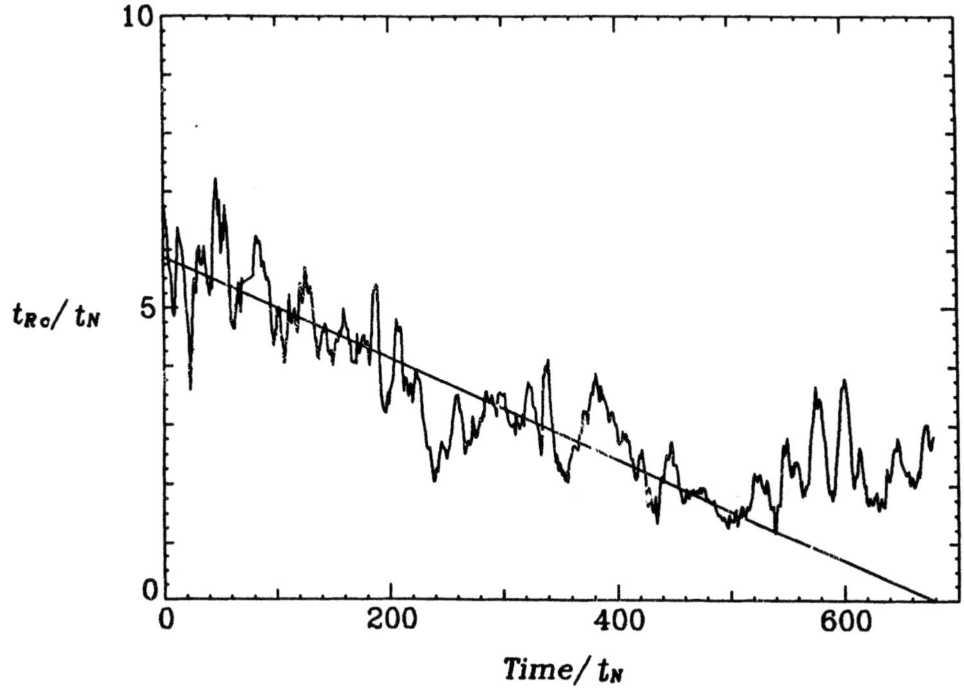

Figure 3: Central relaxation time as a function of time.

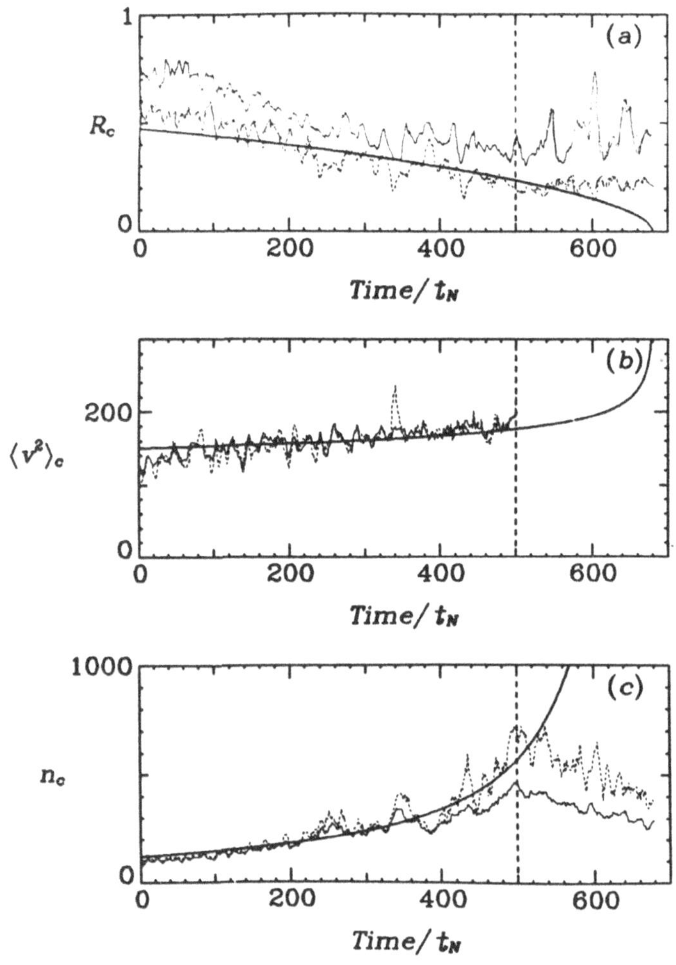

Figure 4: Core radius (a), central velocity dispersion (b), and central density (c) as a functions of time. The solid curves the predictions of the "large-N" theory of collapse.

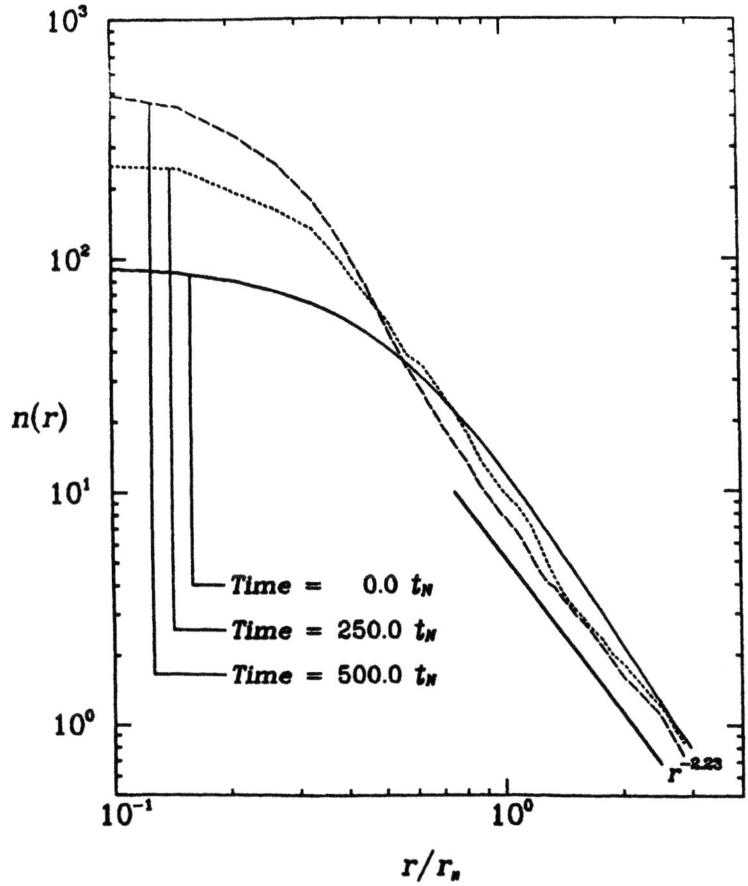

Figure 5: Density profile at three different times, ending at core collapse.

3.4 Expansion and Recollapse

The expansion phase was driven by the formation and heat input of a central binary as can be seen in Figure 6. As soon as the binary formed and began to harden, the core began to expand. The overall energetics of the expansion can be understood by simply noting that, at the time of maximum density, $N_c \approx 25$ and $\langle v^2 \rangle_c \sim 180\, Gm_*/r_N$, so the total core kinetic energy is $T_c \sim 2250\, Gm_*^2/r_N$. The total change in the binding energy of the pair between $t = 500\, t_N$ and $t = 600\, t_N$ was approximately $3000\, Gm_*^2/r_N$, so, even allowing for considerable heating inefficiency, the energy input was sufficient to cause a substantial change in the core parameters. During the expansion of the core, the density profile followed the same one-parameter sequence of shapes as in the initial collapse.

At $t \approx 625\, t_N$, the hard binary formed a relatively tightly bound ($H \sim 1$) triple system with a third star. Here H is the binary binding energy divided by the mean kinetic energy of a star. The system was hierarchical in nature, that is, the semi-major axis of the binary–single star orbit was initially much greater than the separation of the members of the binary, although the outer orbit rapidly began to harden and had reached $H \sim 2.5$ by the time special steps were taken to deal with it.

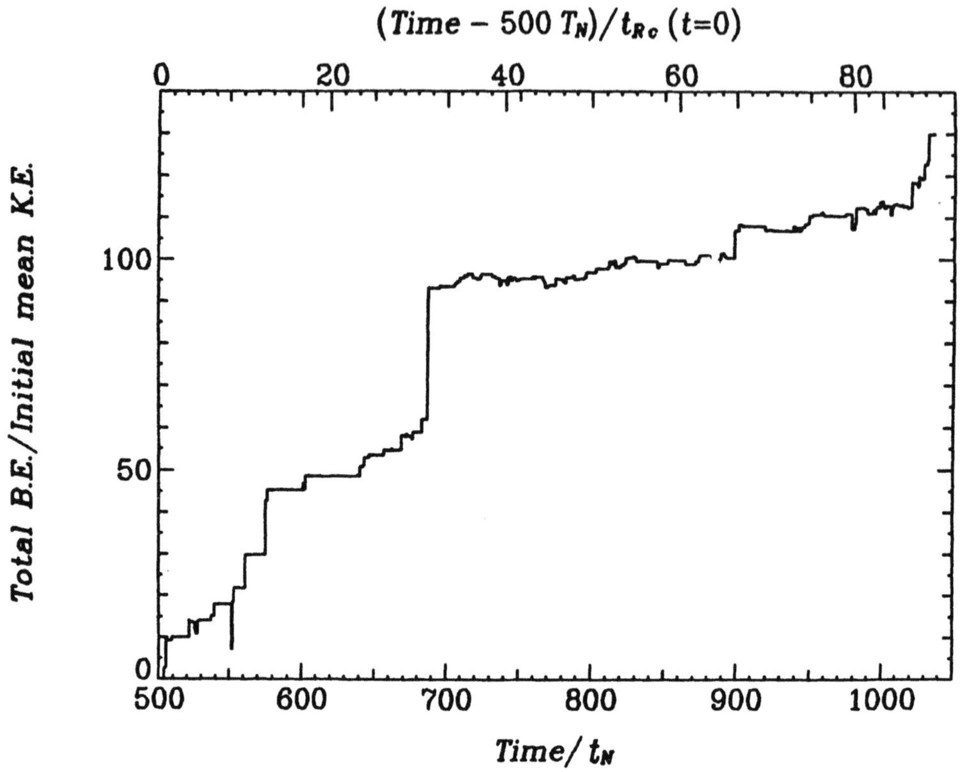

Figure 6: Binding energy in binaries as a function of time.

To enable the integration to proceed, we collapsed the inner hard binary to a single particle of double mass while the outer particle was becoming increasingly bound from encounters with field particles. Eventually, the shrinking orbit of the outer particle forced an end to the approximation. Once a triple loses its hierarchical character, there is strong numerical evidence (Agekian and Anosova, 1968; Harrington, 1972; Szebehely, 1972) that it becomes unstable. Harrington finds empirically that this happens when the pericenter of the outer orbit is a few times the scale of the inner one. Apportioning the recoil energy among the decay products gave the binary center of mass a velocity of 1.8 times the root mean square core velocity, so we simply ejected the binary from the system at this point.

At $t \approx 680\, t_N$, with the central binary gone, the core began recollapsing. During the remainder of the run three more binaries formed and, near the end, two of them were hardening at a rate sufficient to begin a new expansion.

4. SUMMARY OF RESULTS AND CONCLUSIONS

4.1 Summary of Results

In brief, our results are as follows: (1) The homologous collapse continues into the "small-N" regime, all the way until the core bounces. (2) After the first core collapse, the core undergoes a series of expansions and contractions, forming a couple of new binaries in each cycle. Eventually, binary-binary interactions must become important. We believe this behavior is inevitable, given the small number of particles in the core at the time of core bounce. (3) The density profile is unchanged in the outer regions, at several core radii and beyond, and follows a 1-parameter regular isothermal form in the inner regions.

4.2 The Long Term Post-Collapse Evolution

The long-term evolution of the system in post-collapse expansion has been recently studied in the fluid approximation by a number of investigators. Heggie (1984) finds that the inner regions form a regular isothermal sphere, out to a radius $R_*(t) \propto (t - t_{coll})^{2/(6-\alpha)}$, at least until this radius expands to the half-mass radius of the system. The density at R_* has the form $n_* \propto R_*^{-\alpha}$. The above results can be quantified completely with the additional relation $t_R(R_*) \approx (t - t_{coll})$ at all times during the expansion. The core quantities obey the relations $n_c \propto n_*^{(7\alpha-12)/3\alpha}$ and $R_c/R_* \approx (n_c/n_*)^{-1/2} \propto n_*^{(6-2\alpha)/3\alpha}$. The outward luminosity necessary to maintaia this expansion, at the radius R_*, satisfies $L(t, R_*) \equiv L_* = 0.4\, n_*\, R_* \langle v^2\rangle_*^{1/2}\, Gm_*^2 \left(\frac{300}{\gamma}\right) \ell n\, \Lambda$, where γ is the number of central relaxation times to collapse.

The dashed line in Figure 6 gives the integral of the above luminosity expression, assuming all of this energy is provided by binaries. The shape is not in bad agreement with our numerical results and the overall amplitude is also in agreement to within a factor of two.

If the long-term fluid calculations are valid, we can predict that the core oscillations observed shortly after core bounce will eventually damp out. Core oscillations are driven by the discrete formation and ejection of single binaries in the core. When the time to form a new binary, t_{bin}, is much shorter than the core evolution time scale, τ_c, we expect the oscillations to cease – the core as a whole then responds to a time-averaged binary heat input, just as does the halo, and a steady-state of secular core expansion is possible. The time scale for forming a new binary, ignoring the logarithmic term, satisfies $t_{bin} \sim N_c\, t_{Rc}$. The core evolution time scale can be defined as $\tau_c = N_c \langle v^2\rangle_c / L_*$ (a definition that keeps $\tau_c \propto t_{coll}$ during the collapse phase, when $R_* \propto R_c$). Finally, we write everything in terms of R_*, using the results $\langle v^2\rangle_* \approx \langle v^2\rangle_c$, the virial theorem $\langle v^2\rangle_* \propto n_*\, R_*^2$, $n_* \propto R_*^{-\alpha}$, we obtain

$$\tau_c/t_{bin} \propto R_*^{\frac{3-\alpha}{3}} \propto R_*^{0.26}\,. \tag{9}$$

As R_* increases with time in the post-collapse evolution, t_{bin} monotonically decreases relative to τ_c. Putting in the numerical factors, it is straightforward to show that this ratio is of order unity at the moment of the initial core collapse, when the oscillations are observed in the hybrid code. We thus expect these ocillations to cease after R_* has increased by a factor of a few.

In conclusion, discrete N-effects are quite visible in our calculations, but absent in the "large-N" fluid approximations. Our results are consistent with the "large-N" post collapse evolution, but do not extend far enough in time to confirm them. If a constant density, secularly expanding core is confirmed, theory and observations of globular clusters may be brought into agreement.

This work was supported in part by NASA grant NAGW-246.

REFERENCES

Aarseth, S. J.: 1968, *Bull. Astron.*, **3**, p. 105.
Aarseth, S. J.: 1974 *Astron. Astrophys.*, **35**, p. 237.
Agekian, T. A. and Anosova, Zh. P.: 1968, *Astrophys.*, **4** , p. 11.
Antonov, V. A.: 1962, *Vestn. Leningr. Gros. Univ.*, **7** , p. 135.
Cohn, H.: 1979, *Ap. J.*, **234**, p. 1036.
Cohn, H.: 1980, *Ap. J.*, **242**, p. 765.
Goodman, J.: 1984, *Ap. J.*, **280**, in press.
Harrington, R. S.: 1972, *Celest. Mech.*, **6**, p. 322.
Heggie, D. C.: 1975, *M. N. R. A. S.*, **173**, p. 729.
Heggie, D. C.: 1984, *M. N. R. A. S.*, **206**, p. 179.
Heggie, D.C.: 1985, this volume.
Hénon, M.: 1964, *Ann. Astrophys.*, **27**, p. 83.

Hoerner, S. von: 1963, *Z. Astrophys.*, **57**, p. 47.
Inagaki, S. and Lynden-Bell, D.: 1983, *M. N. R. A. S.*, **205**, p. 913.
Kustaanheimo, P. and Steifel, E. J.: 1965, *J. Math.*, **218** , p. 204.
Lightman, A. P.: 1982, *Ap. J. Lett.*, **263**, p. L19.
Lightman, A. P. and Fall, S. M.: 1978, *Ap. J.*, **221**, p. 567.
Lynden-Bell, D. and Wood, R.: 1968, *M. N. R. A. S.*, **138**, p. 495.
Lynden-Bell, D. and Eggleton, P. P.: 1980, *M. N. R. A. S.*, **191**, p. 483.
Shapiro, S. L. and Marchant, A. B.: 1978, *Ap. J.*, **225**, p. 603.
Spitzer, L. and Hart, M. H.: 1971, *Ap. J.*, **164**, p. 399.
Spitzer, L.: 1985, this volume.
Stodolkiewicz, J. S.: 1982, *Acta Astron.*, **32**, p. 63.
Sugimoto, D. and Bettwieser, R.: 1983, preprint.
Szebehely, V.: 1972a, *A. J.*, **77**, p. 169.
Szebehely, V.: 1972b, *Celest. Mech.*, **6**, p. 84.

DIRECT N-BODY SIMULATIONS WITH A RECURSIVE CENTER OF MASS REDUCTION AND REGULARIZATION

J.G. Jernigan
University of California
Space Sciences Laboratory
Berkeley, Ca. 94720

ABSTRACT. We report on an investigation of the core collapse of an initially bound system of 1024 point masses based on a new numerical method which incorporates a recursive center of mass reduction of the N-body system and includes regularization of the two body encounters. A tree which organizes the point masses into a logarithmic hierarchy is the primary feature of the numerical algorithm. The spatial structure of the tree allows for rapid force calculation while the associated temporal structure allows for variable time steps. Simulations are calculated for several core collapse times in an effort to follow the end point evolution of the system. The point-like nature of the interactions are maintained throughout the simulations.

1. INTRODUCTION

The dynamical evolution of star clusters has been studied by a variety of numerical methods including direct integration of the gravitational N-body equations for point-like masses. Direct methods provide realistic description of the system without any simplifying assumptions about the evolution. In contrast, the alternative methods of simulation including the fluid dynamical approach, the Fokker-Planck approximation, Monte Carlo models, and simple analytic models require certain approximations which fail for "small N" systems in which discreteness effects dominate. In particular, the nature of the end point evolution of an initially spherically symmetric system of N point-like masses is an important subject in stellar dynamics. Astrophysical systems such as open clusters ($N = 1000$) and Globular clusters ($N = 1{,}000{,}000$) are the principal motivation for the study of idealized point-like N-body models (see Lightman and Shapiro 1978). We present results of a numerical experiment in which a 1024 body system is evolved past core collapse utilizing a new method for the direct integration of the gravitational N-body equations. The results of this single experiment are compared to the results obtained by the alternative methods.

2. NUMERICAL METHOD

The new method for the direct numerical integration of the N-body equations is based on a coordinate transformation in which masses are grouped in clusters recursively and are thereby organized in a tree-like structure. Each node in the tree is associated with the center of mass and momentum of the members of that cluster. This analytic transformation generates a new set of N-body equations which describe the dynamics of the nodes of the tree instead of the

J. Goodman and P. Hut (eds.), Dynamics of Star Clusters, 275–284.

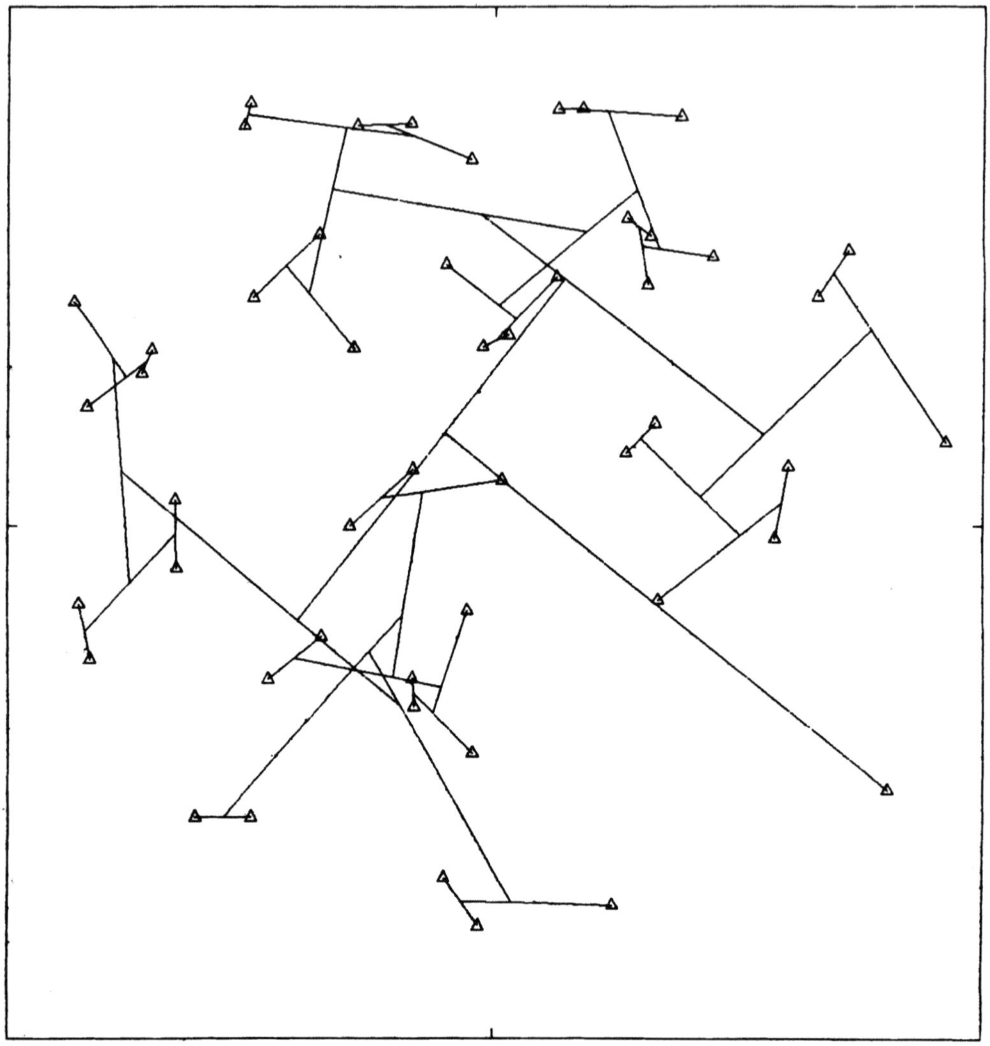

Figure 1. Two dimensional example of a tree generated by a recursive center of mass reduction. Triangles are mass points. Intersections of lines mark the center of mass of all mass points which are members of that node. The tree is arranged to match the clustering of the mass points.

particles. The system of equations is completely equivalent to the original N-body equations. Figure 1 illustrates the tree structure for a two dimensional example. Every possible tree generates a correct set of equations, but only those trees which are arranged to mimic the clustering of the particles lead to a numerically tractable scheme. The equation of each node of the tree is in the form of a perturbed two body problem, and therefore amenable to Kustaanheimo-Stiefel (KS) regularization (Kustaanheimo, 1964 and Stiefel, 1965). This leads to a numerical scheme which handles binaries in a natural manner. The properties of the integration scheme are summarized as follows; (1) rapid force calculation with tree structure, (2) rapid numerical integration with individual time steps for nodes, (3) regularization of all two body encounters, (4) accurate integration of "small N" subsystems such as binary binary collisions.

Several numerical experiments have been calculated in order to test the veracity of the code. These include strongly interacting triple systems in which tens of close encounters and binary/single exchanges occur before ultimate disassociation of the system. Such resonant "small N" systems are the most difficult to integrate since the harmonic oscillator of the KS regularization for each node is strongly perturbed and therefore small time steps are required. Such interactions are rare and of short duration in "large N" systems.

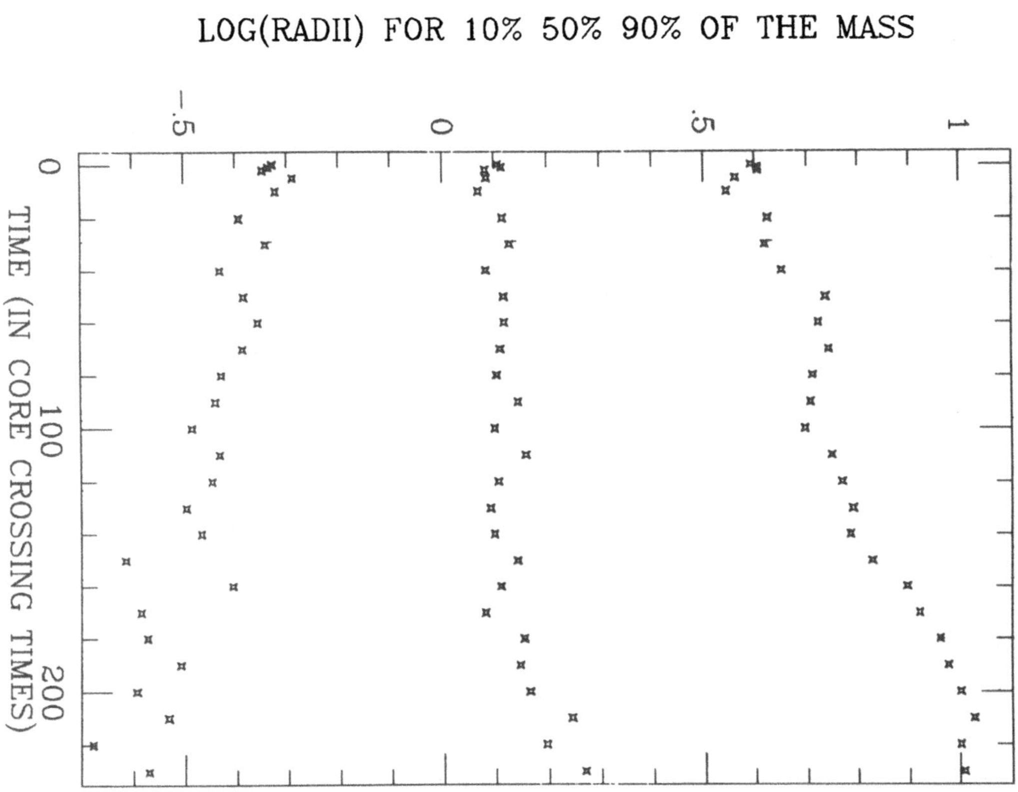

Figure 2. Display of the radii containing 10%, 50%, and 90% of the particles of an approximately spherically symmetric 250 body problem. The initial condition is a Plummer model.

Also the code was tested for a sample "large N" problem which has been independently calculated with alternative methods. A Plummer model with 250 equal masses was simulated for several hundred crossing times and compared to previous experiments (Aarseth, Henon, Wielen 1974). The results shown in figure 2 are in excellent agreement with previous efforts and clearly show the expected core contraction, halo expansion, and formation of a hard binary during the late stages.

3. CORE COLLAPSE EXPERIMENT

A single numerical experiment with 1024 point-like masses was evolved for ~120 initial crossing times. The initial distribution was a two component Plummer model in which the heavy particles are three times the mass of the light particles and the ratio of the total mass in heavies to the total mass in lights is 0.3. The velocities and spatial coordinates of each component (931 lights and 93 heavies) are generated according to the algorithm in Aarseth, Henon, and Wielen (1974) for a single component Plummer model. The initial core radii and velocity dispersion are the same for the heavy and light components as expected just after violent relaxation. This configuration corresponds to the analytic model E of Lightman and Fall (1978) and was selected to facilitate a direct comparison. The mass equipartition parameter Γ as defined by Lightman and Fall is equal to 3.5 for the initial condition, therefore since $\Gamma > 1$ the system is unstable to mass segregation as would be expected for any realistic stellar system. The criterion for mass segregation instability as noted here is essentially the same as that derived by Spitzer (1969). The left panel of figure 3 is a display of the initial condition in projection. The region shown is ten Plummer radii across and clearly indicates the approximate spherical symmetry and central concentration. The numerical model is scaled such that in one unit of time a typical particle of the initial Plummer distribution will travel one Plummer core radius. The right panel of figure 3 is a display of the system after 100 units of time (initial crossing times). The linear scale is the same as for the left panel clearly indicating the increased central density of the system following core collapse.

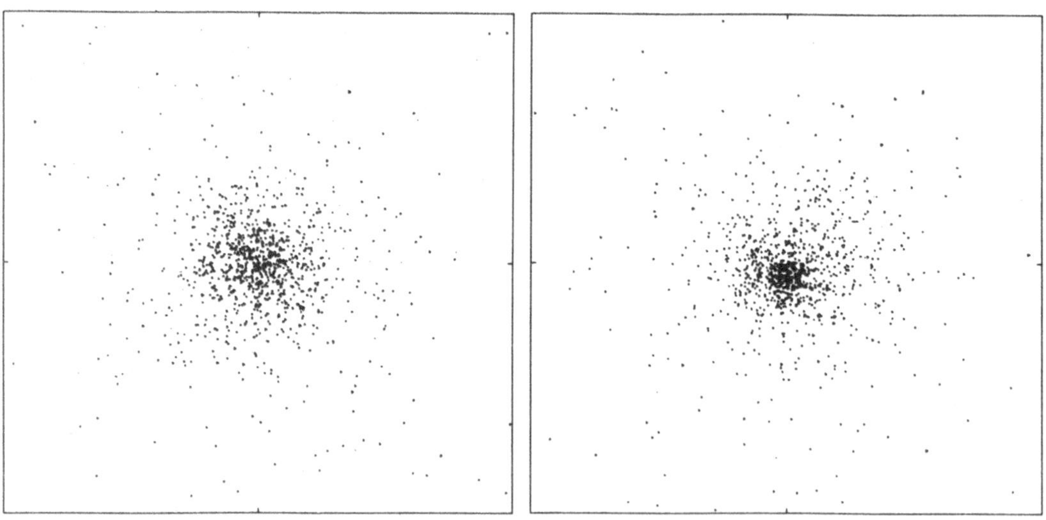

Figure 3. Two dimensional projection of a 1024 body Plummer model with two mass components: initial configuration (left panel) and after 100 initial crossing times (right panel). The linear scale across each panel is ten initial Plummer core radii.

The evolution of the numerical system follows closely the stages of development of the model of Lightman and Fall (1978) outlined as follows. During the conductive regime the heavy particles transfer energy to the light particles and thereby cool with a subsequent decrease in central velocity dispersion and a decrease in core radius since approximate virial equilibrium is maintained. The process continues for about 30 crossing times when thermal equilibrium is reached between the heavy and light components. Even though the two components are strongly gravitationally interacting, the simple scheme of a slow approach to thermal equilibrium of two separate components each in virial equilibrium is accurate. During this period the evolution of the lighter component is slow since the relaxation time is much longer than for the heavy component. From 30 to 60 crossing times the heavy component evolves with a slowly contracting core nearly in thermal equilibrium with the lights. During this period the heavy component evaporates mass into an intermediate halo thereby decreasing the mass of the inner core composed predominately of heavies. Eventually the heavy component becomes sufficiently compact and therefore self gravitating that it undergoes core collapse with an associated increase in central velocity dispersion and a further decrease in core mass. This runaway core collapse is similar to single component core collapse (see figure 2) since the heavy component is now thermally decoupled from the light component. This last phase is rapid, occurring in less than ten initial crossing times and corresponds to the singular event which terminates the analytic model. In the numerical experiment this phase ends with the formation of a single dominate hard binary which halts and reverses the core contraction. Figure 4 tracks the evolving central density of the heavy component. Unfortunately there is no unique model independent indicator of

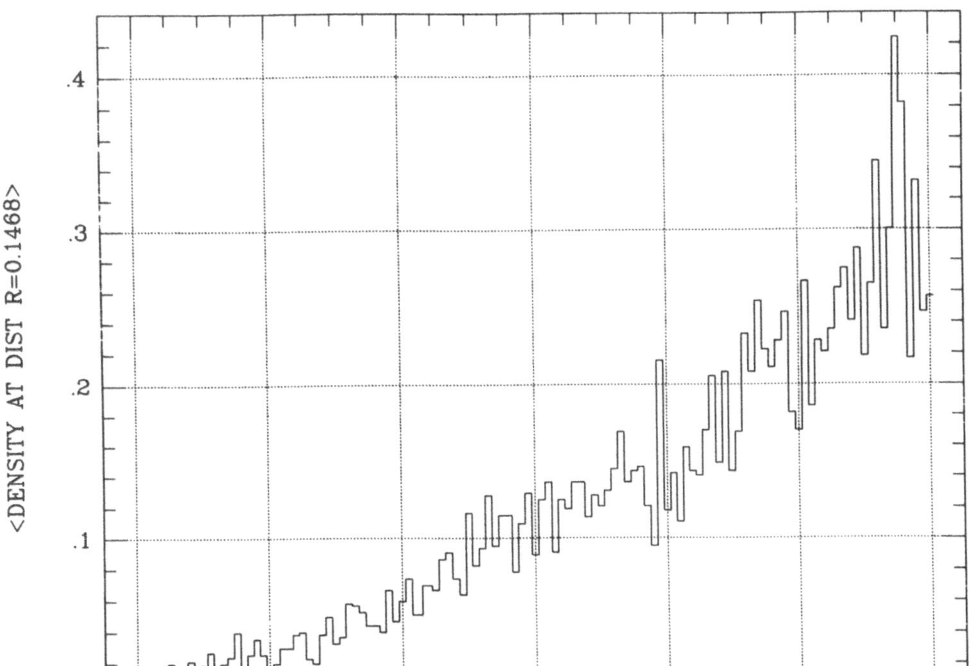

Figure 4. Central density of the heavy mass component as a function of time determined by counting all pairs of heavy point masses with separations less than ~.15.

core collapse, but any function of the distribution which depends on the relative rather than absolute coordinates is numerically suitable. In figure 4 the central density is computed by counting all pairs of heavies with separations less than ˜.15. Any other function of relative coordinates such as potential energy would be desirable, in contrast to model fitting a spherical profile with an assumed point of symmetry. Any core parameter is likely to be ill defined after the formation of hard binaries (after time 60 in this example).

The best indicator of the onset of core collapse is the formation of hard binaries which then dominate the immediate post-collapse phase. Figure 5 is a display of the binding energy of the hardest binary and the sum of the five hardest binaries. The endpoint core collapse is clearly indicated by the rapid growth in the hardness of one or perhaps two binaries. The process is highly stochastic since the identity of the dominate binary is not the same throughout the post collapse evolution. Episodes of hardening and softening of the binaries are correlated with rapid fluctuations of the velocity dispersion of a core of ˜30 predominately heavy point masses.

The experiment required about one hour of CPU time per crossing time on a VAX11/780 with a floating point accelerator. Approximately half of the time is required for the post collapse phase. The clear advantage of a direct simulation is the possibility of investigating the post-collapse phase without the encumbrance of special assumptions.

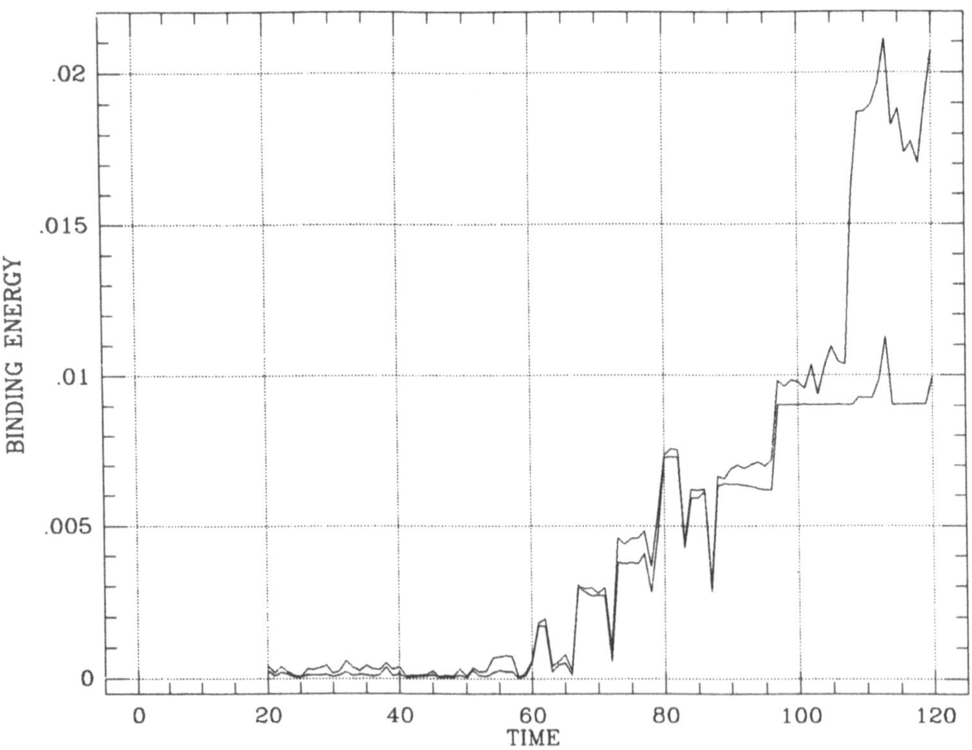

Figure 5. Binding energy of the most tightly bound binary and the sum of the five most tightly bound binaries as a function of time in units of initial crossing time.

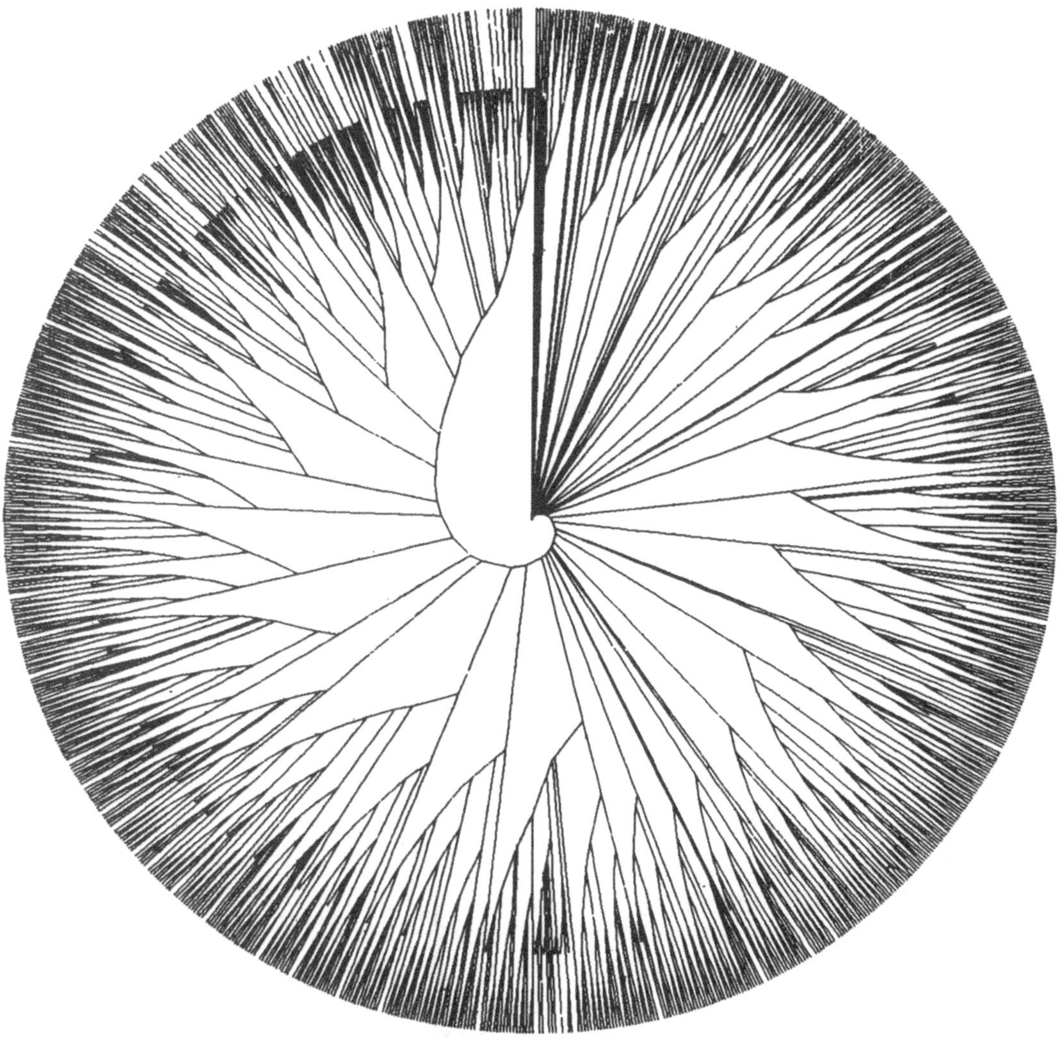

Figure 6. Unfolded display of the node tree for the 1024 body Plummer model after 100 initial crossing times. The radial position of a node is proportional to the logarithm of its mass with the center corresponding to the entire system and the outer perimeter corresponding to an individual light mass point. The order of the mass points on the outer rim is selected so that no interior lines cross. The angular position of a node or mass point indicates the radial position in the three dimensional system with the distant halo at the top right, the core at the top left, and with clockwise rotation corresponding to decreasing radial coordinate.

The evolution of the recursive tree for the system provides an alternative description of the dynamics of the numerical experiment. Since each node of the tree has an associated dynamical equation of the KS regularized form, the system can be fully described by the properties of each node including the spatial displacement, relative velocity, and binding energy or frequency. Figure 6 is a two dimensional unfolded display of the three dimensional tree structure of the system after 100 initial crossing times. The radial position of a node is proportional to the logarithm of the mass of the node. The angular position is selected such that the lines connecting nodes to subnodes do not cross and nodes (and mass points) are organized approximately according to distance from the center of the system with the most distant components at the top right, clockwise for decreasing radial distance, reaching the core at top left. The state of each node combined with the global structure of the tree (figure 6) are a complete dynamical description of the system.

The advantage of the tree-like description of the system is illustrated in a scatter plot of the mass versus binding energy per unit mass for each node (see figure 7). Unlike a display of particle properties, all the major dynamical components of the system are clearly visible in the plot, including escapers, the formation of a two component core, and the existence of a single dominate binary. This representative scatter plot of the properties of nodes illustrates the natural way in which dynamical constructs of all spatial and mass scales are represented in the tree structure.

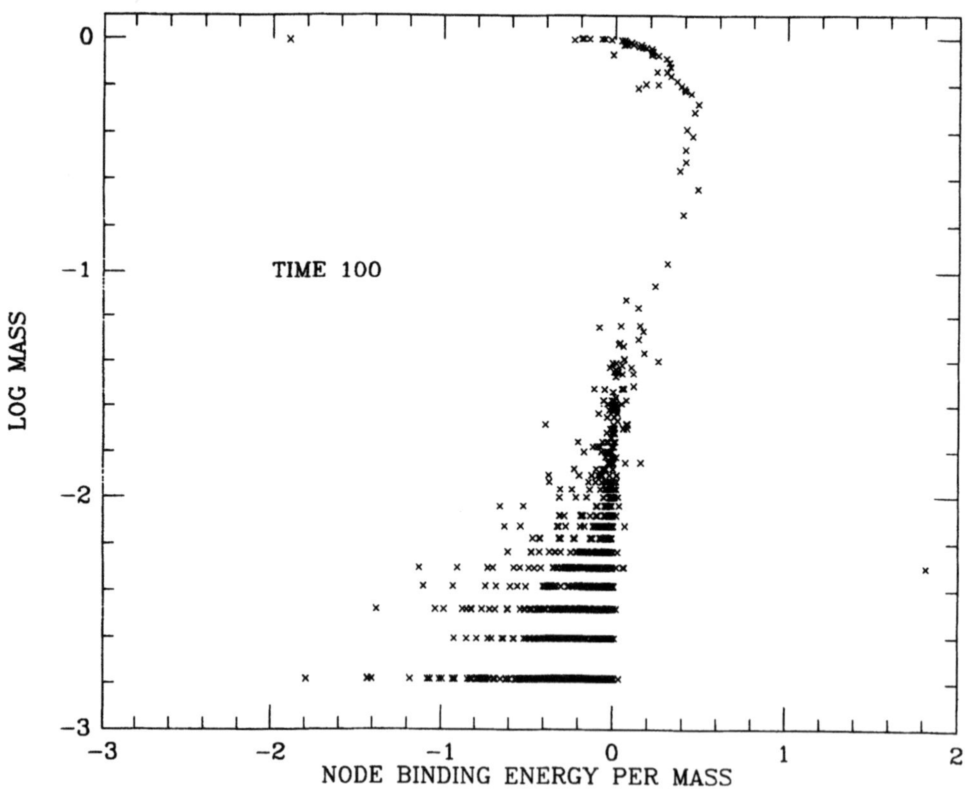

Figure 7. Scatter plot of node mass versus binding energy per unit mass after 100 crossing times. Positive (negative) binding energy corresponds to a bound (unbound) node. The most massive unbound nodes indicate escapers. The core of the system appears as the curved portion at the top of the diagram. The single dominate hard binary is in the lower right corner.

3. SUMMARY

A recursive and regular transformation of the gravitational N-body equations is derived in the form of a hierarchical tree of nodes, each individually associated with a KS regularized two body equation or perturbed harmonic oscillator. This transformation provides both an efficient numerical method for the simulation of a large system and a new approach for diagnosing the nature of the dynamical evolution of the system under study. The specific numerical experiment, the core collapse of 1024 body point mass system is intended as a representative and idealized example of the methodology. Any direct N-body technique (also see Aarseth in this volume) has the advantage of assumption free simulation of the "small N" core of the late and post-collapse phases. The disadvantage of the direct approach is the limitation on the size of N that is practical. Alternative methods utilizing Fokker-Planck or Monte-Carlo techniques readily deal with the large N boundary associated with a small collapsing core. McMillan and Lightman (1984 and in this volume) have developed a "hybrid code" which combines the advantages of both types of methods. Clearly a variety of efficient hybrid are needed to fully explore the post-collapse evolution of large systems.

Acknowledgements

We thank David Porter for his contributions to the implementation of the numerical integrator, the Astronomy Department of the University of California at Berkeley for the use of the VAX11/780, and Theda Crawford for the preparation of this manuscript. This work was supported in part by the California Space Institute.

References

Aarseth, S.J., Henon, M., and Wielen, R., 1974, *Astron. Astrophys., 37,* 183.

Kustaanheimo, P., 1964, *Ann. Univ. Turku. Ser. AI., 73.*

Lightman, A.P. and Fall, S.M., 1978, *Ap. J., 221,* 567.

Lightman, A.P. and Shapiro, S.L., 1978, *Rev. Mod. Phys., 50,* No. 2, 437.

McMillan, S.L. and Lightman, A.P., 1984, *Ap. J.,* in press.

Spitzer, L., 1969, *Ap. J. (Letters), 158,* L139.

Stiefel, E., 1965, *J. Reine Angew. Math., 218,* 204.

DISCUSSION

GOODMAN: This is very clever and interesting work, but could you say something about what seems to be the hardest part of the calculation, namely how to choose the tree in the optimal manner? I should think that there are of the order of N! trees possible for N particles, and that the optimal tree for one particle pair would not be optimal for another pair.

JERNIGAN: First of all, every possible tree produces a new set of analytic equations which are correct. However, only a small fraction of the trees produce a set of numerically stable equations. One way to generate a numerically useful tree is to build the tree recursively by first connecting the closest two bodies in the system and then replacing the two bodies by a particle at their center of mass, thereby reducing a N body system to an N-1 body system. This process is recursed until the entire tree is built. This process can also be shown to require only NlogN steps, since one can also recursively add a new particle to a tree for N particles and produce a tree for N+1 particles. In the actual integrator the process is easier since one only modifies the tree at a previous step to generate the tree for the next step.

TERLEVICH: I run models of 1000 bodies (Aarseth regularized code) to compare them with open clusters (therefore they have included mass loss through stellar evolution, galactic tidal field, a power low mass function (α = 2.75)). I do not find signs of core collapse, which I attribute to the combined action of binaries and mass-loss (which occur frequently when the stars are binary components).

JERNIGAN: This is an interesting comment and indicates the large difference between open and globular clusters.

TOOMRE. Would you care to compare the speed or cost of using your algorithm, for instance on some 1000-body problem, versus whatever Aarseth's or anyone else's code would have yielded in the same circumstances?

JERNIGAN: My code requires $\sim$1 hour of VAX780 CPU time per crossing time for a 1000 body Plummer model. I have run problems for N up to 10,000 and see a $N^{1.3}$ dependence for the force calculation which is the dominant part of the computation. This is somewhat faster than the $N^{1.6}$ for the Ahmad-Cohen method that Aarseth uses. The $N^{1.3}$ dependence over the range N=100 to N=10,000 is consistent with NlogN dependence asymptotically. The particular 1000 body problem which Aarseth presented earlier today runs about 3 times slower than the problem that I have presented. However, it is not exactly the same problem, therefore a definite comparison is not possible. For large N, I would predict that my code would scale NlogN whereas the Ahmad-Cohen scheme would ultimately scale N^2. However, for any numerically realistic N the asymptotic behavior may not be relevant.

ROTATION AND FLATTENING OF GLOBULAR CLUSTERS

S. Michael Fall[1,3] and Carlos S. Frenk[2,3]
1) Institute of Astronomy, Madingley Road, Cambridge CB3 OHA England
2) Astronomy Centre, University of Sussex, Falmer, Brighton BN1 9QH England
3) Institute for Theoretical Physics, University of California, Santa Barbara, California 93106 USA

1. INTRODUCTION

Pease and Shapley (1917) first remarked on the apparent flattening of several Galactic globular clusters, a view that has been confirmed by many subsequent studies. Tidal stresses, internal rotation, and velocity anisotropies can cause deviations from sphericity in stellar systems. In general, we might expect globular clusters to have some angular momentum at the time of formation and, if they collapsed from flattened initial conditions, to have anisotropic pressure support. Since the velocity distributions within the clusters can be altered by a variety of internal and external processes, their shapes are expected to evolve. In this article, we review the methods for measuring ellipticities and the results that have emerged from such studies. Our main purpose, however, is to discuss the processes that determine the shapes of globular clusters and the ways in which they change with time.

Dynamical evolution may be observable in the rich star clusters of the Magellanic Clouds. Searle, Wilkinson, and Bagnuolo (1980, hereafter SWB) have classified these objects into seven types on the basis of their spectrophotometric properties. An approximate age-calibration of the SWB sequence, using Hodge's (1983) compilation of color-magnitude diagrams, gives

$$\log(\text{age/yr}) \approx 0.5\ (\text{SWB type}) + 6.6 \qquad (1)$$

The clusters in the Large Magellanic Cloud (LMC) with SWB types I-IV certainly are, and those with SWB types V-VII probably are, members of the disk population (Freeman, Illingworth, and Oemler 1983). However, the richest clusters of all ages are considerably more massive than the open clusters in our Galaxy. As is now customary, we refer to these objects as globular clusters even though very few of them may be members of the halo population in the Magellanic Clouds.

J. Goodman and P. Hut (eds.), Dynamics of Star Clusters, 285–296.

2. MEASUREMENTS OF ELLIPTICITY

The shapes of globular clusters are usually quantified in terms of apparent ellipticities, $\varepsilon_a \equiv 1 - b/a$, where a and b are respectively the major and minor axes of the projected images. Since most of the light comes from a relatively small number of bright stars (red giants in old clusters and early-type stars in young clusters), the accuracy of the measurements is limited mainly by sampling errors. The ellipticities derived from star counts are consequently more reliable than those derived from smooth surface brightness maps. Kholopov (1953) used star counts for 19 Galactic globular clusters and we used them for 12 young LMC clusters (Frenk and Fall 1982). Geisler and Hodge (1980) estimated the ellipticities of 25 old LMC clusters from microdensitometer scans but, by undersampling the images, they appear to have introduced large systematic errors. Geyer and Richtler (1981) used Agfa contourfilm for 25 clusters of all ages in the LMC and Geyer, Hopp, and Nelles (1983) used the same method for 20 globular clusters in our Galaxy and 4 in the Magellanic Clouds.

The simplest way to measure ellipticities is by eye. Shapley and Sawyer (1927) used this method for 75 Galactic globular clusters and we used it for 93 globular clusters in the Galaxy and 52 in the LMC (Frenk and Fall 1982). Our measurements, which were made from polaroid enlargements of the Palomar and SRC Sky Surveys, apply to the regions near (1-2) r_h where r_h is the median radius of a cluster (i.e. the radius containing half of the light in three dimensions). Since the results are somewhat subjective, they need to be checked by other methods. Figure 1 is a comparison between our eye-measurements and determinations from star counts by Kholopov and by us in the same regions of the clusters. The standard deviation for individual ellipticities is about 0.05 and there is no systematic difference between the results from the two methods.

Our eye-measurements are compared in Fig. 2 with the ellipticities derived from Agfa contourfilm by Geyer and Richtler (1981) and Geyer, Hopp, and Nelles (1983). The agreement for most of the sample is reasonable but the systematic deviations for several of the flattest

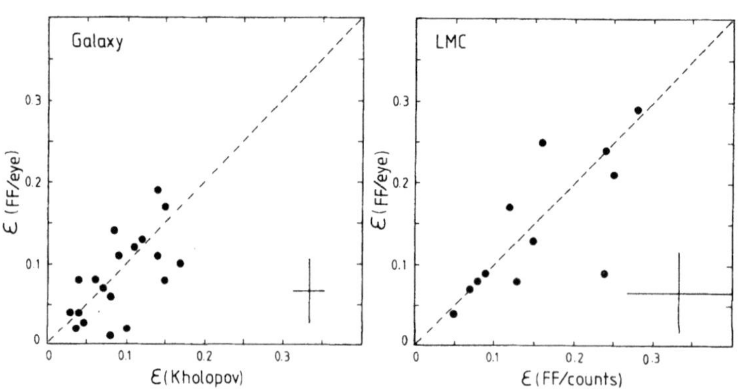

Figure 1. Comparison of ellipticities from eye-measurements and star counts in the same regions of the clusters.

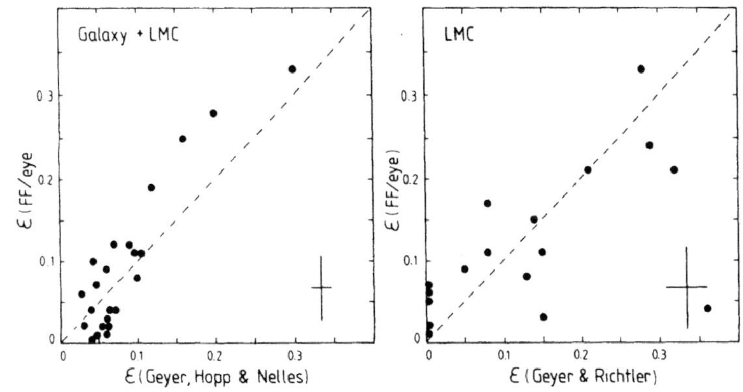

Figure 2. Comparison of ellipticities from eye-measurements and from Agfa contourfilm. The left-hand panel differs slightly from Fig. 3 of Geyer et al. because we have plotted the means of their measurements $\bar{\varepsilon}$ rather than their extrapolated values $\langle\varepsilon\rangle$.

clusters led Geyer et al. to question the validity of our results. In fact, most of the difference arises because their ellipticities are averages over large regions of the clusters whereas the eye-measurements apply to a smaller range of radii. The discrepant clusters in the Galactic sample are 47Tuc (NGC104), ωCen (NGC5139), and M19 (NGC6273); and, in at least two of them, Geyer et al. found strong ellipticity gradients. At the radii where the eye-measurements were made, the agreement with the results derived from Agfa contourfilm is excellent. This is shown in Fig. 3 for the four clusters with the most extensive data, including two of the discrepant clusters.

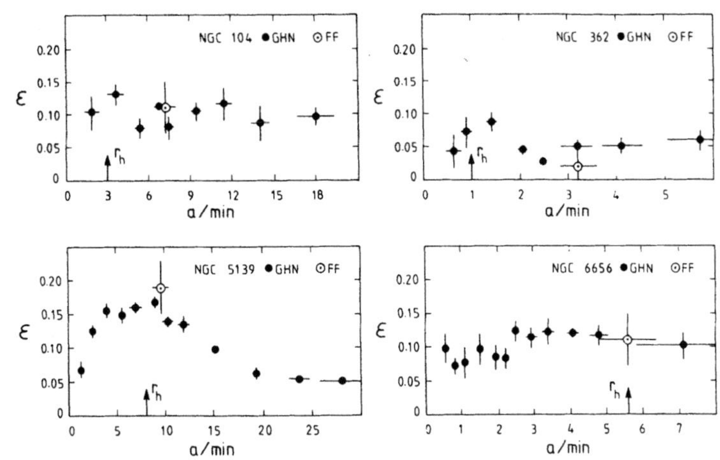

Figure 3. Radial dependence of ellipticity derived from Agfa countourfilm by Geyer et al. in four Galactic globular clusters. The open circles represent our eye-measurements.

While improved determinations of ellipticities would be highly desirable, the eye-measurements have allowed us to explore the statistical properties of large samples (Frenk and Fall 1982; Fall and Frenk 1983). The main results can be summarized as follows. In both the Galaxy and the LMC, globular clusters appear to be oriented at random. After deprojecting the apparent shapes of the clusters, the mean of the true ellipticities is $\langle\varepsilon\rangle = 0.12 \pm 0.01$ for the Galactic sample and $\langle\varepsilon\rangle = 0.18 \pm 0.02$ for the LMC sample. There is a general tendency for the ellipticities of the LMC clusters to decrease along the SWB sequence. Most of the change occurs near types III and IV, which corresponds to ages of a few times 10^8 yr by Eq. (1). On average, the youngest LMC clusters are nearly as flat as elliptical galaxies and the oldest LMC clusters are nearly as round as the globular clusters in our Galaxy.

van den Bergh (1983) found that the flattest clusters in the LMC are generally among the brightest. Since the clusters become fainter as they get older, this may be in part a reflection of the correlation between ellipticity and age. Without a theoretical model for the origin and evolution of the clusters, it is not possible to say which relation has the greatest physical significance. van den Bergh (1984) also found a correlation between the visual extinction of Galactic globular clusters and their apparent ellipticities. He suggests, rather arbitrarily, that the clusters with $A_V > 1.0$ are distorted by patchy obscuration. This is true for M19, which appears kidney-shaped at visual wavelengths and suffers differential redding across the image (Shapley 1930; Harris, Racine, and deRoux 1976). A smooth contribution to A_V would not affect the shape of a cluster and, apart from M19, there is no evidence that the other clusters in our sample are distorted by patchy obscuration.

The absence of alignment with the structural features of the Galaxy and the LMC rules out tidal stresses as the cause of flattening in globular clusters. This is not surprising since the densities are too high in the regions where ellipticities are measured for tidal stresses to be important (King 1961). The remaining possibility is that globular clusters owe their non-spherical shapes to internal rotation, velocity anisotropies, or some combination of both effects. Rotation appears to be the dominant cause of flattening in four of the Galactic globular clusters for which stellar radial velocities are available: ωCen (Freeman and Seitzer 1984; Meylan and Mayor 1984), 47Tuc (Mayor et al. 1984; Freeman and Da Costa 1984), M13 and M92 (Lupton, Gunn, and Griffin 1984). The other cluster with extensive data, M3, is nearly round and shows no signs of rotation. Using proper motions, Cudworth (1979) has found direct but weak evidence for velocity anisotropies in the outer parts of this cluster. The radial velocities measured by Gunn and Griffin (1979) are also fitted best by a distribution function that is elongated toward the center of M3.

3. THE TENSOR VIRIAL THEOREM

The tensor virial theorem provides a useful connection between the global rotation, velocity anistropy, and shape of a self-gravitating system (Chandrasekhar 1969). In a steady state, this takes the form

$$2T_{kl} + \Pi_{kl} + W_{kl} = 0 , \tag{2}$$

where T_{kl}, Π_{kl}, and W_{kl} are respectively the kinetic energy, pressure, and potential energy tensors. For an axisymmetric system with no radial or vertical streaming motions, the only nontrivial components of Eq. (2) are

$$\begin{aligned} &M(v^2 + \sigma_r^2 + \sigma_\phi^2) + W_{11} + W_{22} = 0 , \\ &M\sigma_z^2 + W_{33} = 0 , \end{aligned} \tag{3}$$

where M is the total mass, v is the root mean square rotation velocity, and σ_r, σ_ϕ, and σ_z are the velocity dispersions in the radial, azimuthal, and vertical directions when integrated over all positions.

To proceed further, we impose the additional restriction that the system be stratified on similar oblate ellipsoids with an axial ratio $q \equiv 1 - \varepsilon$ but with an otherwise arbitrary density profile. Equations (3) can then be combined into the form

$$(3 + \alpha + \beta)(T_{rot}/T_{rand}) + (\alpha+\beta) = \frac{(1+2q^2)\cos^{-1}q - 3q(1-q^2)^{1/2}}{q(1-q^2)^{1/2} - q^2\cos^{-1}q}$$

$$\approx \frac{8}{5}\,\varepsilon \text{ for } \varepsilon \ll 1, \tag{4}$$

where, by definition,

$$\alpha = (\sigma_r/\sigma_z)^2 - 1, \quad \beta = (\sigma_\phi/\sigma_z)^2 - 1,$$

$$T_{rot} = \frac{1}{2} Mv^2, \quad T_{rand} = \frac{1}{2} M(\sigma_r^2 + \sigma_\phi^2 + \sigma_z^2) \tag{5}$$

(Binney 1978). For a fixed value of T_{rot}/T_{rand}, the ellipticity depends on anisotropic stresses only through the combination $\alpha+\beta$ and the case $\alpha=\beta=0$ corresponds to a rotationally supported system. [The approximation suggested by King (1961) differs from Eq. (4) by a factor of 3 in this limit.] If the phase space distribution is a function only of the energy E and the angular momentum J_z about the symmetry axis, then $\sigma_r = \sigma_z$. The converse is not necessarily true for global velocity dispersions but in practice $\sigma_r \neq \sigma_z$ will usually imply the existence of non-classical integrals. The price we pay for the simple description given by the tensor virial theorem is the loss of information about local variations in the ellipticity, rotation, and velocity anisotropy.

Two-body diffusion causes a system with initial anisotropies to evolve quasistatically toward a state with isotropic pressure support. The characteristic time-scale for this process is the deflection time, defined as $\tau_D \equiv \langle u^2\rangle / \langle(\Delta u_\perp)^2\rangle$ where $\langle u^2\rangle^{1/2}$ is the local velocity dispersion and $\langle(\Delta u_\perp)^2\rangle$ is the diffusion coefficient for a test star in the direction perpendicular to its motion (Spitzer 1956). This is comparable with the relaxation time $\tau_r \equiv \langle u^2\rangle/3\langle(\Delta u_{\parallel})^2\rangle$ where $\langle(\Delta u_{\parallel})^2\rangle$ is the diffusion coefficient for a test star in the direction parallel to its motion. Thus, we expect the velocity distribution at each point in a cluster to isotropize at nearly the same rate as it relaxes. Similarly, we expect the global anisotropy parameters α and β and the ellipticity near the median radius to change on small multiples of the reference relaxation time τ_{rh}, which is defined as the value of τ_r at the mean density within r_h (Spitzer 1975). The N-body models and the Fokker-Planck integrations discussed below confirm this expectation.

Since most of the old LMC and Galactic globular clusters have ages greater than τ_{rh}, they should have nearly isotropic velocity distributions in the regions where ellipticities are measured. Setting $\alpha = \beta = 0$ in Eq. (4) and deprojecting the ellipticities measured by eye, we find $\langle(v/\sigma)^2\rangle^{1/2} = 0.47 \pm 0.03$ for Galactic globular clusters and $\langle(v/\sigma)^2\rangle^{1/2} = 0.56 \pm 0.06$ for the LMC clusters with SWB types V-VII (Fall and Frenk 1983). As the result of residual anisotropies, these estimates may be slightly too large but they are in rough agreement with the values of v/σ derived from stellar radial velocities in the four clusters mentioned in the previous section. If globular clusters form by aspherical collapse and incomplete violent relaxation, we would expect them to have anisotropic velocity distributions when they first reach dynamical equilibrium (Aarseth and Binney 1978). This may account for the flattened shapes of the LMC clusters with ages smaller than their relaxation times. Unfortunately, a direct test of this conjecture, which would require proper motions for inidividual stars, is virtually impossible.

4. N-BODY MODELS

To simulate the evolution of globular clusters with initially anisotropic velocity distributions, we computed a series of models with equal-mass particles using Aarseth's (1984) code. The validity of this approach can be justified as follows. In a typical globular cluster, the dynamical, relaxation, and close-encounter times are in the ratios $\tau_{dh}:\tau_{rh}:\tau_{ch} \approx 1:7 \times 10^3:4 \times 10^5$. In an N-body model, there is some freedom to alter the characteristic time-scales by adjusting the number of particles and the softening parameter in the interaction potential $\phi(r) = -G/(r^2 + s^2)^{1/2}$. We chose $N = 400$ and $s = r_h/16$, which implies $\tau_{dh}:\tau_{rh}:\tau_{ch} = 1:14:700$. Although these time-scales do not differ by as much as in globular clusters, they are in the correct sequence and are fairly well separated. Thus, the global properties of a globular cluster are modelled accurately by integrating the equations of motion for many fewer particles.

As the initial conditions for the N-body simulations we chose a slowly rotating, oblate spheroid with a density profile resembling that of a King model. The initial concentration parameter we adopted, $c \approx 1.3$, is typical of the globular clusters in the LMC (Chun 1978). The initial velocities of the particles were assigned randomly in such a way that the distribution function was the same throughout the system and approximately in a steady state. We computed five models with different statistical realizations of two sets of initial conditions and then combined the results into two ensembles. Ensemble A (with 3 models) has an initial ellipticity of $\varepsilon_i = 0.4$ and initial anisotropy parameters of $\alpha_i = 0.3$ and $\beta_i = 0.5$; ensemble B (with 2 models) has $\varepsilon_i = 0.6$, $\alpha_i = 0.8$, and $\beta_i = 1.2$. In both cases, the initial streaming motion consists of solid-body rotation about the z axis with $T_{rot}/T_{rand} = 0.05$, which corresponds to $v/\sigma \gtrsim 0.4$. Thus, rotation plays only a minor role in the early stages but it becomes more important as the velocity anisotropies decay.

The N-body simulations were stopped after a time equal to $175\tau_{dh}$ or equivalently $13\tau_{rh}$, which is not long enough for the core to collapse in isolated systems with equal-mass particles (Cohn 1980). Since only one particle escaped from one of the five models, evaporation has no influence on their evolution. The models remained axisymmetric at all times and we obtained a global estimate of the ellipticity ε_{ML} by a maximum likelihood method and an estimate for the inner half of the mass ε_h by White's (1978) ranking method. These techniques and other properties of the simulations are described in detail in a separate paper (Fall and Frenk 1984). Figure 4a shows the ellipticities as a function of time in units of the instantaneous value of τ_{rh}. In both ensembles, ε_{ML} and ε_h decrease by a factor of 2 on time-scales of about $5\tau_{rh}$. After that, the evolution is slower and, by the end of the simulations, the ellipticities appear to be close to their asymptotic values. The arrows in Fig. 4a indicate the times equal to the local relaxation time for radii containing 30, 50, and 70 percent of the mass. As successive shells relax, weak ellipticity gradients develop, which are reflected in the differences between ε_{ML} and ε_h.

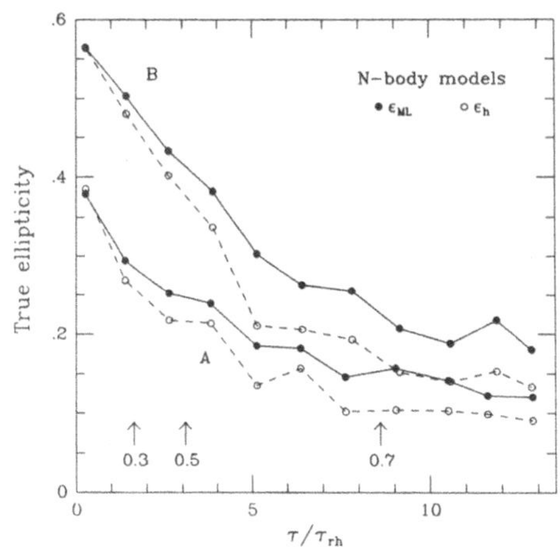

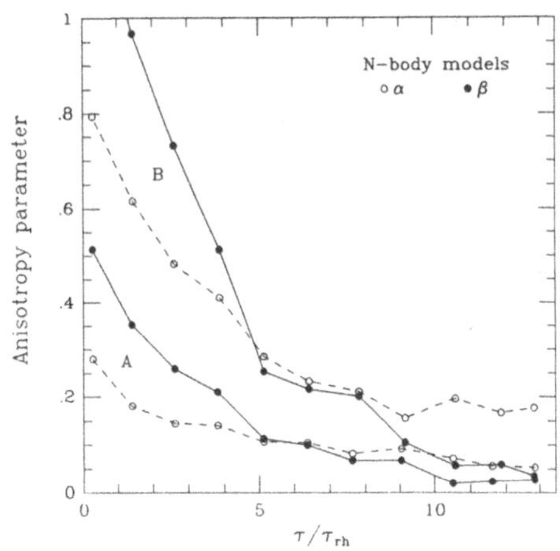

Figure 4. Evolution in the two ensembles of N-body models: (a) ellipticities ε_{ML} and ε_h from the maximum likelihood and ranking methods, (b) anisotropy parameters α and β in the radial and azimuthal directions. Each symbol represents an average over 24 different times separated by $0.5\tau_{rh}$.

The evolution of the anistropy parameters for both ensembles is shown in Fig. 4b. As expected, α and β decrease at nearly the same rate as the clusters become rounder although some residual anisotropy remains at the end of the simulations. This probably explains why the two ensembles, which have the same total angular momentum, do not have exactly the same final shapes. Despite the development of ellipticity gradients, the tensor virial theorem for similar ellipsoids provides a remarkably good description of the N-body models. The ratio v/σ remains approximately constant but, as a result of residual anisotropies in the final state, it is smaller than the value that would be inferred from Eq. (4) with $\alpha = \beta = 0$. The discrepancy of 30 percent for ensemble A and nearly 100 percent for ensemble B gives a rough indication of the possible systematic errors in the estimates of $\langle(v/\sigma)^2\rangle^{1/2}$ mentioned in the previous section.

5. SIMPLE EVAPORATION MODEL AND FOKKER-PLANCK INTEGRATIONS

Agekian (1958) suggested that the shape of a rotating cluster would change as stars escape by two-body diffusion. As an approximation, he assumed that all of the stars in the unbound tail of a Maxwellian distribution would leave the system with zero energy in each relaxation time. Since the low-velocity part of the distribution is isotropic in the rotating frame, more stars escape in the direction of rotation and the angular momentum per unit mass of the cluster decreases. Agekian applied this simple evaporation model to a Maclaurin spheroid and found that it becomes rounder if the initial ellipticity is less than 0.74. Shapiro and Marchant (1976) obtained the same result and suggested that globular clusters could have been much flatter at the time of formation than they are at present.

The previous arguments can be applied even more simply to a system with an arbitrary density profile if the ellipticity is small. This limit is more appropriate for globular clusters than Agekian's assumptions of arbitrary ellipticity and uniform density. For an oblate system with $\varepsilon \ll 1$ and $\alpha = \beta = 0$, the tensor virial theorem in the form of Eq. (4) gives $\varepsilon \approx 15T_{rot}/8T_{rand}$. In general, we can write

$$T_{rot}/T_{rand} = g(\varepsilon)J^2(-E)G^{-2}M^{-5} \quad , \qquad (6)$$

where M, E, and J are respectively the total mass, energy, and angular momentum of the system and $g(\varepsilon)$ is some function that approaches a constant value in the limit $\varepsilon \to 0$. Thus, at constant energy, the evolution is governed by

$$d(\ln\varepsilon)/dt \approx 2d(\ln J)/dt - 5d(\ln M)/dt \quad , \qquad (7)$$

to lowest order in the ellipticity. By definition, we can also write

$$d(\ln M)/dt = - f_M(\varepsilon)/\tau_r \ , \quad d(\ln J)/dt = - f_J(\varepsilon)/\tau_r \ , \qquad (8)$$

where $f_M(\varepsilon)$ and $f_J(\varepsilon)$ are respectively the fractions of the mass and angular momentum lost in each relaxation time τ_r. These functions tend to constant values in the limit $\varepsilon \to 0$ and, in the simple evaporation model, are given by $f_M \approx 7 \times 10^{-3}$ and $f_J \approx 3 \times 10^{-2}$ (Agekian 1958). Thus, the e-folding time for local changes in the ellipticity is

$$\tau_\varepsilon \equiv - dt/d(\ln\varepsilon) \approx \tau_r/(2f_J - 5f_M) \approx 40\tau_r \, . \tag{9}$$

For changes in the ellipticity near the median radius, the reference relaxation time would seem appropriate and we therefore expect $\tau_\varepsilon \approx 40\tau_{rh}$. This is an order of magnitude larger than the time-scale for the decay of velocity anisotropies and is also larger than the ages of most globular clusters. Shapiro and Marchant's suggestion was based on central relaxation times, which are generally an order of magnitude smaller than reference relaxation times. They therefore overestimated the influence of evaporation on the overall shapes of globular clusters by a similar factor.

A more accurate description of the effects of two-body diffusion in a stellar system is given by the orbit-averaged Fokker-Planck equation. Goodman (1983) used this method to study the evolution of clusters supported either by rotation or pressure. He assumed that the distribution function depends only on E, J_z, and time, which reduces the number of dimensions in the problem from seven to three. Thus, in contrast to the N-body simulations, his method neglects the effects of non-classical integrals. Goodman calculated the evolution of two non-rotating models with anisotropic velocity distributions and equal-mass particles. They were tidally truncated with an initial concentration parameter of $c = 1.9$ and were stopped when the cores collapsed at times equal to $(2\text{-}3)\tau_{rh}$. Over this limited interval, the ellipticities in the Fokker-Planck integrations decreased at roughly the same rate as in the N-body simulations, despite the different assumptions.

Goodman also calculated the evolution of two rotating models with isotropic velocity distributions. In this case, the shapes changed more rapidly than predicted by the simple evaporation model. However, Goodman attributed most of the evolution to two effects other than the escape of stars. The first, which has also been discussed by Aarseth (1966), is the relaxation-driven transfer of angular momentum from the inner to the outer parts of a cluster when the angular velocity decreases with radius. This is the analog of viscous transport in a rotating fluid but is not a local phenomenon because the mean-free path is much longer in a stellar system. The second reason for the faster evolution in Goodman's models is that their density profiles changed. As the cores contract, they release energy and rotate more rapidly while the envelopes, which absorb the energy and expand, rotate more slowly. These effects are essentially absent in our N-body models because they have solid-body rotation and nearly constant density profiles.

6. CONCLUSIONS

The arguments given here suggest that globular clusters become rounder as they get older. Two-body diffusion causes the decay of velocity anisotropies, the evaporation of stars, the viscous transport of angular momentum and the redistribution of matter during core collapse. We therefore expect the ellipticities near the median radii of the clusters to be functions of τ/τ_{rh} where τ is the age and τ_{rh} is the reference relaxation time. For the Galactic globular clusters, which have $\tau \approx 1.5 \times 10^{10}$ yr, the values of τ_{rh} range from 10^8 yr to 10^{10} yr (Spitzer 1975). The N-body and Fokker-Planck models discussed above have equal-mass particles and evolve on small multiples of τ_{rh}; the inclusion of a realistic mass spectrum would probably lead to somewhat faster rates. Thus, globular clusters could be much rounder now than they were at the time of formation, especially if their initial velocity distributions were anisotropic. Evaporation has almost certainly played a less important role since the characteristic time-scale for changes in the overall shapes by this process alone is about $40\tau_{rh}$.

Unfortunately, the masses and radii needed to derive relaxation times are available for only a few of the brightest clusters in the LMC. From the data given by Chun (1978) and Elson and Freeman (1984), we estimate values of τ_{rh} between 10^8 yr and 10^9 yr but the dispersion in the whole LMC sample is likely to be at least as large as in the Galactic sample. Since the ages of the clusters with SWB types III and IV are a few times 10^8 yr, they correspond, very roughly, to $\tau \approx \tau_{rh}$. Thus, the observed decrease in the ellipticities of LMC clusters could be explained by the decay of initial velocity anisotropies; viscous transport of angular momentum could have also played a role. We cannot, however, rule out the possibility that the initial distribution of shapes has varied over the life-time of the LMC. There is some evidence that the youngest clusters are less massive than the oldest and variations in their other properties would not be surprising. Nevertheless, the young LMC clusters will become rounder with time and some contribution to the relation between ellipticity and age from the processes discussed here seems almost inevitable. More accurate estimates of the ages, ellipticities and relaxation times of the LMC clusters would undoubtedly lead to a better understanding of these problems.

ACKNOWLEDGMENTS

This material is based upon research supported in part by the National Science Foundation under Grant No. PHY77-27084, supplemented by funds from the National Aeronautics and Space Administration.

REFERENCES

Aarseth, S.J. 1966, M.N.R.A.S., **132**, 35.
Aarseth, S.J. 1984, in Methods of Computational Physics, eds. J.U. Brackbill and B.I. Cohen (New York: Academic Press), p. 1.

Aarseth, S.J. and Binney, J. 1978, M.N.R.A.S. 185, 227.
Agekian, T.A. 1958, Soviet Astr. - A.J., 2, 22.
Binney, J. 1978, M.N.R.A.S., 183, 501.
Chandrasekhar, S. 1969, Ellipsoidal Figures of Equilibrium (New Haven: Yale University Press).
Chun, M.S. 1978, A.J., 83, 1062.
Cohn, H. 1980, Ap. J., 242, 765.
Cudworth, K.M. 1979, A.J., 84, 1312.
Elson, R.A.W. and Freeman, K.C. 1984, Ap. J., submitted.
Fall, S.M. and Frenk, C.S. 1983, A.J., 88, 1626.
Fall, S.M. and Frenk, C.S. 1984, in preparation.
Freeman, K.C. and Da Costa, G. 1984, in preparation.
Freeman, K.C., Illingworth, G., and Oemler, A. 1983, Ap. J., 272, 488.
Freeman, K.C. and Seitzer, P. 1984, in preparation.
Frenk, C.S. and Fall, S.M. 1982, M.N.R.A.S., 199, 565.
Geisler, D. and Hodge, P. 1980, Ap. J., 242, 66.
Geyer, E.H., Hopp, U., and Nelles, B. 1983, Astr. Ap., 125, 359.
Geyer, E.H. and Richtler, T. 1981, in Astrophysical Parameters for Globular Clusters, eds. A.G.D. Philip and D.S. Hayes (Schenectady: L. Davis Press), p. 239.
Goodman, J. 1983, Ph.D. Thesis, Princeton University.
Gunn, J.E. and Griffin, R.F. 1979, A.J., 84, 752.
Harris, W.E., Racine, R., and deRoux, J. 1976, Ap. J. Suppl. 31, 13.
Hodge, P.W. 1983, Ap. J., 264, 470.
Kholopov, P.N. 1953, Publ. Astr. Sternberg Inst., 23, 250.
King, I. 1961, A.J., 66, 68.
Lupton, R., Gunn, J.E., and Griffin, R.F. 1984, in preparation.
Mayor, M. et al. 1984, Astr. Ap., 134, 118.
Meylan, G. and Mayor, M. 1984, preprint.
Pease, F.G. and Shapley, H. 1917, Contr. Mt. Wilson Obs., 129.
Searle, L., Wilkinson, A., and Bagnuolo, W.G. 1980, Ap. J. 239, 803.
Shapiro, S.L. and Marchant, A.B. 1976, Ap. J., 210, 757.
Shapley, H. 1930, Star Clusters (New York: McGraw-Hill).
Shapley, H. and Sawyer, H. 1927, Harvard Obs. Bull., 852.
Spitzer, L. 1956, Physics of Fully Ionized Gases (New York: Inter-Science).
Spitzer, L. 1975, in Dynamics of Stellar Systems, ed. A. Hayli (Dordrecht: Reidel), p. 3.
van den Bergh, S. 1983, P.A.S.P., 95, 839.
van den Bergh, S. 1984, Observatory, in press.
White, S.D.M. 1978, M.N.R.A.S., 184, 185.

DISCUSSION

MAYOR (Comment to Fall): The v_0 / σ_0 values derived from our measurement of 47 Tuc and ω Cen are strongly in favor of a quasi-isotropic distribution of velocities in these two clusters. In spite of its large relaxation time ω Cen does not show any global evidence of anisotropy.

DYNAMICAL INSTABILITIES IN SPHERICAL STELLAR SYSTEMS

Joshua Barnes

Astronomy Department,
University of California,
Berkeley, CA 94720 USA

Abstract. Equlibrium spherical stellar systems exhibiting instabilities on a dynamical timescale were first studied by Henon (1973), using a spherically symmetric N-body code. We have re-examined Henon's models using an improved code which includes non-radial forces to quadrupole order. In addition to the radial instability reported by Henon, two new non-radial instabilities are also observed. In one, found in models with highly circular orbits, the mass distribution exhibits quadrupole-mode oscillations. In the other, seen in models with highly radial orbits, the system spontaneously breaks spherical symmetry and settles into a tri-axial ellipsoid. These instabilities, which are driven by fluctuations of the mean field, offer some analogies to the well-known dynamical instabilities of a cold disk of stars. While our models are rather artificial, they indicate that dynamical instabilities may be more common in spherical systems than had been thought.

Henon's generalized polytropic models are equlibrium spherical stellar systems based on the distribution function

$$f(r,v) \propto \max(E_1-E,0)^{n-3/2}\ J^{2m},$$

where E_1, n, and m are constants, $E = \phi(r)+\frac{1}{2}v^2$, and $J = rv_t$. Here $n \geq \frac{1}{2}$; if $n = \frac{1}{2}$ then all stars have the same E. The stellar density of these models vanishes at

$$r_1 = (4m+6)/(3m-n+5)$$

in units where G = 1, total mass $M_1 = 1$, and binding energy $T+U = -\frac{1}{4}$. The velocity anisotropy is independent of radius. Instead of m, we use the parameter

$$\kappa = \langle v_t^2\rangle/(\langle v_v^2\rangle+\langle v_r^2\rangle) = (2m+2)/(2m+3).$$

The cases κ = 0, 2/3, and 1 correspont to radial, isotropic, and circular orbits respectively.

J. Goodman and P. Hut (eds.), Dynamics of Star Clusters, 297–299.

We numerically solve Poisson's equation for a given choice of n and κ, obtaining the potential Φ and radius r as functions of the enclosed mass M. To produce an N-body realization of a generalized polytrope, particle radii are selected by applying r(M) to a series of uniformly distributed random numbers, while the radial and transverse velocity components are selected by a von Neumann rejection procedure.

Most of our results were obtained with a "reduced" N-body code (White 1983) in which the mean force field is expressed as a sum of monopole, dipole and quadrupole terms. Individual 2-body interactions are not included, so this code is collisionless. Our key results have been checked with a direct-summation Aarseth code. In general, the two codes yield similar results, although the more expensive direct-summation experiments also show signs of 2-body relaxation.

The first non-radial instability occurs in models with orbits circular enough to initially confine most of the mass to a spherical shell. Figure 1 shows a snapshot of such a model; particles with high angular momentum have been plotted as circles to bring out the quadrupole form of the instability. The net ellipticity of the model, derived from the quadrupole moment, oscillates with a period of approximately half the mean orbital time. In the limit $\kappa = 1$, particles are confined to the surface of a sphere; since all have the same orbital period, any initial perturbation will recur at regular intervals. The self-gravity of a perturbation attracts unperturbed particles, so its amplitude grows with time.

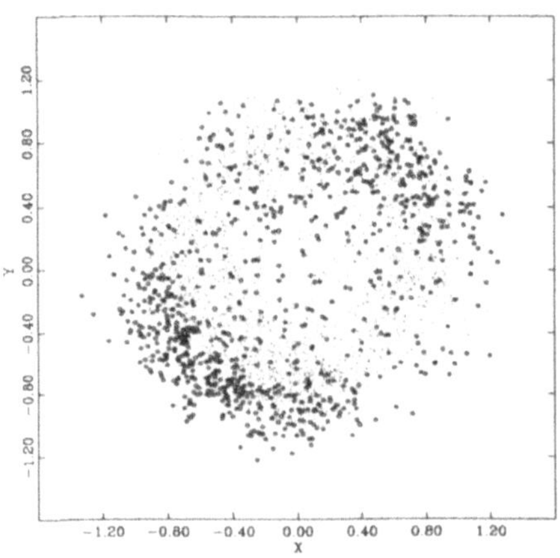

Figure 1. Snapshot of an N-body model of the polytrope $n = \frac{1}{2}$, $\kappa = 11/12$ after 13 time units. Particles with $J \geq 0.8$ are plotted as circles.

The second non-radial instability is seen in models with highly radial orbits. In these cases the system does not oscillate, but quickly settles into a highly tri-axial ellipsoid, as shown in figure 2. No further evolution has been seen in the longest experiments to date.

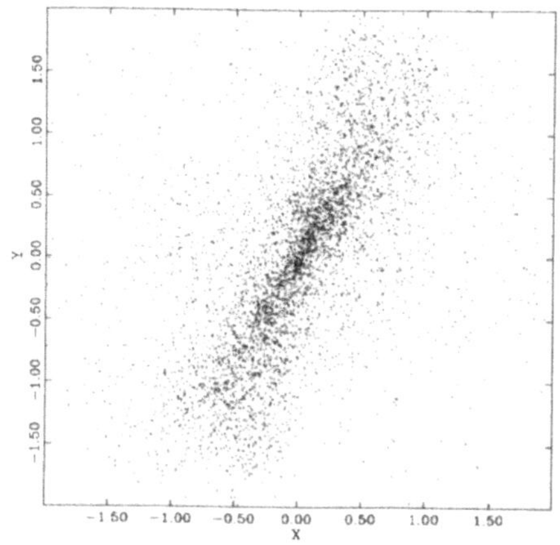

Figure 2. Snapshot of an N-body model of the polytrope $n = 1$, $\kappa = 0$ after 10 time units.

To map out the domain of these instabilities, we have run a grid of 120 experiments in the (n,κ) plane, spanning the range from $n = 5/10$, $\kappa = 2/12$ to $n = 16/10$, $\kappa = 11/12$. We confirm Henon's results for the radial instability, which appears to be confined to models with small values of n and κ. The non-radial instabilities occur over a much wider range of parameters; there is little indication that either instability becomes weaker with increasing n up to $n = 16/10$.

Antonov (1962) and Lynden-Bell and Sanitt (1969) have shown that no instabilities exist if the distribution f is a function of the energy E only and $df/dE < 0$. The latter condition alone is sufficient for stability if only radial modes are allowed (Gillan et. al. 1976). Our results show that the condition $\partial f/\partial E < 0$ is not sufficient for stability against non-radial perturbations. The unstable models presented here may shed some new light on the general stability problem for stellar systems.

References

Antonov, V. A., 1962, Vest. Leningrad gos. Univ., 19, 96.
Gillon, D., Cantus, M., Doremus, J. P. and Baumann, G., 1976, Astron. Astrophys., 50, 467.
Henon, M., 1973, Astron. Astrophys., 24, 299.
Lynden-Bell, D. and Sanitt, N., 1969, MNRAS, 143, 167.
White, S. D. M., 1973, Ap. J., 274, 53.

WHY ARE STELLAR SYSTEMS ANISOTROPIC?

E. Bettwieser, K. J. Fricke, R. Spurzem
Universitätssternwarte, Geismarlandstraße 11
D-3400 Göttingen, Federal Republic of Germany

Spherical stellar systems show during their secular evolution the development of velocity anisotropy in their halo (cf. e.g. Hénon, 1971). The present study examines the general reasons for generation of anisotropy in stellar systems by means of a gaseous star cluster model including anisotropy. Moment equations of the Boltzmann equation are considered for spherical symmetry in coordinate space but not in velocity space closed in third order by a heat flux equation. The coefficient of heat conductivity is tailored to describe the flux of energy due to the cumulative effect of distant gravitative encounters and generalized to include effects of anisotropy and external gravitation by a massive central object (Bettwieser et al., 1984).

Anisotropy in velocity space leads to additional equations and terms compared to a usual gasdynamical approach (Heggie, 1984). Thus for example anisotropic hydrostatic equilibrium has the form

$$dp/dr = -\rho g - 2p_a/r \; . \qquad (1)$$

Here the stellar mass density, gravitational acceleration and radius are denoted with ρ, g, r, resp. and the definitions $p=\rho\sigma_r^2$, $p_a=\rho(\sigma_r^2-\sigma_t^2)$ are used, where σ_r^2 and σ_t^2 denote the one-dimensional mean squared radial and tangential velocity dispersions. $p_a=0$ means isotropy. Note that in hydrostatics the anisotropy is another degree of freedom which is not a priori physically determined. The development of anisotropy is described by an anisotropy balance equation yielded from second order moment equations. It is given here in a simplified form for nearly isotropic systems ($p_a \ll p$):

$$\frac{dp_a}{dt} = -\,2\rho\sigma_r^2 \frac{d}{dr}\left(\frac{u}{r}\right) - \frac{4}{5}\frac{d}{dr}\left(\frac{F}{r}\right) \quad . \qquad (2)$$

u and F are the mean radial velocity and the energy flux, resp. Due to hydrostatic equilibrium in homogeneous cores of self-gravitating systems it is in first order $\nabla\sigma_r^2 \propto r$ and the coefficient of heat conductivity is constant, hence $F \propto r$. A similar argument can be given for u. So

J. Goodman and P. Hut (eds.), Dynamics of Star Clusters, 301–303.

there is no anisotropy generation. In general non-linear velocity or flux are a source of anisotropy according to equation (2) even for an initially isotropic configuration. The anisotropy is limited by a collision term proportional to p_a over stellardynamical relaxation time, which describes the isotropization of the velocity distribution by distant encounters. The general form of eq. (2) for appreciable degree of anisotropy ($p_a \approx p$) is more complex and does not allow for a simple discussion as in the special case of small anisotropy above.

We also considered the problem of anisotropy generation from the following point of view. The Boltzmann entropy is extremized by stationary solutions of the hydrostatic equilibrium. But, allowing for the anisotropy degree of freedom, is it in fact a local maximum? We know that it is not if the stationary configuration is a bound isothermal sphere with a density contrast above the critical value of 709. Besides this normal gravothermal runaway there exist different perturbations, which increase the total entropy. The present stage of our results is the following: we considered for a singular isothermal sphere (SIS, $\rho \propto r^{-2}$) perturbations in anisotropy δp_a but did not allow for redistribution of heat. Assuming $\delta p_a(r=0)=0$ as a boundary condition yields an increasing specific entropy for anisotropy perturbations $\delta p_a > 0$. Thus even a perturbation which suppresses the gravothermal effects makes the hydrostatic solution unstable. We guess that if one considers the more realistic problem including the thermal perturbation effects of the anisotropic catastrophe will be enhanced.

Numerical simulations were carried out by using a fully implicit code solving the set of discretized time-dependent moment equations in a Eulerian grid. Effects of a massive central black hole, which is assumed to sit fixed at the very centre and accretes stars by tidal disruption, are included. Without central object and a regular Plummer's model as initial model we find that during the secular core collapse the halo develops more and more anisotropy, whereas the shrinking core remains strictly isotropic. The halo anisotropy increases outwards. SIS taken as another initial model developed in contrast to the regular one extreme anisotropy ($\sigma_r^2 \gg \sigma_t^2$) in the post-collapse phase, which has a spatial maximum in the central region and decreases outwards. Note that the gravothermal instability of SIS (cf. Bettwieser and Sugimoto, 1984) is here accompanied by an anisotropy instability as it is proposed by the entropy considerations above. For simulations of the star cluster with star-accreting central black hole the accretion of bound stars in its vicinity and the related energy release are taken into account. The loss cone process was neglected (cf. for details Frank and Rees, 1976). After 16.1 Spitzer-Hart reference times the core contraction was halted due to the enhanced energy release by the star accretion and turned into reexpansion. The figure next page refers to a post-collapse model with an age of 71.4 reference times. Plotted are the r.m.s. radial (solid line) and tangential (dashed line) one-dimensional velocity dispersions in km/s versus radius in pc. There is appreciable anisotropy not only in the halo but also in the central velocity cusp surrounding the black hole. The radius inside which the hole's gravity dominates is marked.

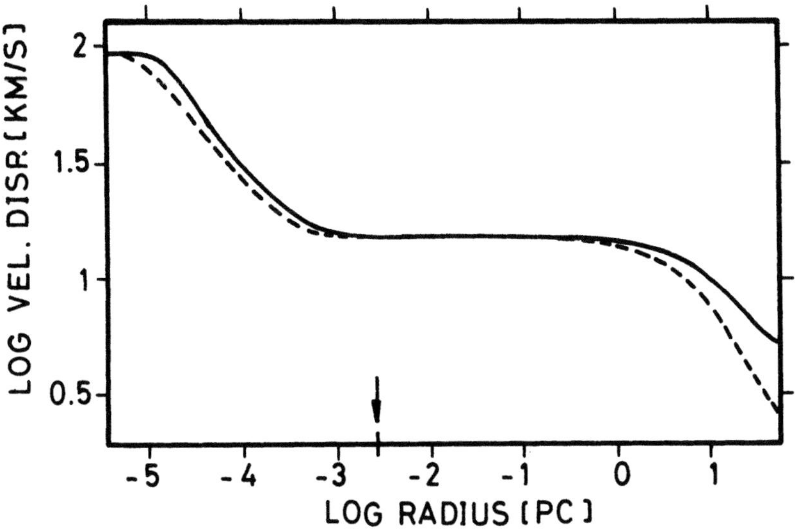

The onset of anisotropy generation in the central region becomes evident from the anisotropy balance eq. (2): for an isotropic steady state under external gravitation the fluxes of energy and stars are source free ($F \propto r^{-2}$, $u \propto r^{-1/4}$). Hence the right hand side of eq. (2) causes the generation of anisotropy. Thus an external gravitational field makes anisotropy by non-linearity of fluxes. Under this viewpoint the anisotropy generation in the halo is explainable, too, because the core's gravity acts on it like an external gravitational field.

Conclusions: we have outlined the essential reasons for anisotropy to develop in stellar systems. External gravity causes non-linear odd-order moments. Hence e.g. the core region around a massive central black hole and the halo of a star cluster become anisotropic in the course of evolution. Anisotropy generation is considered to be a secular instability of such systems. Isotropy is unstable, too, for the singular isothermal sphere which even suffers from an anisotropic catastrophe. Not that also in this case there is a singular gravitational acceleration ($g \propto 1/r$) simulating the 'external' gravitational field. For real stellar systems containing a massive central black hole however, the loss-cone accretion has to be included and it will influence the anisotropy, too (cf. Cohn and Kulsrud, 1978).

REFERENCES

Bettwieser, E., Fricke, K.J., Spurzem, R.: 1984, to be subm. to Mon.Not.R.astr.Soc.

Bettwieser, E., Sugimoto, D.: 1984, Mon.Not.R.astr.Soc. 208, 493.

Cohn, H., Kulsrud, R.: 1978, Astrophys.J. 226, 1087.

Frank, J., Rees, M.: 1976, Mon.Not.R.astr.Soc. 176, 633.

Heggie, D.C.: 1984, Mon.Not.R.astr.Soc. 206, 179.

Hénon, M.: 1971, Astrophys. Space Sc. 13, 284.

A STATISTICAL TREATMENT OF LOW-N SYSTEMS

Stefano Casertano
Institute for Advanced Study
Princeton, New Jersey

The long-term evolution of N-body systems with high N ($>10^4$) cannot yet be studied by direct integration of the equations of motion. Instead, the study of systems with smaller N can give insight into the higher N limit in two ways. First, we can measure the dependence on N of all quantities of interest for a range of N-values, and extrapolate the results to higher N. Second, even a very high-N system will eventually undergo core collapse, at which point the central dynamics will be dominated by a subset of stars with small N (10-100).

For both reasons it is interesting to explore the extent to which we can meaningfully define statistical quantities, even for very low N. As a first step in this direction, we report here some results obtained with $N = 8$, using a regularized code written by S. Aarseth. The initial conditions have been drawn from a Plummer model, a polytrope of index $n = 5$. We have now 35 realizations of this system. We are currently extending these explorations to higher N, and we hope to study the different characteristics of the evolution when approaching the high-N (Fokker-Planck) limit.

Our results indicate that multi-body encounters, as opposed to two-body scattering, play an important role in the dynamical evolution of systems with small N, especially in their early stages (cf. Aarseth & Lecar 1975). Multi-body encounters favor the formation of binaries, which - during their formation and subsequent hardening - can release enough energy to unbind the cluster (see Fig. 3).

After the formation of a hard binary the evolution of the system can be sketched as follows. The binary will gain binding energy in discrete steps, corresponding to encounters with field stars. As its binding energy increases, encounters will become less frequent - because of the smaller geometrical cross-section - but more energetic (Hut 1984). If, after one of these encounters, the binary acquires enough energy to unbind the cluster, the system will be disrupted. If, on the other hand, the separation becomes too small, the binary is "frozen" and will no longer play a role in the dynamics of the system.

J. Goodman and P. Hut (eds.), Dynamics of Star Clusters, 305–308.

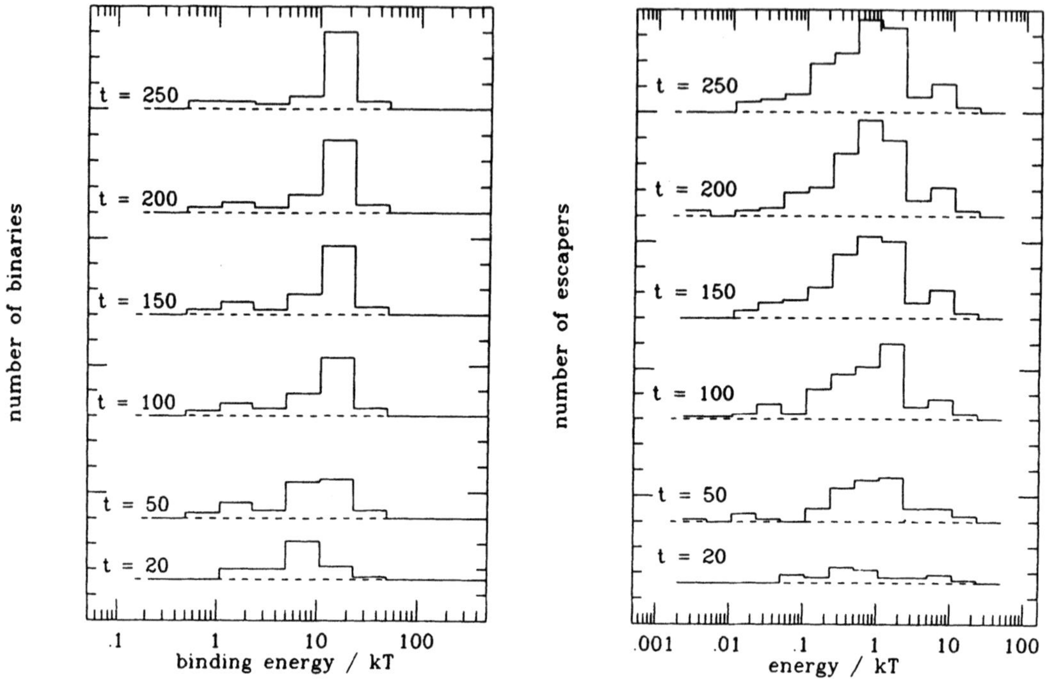

Figure 1: Spectrum of (a) binding energy of binaries and (b) energy of escapers at different times. Tickmarks are in units of ten. In all figures, a time unit is used in which the crossing time $t_{cr} = 2\sqrt{2}$.

This schematic description is illustrated in Fig. 1a, where we show the variation with time of the cumulative spectrum of binaries, summed over all 35 runs. The distribution of binaries evolves in time towards higher energies, until - at the end of the time interval used (approximately 90 crossing times) - almost every system has a binary in the range 10 - 25 kT, where T is defined in terms of the <u>initial</u> velocity dispersion. Fig. 1b shows the corresponding evolution in the spectrum of escapers.

In Fig. 2 we show the number of surviving systems as a function of time (the system is disrupted when four particles of the initial eight escape). The probability of disruption seems to decrease after the initial period. This agrees with the fact that binaries become less effective in disrupting the system as they harden. Note that, even for N=8, the average survival time is orders of magnitude larger than the crossing time. For our sample, the harmonic mean of individual lifetimes is about 100 (35 t_{cr}).

A more detailed diagnostic of the evolution is in Fig. 3a, where we give the energy balance of the system as a function of time. There

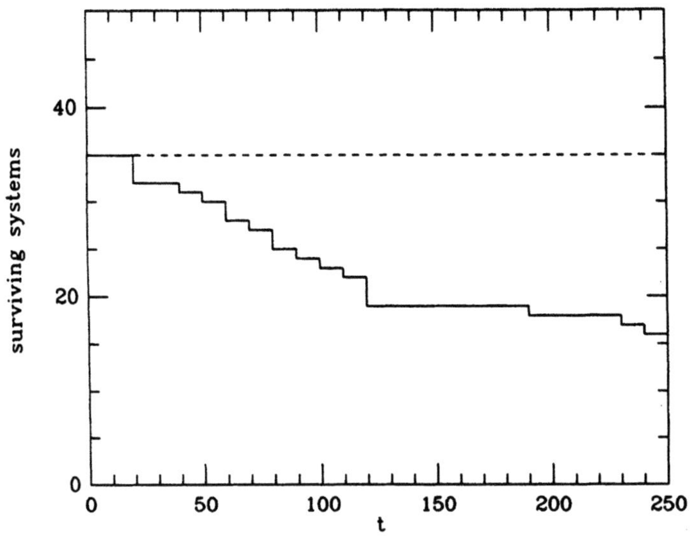

Figure 2: Number of surviving systems as a function of time.

are three types of energy: the internal binding energy of the binaries (thick solid line); energy carried away by the escapers (dotted line); and energy available to the system. The latter is represented by the difference between the solid and the dotted lines, since the binding energy has been offset vertically by a quantity equal to the total energy of the system (constant; dashed line). The energy available to the system increases with time, as energy is poured off by the hardening of binaries. When it becomes positive (i.e., the solid and dotted lines in Fig. 3a cross each other), the system is unbound.

In Fig. 3b we show the variation with time of the core radius, averaged over all 35 runs. The core radius is defined here as in von Hoerner (1963). There seems to be no indication of core collapse in the initial phase of the evolution, and the core radius increases steadily with time. This effect is probably related to the small number of particles, as it is known (Aarseth et al 1974) that for $N > 100$ the core radius decreases initially. We are currently extending our study to larger values of N, hoping to be able to set limits on the minimum number of particles for which the core collapse can take place.

Although general patterns of behaviour can be found, there is a large scatter when we consider individual runs. For this reason we emphasize the need for good statistics, especially for small values of N. In the forthcoming work - with N in the range 10-100 - we are also trying to draw the initial conditions from the same ensemble as for N=8, in order to compare more readily the results for different values of N.

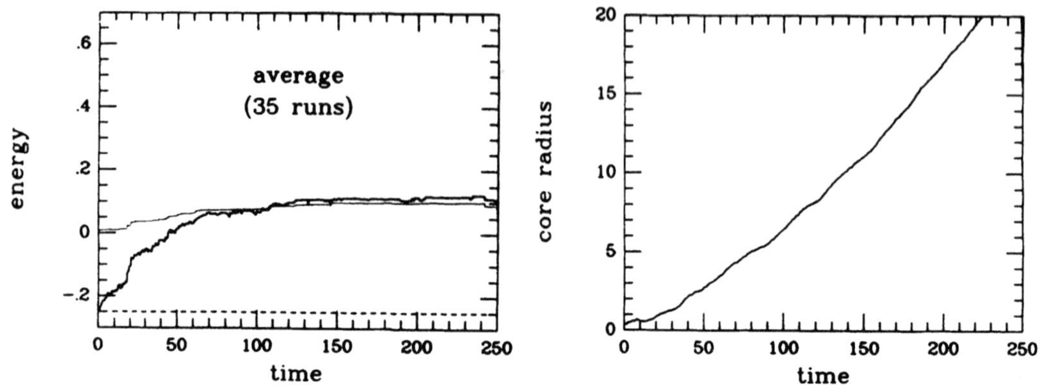

Figure 3: (a) Energy balance as a function of time, averaged over 35 runs with 8 equal mass particles. Units are such that $G = 1$, $M_{tot} = 1$ and $E_{tot} = -0.25$. The difference between the solid and the dotted line is the (negative) total energy of the system, except that frozen in internal degrees of freedom of the binaries or carried away by the escapers.
(b) Evolution of the core radius, averaged over 35 runs. There is no indication of a collapse phase before the expansion.

This work is part of an ongoing project in collaboration with Piet Hut and Bob Mathieu. It was supported in part by National Science Foundation through grant PHY-8440263.

REFERENCES

Aarseth, S. J., Hénon, M. & Wielen, R., 1974: Astron. Astrophys. 37, 183.
Aarseth, S. J. & Lecar, M., 1975: Ann. Rev. Astron. Astrophys. 13, 1.
Hut, P., 1984: present symposium.
von Hoerner, S., 1963: Z. Astrophys. 57, 47.

MINIMUM OF THE ECCENTRICITY OF THE GALACTIC GLOBULAR CLUSTER ORBITS

Jacques COLIN
Besançon Observatory
25000 BESANCON (France)

ABSTRACT. We determine if the orbits of the galactic globular clusters can be circular, using a spherical logarithmic potential. If they are not, we calculate their perigalactic distances and we deduce a pseudo-eccentricity. Using a numerical and geometrical method, we obtain the minimum of all the possible pseudo-eccentricities for each cluster. The corresponding perigalactic distances are compared with the values given by Innanen et al. (1983) with a different method.

In this paper we consider that the galactic potential can be represented by a logarithmic function. This is consistent with the observed flat rotation curve, and the existence of a massive halo. In such a model, the visible disk will not, to first order, modify the orbital characteristics of the globular clusters.

Using this potential and the radial velocity V_R given by Harris and Racine (1979),we consider all the orbits compatible with the model and we select those which have the minimum pseudo-eccentricity.

Thus, we obtain a sample of globular clusters whose orbits are not circular and we compare the corresponding perigalactic distances with those given by Innanen et al. (1983).

If V_{Rot} is the constant rotational velocity corresponding to the flat rotation curve, V_T, the unknown tangential velocity and V_S the unknown spatial velocity, the orbit is circular if:

$$V_S = V_{Rot}$$

This is possible only if $V_{Rot} > (V_R / \sin \beta)$ where β is the angle under which is seen the distance Sun-Galactic center.

If $V_{Rot} < (V_R / \sin \beta)$, the orbit cannot be circular, thus there exists a perigalactic and an apogalactic distance r_p and r_a, and we define a pseudo-eccentricity:

$$e = (r_a - r_p) / (r_a + r_p)$$

J. Goodman and P. Hut (eds.), Dynamics of Star Clusters, 309–312.

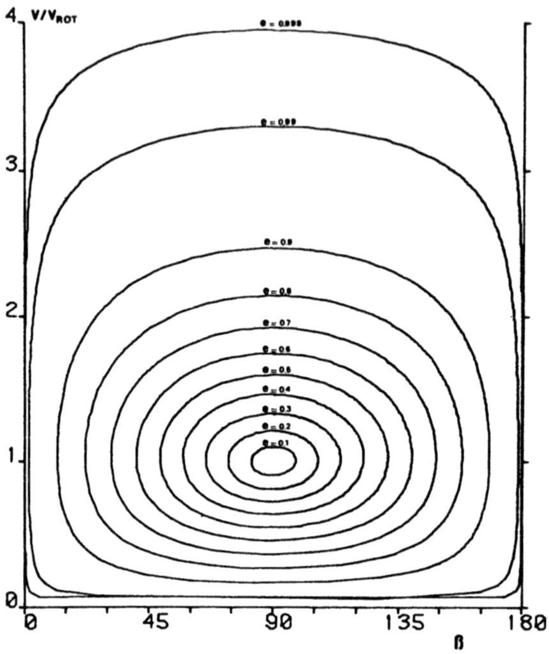

Figure 1 - Eccentricity of an orbit in a logarithmic potential as a function of the velocity vector define by the modulus V/V_{Rot} and the direction β. This diagram is independant of the distance to the galactic center.

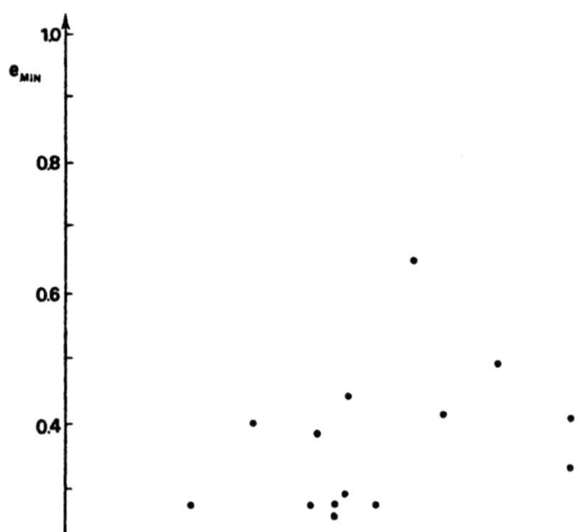

Figure 2 - Minimum of the eccentricity for 30 globular clusters as a function of the distance r. For the other clusters, the minimum is zero.

By varying V_T we explore all the possible orbits and the corresponding pseudo- eccentricities. For that, we use two methods, a geometrical and a numerical one.

NUMERICAL METHOD

For each cluster we compute a sample of possible orbit by varying V_T. We represent the variation of e relative to V_T and we obtain a minimum for the eccentricity.

GEOMETRICAL METHOD

In a logarithmic potential it is possible to determine the eccentricity of the orbit of a test particle, when its velocity vector is known (Colin and Puel, unpublished).

On Figure 1 we show the relation between the components of the velocity vector (modulus V and direction ß) and the ellipticity of the orbits. This diagram must be compared to the Bottlinger diagram but here the distance to the galactic center is ignored. With this diagram the radial velocity and the position of the cluster it is possible to determine the minimum of the pseudo-eccentrictry of a globular cluster. But we don't obtain r_a and r_p with this method.

We obtain 30 globular clusters for which the pseudo-eccentricity is different from zero. We also obtain the ratio (R_p/R) of the pericenter distance to the present distance of the clusters. These values may be compared with those of Innanen et al. (1982) but the significance is not exactly the same. For Innanen et al. the values are deduced from the observations and must correspond to the true orbits. In our case the given values correspond to the maximum possible value, but not necessarily to the real one. Thus, the values of Innanen et al. (1982) must not be greater than ours.

DISCUSSION

From a geometrical and a mechanical point of vue, 60% of the globular clusters may have a circular orbit, if the galaxy is represented by a spherical logarithmic potential. If we exclude the $R_p/R > 1$, most of the values of R_p/R given by Innanen et al. (1982) agree with ours, they don't exceed our values that normally are maximum values.

It also seems that the minimum eccentricity grows with distance. Figure 2 shows this peculiarity already pointed out by Rastorguev and Surdin (1980) for a different parameter ($e_{min} = (R - R_p) / (R + R_p)$.

REFERENCES

Harris, W.E. and Racine, R. (1979) Ann. Rev. Astron. Astrophys. **17**, 241.
Innanen, K.A., Harris, W.E., Webbink, R.F. (1983) Astron. J. **88**, 338.
Rastorguev, A.S. and Surdin, V.G. (1980) Astron. Tsirk., N° **1102**.

FUNCTIONAL INTEGRALS IN STELLAR DYNAMICS

ELDAD DAGAN
THE HEBREW UNIVERSITY OF JERUSALEM
THE RACAH INSTITUTE OF PHYSICS
JERUSALEM, ISRAEL

Abstract: In the late 70s, the functional integral formulation equivalent to the Fokker-Planck equation was worked out (Graham). We apply this functional integral formulation to gravitationally interacting systems, whose dynamics may be analyzed by separating the forces operating on a particle into a mean field force and fluctuations due to random collisions at intermediate range (scattering at small angles). In this poster the formalism is presented for short periods (in the stochastical meaning) to systems with isotropic distribution background in velocity space (different spatial densities are possible). Later the functional integral for the local change for the distribution function is evaluated in the steepest descent approximation. In the end we point out the applicability of the method to slowly evolving globular star clusters near thermal equilibrium. In conditions of slow evolution we can express the evolution of the orbits in terms of local deviation from equilibrium.

I. The functional integral formalism for Nartonian gravity

R. Graham treated Fokker Planck Equation of the type:

$$(1)\quad \frac{\partial f}{\partial t} = \frac{\partial}{\partial q^{\nu}}\left(K^{\nu}(\vec{q})f\right) + \frac{1}{2}\frac{\partial^2}{\partial q^{\nu}\partial q^{\mu}}\left(Q^{\nu\mu}(\vec{q})f\right) + V(\vec{q})f$$

He had built functional integral solution for the density distribution function at later times for an initial distribution of the delta function type. The functional integral has the form:

$$(2)\quad f(\vec{q}\,t\,|\,\vec{q}_0\,0) = \int_{\vec{q}(0)=\vec{q}_0}^{\vec{q}(t)=\vec{q}} \mathcal{D}\mu([\vec{q}(\tau)])\, \exp\left(-\int_0^t L(\dot{\vec{q}}(\tau), \vec{q}(\tau))\,d\tau\right)$$

where $L(\dot{\vec{q}}, \vec{q})$ - the Lagrangian in the problem is given by:

J. Goodman and P. Hut (eds.), Dynamics of Star Clusters, 313–315.

$$(3)\quad L(\dot{\vec{q}},\vec{q}) = \frac{1}{2} Q^{-1}_{\nu\mu}(\dot{q}^{\nu}-h^{\nu})(\dot{q}^{\mu}-h^{\mu}) + \frac{1}{2}[Q^{-1}]^{-\frac{1}{2}}\frac{\partial}{\partial q^{\nu}}\{[Q^{-1}]^{\frac{1}{2}}h^{\nu}\} - V(\vec{q}) + \frac{R}{n}$$

where

$Q_{\nu\mu} = (Q^{-1})^{\nu\mu}$ - the diffusion matrix

$Q^{-1}_{\nu\mu}$ - the inverse of the diffusion matrix

K^{ν} - the original drift

h^{ν} - the covariant drift: $h^{\nu} = K^{\nu} - \frac{1}{2}[Q^{-1}]^{-\frac{1}{2}}\frac{\partial}{\partial q^{\mu}}\{[Q^{-1}]^{\frac{1}{2}}\{Q^{-1}\}^{\nu\mu}\}$

$Q^{-1} = \det Q^{-1}_{\nu\mu}$

n - integer taking the values: 6, 8, or 12.

R - the scalar curvature, built on a metric which is the inverse of the diffusion matrix.

The Lagrangian has the character of a scalar, which is covariant to general coordinate transformations, with the role of the metric being played by the reciprocal diffusion matrix.

For one component, gravitationally interacting systems we get the following expression for the distribution function after short time τ, where the initial distribution function was $f(\vec{r}', \vec{v}'; t)$

$$(4)\quad f(\vec{r},\vec{v},t+\tau) = \int \frac{1}{(2\pi\tau)^{3/2}D^{1/2}} \exp\Big[-\Big\{\frac{1}{2\tau}D^{-1}_{\nu\mu}(v^{\nu}-v^{\nu'}-\tau h^{\nu}-\tau g^{\nu}) \times (v^{\mu}-v^{\mu'}-\tau h^{\mu}-\tau g^{\mu}) + \frac{1}{2}[D^{-1}]^{-\frac{1}{2}}\frac{\partial}{\partial v^{\mu}}\{[D^{-1}]^{\frac{1}{2}}h^{\nu}\}\tau + \frac{R}{n}\tau + \ln f(\vec{r}',\vec{v}';t=0)\Big\}\Big] \times \delta(\vec{r}-\vec{r}'-\vec{v}\tau)\,d\vec{r}'\,d\vec{v}'$$

where: $K^{\nu} = \langle\Delta v^{\nu}\rangle$, $D_{\nu\mu} = \langle\Delta v^{\nu}\Delta v^{\mu}\rangle$, $\vec{g} = \frac{\partial\phi}{\partial\vec{r}}$, the mean force.

In the case of locally isotropic background the transport coefficients in Cartesian coordinates take the form:

$$(5)\quad D_{ij} = \frac{D_{\parallel}(v)-D_{\perp}(v)}{v^2}v^i v^j + \delta^{ij}D_{\perp}(v);\quad D^{-1}_{ij} = \frac{D^{-1}_{\parallel}-D^{-1}_{\perp}}{v^2}v^i v^j + \delta^{ij}D^{-1}_{\perp}$$

$$K^i = \eta(v)\frac{v^i}{v}\quad [D_{\perp} = \langle\Delta v_{\perp}\Delta v_{\perp}\rangle,\ D_{\parallel} = \langle\Delta v_{\parallel}\Delta v_{\parallel}\rangle,\ \eta(v) = \langle\Delta v_{\parallel}\rangle]$$

We can diagonalize the diffusion matrix into the form $D_{ij} = \begin{pmatrix} D_{\parallel} & & 0 \\ & D_{\perp} & \\ 0 & & D_{\perp}\end{pmatrix}$ and then the christoffels are given by:

$$\Gamma_{11}' = -\frac{1}{2}(\ln D_{\parallel})', \quad \Gamma_{22}' = \Gamma_{33}' = -\frac{1}{2} D_{\parallel}(D_{\perp}^{-1})' + D_{\parallel}\frac{D_{\parallel}^{-1} - D_{\perp}^{-1}}{v}$$

(6) $$\Gamma_{21}' = \Gamma_{31}' = -\frac{1}{2}(\ln D_{\perp})'$$

The transport coefficients for that case are given by Hénon.

II. The steepest descent evaluation near equilibrium states

For systems near thermodynamic equilibrium we use the steepest descent method to evaluate the integrals in question.

In the limit of small velocities the transport coefficients takes the asymptotic form:

(7) $D_{\perp} \to D_{\parallel} = Q_0 = \text{const}, \quad \eta(v) \to \eta_0(v)\,[\eta_0 = \text{const}]$

In that limit the method gives exact results for maxwellian type distributions:

(8) $$f(\vec{r}, \vec{v}, \tau) = A\left(\frac{1}{1+4S\tau}\right)^{3/2} e^{-\frac{\beta}{(1+4S\tau)}\frac{v^2}{2} - \beta\phi(\vec{r})}$$

where A is a normalization factor and $\Delta S = \beta Q_0 - 2\eta_0$

In Equilibrium $\Delta S = 0$ and so the form of the distribution is conserved.

In Wooley model $S = -\frac{2}{3}C$, where $C = 16\pi^2 G^2 \ln N$ (defined in Hénon paper).

The local changes in the distribution function has no meaning in globular clusters. However, if the evolution is slow, we can calculate the change in the distribution of orbits by first solving for the local changes and then averaging on the orbits.

Relating to this matter, we have found that in the Woolley models, particles below the energy curoff, lose energy at constant rate.

References

1. R. Graham, Z. Phyzik B 26, 397-405, 281-290.

2. S. Chandrasehur, Principles of Stellar Dynamics, Dover 1960.

3. M. Hénon, 'Collisional Dynamics in Spherical Stellar systems' in Dynamical Structure and Evolution of Stellar Systems, Saas Fec, 1973.

MULTI-COMPONENT MODELS FOR THE STRUCTURE AND EVOLUTION OF SPHERICAL STELLAR SYSTEMS

Emmanuel Davoust
Observatoire de Besançon
25044 BESANCON CEDEX
France

ABSTRACT. We present stationary analytical models for the structure of spherical stellar systems with several mass components. Evolution is simulated by increasing the central velocity of escape. We compare our models to the results of numerical simulations. An observable kinematical effect of mass segregation is pointed out.

1. MULTI-COMPONENT ANALYTICAL MODELS

This paper is an investigation of the structure and evolution of spherical stellar systems with components of different masses. One expects mass segregation to occur in these systems. In the course of evolution, the more massive stars (or galaxies) concentrate toward the center, and the lighter ones form an extended halo. Multi-component dynamical models are then required to interpret their structure and kinematics.

Our models are based on two assumptions. The phase density distribution is a sum of pseudo-maxwellian isotropic distributions. The precise functional form of the distribution has been tested on one-component models (Davoust, 1977; hereafter paper I). Each of the components is in hydrostatic equilibrium, but there is no equipartition of energy between them (not even at the very center). The evolution is simulated by increasing the central velocity of escape, k. The sole limitation of the models, in view of recent observational developments, is that they are not rotating.

The phase density distributions are :

$$f_i = K_i \left[e^{-\frac{m_i(E-E_e)}{\sigma^2}} - 1 + \frac{m_i(E-E_e)}{\sigma^2} \right]$$

where E is the energy, E_e the energy of escape and m_i the masses of individual stars. The space densities and velocity dispersions are derived by straightforward integrations, as in paper I.

J. Goodman and P. Hut (eds.), Dynamics of Star Clusters, 317–319.

The only free parameter of our models is k, the normalized central velocity of escape (see paper I). The central density ratio Cij between components i and j must be adjusted so that the ratio of total mass of components i and j remains constant.

As our models evolve (as k increases), the Cij's increase and the deviations from equipartition between the components, Sij, decrease. The heavier components are closest to equipartition. The deviation is always in the same sense: the heavier component has a larger velocity dispersion than required for equipartition.

The core structure of our 3-component isotropic model with the mass function of Spitzer and Shull (1975) is given in the table below for different values of k. In the lower part of the table, we give the same parameters derived from the results of Spitzer and Shull's (1975) numerical simulations. The main features of the evolution of the numerical models are well reproduced in the isotropic model. This validates our two basic assumptions. One discrepancy, the fact that the mass ratios Mij in the core (containing 4% of the total mass of the system) decrease after a maximum in our isotropic 3-component model, is weakened by the fact that the corresponding Cij's do increase mmonotonically. This discrepancy is due to the fact that the halo density of the isotropic model falls off more rapidly than in the numerical simulations. This affects the total mass and in turn the core radius.

k	C21	C31	M21	M31	S21	S31	S32
1.0	.822	.863	.805	.773	2.293	4.186	1.825
1.4	.959	1.765	.923	1.435	2.181	3.338	1.531
1.8	1.230	5.145	1.142	3.473	2.055	2.715	1.321
2.2	1.838	39.70	1.430	11.82	1.924	2.286	1.188
2.6	2.640	265.	1.307	10.43	1.793	1.986	1.108
3.0	3.890	1545.	1.253	8.05	1.669	1.770	1.060
t			**M21**	**M31**	**S21**	**S31**	**S32**
260			1.069	1.143	1.826	3.519	1.928
860			1.407	3.479	2.061	2.646	1.284
1460			1.686	10.43	2.077	2.448	1.179
1860			2.135	32.68	1.422	1.769	1.243

2. KINEMATICAL EFFECTS OF MASS SEGREGATION

The velocity dispersion decreases monotonically with increasing distance in one-component globular cluster models. This is not the case in multi-component models with important mass segregation (Kondrat'ev and Ozernoy, 1981).

We have determined the radial profile of the total velocity dispersion in our 3-component isotropic model and in the 3-component Monte-Carlo models of Spitzer and Shull (1975) and the 4-component ones of Hénon (1975). The results are very similar. At the beginning of the evolution the profiles are monotonously decreasing. As evolution proceeds, the total velocity dispersion goes through a maximum, at a radius containing about 10% of the total mass in our and Hénon's (1975) models, and 12 to 20% in Spitzer and Shull's (1975) model. In this region, the velocity distribution is still isotropic in the numerical simulations, and justifies using our isotropic model.

While the three models show distinctive features, the maximum in the central region is always present at later evolution times. It should be indicative of mass segregation in stellar systems with flat central density, such as globular clusters and clusters of galaxies, but not in systems with sharply peaked mass concentration, such as elliptical galaxies or systems of globular clusters, where it may be the result of strong velocity anisotropy (Bailey and MacDonald, 1981).

This maximum should be observable in globular clusters of our Galaxy, provided the luminosity of the stars scales with their mass, i.e. that there is no large population of heavy remnants, such as white dwarfs and/or neutron stars. The radius at which the hump occurs depends on the concentration and the total mass of the cluster, but it is comparable to, if not larger than, the core radius of the cluster. This radius is about 6 arc sec. in the worst case, and several arc min. in the most favorable ones. The maximum is only 1 or 2 kms^{-1} above the central value and might be difficult to detect with present techniques.

Finally, this feature is present in the Coma (Kent and Gunn, 1982) and Perseus (Kent and Sargent, 1983) clusters of galaxies and reveals mass segregation in these clusters.

A more detailed version of the paper is available from the author.

ACKNOWLEDGEMENTS

We would like to thank Dr. M. Hénon and L. Spitzer for providing unpublished details of their Monte-Carlo simulations. This research was supported by a grant of the CNRS to ERA 07904.

REFERENCES

Bailey, M.E., MacDonald, J. : 1981. Mon. Not. R. astron. Soc. **194**, 195.
Davoust, E. : 1977. Astron. Astrophys. **61**, 391.
Hénon, M. : 1975. IAU Symp. **69**, 133.
Kent, S.M., Gunn, J.E. : 1982. Astron. J. **87**, 945.
Kent, S.M., Sargent, W.L.W. : 1983. Astron. J. **88**, 697.
Kondrat'ev, B.P., Ozernoy, L.M. : 1982. Astrophys. Sp. Sci. **84**, 431.
Spitzer, L., Shull, J.M. : 1975. Astrophys. J. **201**, 773.

THE ROLE OF HARD BINARIES IN CLUSTER EVOLUTION

G. Giannone, D. Molteni
Istituto di Fisica
Via Archirafi 36
90123 Palermo
Italy

ABSTRACT. The presence of an initial binary population in a stellar cluster can have a dominant effect on the dynamical evolution of the whole system. We discuss simulations, carried out with a direct N-body method for several models with N = 300 equal mass objects, with 20% of all the stars as initial binaries having different binding energy. These calculations show that binaries with values of the binding energy, h, in the range 5 - 10 times the mean kinetic energy, x, are the most strongly interacting ones with the field stars and between themselves. The main result of these interactions is an enhanced expansion of the core and also of the whole cluster. The binary heating prevents the gravitational collapse for a time longer than the collapse time for systems without binaries. Energetic binaries with h greater than 25x concentrate at the center, due to mass segregation but these binary-binary interactions produce disruption and escape of the components, with smaller heating effect. Binary-binary interactions occur predominantly in the inner regions and are mainly responsible for the core heating and for the escaping stars. Single star-binary encounters, instead, occur more frequently in the outer regions, and only contribute to the strong expansion of the halo. In all the simulations there was no formation of new persistent hard binaries.

Further details of these calculations will be published elsewhere.

J. Goodman and P. Hut (eds.), Dynamics of Star Clusters, 321.

COLLISIONAL STELLAR DYNAMICS IN GENERAL RELATIVITY: AN OVERVIEW

Henry E. Kandrup
Center for Relativity, Center for Theoretical Physics, and
Center for Studies in Statistical Mechanics
University of Texas, Austin, TX 78712 U.S.A.

ABSTRACT. Recently, Israel and Kandrup (1984; Kandrup 1984 a,b,c,d) have formulated a new, manifestly covariant approach to non-equilibrium statistical mechanics in classical general relativity. The object here is to indicate how that formalism may be used to construct a theory of 'collisional' stellar dynamics, valid for a collection of point mass stars in the limit that incoherent radiative effects may be neglected.

The idealisation of point masses is of course a gross oversimplification which, however, at least in principle, should not be that difficult to overcome. The neglect of radiative effects is more important, since it implies ultimately that deviations from some 'average' mean field conditions may be modeled by a comparatively simple direct interaction. This means that one need not consider explicitly the degrees of freedom of the gravitational 'field'. It should, therefore, be stressed that this neglect really is legitimate: the time scale for energy loss via radiation reaction will exceed the evaporation time scale and all other relevant scales provided that the central redshift $z \lesssim 1$, a condition believed necessary to ensure dynamical stability.

Whether there exist realistic situations in which relativistic effects actually are important is at present an unanswered question. The object here is simply to understand how one might describe such relativistic systems if they do in fact exist. In this sense, the problem may, as suggested by Zel'dovich and Podurets (1966), be termed one of 'methodological interest.'

In order to see what is required, it is useful to recall the Newtonian theory. In its most rigorous form, the starting point for the description of a collection of N identical, point mass stars is an N-particle distribution function μ, the probability density for finding each star i at a given point $\vec{x}_i$ with momentum $\vec{p}_i$ at time t. The one-particle f, which is of more direct physical relevance, is constructed as a reduced distribution. Given this f, one may generate an average mass density ρ and other quantities of interest. The evolution of μ is of course governed by the N-particle Liouville equation. What one really wants, however, is an equation for the evolution of f.

Physically, one anticipates that any such equation will allow for two sorts of effects, namely (i) the 'average' gravitational force $-\vec{\nabla}\Phi$ associated with ρ, and (ii) 'fluctuating' forces that arise because the 'true' and 'average' potentials do not coincide. The important point is that such

J. Goodman and P. Hut (eds.), Dynamics of Star Clusters, 323–325.

a decomposition of the total force can be implemented rigorously, so that one can derive an exact, closed, non-linear, non-local equation for the evolution of f (Kandrup 1981). In the limit that the fluctuating components are neglected, one recovers the standard 'collisionless' theory. In the limit that their effects may be evaluated as if the stars were following linear trajectories through the center of an infinite, homogeneous system, one recovers the Fokker-Planck equation of 'collisional' stellar dynamics.

The question is of course: Could one formulate analogous relativistic theories of 'collisionless' and 'collisional' stellar dynamics? An heuristic 'collisionless' theory is easy enough to obtain (see, e.g., Ipser and Thorne 1968). One again considers a one-particle f, now defined covariantly, and, given this f, one may construct an average stress-energy $T_{\mu\nu}$ which in turn serves as a source for an average spacetime metric $g_{\mu\nu}$. To the extent that 'fluctuations' may be neglected, one might suppose that the stars may be approximated as following geodesics in this average spacetime, and, in this limit, one recovers a simple relativistic generalisation of the 'collisionless' Vlasov equation.

Transcending a 'collisionless' description is more difficult. One obvious complication is the fact that one really ought to incorporate explicitly the degrees of freedom of the gravitational field. Thus, in analogy with relativistic plasma physics, one ought to start from a more complicated distribution function involving both particle and field variables (Kandrup 1984d). This complication can, however, be circumvented if one is concerned only with formulating an approximate kinetic theory. The important point is that, to the extent that incoherent radiative effects may be neglected, one can model the gravitational 'forces' by a direct interaction involving only the coordinates and momenta of the stars (Israel and Kandrup 1984, Kandrup 1984a,b).

Another complication arises when one tries to specify what is really meant by a gravitational 'force'. After all, the guiding principle of general relativity is that, in the absence of non-gravitational effects, the stars follow geodesics in the spacetime which they generate. To isolate upon a sensible gravitational 'force' one must re-express the 'true' physics — stars following geodesics in the irregular, rapidly fluctuating spacetime — in terms of another, smoothed background spacetime chosen to satisfy some appropriate field equations, e.g., the equations of the collisionless theory. The tensor $\delta\Gamma^{\lambda}_{\mu\nu}$, the difference between the Christoffel symbols for the true and background spacetimes, serves to generate a 'fluctuating force', the existence of which is manifest by the fact that the stars no longer appear to follow geodesics (Israel and Kandrup 1984). The form of this $\delta\Gamma^{\lambda}_{\mu\nu}$ must in general be determined by solving non-linear field equations which are virtually intractable analytically. To the extent, however, that the difference between the two spacetimes is small, so that the 'fluctuating forces' are weak, it will suffice to linearise the field equations about the background solution. A similar argument then suggests that, when solving these equations, one may proceed in a first approximation as if the stars were following geodesics in the background spacetime: this is the relativistic analogue of the standard 'impulse' approximation. What this means is that the gravitational forces will be generated from simple retarded potentials involving only the coordinates and momenta of the stars: one has achieved the desired direct interaction approximation.

Given these sorts of expressions for the inter-stellar forces, it is straightforward to proceed in comparatively close analogy with the Newtonian description to introduce an N-particle distribution function μ and a reduced one-particle f; and, starting from this 'complete' description, it is possible to derive an exact, closed equation for the evolution of f. What remains then is to approximate that exact equation by a tractable kinetic equation. This entails four general assumptions, the justification of which has been discussed elsewhere (Israel and Kandrup 1984, Kandrup 1984a,b,f). (i) It is assumed that any initial correlations have a negligible influence upon the evolution of f. (ii) It is assumed that there exists a 'collision' time scale short compared with the dynamical and relaxation time scales (in the context of an impulse approximation, this only makes sense if one introduces a cut off in the potential at large impact parameter to avoid the standard logarithmic divergences). (iii) It is assumed that the stars may be approximated as following geodesics in the average spacetime. (iv) It is assumed that, in evaluating the effects of the fluctuating forces at some point x^{μ}, one may introduce a local Lorentz frame and then proceed as if the spacetime were really homogeneous and flat ('localised fluctuations'). This final assumption enables one to express the forces in terms of their Fourier transforms, and one obtains thereby the kinetic equation (7) of Kandrup (1984c).

Various properties of this kinetic equation have been discussed elsewhere (Israel and Kandrup 1984, Kandrup 1984a,c,f) but certain implications should be reiterated here. (i) In the obvious Newtonian limit, one recovers the well-known Landau equation for a 1/r potential, and, if one performs the $\vec{k}$-space integrations by introducing suitable cut offs, one obtains the standard Fokker-Planck equation. (ii) Provided that one introduces a cut off appropriate for large impact parameter, one can conclude that the relativistic isothermal distribution constitutes an exact stationary solution. If, however, one does not introduce a cut off, this is no longer true. Just as in the Newtonian limit (Kandrup 1984e), one anticipates deviations from an isothermal! (iii) The kinetic equation guarantees energy-momentum conservation, i.e., it implies that the mean field $T^{\mu\nu}$ is intrinsically divergence-free: $\nabla_{\mu} T^{\mu\nu} = 0$. (iv) The kinetic equation implies an H-theorem inequality, i.e., the divergence of the Boltzmann entropy flow S^{μ} associated with f is non-negative: $\nabla_{\mu} S^{\mu} \geq 0$, with equality holding only for a locally Maxwellian distribution of velocities.

REFERENCES

Ipser, J. R. and Thorne, K. S. 1968, Ap. J. 154, 251 - 270.
Israel, W. and Kandrup, H. E. 1984, Ann. Phys. (NY) 152, 30 - 84.
Kandrup, H. E. 1981, Ap. J. 244, 316 - 330.
____. 1984a, Ann. Phys. (NY) 153, 44 - 102.
____. 1984b, Physica A (in press).
____. 1984c, Ap. J. 282 (in press).
____. 1984d, J. Math. Phys. (in press).
____. 1984e, 'Should a Self-Gravitating System Relax towards an Isothermal Distribution?' preprint.
____. 1984f, 'Collisional Stellar Dynamics in General Relativity,' preprint.
Zel'dovich, Ya. B. and Podurets, M. A. 1966, Sov. Astron. AJ 9, 742 - 749.

3-INTEGRAL MODELS FOR GLOBULAR CLUSTERS

Robert Lupton
Princeton University Observatory
Princeton, NJ 08544

This work was done in collaboration with Jim Gunn.

In order to represent rotating clusters with an anisotropic velocity dispersion we took

$$f = C(e^{-\beta E}-1)e^{-\beta\gamma I3}e^{-\beta\tilde{\Omega}J3} \quad (1)$$

and approximated the third integral I_3 by the total angular momentum, J^2.

The term in J_3 leads to solid body rotation in the core, with the rotational velocity dropping to zero at the tidal radius. This rotation leads to flattened isopotentials which means that J^2 is not an integral of the motion. This approximation is discussed below. We further extended equation 1 by introducing a number of mass classes, with masses m_j.

MAKING THE MODELS

Introducing dimensionless variables with substitutions

$$M = \frac{\sum_i M_j \rho_{jo}}{\sum_i \rho_{jo}}, \quad W = \beta \bar{M} \psi, \quad v_o^2 = \frac{1}{\beta M}, \quad u^2 = \frac{v^2}{{v_o}^2},$$

$$\mu_j = \frac{M_j}{\bar{M}}, \quad {r_c}^2 = \frac{9{v_o}^2}{4\pi G\rho_o}, \quad \xi = r/r_c, \quad \Omega = \frac{3}{(4\pi G\rho_o)^{1/2}}\tilde{\Omega}, \quad (2)$$

$$\alpha_j = \frac{\rho_{jo}}{\rho_j}, \quad \sigma = \frac{\rho}{\rho_{jo}},$$

We have

$$f_j = C_j(e^{-1/2\mu_j u^2+\mu_j W}-1)e^{-1/2(\xi/\xi_t)^2\mu_j(u^2-{u_r}^2)}e^{\xi\sin\theta\Omega\mu_j u_\phi} \quad (3)$$

J. Goodman and P. Hut (eds.), Dynamics of Star Clusters, 327–330.

The C_j are found by the condition that

$$\rho_{jo} = \int f_j(\underline{x}=0)d^3x$$

To find the potential W, which is needed to calculate the density and velocity moments, we must solve Poisson's equation which takes the form

$$\frac{1}{\xi^2}\frac{\partial}{\partial\xi}\left(\xi^2\frac{\partial W}{\partial\xi}\right) + \frac{1}{\xi^2\sin^2\theta}\frac{\partial}{\partial\theta}\left(\sin\theta\frac{\partial W}{\partial\theta}\right)=-9\sigma \tag{4}$$

As the angular part of this equation is simply the legendre operator, we expand W and σ in legendre polynomials in $\cos\theta$, and end up with a set of ordinary differential equations for the coefficients $W^{(i)}$ and $\sigma^{(i)}$, $i=0,2,4,\ldots$ which may be solved using Green's functions.

In practice, we truncate at $P_4(\cos\theta)$, evaluate the density and the velocity moments, weight by the mass-to-light ratios of the different mass classes, and project onto the sky.

Some of these models for M13 and M92 are given in the talk by Lupton and Gunn at this conference. For M13, both anisotropy and rotation play an important role.

USE OF J^2 AS AN INTEGRAL

We have proceeded along two paths to justify our use of J^2 as an approximate integral. Firstly, by the use of an Eddington potential to approximate our true potential, and secondly by integrating f around orbits in our model potentials.

EDDINGTON POTENTIALS

For the class of potentials given by

$$\psi(r,\theta)=\zeta(r)+\frac{\eta(r)}{r^2} \tag{5}$$

$I_3=J^2-2\eta(r)$ is an exact third integral. As may be seen from figure 1, our potential approximates this form in the outer parts, with

$$\eta(r)=\lambda m_j{}^2 r_c{}^2 v_o{}^2 P_2(\cos\theta) \tag{6}$$

and

$$W=W^{(0)} + \frac{\lambda P_2}{\xi^2} \tag{7}$$

To the extent that this is correct, the only difference between our (J^2) models and a model using the true third integral I_3 is a factor

$$k=e^{\beta\gamma(J^2-I_3)}=e^{\lambda\mu_j P_2/\xi_t^2} \tag{8}$$

For the case shown, $\lambda \approx 0.8$ and the μ_i = 1.75, 1.17 and 0.58. These lead to values of k = 1.057, 1.038 and 1.019 at $\theta=0$. Due to mass segregation the first mass class is negligible this far out ($\xi>\xi_t$) in these models.

It is clear that the use of J^2 leads to errors of a few percent that are proportional to P_2.

DIRECT INTEGRATION

We have integrated orbits in our model potential and looked at the variation of f, results are shown in figures 2 and 3. For the orbit in figure 2 the period is about 80 time units, and the ap- and peri-clusticon distances are about 3 and 27 core radii. J^2 was constant at about 3%, and f to about 2%. For the orbit in figure 3, the period was around 2 units, and it spent all of its time in the region from 0.8 - 1 core radii. J^2 varied by about 7%, although f varies by only 0.1% due to the factor $e^{(\xi/\xi t)^2}=1.02$.

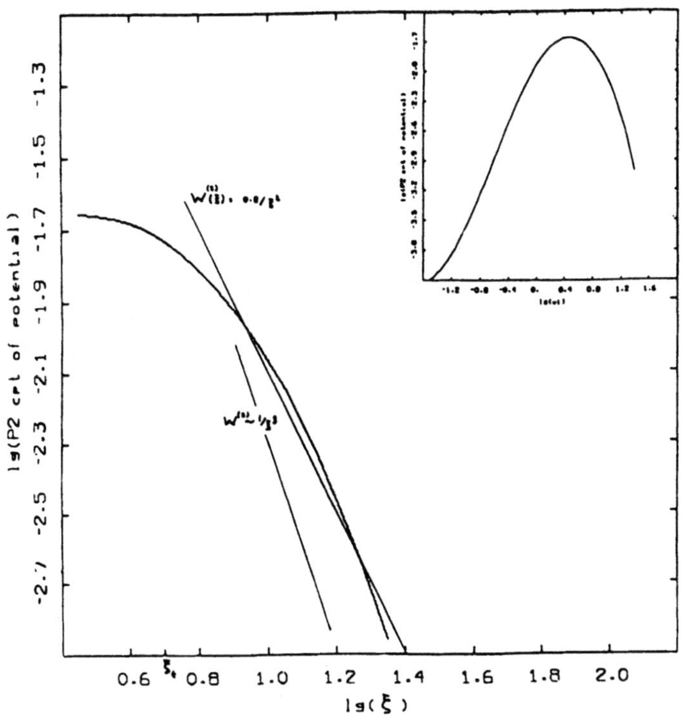

Figure 1. The P_2 component of the potential for a cluster model as a function of dimensionless radius. The insert is for the whole model, the main body of the figure shows only the outer parts. The two straight lines show two power law fits, $1/\xi^2$ and $1/\xi^3$. The $1/\xi^2$ is an Eddington potential.

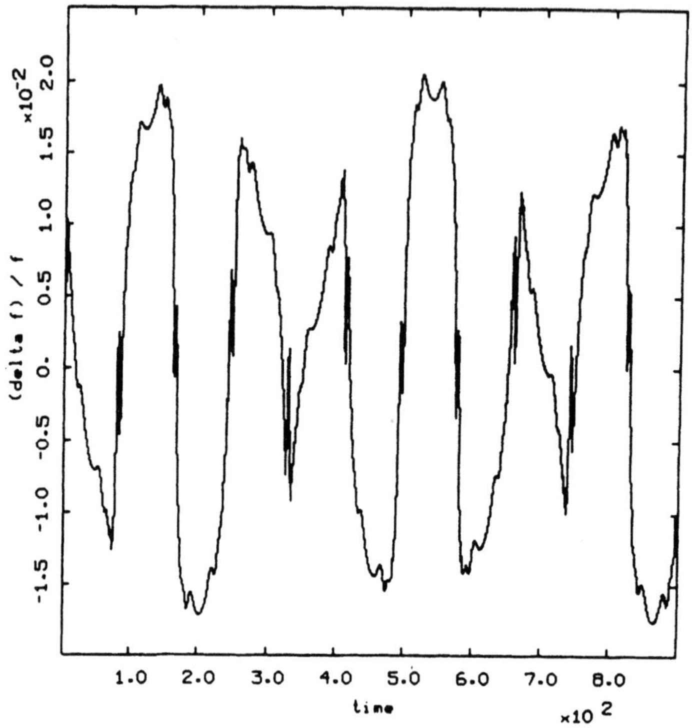

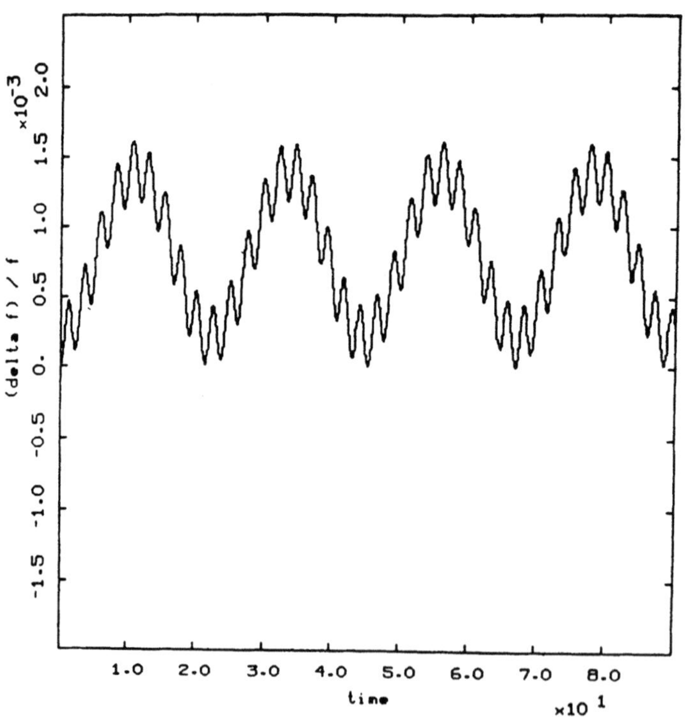

Figures 2 & 3. The fractional variation of the distribution function f as a function of time for two orbits, as described in the text.

THE EARLY STAGES OF POST-COLLAPSE CLUSTER EVOLUTION

Stephen L. W. McMillan
Department of Physics
University of Illinois at Urbana-Champaign
1110 W. Green Street
Urbana, Illinois 61801, U.S.A.

I. INTRODUCTION

We present here some recent (and very preliminary) findings from a study of the early stages of the post-core-collapse evolution of an isolated cluster of identical point "stars". The method used to follow the behavior of the system is the unified N-body/statistical treatment described in detail by McMillan and Lightman (1984a) and by Lightman and McMillan elsewhere in this volume. Briefly, the method combines the standard "large-N" and "small-N" approaches to the problem in the régimes where they are appropriate by treating the inner regions ($r < r_N$) exactly with a regularized Aarseth N-body code (Aarseth, 1972), while permitting stars at greater and greater radii to retain less and less of their individual identities, ultimately treating the outer portions of the system ($r > Kr_N$) in an almost purely statistical fashion.

The calculations described here may be viewed as an extension of those reported by McMillan and Lightman (1984b, hereafter referred to as ML), although this is a completely separate run with somewhat different initial conditions and operating parameters and a number of significant improvements to the computer program. The most important improvement is in the handling of close triple and higher-order encounters within the N-body code. While the regularization procedure adopted deals effectively with close two-body interactions, situations in which there is no dominant pair to regularize within a high-density "clump" of three or more stars can lead to substantial integration errors. This problem is virtually eliminated by the establishment of a "group" structure on the N-body stars. Basically, the definition of a group is such that the critical clumps of stars just mentioned can be identified and their members given special attention. The single most important step taken with group members is the shortening of their timesteps; in addition, specific troublesome configurations can be checked for and appropriate action taken. More detailed definitions and procedures for group initialization and termination will be presented in a future paper.

For this run, the system was initialized from the late-time distribution reported by Cohn (1980) with core radius equal to 0.5 r_N and around fifty stars in the core (125 N-body stars altogether). The

J. Goodman and P. Hut (eds.), Dynamics of Star Clusters, 331–333.

interface between the exact and purely statistical radial zones had K = 6 and consisted of approximately 450 mass shells initially. The run consumed a total of 31 hours of CPU time on an FPS-164 Array Processor, almost all of the time being taken up by the N-body integration. With the above-mentioned group handling in place, typical N-body integration errors were less than two parts in 10^5 per central relaxation time.

II. RESULTS

The variation with time of the central density and the total energy (in units of the initial central mean stellar kinetic energy, E_0) "locked up" in the form of tightly bound binaries are shown in Figures 1 and 2. In each case, the time unit, t_N is the dynamical timescale, roughly one-fifth of the initial central relaxation time, and the results have been averaged over twenty dynamical times so that the long-term behavior of the system is not obscured by noise.

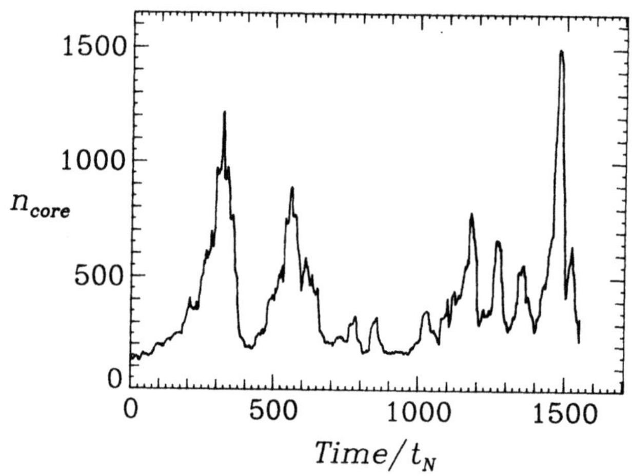

FIGURE 1. Mean density within one characteristic core radius (as defined by Aarseth, 1974) of the density center.

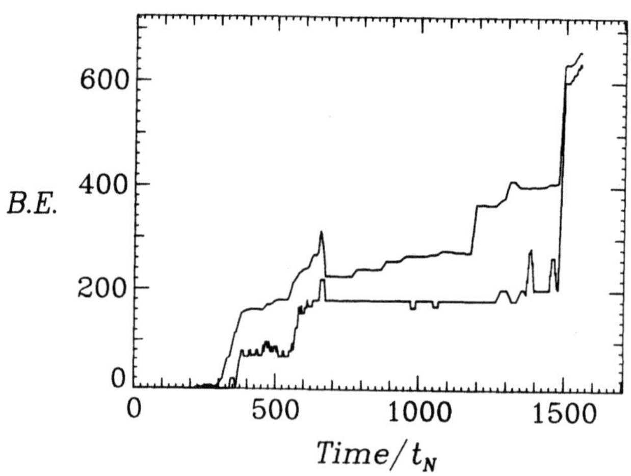

FIGURE 2. Binding energy (defined to be positive for a bound pair) stored in the form of hard binaries. The upper and lower lines represent the total energy and the energy outside the N-body region, respectively.

These figures may be compared with the results, and discussion thereof, given by ML. Once again, we find that the projected time to complete initial collapse is around one hundred initial central relaxation times. Several cycles of the core oscillations reported by ML can be seen. The claim that the oscillations are regulated by binary formation, hardening and ejection from the core is strongly supported by the facts that (i) with one notable exception (at around 1350 t_N), the termination of each collapse phase is associated with a significant increase in total binary binding energy, and (ii) the magnitude of each increase may be shown to be comparable with the core kinetic energy around the time of the density maximum. In addition, in many cases, the subsequent ejection of the binary from the N-body region can easily be seen in Figure 2.

At the end of the run, there were five very hard binaries in the system, with binding energies ranging from 15 E_o to 400 E_o. Again as conjectured by ML, binary-binary (and even binary-binary-binary!) interactions became quite frequent at late times; however, the effects of these encounters have not yet been analyzed in any detail.

III. DISCUSSION

It is not clear whether the apparent decay of the oscillations is real or just a chance phenomenon. The edge of the central isothermal region (Heggie's (1984) R_*) is rather hard to determine precisely, but if we assume that R_* lies between ten and twenty core radii at the time of maximum collapse, it may be shown (ML) that the decay timescale associated with the increase in R_* is between 1500 and 6000 t_N, suggesting that we might indeed be seeing the precursor to a steady reexpansion phase. Two other interesting findings are (i) binary lifetimes and maximum energies seem to be much larger than expected (indeed, the hardest binary in the system at the end was formed in the initial collapse, but was ejected only during the final oscillation) and (ii) there appears to be a substantial "non-diffusive" flux of stars ejected directly from the core to quite large radii. An explanation of the longevity may be that many energetic encounters seem to occur within groups of stars; by absorbing some of the liberated energy, these groups could reduce the binary recoil velocities. These facts clearly affect the arguments given in ML and warrant considerably closer scrutiny.

This work was supported by NSF grant PHY80-25605.

REFERENCES

Aarseth, S. J., 1971, in The Gravitational N-body Problem, ed. M. Lecar (Reidel, Dortrecht), p. 373.
Aarseth, S. J., 1974, Astron. Astrophys. **35**, 237.
Cohn, H., 1980, Ap. J. **242** 765.
Heggie, D. C., 1984, M.N.R.A.S. **206**, 179.
McMillan, S. L. W. and Lightman, A. P., 1984a, Ap. J. **283**, in press.
McMillan, S. L. W. and Lightman, A. P., 1984b, Ap. J. **283**, in press.

NUMERICAL SIMULATIONS OF ENCOUNTERS OF HARD BINARIES

Seppo Mikkola
Turku University Observatory,
Itäinen Pitkäkatu 1
SF-20520 Turku 52
Finland

ABSTRACT. Results from numerical integrations of random binary-binary encounters have been used to obtain various cross-sections and outcome distributions for the four-body scattering. The initial orbital elements were chosen randomly except the Kepler-energies for which various selected values were used. Rough estimates for mass effects were obtained by simulating encounters of binaries with unequal component masses.

We developed a semi-analytical theory for obtaining the types and energies of the outcome configurations. The theory contains some adjustable parameters, the values of which we deduced by comparing the theory and experiments.

The energy transfer rate by collisions (=outcome is not two binaries) dominates over that due to fly-by's by an order of magnitude, provided that the binaries are hard. The formation of a hierarchical three-body system is fairly common. In a collision of energetically similar very hard binaries the probability is about 20 percent, while it is greater than 50 percent if the binding energies differ by a factor of more than four.

1. INTRODUCTION

Many-body systems with more than two bodies are usually unstable against disruption. Therefore binaries are the most important subsystems in a star cluster. However, binaries may make close encounters with single stars and with other binaries. The problem of three-body interactions has been widely studied by many authors (e.g. Saslaw, Valtonen and Aarseth 1974, Hills 1975a,b,1980, Hut & Bahcall 1983, Heggie 1975, and others). However, the binary-binary scattering has been studied rather little. A few papers in this field to be mentioned are Saslaw et al (1974) and Hoffer (1983). This paper is a short summary of the work done by the author (Mikkola 1983,1984a,b) and includes also some new preliminary results concerning the effect of unequal masses.

J. Goodman and P. Hut (eds.), Dynamics of Star Clusters, 335–338.

2. NUMERICAL EXPERIMENTS

The Keplerian elements ω, Ω, M_0, cos(i), and e^2 of the binaries and the semiparameter p of the relative orbit were drawn from uniform distributions. Selected values were used for the Kepler energy T_∞ of the relative motion (which we call the 'impact energy') and for the ratio of the initial binding energies. Many experiments which were made with equal masses have been reported elsewhere (Mikkola 1983,1984a,b). Here we include also some results obtained from 9000 experiments with unequal masses. Both fixed mass families and certain random mass families were studied.

3. CLASSIFICATION OF THE FINAL CONFIGURATIONS

When the binaries are hard i.e. their binding energy exeeds the impact energy, only the following outcomes are possible:

(1) two binaries in a hyperbolic relative orbit
(2) one escaper and a hierarchical three-body system
(3) one binary and two escapers

A type (1) encounter is called a 'fly-by', while types (2) and (3) are classified as 'collisions'. The fraction of type (2) outcomes from all the collisions increases when the ratio of the initial semi-major axes increases. In the equal mass case we may approximate this fraction by

$$f(2) = 20\% + 15\% * \log_2(a_>/a_<) \tag{1}$$

where $a_>$ and $a_<$ are the larger and smaller semi-major axes. This is valid for very hard binaries and has been obtained from samples in which $\log_2(a_>/a_<) < 2.5$. We conclude that this outcome dominates for $(a_>/a_<) > 4$.

4. A MODEL FOR THE COLLISIONS

4.1. Cross-section

We have attempted to obtain formulae for the collision cross-section. It was found (Mikkola 1984a) that a reasonable approximation may be constructed as follows: We assume that a collision occurs if the perturbation of the Kepler-energy of the relative motion is such that this energy is negative after the close approach. We suppose further that the cross-section satisfies the relation

$$\sigma \propto s_x^{\,2} \tag{2}$$

where s_x is the maximum possible value of impact parameter for a collision. The dependence of s_x on the parameters of the system was obtained by writing the perturbation y proportional to T_∞ plus a cut-off energy ε. We have thus an equation for s_x:

$$y(s_x) \propto T_\infty + \varepsilon \tag{3}$$

When deriving the expression for y the corresponding three-body result by Heggie (1975) was used. Finally the coefficients of proportionality in (2) and (3) was determined by fitting the experimental data. We noted that the data can be fitted within the expected statistical uncertainty.

4.2. Outcome energies

A simple statistical model for the disruption of the four-body system formed during a collision has been discussed in Mikkola (1984a). Here we give only a short description of this theory: At first the system ejects one of its stars. Depending on the configuration at the moment of ejection the remnant three-body system may disrupt or form a stable configuration. We consider in more detail only the type (3) outcomes. Let the first ejected star (number 4 say) have a Kepler energy E_4 with respect to the centre of mass of the other stars. Later the remnant three-body system disrupts with disruption energy E_3. These disruptions are independent (except that the energy of the first disruption affects the energy of the second) and thus the angle between the ejection directions is random (cosine of the angle has uniform distribution). The various energies may be written in the form

$$E_4 = \tilde{E}_4(\xi_1^{-2/7}-1) \;\; ; \;\; \tilde{E}_4 = 1.8((E_{12}+T_\infty)^{-1}+(E_{34}+T_\infty)^{-1})^{-1} \tag{4}$$
$$E_3 = \tilde{E}_3(\xi_2^{-2/7}-1) \;\; ; \;\; \tilde{E}_3 = 0.82(E_4+E_{12}+E_{34}-T_\infty)$$

where E_{12}, E_{34} and T_∞ are the initial binding energies and the impact energy. The ξ's are two random numbers uniform in the interval (0,1).

A treatment of the fly-by's in the case of equal masses may be found in Mikkola (1984b). We state here the main results: The fly-by transfer of energy between the binding energy of the binaries and the kinetic energy of their relative motion is typically small. For hard binaries the collisional rate dominates by an order of magnitude. In most situations the fly-by's are thus negligible as compared with collisions.

5. ON MASS EFFECTS

In our formula for the collision cross-section the mass effects are included (approximately at least). Consideration of the energy distributions from our various unequal mass samples do not show significant differencies against the equal mass case. A possible exception is a sample in which we have very large mass ratios (typically 10). In this case the mean values of the energy transfer are about 1.5 times greater than in other cases. We thus conclude that the energy distributions are not sensitive to the masses.

An additional interesting question is: which masses are likely to be ejected and which finally form the remaining binary? More precicely we may ask for the probability of a given mass to be ejected, first from the four-body system and then from the remaining three-body system. When we consider the outcome type (3) we in fact restrict the consideration to a sample in which it may be reasonable to take the probability to be a function of the masses only. For a restricted mass range a simple power law is a plausible approximation. We assume that for three- and four-body systems the escape probabilities are, respectively, proportional to

$$P_j \propto m_j^{-n} \ ; \ P_j \propto m_j^{-k} \qquad (5)$$

We wrote down the expressions for the probability that a given pair forms the remaining binary and then we used the maximum likelihood method to determine the most likely values for the exponents. The result was $n = 2.5$, $k = 1.9$. The goodness of this law was also studied by calculating quantities like the mean mass ratio and mass sum of the remaining binary. A good agreement with experiments was found.

6. CONCLUSION

A rough statistical understanding of the binary-binary scattering has been reached. We have found that the mass effects enter the collisional interaction mainly through the cross-section and the escape probability while the energy distributions do not depend sensitively on the masses. However, especially for the case of very large mass ratios the situation is yet unclear. We must also emphasize that the bad statistics due to small number of experiments makes these results only preliminary. We hope to publish elsewhere the details for the unequal mass case.

REFERENCES

Heggie, D.C., 1975. Mon.Not.R.astr.Soc. 173,729.
Hills, J.G., 1975. Astr.J. 80,809.
ibid 1975b. Astr.J. 80,1075.
ibid 1980. Astr.J. 85,1281.
Hoffer, J.B., 1983. Astr.J. 88,1420.
Hut, P. & Bahcall, J.N., 1983. Ap.J. 268,319.
Mikkola, S., 1983. Mon.Not.R.astr.Soc. 203,1107.
ibid 1984a. Mon.Not.R.astr.Soc. 207,115.
ibid 1984b. Mon.Not.R.astr.Soc. 208,75.
Saslaw, W.C., Valtonen, M.J. & Aarseth, S.J., 1974. Ap.J. 196,253.

COLLISIONAL RELAXATION : A NEW APPROACH

G. Severne and M. Luwel
Physics Dept., U. of Brussels (V.U.B.),
Pleinlaan 2, 1050 Brussels, Belgium

ABSTRACT. Recent numerical simulations for 1-dimensional systems have shown that the relaxation time due to encounters is far shorter than the generally accepted estimate. To account for this, a new approach to the theory is necessary. The analysis of encounters presented here is characterized by the retention of periodic trajectories in the mean field. The kinetic equation obtained yields a relaxation time scale in qualitative agreement with the simulations. The analysis can be extended to the 3-dimensional case, and preliminary results predict here also a reduction of the relaxation time.

Our present understanding of the relaxation process in gravitational systems has not significantly progressed with respect to Chandrasekhar's original formulation (1) in 1941. The various refinements and reformulations introduced over the years have left his results essentially unchanged. All existing theories have in common one important approximation : the influence of the mean gravitational field is not taken into account in the analysis of encounters. These are described as perturbations of straight trajectories, and one obtains, after the ad hoc suppression of a logarithmic divergence, a relaxation time t_R proportional to N t_D/log N, where N is the number of particles in the system and t_D a characteristic dynamical time (crossing time).

Applied to 1-dimensional systems, this approach yields $N^2 t_D$ as minimum estimate for t_R : indeed in 1-d, simple binary encounters cannot give rise to energy exchange, as a result of the constraints imposed by the conservation laws. Recent numerical simulations (2), (3), have however established that collisional relaxation develops on a far shorter time scale t_R < N t_D. A typical example is provided by Fig. 1, giving the evolution in a 2-mass system of the ratio k of the average kinetic energies for the 2 masses : one sees that k(t) effectively relaxes towards the equipartition value k = 1 on a time scale much shorter than N t_D. Moreover, the results exhibited Fig. 2 suggest that the relaxation time for equipartition increase linearly with N.

These simulation results make it imperative to reconsider the classical analysis of encounters. For the 1-d problem one is dealing

J. Goodman and P. Hut (eds.), Dynamics of Star Clusters, 339–341.

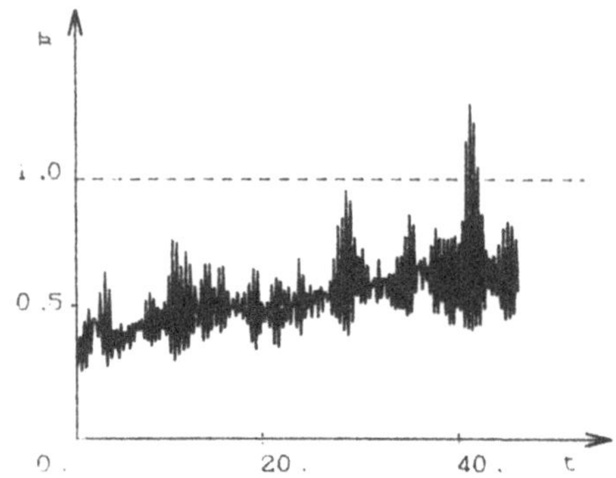

Fig. 1 Evolution of $k = \langle m_1 v_1^2/2\rangle\langle m_h v_h^2/2\rangle$, with $m_1/m_h = 1/3$, $N_1/N_h = 3$ and $N = N_1 + N_h = 200$. The time t is measured in units t_D.

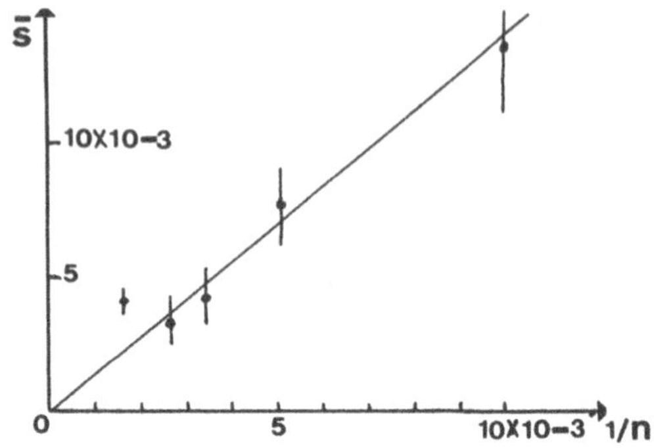

Fig. 2 The slope s of the linear fit to the k(t) plots, in terms of 1/N. Since $s = (t_D/t_R)(m_h-m_1)/m_1$, t_R varies linearly with N.

with a system of mass sheets, interacting through binary forces $ma_{12}=-2\pi Gm^2\,\mathrm{sgn}(x_1-x_2)$; for simplicity, we take particles (sheets) of the same mass (density) m. The particles are confined in the self-consistent mean field mA(x,t), which, contrary to the classical analysis, we retain as a dominant effect. The feasibility of the problem results from the fact that, once the violent relaxation phase is over, mA(x,t) can quite realistically be modelled by the constant harmonic oscillator field $-m\omega^2 x$, with $\omega=2\pi t_D^{-1}$.

The dynamical analysis for the N-body system proceeds from the BBGKY hierarchy and, with the usual approximations on the higher order and initial correlation functions, leads, see ref. (4), to the following equation for the 1-particle distribution function $f_1(t)\equiv f(x_1,v_1,t)$:

$$(\partial_t+L_1)f_1(t) = \int dx_2 dv_2 \int_0^t d\tau\, \partial_{12} a(x_{12}) a(x_{12}\cos\omega\tau-\omega^{-1}v_{12}\sin\omega\tau)$$
$$\left[\partial_{12}\cos\omega\tau + \omega^{-1}\nabla_{12}\sin\omega\tau\right]\left[\exp-(L_1+L_2)\tau\right] f_1(t-\tau)f_2(t-\tau) \quad (2)$$

$$L_i = v_i\, \partial/\partial x_i - \omega^2 x_i\, \partial/\partial v_i,$$

$$x_{12}=x_1-x_2,\ v_{12}=v_1-v_2,\ \partial_{12}=\partial/\partial v_1-\partial/\partial v_2,\ \nabla_{12}=\partial/\partial x_1-\partial/\partial x_2.$$

To focus attention on the time integration, we rewrite Eq.(2) as

$$(\partial_t+L_1)f_1(t) = \int_0^t d\tau F(\tau)\left[\exp-(l_1+L_2)\tau\right] f_1(t-\tau)f_2(t-\tau). \quad (3)$$

$F(\tau)$ is a periodic function of τ (and an operator on position and velocity). The exponential operator advances $f(t-\tau)$ along the unperturbed orbit so that one can approximate

$$\left[\exp-(L_1+L_2)\tau\right] f_1(t-\tau)f_2(t-\tau) \cong f_1(t)f_2(t) \quad (4)$$

as long as the perturbation due to the encounters remain small. If we ignore this restriction and take the usual time limit $t\to\infty$, we obtain, writing the integral as a sum over successive periods $t_D=2\pi\omega^{-1}$,

$$(\partial_t+L_1)f_1(t) = \sum_{n=0}^{p\to\infty} \int_{nt_D}^{(n+1)t_D} d\tau\ F(\tau)f_1(t)f_2(t) \tag{5}$$

$$= \lim_{p\to\infty} p\left[\int_0^{t_D} d\tau F(\tau)\right] f_1(t)f_2(t) \tag{6}$$

The manifest divergence of this result comes from the approximation of Eq. (4), which must break down over times τ of the order of the relaxation time t_R. However the assumption of perfect harmonic motion clearly becomes unrealistic well before that : non-isochronic trajectories, as also fluctuations of the mean field, will introduce a progressive loss of coherence between the successive contributions to Eq. (5). Heuristically, we can proceed by retaining only the first p terms of the sum, with $pt_D<t_R$, thus obtaining for Eq.(2) :

$$(\partial_t+L_1)f_1(t) = 4\pi G^2m^2p \int dx_2dv_2\partial_{12}\mathrm{sgn}(x_{12}v_{12})$$
$$\left[\sin\omega\tau_{12}\partial_{12} - \omega^{-1}(\cos\omega\tau_{12})\nabla_{12}\right]f_1(t)f_2(t) \tag{7}$$

with τ_{12} = Arc tan $\omega x_{12}/v_{12}$.

Eq. (7) is a generalized Fokker-Planck equation : in the collision integral, velocity diffusion is described by the $\partial_{12}\partial_{12}$ contribution and gives rise to a monotonous growth in entropy, while the ∇_{12} term ensures the stationary of the equilibrium distribution exp $-m(v^2+\omega^2x^2)/2$. In order of magnitude, Eq. (7) gives for the relaxation time : $t_R^{-1} \propto p.G^2m^2N/\langle v^2\rangle$. On the other hand, one has $\omega^2 = 4\pi GmN/L$, with $L \cong \sqrt{\langle v^2\rangle}/\omega$. Since $t_D \propto \omega^{-1}$, it follows that $t_R \propto Nt_D/p$ with the coherence factor p to be determined from simulations. This result constitutes a radical reduction, by a factor Np, with respect to the standard estimate, and is in qualitative agreement with our simulations, refs. (2), (3).

The 1-d analysis can be extended (4) to three dimensions, but here the time integration has not yet been worked out. Nevertheless it is clear that there is no longer a divergence at large particle separations and that, in qualitative agreement with recent 3-d simulations (5), a moderate reduction of the relaxation time (by the coherence factor p) is to be expected.

REFERENCES

(1) S. Chandrasekhar : 1942, "Principles of Stellar Dynamics",Dover,N.Y.
(2) M. Luwel, G. Severne, and P. Rousseeuw : 1984, Ap. Space Sci. **100**, 261.
(3) G. Severne, M. Luwel, and P. Rousseeuw : 1984, Astr. Ap. in press.
(4) G. Severne and M. Luwel : 1984, Phys. Lett. in press.
(5) R. Farouki, G. Hoffman, and E. Salpeter, Ap. J. **271**, 11.

TIDAL EFFECTS ON GLOBULAR CLUSTERS

P. Seitzer
Kitt Peak National Observatory
National Optical Astronomy Observatories*
P. O. Box 26732
Tucson, Arizona 85726

It was von Hoerner (1957) who pointed out that the tidal field of the Galaxy imposed a boundary condition on the size of globular clusters, a fact that had observational consequences when the first deep star counts became available (King 1962). More recently, though, dispute (cf Innanen, et al. 1983) has arisen over the identification of the observed limiting radius from star counts with that predicted by simple dynamical theory (King 1962, 1966; Szebehely 1967). Straightforward application of the classical three body problem seemed to imply that no stable stellar orbits would be found beyond the collinear Lagrangian points. Thus, to first order, the observed limiting radius would be just the radius of the inner Langrangian point L_2 from the cluster center. For clusters in eccentric orbits, the tidal radius would be set at perigalacticon, where the tidal force is strongest. Numerical models of stellar orbits about a cluster in a realistic galactic potential by several authors (Jefferys 1976; Keenan and Innanen 1975; Keenan 1981) showed that stable orbits can exist with apocluster distances greater than r_{L_2}, due to the presence of non algebraic integrals of the motion that constrained the stars to move in a restricted region about the cluster rather than escaping.

I undertook a study of the effects of the tidal field on globular clusters in order to quantify the above (Seitzer 1983). I followed 1000 test particles orbiting in the outer half of a cluster (represented by a Plummer potential) for 10 cluster orbits about the galaxy. Escape was arbitrarily defined as when a particle went beyond 3 times the expected tidal radius as computed at apogalacticon. Models were computed for different cluster orbital eccentricities e = 0.0, 0.3, 0.5, and 0.7 to check for the dependence of observed tidal radius on eccentricity. The conclusions are summarized below:

*Operated by the Association of Universities for Research in Astronomy, Inc., under contract with the National Science Foundation.

J. Goodman and P. Hut (eds.), Dynamics of Star Clusters, 343–346.

1. The observed limiting radius after ten orbits of the cluster about the galaxy was very close to the predicted inner Lagrangian point. A very few stars had apocluster radii outside this limit. This means that the observed tidal radius can be used to estimate the eccentricity of the cluster's orbit about the galaxy, or (to a much lower accuracy) the mass of the galaxy inside the cluster's orbit (Peterson 1974).

2. Most of the stars escaped in the first eight cluster orbits. One model (e = 0.5) was followed for 50 cluster orbits: very few stars escaped after 10 orbits. This suggests that one has to be very careful about interpreting the observed limiting radius of distant systems such as dwarf spheroidals, who probably have completed fewer than 10 orbits about the galaxy since their formation. Conditions today are probably an unknown mixture of initial conditions and tidal effects, and hence tidal radii of these distant systems can not be used as probes of the galactic potential.

3. Something like 50 cluster orbits will be necessary for a state of tidal relaxation. Thus, only the inner clusters will have their outer parts tidally relaxed, the outer halo systems will still reflect (to some level) their initial conditions.

4. There was a slight preference for stars with retrograde orbits about the cluster (compared to the direction of the cluster's orbital angular momentum vector) to be more stable than stars in a direct orbit. However, the effect is not going to be large enough to set up a net retrograde rotation of the outer half of a cluster. In Omega Centauri, the one cluster where the kinematics has been followed beyond half the tidal radius (Freeman et. al. 1984), no rotation is observed.

It must be emphasized that my calculations were not a true N body simulation of a self gravitating system, and therefore can be applied only to very centrally condensed systems such as globular clusters where tidal effects will be neglible in the center. Dwarf galaxies will have to be followed with true N body codes.

REFERENCES

Freeman, K. C., Seitzer, P. and Norris, J. 1984, in preparation.
Innanen, K. A., Harris, W. E. and Webbink R. F. 1983, A. J., **88**, 338.
Jefferys, W. H. 1976, A. J., **81**, 983.
Keenan, D. W. 1981, Astron. and Astrophys., **95**, 340.
Keenan, D. W. and Innanen, K. A. 1975, A. J., **80**, 290.
King, I. R. 1962, A. J., **67**, 471.
King, I. R. 1966, A. J., **71**, 64.
Peterson, C. J. 1974, Ap. J. Letters, **190**, L17.
Szebehely, V. 1967, Theory of Orbits (New York: Academic Press).
von Hoerner, S. 1957, Ap. J., **125**, 451.

DISCUSSION

OSTRIKER: There is a possible explanation for the correlation you pointed out between radius and central relaxation time based on work done at Princeton several years ago. As galactocentric radius decreases, the effects of tidal shocks increase and evolution of the core to larger densities and shorter relaxation times becomes more rapid. This is due to depopulation of orbits at intermediate energies by tidal phenomena which speeds diffusion from the center to these intermediate regions.

WHITE: I'd like to comment on two points. I think that for many clusters it may be important to include two-body relaxation effects in a discussion of limiting radii since their core relaxation times are short enough that significant repopulation of a depleted halo may occur within one or two orbits about the galaxy. I'd also like to reemphasize that it may take a number of orbits to establish a tidal cut-off in binding energy. I have carried out some self-consistent simulations of the Milky Way - Draco system. These simulations show that in a single eccentric encounter, tides do not impose a cut-off in binding energy - many weakly bound stars survive while a number of much more strongly bound stars are lost. Whether a star is lost or not clearly depends on its orbital phase and orientation as well as on its binding energy. This suggests that it may be extremely risky to use the tidal radii of dwarf spheroidals to estimate their mass. If I try to apply this technique to my simulations of Draco I find that the present profile of the model shows no evident limiting radius and that attempts to derive such a radius using standard observational techniques can lead to overestimates of the mass by factors as large as 500. I urge the N-body people to run more self consistent simulations of loosely bound systems.

SHAPIRO: Do I gather, then, that estimates of dwarf spheroidal galaxy masses from measurements of their tidal cut-off radii are completely invalid?

SEITZER: I do not put much faith in these, due to both the observational errors and the fact that these clusters may not have reached a tidally relaxed state.

COHN: Concerning your final figure, which relaxation time was plotted - the core relaxation time or the half-mass relaxation time?

SEITZER: The central relaxation time.

TREMAINE: If the cluster is on a circular orbit the tidal radius is located at the Lagrange points. In the elliptic restricted three-body problem the Lagrange points move closer or further from the cluster as it moves closer or further from the galactic center, keeping at a fixed fraction of the distance from the galactic center. This suggests that the tidal radius may be a fixed fraction of the present distance rather than the pericenter distance. At what phase in the cluster orbit did you measure the tidal radius and would you get the same answer if you measured it at a different phase of the orbit?

SEITZER: I measured the density profiles at apogalacticon, and found that the observed limiting radius is not too far from the calculated radius at perigalacticon for tightly bound systems. For loosely bound system (dwarf galaxies), this may not be true.

KING: I would like to emphasize your point that many of the "observed" limiting radii are rather weak. When Peterson and I determined them, we began with the cases where the models were a good fit and the observations were good too. But then we went on and estimated a limiting radius for any cluster where we thought it could be done at all. These should surely be redone from better observational material that can be obtained today.

SEMENZATO: 1) What is the agreement of your results with those of Keenan's numerical experiments? 2) I have seen reports of evidence of observations of retrograde rotation in M13 and ω Cen. Can you explain this rotation with only the slight surplus of retrograde orbit stars that you get from your numerical simulations?

SEITZER: 1) My models produced nearly the same results as those of Keenan. However, Keenan would say that because there are unstable orbits at $r < r_t$, the observed limiting radius will be less than the collinear Lagrangian radius. But the King models do not have all possible orbits in phase space at $r < r_t$. Considerable volumes of phase space are not inhabited in a spherically symmetric King model at $r < r_t$: the observed cut-off is very gradual with continuous derivatives. 2) Rotation in M13 and ω Cen is almost certainly due to initial conditions, since tidal forces do not penetrate very far into the cluster.

FREEMAN: Seitzer, White and van Albada have all pointed out that the tidal radius of a satellite or globular cluster takes several galactic orbits to be established. There is at least one situation where this is not so. The young globular clusters in the LMC have ages that are only a fraction of their orbital time around the LMC. However, the masses that one estimates for these clusters from their tidal radii are in excellent agreement with the masses estimated from their measured luminosity functions and from a recent direct measurement of velocity dispersion (for one somewhat older cluster). The tidal estimates work because these clusters have nearly circular orbits: the tidal radii are probably associated more with the clusters' *formation* than with the subsequent dynamical evolution. If we knew that the dwarf spheroidal satellites of our galaxy were also born in low eccentricity orbits, then their tidally estimated masses could also be reliable, despite the low number of galactic orbits that they have made.

SEITZER: Agreed. A knowledge of initial conditions when clusters and dwarf galaxies formed is essential to understand their current tidal state.

PHYSICAL INTERACTIONS BETWEEN STARS

Jeremiah P. Ostriker
Princeton University Observatory
Princeton, New Jersey 08544 USA

ABSTRACT. Physical arguments are presented to show that two-body, tidal-capture binaries should form in abundance ($N_b \sim 10^3$) during the evolution of globular clusters by the time that core collapse begins. Interactions amongst these binaries and with core single stars will cause ejections from the cluster which pump energy into the system producing a bounce and re-expansion. Detailed numerical Fokker-Planck evolutionary calculations presented here confirm this scenario and indicate that this process is likely to be the dominant energy input for most clusters. During the re-expansion phase $r_{core} \propto t^{1/3}$, $r_{half} \propto t^{2/3}$ with the core containing several hundred very close binary star systems.

1. INTRODUCTION

The mathematical treatment of globular cluster evolution has almost always idealized the interacting stars, treating them as point masses. Most of the work presented at this conference has also adopted this usually excellent approximation. The purpose of this paper is to show that, at critical stages in the life of a cluster, finite size effects will have an important, even dominant, influence on its subsequent evolution. This possibility has been noted by several authors, including: Lightman and Fall (1978), Milgrom and Shapiro (1978), Alexander and Budding (1979), Ozernoy and Dokuchaev (1982a,b), Inagaki (1984a,b), Krolik (1983) and Stodolkiewicz (1985). Here I will review the physical situation and then present the first detailed model evolution (through and past core collapse) for a stellar system which is forming and ejecting binaries produced by the two-body tidal capture process. It appears from this work that such physical interactions are the primary mechanism for arresting and reversing core collapse and also for driving the late asymptotic expansion phase.

To develop the physical framework required to understand the problem, we will consider a simple one-component system and begin by comparing three time scales in the central regions of the cluster. The inverse relaxation time ($\nu_r \equiv t_r^{-1}$) is conventionally

$$\nu_r \equiv 15.4\ G^2 m_*^2 n \ell n \Lambda / v^3 \quad , \tag{1}$$

J. Goodman and P. Hut (eds.), Dynamics of Star Clusters, 347–360.

where v is the three-dimensional local velocity dispersion, (m_*, n) the stellar mass and local number density and Λ typically approximated by 0.4 N. All quantities, including N, may be thought of as locally defined; e.g., $N \equiv N(r)$ the number of particles interior to r. Close two-body encounters locally occur at a rate $\nu_2 = n\sigma v$ where the cross section can be written in terms of minimum orbital separation R_{min} as

$$\sigma = \pi\left(R_{min}^2 + R_{min} \frac{2Gm_*}{u^2}\right) \approx \pi\; R_{min}R_* \left(\frac{v^*}{u}\right)^2 \quad , \qquad (2)$$

where v_* is the escape velocity from a star ($v_*^2 \equiv 2Gm_*/R_*$) and u the relative velocity. For physical (including tidal) encounters we will find that, as expected, R_{min} cannot be much larger than R_*, so we write $R_{min} \equiv aR_*$ and note the weak dependence of a on v. Thus we have

$$\sigma = \pi a(v)R_*^2(v_*/v)^2, \text{ where } \qquad a(v) \equiv 0.54(v_*/v)^{0.2}; \qquad (3)$$

giving

$$\nu_2 = 16.1\; naR_*^2 \left(\frac{v_*}{v}\right)^2 v \quad , \qquad (4)$$

where integration over an assumed local Maxwellian has been performed ($2\langle u^2\rangle = v^2$) and, for a(v) in equation (3), we have used the calculations of Press and Teukolsky (1977) for tidal capture of n = 3 polytropes. Finally, we adopt (from Hut, 1985) for the rate of production of three-body binaries hard enough to survive

$$\nu_3 = 126\; n^2G^5m_*^5/v^9 \quad . \qquad (5)$$

Let us first compare the rates of relaxation and two-body capture from equations (1) and (4):

$$\frac{\nu_r}{\nu_2} = 0.24\; \frac{\ell n\Lambda}{a} \left(\frac{v_*}{v}\right)^2 \quad . \qquad (6)$$

Since (v_*/v) is large in normal globular clusters and the other numerical factors together are close to unity, we would not normally expect two-body processes to be important. As an example, if we adopt the parameters (v_* = 600 km/s, v = 20 km/s, $N_0 = 10^{5.0}$) we find that (ν_r/ν_2) = 1,200 and two-body processes should be utterly negligible in importance. However, as core collapse proceeds, the central velocity dispersion increases, $v^2 \propto \rho^{0.10}$ (Cohn, 1980), the number of stars in the core decreases, $N_c \propto \rho^{-0.35}$ leading to a decrease in $\ell n\Lambda$ and, most important, the evolution becomes slower and slower per relaxation time as isothermality becomes better in the central regions with, at late times, $\nu_{r,eff} \equiv (\dot{\rho}/\rho) = 10^{-2.5}\; \nu_r$ (Cohn, 1980). Thus, by the time the central density has increased by six orders of magnitude as core collapse approaches, the velocity dispersion has increased by a factor of 2, and the number of stars in the core decreased to 800, giving ($\nu_{r,eff}/\nu_2$) = 1. Well beyond this point the two-body processes will be more important than the heat flow induced by relaxation, so energy input from binaries should be able to reverse the collapse. This is close to the point in our numerical computation (cf. Fig. 1) when re-expansion does in fact, begin.

Second, we may compare the two-body and three-body processes using equations (4) and (5):

$$\frac{\nu_3}{\nu_2} = \frac{0.4}{aN^2}\left(\frac{v_*}{v}\right)^2 \quad . \tag{7}$$

This shows that the three-body processes will not be able to compete with tidal capture, until N_c is quite small ($N_c < 0.7 \times (v_*/v)^{0.9}$, and the core has reached a correspondingly large central density. Another way to express this point is to note that $(\nu_3/\nu_r) = (3.7/N^2 \ln\Lambda)$ so that very few stars (perhaps only a few binaries) will be left in the system when three-body processes become significant regardless of the internal structure of the stars. As a result, the total number of binaries formed by the three-body process will be much smaller than the number of two-body binaries already present. Of course, if most of the stars were "hard" enough such as degenerate dwarfs or neutron stars then, even when the core was reduced to one binary, finite size effects would not have become important. But for normal stars, and stellar systems such as globular clusters or galactic nuclei it appears that two-body processes will almost always be more important than three-body collisions among single stars. We will make this argument more quantitative in the next section.

After tidally captured binaries form, they will affect the system in two ways. First they may interact with singles or other binaries in close encounters. In these an exchange of stars often occurs, and then typically the reaction products leave the encounter at velocities of order the orbital velocity ($\Delta v \approx 1/4\ v_* \approx 10^2$ km/s) which will cause ejection from the cluster. To evaluate the effect on the cluster one need only consider the reverse process, bringing mass Δm in from infinity to the cluster center where it is given a velocity $v_0{}^2$. The energy release is obviously $-(+\phi_0 + 1/2\ v_0{}^2)\Delta m$. The energy input to the cluster caused by mass Δm leaving is, of course, the negative of this and it is distributed according to the integral, $G\Delta m\int dm/r$, which puts equal amounts of energy in equal logarithmic intervals of radius for an isothermal cluster. As a consequence, this form of "heating" is not centrally concentrated but deposited typically far from the core. In addition, more distant encounters of stars with binaries will deposit a comparable amount of energy directly into the central regions by the processes investigated by Spitzer and Mathieu (1980). In the detailed integrations to be reported on shortly, we have neglected this process and we therefore underestimate the importance of the two-body binaries. Finally we note that, ultimately, we can expect an equilibrium density of binaries to be established since, if the density is low, then the ejection rate by the dominant binary-binary interactions is slow but the formation rate by single-single interactions is rapid so the binary number density will increase (and vice versa). We can expect the post-collapse expansion, if one exists, to be similar to that found by Goodman (1984) with the half mass radius increasing as $t^{2/3}$, since as noted in that work and by others (Henon, 1961; Lynden-Bell and Eggleton, 1980), the overall evolution is insensitive to the details of the energy generation process in the post-collapse era.

2. AN IDEALIZED EVOLUTION UNDER THE INFLUENCE OF TIDAL CAPTURE BINARIES

The calculations were performed by T. Statler who has extensively modified the code developed by H. Cohn (1980) for orbit averaged Fokker-Planck evolution to include the various binary rates. Details of the work will be published elsewhere (Statler, Ostriker and Cohn, 1984). The only significant modification made to Cohn's treatment of collisions was to take $\Lambda = 0.4\ N_{core}$ rather than keeping it constant throughout time. For binary creation we use the orbit averaged cross sections calculated by Press and Teukolsky (1977). All binaries are taken to be at the average separation $\langle R_{sep} \rangle = 2.5\ R_*$ and no allowance is made for mergers. The inelasticity of the sticking collisions is specifically included. Energy release during single-binary encounters is calculated via a modified form of the Hut (1984) cross sections.

$$\frac{d\sigma}{d\Delta} = \frac{135}{8}\left(\frac{v_*}{u}\right)^2 \pi R_*^2 (1 + \Delta)^{-4} \Delta^{-0.5} \quad , \tag{8}$$

where $\Delta \equiv |\Delta E/E_{binary}|$. We note how much larger the numerical factor is in this cross section than in the binary formation cross section given by equation (4). In binary-binary interactions Mikkola (1983) is adopted for close encounters ($\Delta > 1$), and a further modified version of the Hut cross sections taken for the distant collisions. After each encounter we check the probability of ejection of any of the components entering into the interaction (sharing the energy released in accordance with momentum conservation). If a star, single or binary, can escape, it is removed from the distribution function but if it cannot, then its energy is not changed (i.e., we ignore heating) and the interaction is ignored. This is done in the "Fokker-Planck step" of the integration. Then, keeping the particle density conserved (in the appropriate action-action space), Poisson's equation is solved iteratively until self-consistency is reached with the new distribution function. In this step the potential change due to the mass lost in the F-P step by ejected stars causes all exterior stars to lose energy and their orbits to expand.

The temporal evolution is presented in the accompanying figures where the dimensionless units are for $[R] = r_0$, $[t] = t_{rh,0}$, $[\rho] = (M/r_0^3)_0$, $[v^2] = (GM/r_0)_0$ where M and r_0 are the mass and scale radius of the initial Plummer model. Physical units displayed assume the initial model to have $r_0 = 1.13$ pc and to contain 3×10^5 stars, each of mass $m_* = 0.7\ M_\odot$ and radius $R_* = 0.57\ R_\odot$. Table I summarizes the physical quantities at various epochs (labeled by arrows in Figure 1). The core radius is defined in the conventional way for isothermal spheres $[4\pi G\rho(0) \equiv 3v^2(0)/r_c^2]$, the core radius used in photometric discussions being three times larger; r_h is the half mass radius, $\sigma_s(0)$ the central one-component velocity dispersion of singles. N_b is the current number of binaries, T_b their mean age at that time and ν_{bf} the current rate of binary formation.

In Figure 1 we see that the central density follows the pre-core collapse evolution nearly identical to that without binaries reaching very high values briefly at $t = 3.3 \times 10^9$ years. The time to peak den-

Table I

Model Cluster at Four Epochs

Epoch:	Initial	Early	Peak	Late
$t/t_{rh,0}$	0	7.84	14.68	76.4
t(yr)	0	1.76(9)	3.303(9)	1.72(10)
r_c(pc)	.799	.295	2.03(-3)	1.98(-2)
r_h(pc)	1.47	1.56	2.16	8.38
$\sigma_s(0)$(km/s)	11.6	11.6	17.3	5.8
N_b	0	335	1462	370
T_b(yr)	0	7.36(8)	8.12(8)	1.19(9)
ν_{bf}(binaries/yr)	1.07(-7)	3.05(-7)	9.29(-6)	2.91(-7)

sity in Cohn's (1980) work was found to be t_c = 15.8 $t_{rh,0}$; binaries accelerate the collapse at late times and peak density is reached at t_c = 14.7 $t_{rh,0}$ in the present calculations. Subsequently the central density decays as the 4/3 power of the time. The central velocity (Fig. 2) is nearly constant until nearly the peak when it climbs briefly to 18 km/s (from an initial value of 12 km/s) and subsequently decays as $t^{-1/3}$. The core and half mass radii behave (Fig. 3) in expected ways prior to maximum concentration and then expands as $t^{1/3}$ and $t^{2/3}$, respectively, in the post-collapse phase. Thus the post-collapse phase is not truly homologous and, although all radii expand, the cluster becomes more and more centrally concentrated.

The post-collapse density distributions (4a,b) show the buildup of an extensive and expanding nearly isothermal structure with, of course, $\rho_b = \rho_{bo}(\rho_s/\rho_{so})^2$; the binaries remaining primarily in the center even as some diffuse into the cluster halo.

We have treated the cluster as an isolated system. Had we taken a tidal cutoff at R_t corresponding to a fixed mean density, then we would have found the mass decreasing linearly (cf. Stodolkiewicz, 1985) at late times as core energy input pushed the expanding cluster beyond the tidal limit.

Figure 5 shows the rates of formation and ejection of stars via various processes. Note that the primary process, binary formation increases only gradually in the precollapse phase and then decreases approximately as $t^{-1.2}$. During and after core collapse the binary-binary ejection process is most important and the three-body binary creation rate (dot dash curve) is never competitively important. The total number of binaries formed during the run is approximately 20,000 (20% of mass) of which 2,000 (2% of mass) had formed prior to the moment of maximum density. However, the lifetime of binaries once made is relatively short, so the number currently in the cluster declines from 2,000 at core collapse to 200 at the end of the numerical evolution. The self-similar nature of the central regions in the post-collapse epoch is evident in the central ratio of binaries to singles which is constant at (2/1) by mass throughout the late phases.

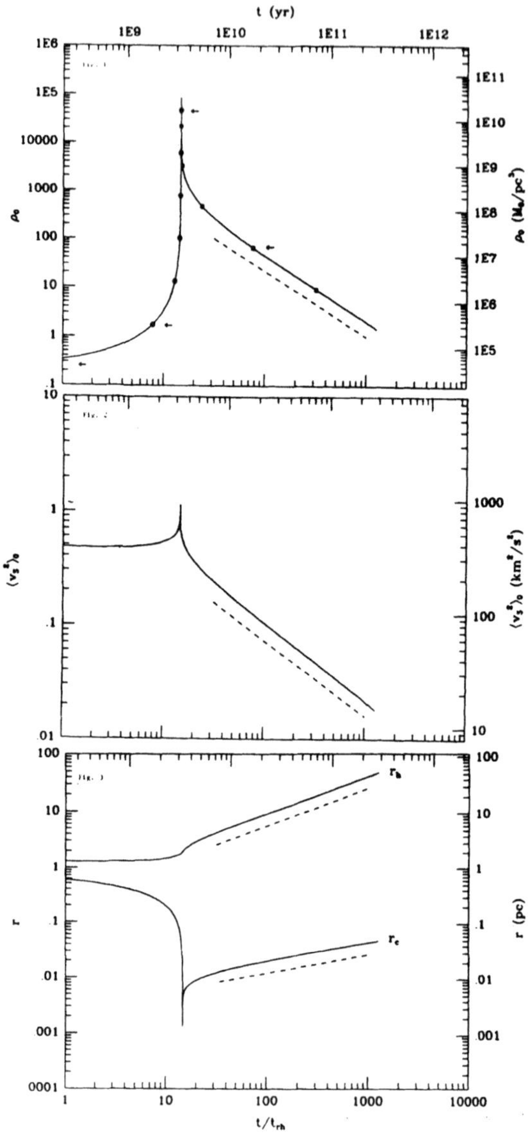

Figure 1. Cluster central density vs time in physical and dimensionless units (cf. §2 for definitions). Arrows mark "early" peak and "late" epochs referred to in Table I. Dashed line shows $\rho_0 \propto t^{-4/3}$ found from analytical treatment eq. (15a).

Figure 2. Central velocity dispersion (squared) of single stars vs time. To high accuracy the binaries always have $\langle v_{b0}{}^2\rangle = 1/2\langle v_{s0}{}^2\rangle$. Dashed line shows $v^2 \propto t^{-2/3}$ dependence of equation (14a).

Figure 3. Half mass radius r_h and core radius r_c vs time with analytical estimates, ($r_h \propto t^{2/3}$, $r_c \propto t^{1/3}$) from equations (14b) and (15b) shown as dashed lines.

Although the calculation is not intended to be realistic in that the cluster tidal radius, a spectrum of masses and other important features have been knowingly neglected; it is of interest to check on the mean age of the binaries as shown in Fig. 6. There we see that up to a time of $t = 1.5 \times 10^{10}$ yrs the mean age of the binaries has not exceeded 9×10^8 years. Thus gravitational radiation, which would cause the stars to merge into singles, would probably not have been effective. If, however, as seems not unlikely, other angular momentum loss processes are much more efficient than gravitational radiation, then at late stages the evolution may be significantly different than what we have found. However, during the important bounce and re-expansion phases the binaries are typically quite young so our treatment should be valid. Other calculations not presented here indicate that neither angular momentum loss nor other causes of stellar fusion are likely to significantly alter the qualitative features of the numerical evolution shown in Figures 1-6 for an isolated cluster.

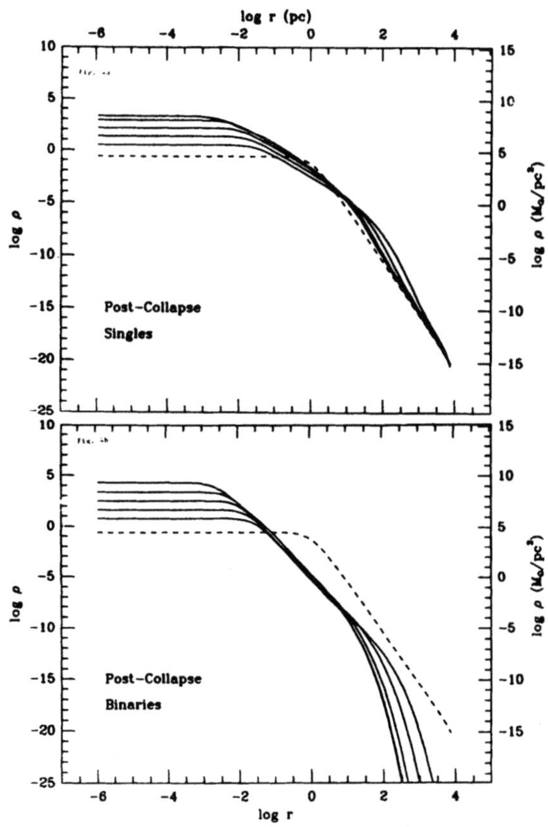

Figures 4a,b. Post-collapse density distributions of singles (a) and binaries (b) shown at various times with the five solid lines corresponding to the five square symbols in Fig. 1. The initial Plummer model is shown as a dashed line.

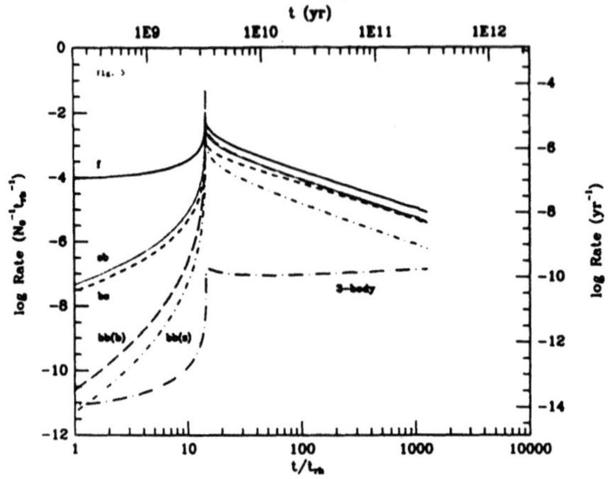

Figure 5. Rates (integrated over the whole cluster) for various processes. Here f designates two-body tidal-capture binary formation rate; sb and bs are the collision rates between binaries and singles that eject singles and binaries, respectively, and bb(b), bb(s) the collision rates between binaries that eject binaries or singles. The rate of formation of binaries by three-body interactions among single stars is always relatively unimportant.

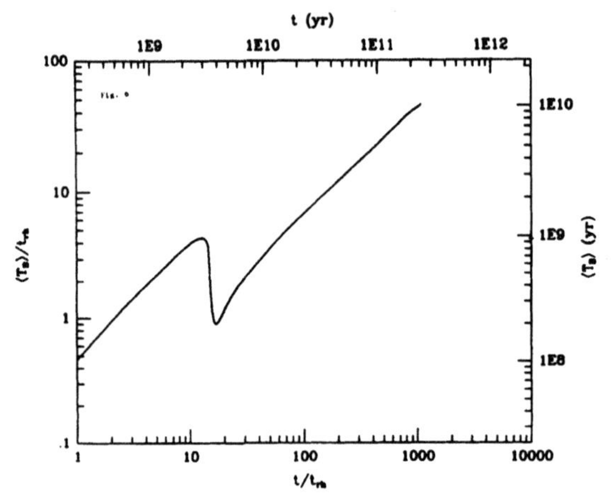

Figure 6. The mean age of the binaries in the cluster at any time vs the time itself. For comparison the timescale for gravitational radiation to bring the stars together would be about $10^{9.3}$ years.

3. ANALYTICAL ANALYSIS OF CLUSTER EVOLUTION

The number of binaries increases only gradually, approximately giving

$$N_b = 10^{1.5} \left(\frac{v_0}{v_*}\right)^{1.8} \frac{N_0}{\ell n \Lambda_0} \left[1 - \left(\frac{\Delta t}{t_c}\right)^{0.29}\right] \quad , \tag{9}$$

as collapse is approached $\Delta t \equiv t_c - t$). During the collapse there is a significant overshoot and, although Figure 5 shows that three-body encounters among the single stars are never important, physical collisions and three-body encounters among binaries would be important--and would presumably lead to core bounce at a somewhat lower density than the one calculated. Now we can check more quantitatively the circumstances under which three-body binaries would be more important than capture binaries. The number of three-body binaries formed in the absence of the two-body processes up to the instant of bounce is approximately $10^{1.5}$ (Hut, Goodman and Cohn, 1984). This implies that if $(N/\ln 0.4N) > (v_*/v_0)^{1.8}$ or, approximately, $N > 1.6 \times 10^4\ v_{0,10}^{-1.8}$, then two-body processes will dominate even at maximum core density. This criterion is satisfied for most clusters.

After core bounce the inner and outer cluster regions are increasingly decoupled. The core evolves on the two-body collisional time scale which implies that

$$\frac{1}{t} = \text{const} \times \nu_2 \approx \text{const} \times n_0 R_*^2 v_*^2 v^{-1} \ . \tag{10}$$

In addition, equilibrium within the core requires that

$$Gm_* n_0 = \text{const} \times v_0^2 r_c^{-2} \ . \tag{11}$$

Now let us consider the outer parts. At the half-mass radius the conductive luminosity must balance energy input

$$\text{const} \times v\ \ln\Lambda \frac{dm_* v^2}{dr} = \frac{d}{dt}(Mv^2) \ , \tag{12}$$

so that the evolutionary time scale in the outer parts is set by the requirement

$$\frac{1}{t} = \text{const} \times \frac{v m_* \ln\Lambda}{M R_h} = \text{const} \times \frac{m_* v^3 \ln\Lambda}{GM^2} \ . \tag{13}$$

Thus, if we can treat $(t\dot{M}/M)$ as a small quantity (a point we shall come back to), then equation (13) tells us that

$$v = K_1 \left(\frac{Gm_*}{\ln\Lambda}\right)^{1/3} N^{2/3}\ t^{-1/3} \quad ; \tag{14a}$$

and

$$R_h = K_2\ [Gm_*(\ln\Lambda)^2]^{1/3} N^{-1/3} t^{2/3} \ . \tag{14b}$$

Where K_1, K_2 are numerical constants to be determined and we have used the virial theorem $(Gm_*N/R_h) \simeq 2.5\ v^2$. Then, with v(t) known, and assuming that (v_0/v) = const, we have, from equations (10) and (11)

$$n_0 = K_3\ v_*^{+2} [N^2/(G^5 m_*^5 \ln\Lambda)]^{1/3}\ t^{-4/3} \ ; \tag{15a}$$

and

$$r_c = K_4\ v_*^{-1} (\ln\Lambda)^{-1/6} (G^2 m_*^2 N)^{1/3}\ t^{1/3} \ . \tag{15b}$$

As shown by the dashed lines in Figures 1-3, the power law dependencies derived here fit our numerical results. A rough quantitative fit is obtained if $K_i = (0.74, 2.0, 0.22, 0.52)$. Then, given these dependencies and the requirement that the mass loss from the center times the central potential equal the energy input into the cluster (const × Mv^2/t), one can show (cf. Statler et al.,1984) that ϕ_0 will be close to that in a truncated isothermal cluster and that $M = \text{const} \times [1 + \alpha \ln (t/t_0)]^{-\beta}$ where α and β positive are constants of order unity, the exact value of which depends on ϕ_0/v_0^2). Thus $(t\dot{M}/M) \ll 1$ as required.

4. DISCUSSION AND OBSERVATIONAL CONSEQUENCES

Many scientists over the years have argued that a significant number of globular clusters must have reached and passed through the stage of "core collapse". From the recently developed theoretical arguments referred to earlier in this paper, these clusters should look like very concentrated isothermal spheres, and there is some evidence (Djorgovski and King, 1984; Djorgovski, 1984) that such objects have indeed been found. The calculations presented here indicate that hundreds of close two-body binaries should be found in the inner parts of such systems concentrated roughly as the square of the surface light intensity). These binaries should be divided into comparable numbers of normal-normal binaries (W Ursa Majoris stars) and normal-degenerate binaries (cataclysmic variables). These last should be seen as soft X-ray sources (cf. Grindlay, 1984) and, of course, since neutron star remnants will be $\sim 10^{-2}$ degenerate remnants, thus of order one X-ray binary per cluster might be anticipated. This last remark of course cannot be taken as a "prediction" since it was the overabundance of X-ray binaries in globular clusters which led Fabian, Pringle and Rees (1975) originally to propose the mechanism of tidal capture in its modern guise.

Finally we note that, if one drops the limitation of isolation, and instead assumes a tidal limit such that $\rho = 0$ outside the radius at which $[3M(R)/4\pi R^3] = \rho_{tidal} = \text{const}$, then we expect the cluster mass to go linearly to zero on the half-mass relaxation time scale. The investigation of this process, which would lead to the total destruction of the cluster, is left to subsequent calculations (Statler et al., 1984).

5. ACKNOWLEDGEMENTS

I would like to thank T. Statler for his intelligent and indefatigable labors and also H. Cohn, J. Goodman, J. Gunn, P. Hut and L. Spitzer for numerous useful comments. This research was supported in part by National Science Foundation grant AST83-17118.

REFERENCES

Alexander, M. E. and Budding, E. 1979, Astron. & Astroph., 73, 227.
Cohn, H. 1980, Ap. J. 242, 765.
Djorgovski, S. 1985 (this conference).
Djorgovski, S. and King, I. R. 1984, Ap. J. (Lett.), 277, L49.
Dokuchaev, V. I. and Ozernoy, L. M. 1982, Astron. & Astroph., 111, 16.
Fabian, A. C., Pringle, J. E. and Rees, M. 1975, Monthly Notices, Roy. Astron. Soc., 172, 15P.
Goodman, J. 1984, Ap. J., 280, 298.
Grindlay, J. 1985 (this conference).
Henon, M. 1961, Ann. Astrophy., 24, 369.
Hut, P. 1984, Ap. J. (Suppl.), 55, 301.
_______ 1985 (this conference).
Hut, P., Goodman, J. J., and Cohn, H. 1984, in preparation.
Inagaki, S. 1984a, M.N.R.A.S., in press.
Inagaki, S. 1984b, preprint.
Krolik, J. H. 1983, Nature, 305, 506.
Lightman, A. P. and Fall, S. M. 1978, Ap. J. 221, 507.
Lynden-Bell, D. L. and Eggleton, P. 1980, Monthly Notices, Roy. Astron. Soc., 191, 483.
Mikkola, S. 1983, Monthly Notices, Roy. Astron. Soc., 203, 1107.
Milgrom, M. and Shapiro, S. L. 1978, Ap. J. 223, 991.
Ozernoy, L. M. and Dokuchaev, V. I. 1982, Astron. & Astroph., 111, 1.
Press, W. H. and Teukolsky, S. A. 1977, Ap. J. 213, 183.
Spitzer, L. and Mathieu, R. D. 1980, Ap. J. 241, 618.
Statler, T. S., Ostriker, J. P., and Cohn, H. 1984, in preparation.
Stodolkiewicz, J. S. 1985 (this conference).

DISCUSSION

HUT: I am very glad to see the first detailed Fokker-Planck treatment of the global effects of dissipative encounters on the evolution of globular clusters. I have two remarks. First, I would recommend to include wide encounters with single stars in your calculations. Although these events change the binding energy of the binary only by a relatively small amount, this change can nevertheless be significant with respect to the much smaller field star energies. Weak encounters with hard binaries can give off $\sim$10 kT, which could turn out to be significant. Second, a treatment of close encounters will be very difficult, since resonant scattering will result in near-central collisions (my numerical orbit calculations show that in at least half of the resonant scattering cases, the centers of two of the three stars approach each other to within ten percent of their radii). Therefore, I expect that close encounters will produce heavy stars by merging (possibly repeatedly); such a heavy star will evolve quickly, and eventually shed its mass which can escape out of the shallow potential well of a globular cluster.

OSTRIKER: I agree with your remarks and since any process which leads to mass loss from the cluster has the same effect on the overall evolution as any other a striking collision perhaps can be neglected.

COHN: I have a question addressed to the observers - what limits do we have on the presence of binaries in the *cores* of globular clusters?

PRYOR: There is no evidence in the cusps, where the stars are unresolved. Our and Griffin & Gunn's observations of the radial velocities of $\sim$100 M3 giants show very little evidence for binaries. If the distribution of periods of the binaries when their components were on the main sequence was that of Abt & Levy, <20% of the stars were in binaries then. Because only giants are observed, there is no direct evidence on the abundance of binaries with separations less than $\sim$0.3AU.

SHARA: White dwarf - main sequence star physical collisions set up very strong shock waves in the non-degenerate star, heating much of that star to $\sim 10^8$K. CNO cycle hydrogen burning then produces $\sim$10 times the main sequence star binding energy in one main sequence star expansion timescale; thus the main sequence star must be totally disrupted. The energy liberated into gas motions (and light?) is supernova-like ($\sim 10^{50}$ erg) and should very effectively sweep globular clusters clear of accumulated gas. Details can be found in Shara & Shaviv, MNRAS 183, 687 (1978).

LARSON: What is the role of red giants and horizontal-branch stars in close interactions? Although they are much less numerous than main-sequence stars, they are also much bigger, and my simple arithmetic suggests that they should be at least as important.

OSTRIKER: If one adopted a simple scaling of the cross sections you would certainly be correct. However, red giant (and to a less extent horizontal branch stars) have relatively weakly bound envelopes which could not absorb sufficient energy to catch stars in a dense globular cluster. We, conservatively, neglect the capture by evolved stars altogether.

SPITZER: In principle the ejection of hard tidal-capture binaries could be averted if the two stars approached each other sufficiently rapidly. Could you comment on the rates of the different processes that will bring the two stars closer together, particularly gravitational radiation and effects of a common gaseous envelope? How do these rates compare with the rate of ejection of hard binaries?

OSTRIKER: Detailed curves in the accompanying paper show that in some cases gravitational radiation can be an important process.

GRINDLAY: Letme comment, first, that I agree with your assertions that two body tidal capture, not exchange collisions, is the dominant mode of formation of the observed X-ray binaries in globular clusters and, second, that the production of tidal-capture binaries in globulars *is* important for the cluster evolution (as evidenced by the observed central cusps and your work, which suggests these may be dominated by binaries). My question is: if the central cusps are indeed dominated by tidal-capture binaries and these are largely main sequence-main sequence or main sequence-giant systems, would you expect them to be bluer than the surrounding cluster core due to the formation of "blue stragglers?" We (Hertz and Grindlay 1984, *Ap.J.*, in preparation) have shown that the central cusp in NGC6624 is in fact bluer than the surrounding core and have speculated that this process may be responsible.

KING: (1) Are these two-body-capture binaries important for the late evolution of the cluster, or do they kick themselves out immediately? (2) In making X-ray sources from them, do you require an exchange encounter with a neutron star, in order to make an ordinary pair into something capable of making X-rays?

OSTRIKER: (1) The two-body binaries are important *because* they are kicked out fairly promptly. This puts energy into the remaining stars and causes re-expansion. (2) Exchange reactions can occur but neutron stars can capture normal stars as well.

LIGHTMAN: Isn't it true that once a small number of physical collisions and coalescences have occurred in a core of two-body-formed binaries, there will be a tendency for a single massive star to grow (by coalescence) very rapidly at the expense of the others - both because of its increased size (increased cross section) and its increased mass (increased gravitational focussing)? Because close encounters with two-body-formed binaries have impact parameters of order the stellar radius, the energy released in encounters may be dissipated through the coalescence process rather than put into recoil that can eject the binary. Thus, we may have to face the issue of a growing massive star in the core, involving a number of uncertainties.

OSTRIKER: I agree but massive stars will have a relatively short lifetime before exploding. Thus the net outcome of mass loss from the cluster is the same and the evolution should be the same. We are doing other calculations to test this point.

WEBBINK: In the picture you have outlined, it is clearly the evaporation of hard binaries from the cluster core in energetic encounters with a third star which is the stabilizing mechanism against cluster core collapse. The question I have (for you or Piet Hut) is whether these energetic encounters necessarily require the third star passing within the binary separation; and wouldn't one then expect three-body collisions (given that the separation is only $\sim 3R_*$) rather than exothermic (hardening) encounters in most cases?

OSTRIKER: Yes, the typical collision is close and involves binaries and singles (binary-binary encounters are also quite important). In any case after a complicated interaction stars are ejected from the tightly interacting subsystem with velocities comparable to the initial binary orbital velocities. These first reaction products will tend to be ejected from the cluster.

KING: Regarding seeing binaries in a central cusp, I think that a post-collapse cusp must be so small that it would be quite hopeless to see into it.

OSTRIKER: Yes, I agree but if it contains special stars, say cataclysmic variables in abundance, then these might leave a signature on the observed spectrum of the central parts.

MONTE-CARLO CALCULATIONS

J.S. STODÓŁKIEWICZ
N.Copernicus Astronomical Center, Warsaw, Poland

ABSTRACT. The evolution of a nonisolated globular cluster is presented. The binaries (both, tidally captured and formed in three-body interactions), outflow of mass from stellar envelopes and shocks are considered as sources of energy in the cluster.

1. INTRODUCTION

Many theories of dynamical evolution of globular clusters consider these stellar agglomerations as systems of separate mass points which move under the influence of their mutual gravitational attraction. This assumption, justified by small stars radii relatively to their distances in the clusters, simplifies discussion of a majority of proprieties of globular cluster structure and their evolution. One of the most important conclusions, obtained under this assumption, states that globular clusters are not stationary systems, but that their evolution leads to a gravothermal catastrophe (Antonov 1962, Lynden-Bell and Wood 1968). The process of a core collapse is accelerated if stars of different mass are allowed (Spitzer 1969). The process of the gravothermal collapse can be interrupted (Hénon 1975), only when a new (different than the core collapse) source of energy appears in the cluster.

Almost all processes which can supply the clusters with energy operate only, if we reject the above mentioned assumption and take into consideration that stars are evolving objects of nonvanishing sizes and can form at least two-component systems interacting with other stars of the clusters. There are following sources of energy in globular clusters considered in literature:

J. Goodman and P. Hut (eds.), Dynamics of Star Clusters, 361–372.

1/ binaries formed in three-body interactions (Gurevich and Levin 1950, Heggie 1975),

2/ binaries produced in two-body tidal interactions (Fabian, Pringle, Rees 1975),

3/ outflow of matter from evolving stars (Angeletti and Giannone 1977, 1980, Stodółkiewicz 1982),

4/ shocks, when the cluster passes the galactic plane (Spitzer 1975, Ostriker, Spitzer, Chevalier 1975),

5/ a massive black hole in the cluster centre (Bahcal and Ostriker 1975, Silk and Arons 1975).

The aim of this paper is to examine the influence of heating (or cooling) of the cluster by three first processes from the above list. The heating by shocks is also taken into consideration; however the cluster is assumed to be far from the galactic centre $R_C \approx 10$kpc and thus this effect is not essential for this cluster.

2. METHOD

A Monte Carlo scheme, which is a version of Henon's (1971) method, described by Stodółkiewicz (1982), was expanded by an incorporation of binary-single star and binary-binary interactions. The whole cluster is divided into 1000 superstars. If the number of superstars declines below 500 during the evolution, each superstar is doubled. So the number of superstars in the cluster is always contained between 500 and 1000.

The most serious problem is to represent correctly the very central region of the core, where, as a result of the collapse, the density in a small volume reaches high values. This cusp of the density is caused by only several hundreds of stars and is represented by a few (3-6) superstars. We cannot expect that it will be possible to elaborate any credible statistical theory which could describe the proprieties of such a small and nonuniform agglomeration of stars interacting with the rest of the cluster. The most promising method to treat the very centre of the core is the hybrid Monte-Carlo-N-body code described by McMillan and Lightman (1983) and presented at this Symposium. It seems that faster methods of globular cluster evolution calculations will incorporate in the future the models of the deep core obtained with this code. In this paper I imposed however very simple requirements on the motion of the innermost superstars. These conditions are the same as described by Stodółkiewicz (1982, Model C1b), moreover, if the two innermost superstars form a bound system the sum of their kinetic energies is divided in equal parts between them and their velocities are "isotropised".

The calculations performed under these assumptions gave reasonable results concerning the evolution of the bulk of the cluster. We shall return to this problem at the end of the paper.

Mass loss from stars is simulated by a single, instantaneous, spherically symmetrical outflow of a stellar envelope at the top of the red giant branch of the evolutionary track. It is assumed that the gas from stellar envelopes immediately leaves the cluster. The data on stellar evolution are taken from Bartelli, Bolton, Chiosi, Nasi (1979) and Alcock, Paczyński (1978). We adopt that all stars more massive than 2.8 solar masses have compact remnants of mass equal to 1.4 $M_{\odot}$. Lighter stars reject half of their mass during the evolution.

The influence of the gravitational field of the Galaxy is considered in two ways. Firstly, the size of the cluster is limited by its tidal radius. Secondly, the cluster is heated by shocks induced by its passages through the galactic plane.

The influence of binaries on the evolution of globular clusters has been investigated by many authors. Most of them have used the analytical formulae in order to estimate the mean values of energy released in the binary-single star interactions. In a few papers only the life of population of binaries was investigated together with the evolution of the cluster: Hills (1975), Spitzer and Mathieu (1980), Aarseth (1980), Stodółkiewicz (1983). McMillan and Lighman (1983) using the "hybrid" code obtained the binaries produced in three-body encounters for the first time. This result is a convincing illustration of formation of binaries in large stellar systems. In this paper we adopt (similarly as in Stodółkiewicz 1983) that binaries are born in three-body encounters in the rate given by Heggie's (1975) formula (5.6). We consider only binaries with initial binding energies greater than $8\beta^{-1}$ (where $\frac{3}{2}\beta^{-1}$ is the local mean kinetic energy per particle).

The binaries produced by tidal capture play also very important role in the life of the cluster. They cool the system because during their formation the energy of relative motion of two interacting stars is transformed into the energy of oscillations of components. Direct collisions of stars will lead to dissipation of kinetic energy of the cluster. But we can expect that these binaries, produced in two-body tidal interactions, play also some role in heating of the system. It is realized in two ways. Firstly, in the very dense central part of the core the probability of interaction between binary and field star can reach high values, even for very tight binaries. Secondly, the mass loss from the close and contact binaries or from single, but massive, stars formed as a result of collisions should be more effective than from separate stars.

The dissipative interactions between two stars lead to formation of two types of objects:

(a) Gingold and Monaghan (1980) showed that, if a star passes the other one at the distance even a little shorter than the radius of its Roche sphere, we can expect the essential deformation of that star which will cause an exchange of angular momentum and loss of energy of orbital motion, or even a disruption of the star. The future of the newly formed object strongly depends on its initial parameters, but its sizes probably will be comparable to the separation of components at the moment of their first periastron passage. The chance of the interaction of field star with such systems, before essential loss of mass from them, is small. In this paper we only count the number of this type of systems appearing in the cluster but do not follow them in further computations. The energy of relative motion of reacting stars we subtract from the energy of the nearest superstar, but we do not take into account any supply of energy from these systems.

(b) When the periastron distance of the first passage grows, the result of interactions is less dramatic. The binary system which can be formed in this case has a circular orbit with semiaxis two times greater than the distance between stars at the moment of the first periastron passage. The system is well detached, both stars lie deep inside their Roche lobes. The cross-section for such capture depends on the structure of both stars. We adopt the results given by Press and Teukolsky (1977) to calculate the probability of formation of binaries, separately for the pairs: unevolved-unevolved star and unevolved-compact star. Each newly produced binary is tested, whether it has enough time for the circu-larization of its orbit before the interaction with a field star; if not, the binary is excluded from the computations.

The binary-single star interactions are computed numerically by integration of three body motion equations. Probability of interaction with a field star is evaluated for each binary in the each step of calculation. If the interaction occurs, the impact parameter, the orientation of the binary orbit and the direction of velocity of the field star are chosen randomly, other characteristics of the interaction are determined by the binary and conditions in the vicinity of this event. We take into consideration only these interactions which are, expected to give a recoli energies greater than $1/2\,\beta^{-1}$. During the interaction all stars are treated as mass points. Results of Mikkola (1984) paper are incorporated in the used scheme in order to evaluate the effects of binary-binary interactions. No stable configurations containing more than two stars are considered.

3. EXAMPLE OF GLOBULAR CLUSTER EVOLUTION

In this chapter we investigate the evolution of globular cluster which has the following initial characteristics:

1/ total mass equals 10^5 solar masses,
2/ continuous distribution of stellar masses, given by Salpeter formula in the interval 0.1 and 3.0 solar masses,
3/ density distribution according to King (1966) model with r_c = 1pc and r_t = 31pc,
4/ no mass segregation.

The initial mean relaxation time in the cluster is equal to $1.7 \cdot 10^9$ years.

The calculations are interrupted after $29.7 \cdot 10^9$ years, when the mass of the cluster declined to 2,000 solar masses and the binding energy of the cluster was reduced to 0.0006 of its initial value.

During the first phase of the evolution we observe the growth of the central concentration of unevolved stars in the cluster. This process is accompanied with the very effective creation of binaries produced by two-body tidal capture mechanism. In the first $8 \cdot 10^9$ years 83 percent of all tidally formed binaries are created. At the same time the most massive stars end their evolution and are transformed into neutron stars or white dwarfs. Both, the mass loss from stars and interactions of binaries with field stars, cause heating of the cluster and lenghtening of this phase of its evolution. After about this time the central density of unevolved stars begins to decrease and when the cluster is about 14 billion years old the density of unevolved stars drops below its initial value.

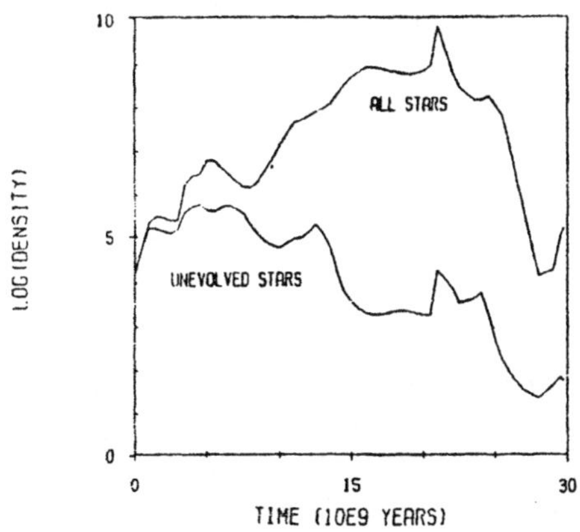

Figure 1. Logarithm of central density (number of stars per pc^3) as a function of time.

The evolution of central density is presented in Fig. 1. From this time, further on, the activity of the cluster declines for external observer: the density of shining stars is low, tidally captured binaries cease their opperation (Fig. 2).

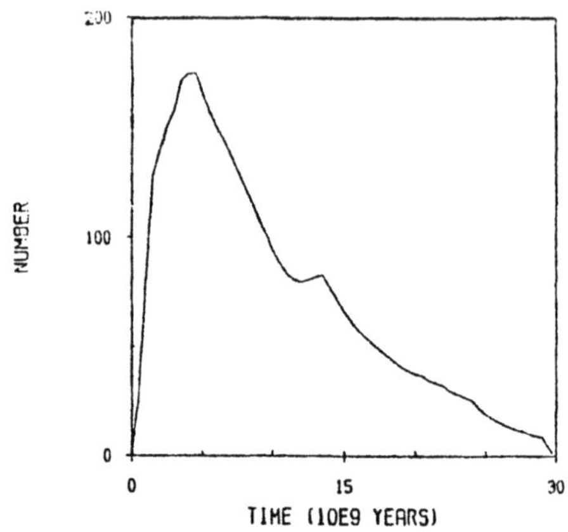

Figure 2. Number of tidally captured binaries actually existing in the cluster as a function of time. Binaries whose components passed the Roche sphere are not presented.

But we are still far from the moment when the collapse is completed!

First binaries formed in the three-body encounters appear in the cluster after $4.5 \cdot 10^9$ years from the beginning of the computations. But the number of these objects is small for a long time. We observed only 10 three-body formed binaries within the first $19 \cdot 10^9$ years. At the end of this time the density of the central core, populated now mostly by evolved stars, crosses the value of 10^9 stars per pc^3. Then a rapid production of the three-body captured binaries and their interactions with field stars stops further collapse of the core. But at the time of the maximum of central density ($21 \cdot 10^9$ years) only about 26 percent of initial mass survives in the cluster and its binding energy decreases tenfold. The last period of the cluster life, after the core collapse, is characterized by decline of density in the whole cluster volume under the influence of still operating three-body captured binaries.

During the evolution of the cluster we observed 94 binaries formed in three-body capture mechanism almost all their components were the evolved stars which took part in 1167 interactions with field stars. The number of stars which were thrown out of the cluster due to encounters with these binaries was 374 (including components of binaries). The tidally captured binaries were much more numerous. There were 1177 such objects (belonging only to the second group), and 3969 interactions of these binaries with field stars. As a result of these interactions 2380 stars were removed from the cluster. During the whole cluster evolution we observed 3658 passages of a star through the Roche sphere of another star.

TABLE 1. Numbers of binaries lost from the system

The way	2-body binaries	3-body binaries
a/	708	72
b/	12	3
c/	417	10
d/	38	8

a/ escape after binary-single star interaction;
b/ disruption after binary-binary interaction;
c/ disruption after binary-single star interaction;
d/ escape in the relaxation process.

The ways of binary loss is presented in Table 1. The interactions of binaries with field stars (the most effective process of loss of binaries) cause that the life-time of tidally captured binaries is short ($2.2 \cdot 10^9$ years) during the first 5 billion years of the cluster evolution. In the latter half of the cluster evolution, when these interactions are much more rare, the life-time of these binaries increases to about $7 \cdot 10^9$ years. The disruption of tidally captured binaries takes place only during the phase of circularization of their orbits, just after the capture of the components and in interactions with the other binaries .

The efficiency of three main sources of the energy in the cluster is presented in Fig. 3. The energy released in the interactions of the two-body captured binaries is much greater than the initial binding energy of the cluster, but only small part (about 10 percent) of this energy is absorbed by the cluster, the rest is carried away by the stars and the binaries leaving the cluster after the interactions. These binaries heat the cluster only during the first phase of the cluster evolution; during first 10 billion years the cluster receives 90% of energy obtained from these systems. It is evident that their action cannot interrupt the core collapse, but, of course, can retard its development. These binaries play also some role in the expultion of unevolved stars from the deep core. After 15 billion years the binaries produced in three-body interactions start to heat the cluster, and later on they become the only source of energy in the cluster. Although the amount of the energy

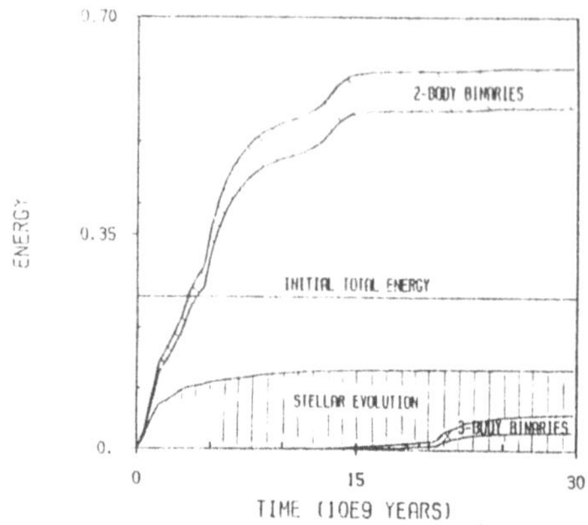

Figure 3. Heating by binaries and outflow of matter from stars. The curves represent the amount of energy integrated up to a given time. Upper curves for binaries represent energy released, lower curves - the energy carried away from the cluster; the curve for stellar evolution describes energy absorbed by the cluster.

received by the cluster from these systems is small comparatively to the initial value of the energy of the cluster , but it is supplied to the much less massive cluster and much loosely bound (except for the deep core) than at the time of its birth. The energy released by these binaries is consumed to prevent the further collapse of the deep core (which contains now most of the binding energy of the cluster) and afterwards to the complete dissolution of the cluster. The most abundant source of energy is the mass outflow from evolving stars. It delivers more than a half of the energy needed to disrupt the cluster. This source it self is sufficient to suppress the effects of collapse of the radius of a sphere containing 10 percent of cluster mass. But it operates effectively only during about the first 5 billion years and, of course, is not able to prevent the collapse of the inner core.

The balance of energy (including all its sources) during the whole evolution of the cluster is presented in Fig. 4 and in Table 2. The energy carried away by the escaping stars is small. But it is a by-product of the most important process enforcing the cluster evolution, the process of relaxation, in which the energy is transported from the core, through the bulk of the cluster, to its envelope, and which causes the steady evaporation of stars. The energy gained from the shocks is small too; it is a consequence of our assumption that the cluster moves far from the galactic centre.

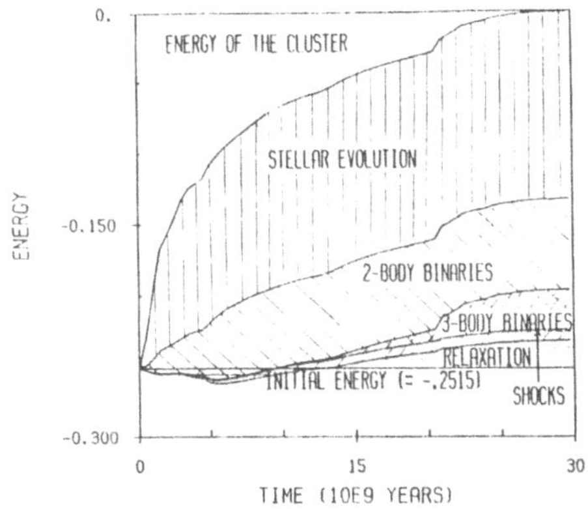

Figure 4. Balance of energy. Amounts of energy absorbed by the cluster until a given time.

TABLE 2. Energy balance

Source / Age 10^9y	10	20	29.7
stellar evolution	.1254	.1316	.1321
2-body binaries:			
energy released	.5354	.6136	.6181
energy absorbed	.0577	.0637	.0637
3-body binaries:			
energy released	.0007	.0143	.0583
energy absorbed	.0004	.0075	.0284
shocks	.0044	.0071	.0076
relaxation	.0035	-.0113	-.0195
mass of the cluster	.6174	.3064	.0197

The role of the same mechanisms for the balance of mass (Fig. 5) is quite different. The relaxation is the most effective process of the mass loss. 82 precent of the cluster mass (but 98.6 percent of stars) escapes due to encounters with field stars. It should be noted that this loss is practically constant during the whole cluster life.

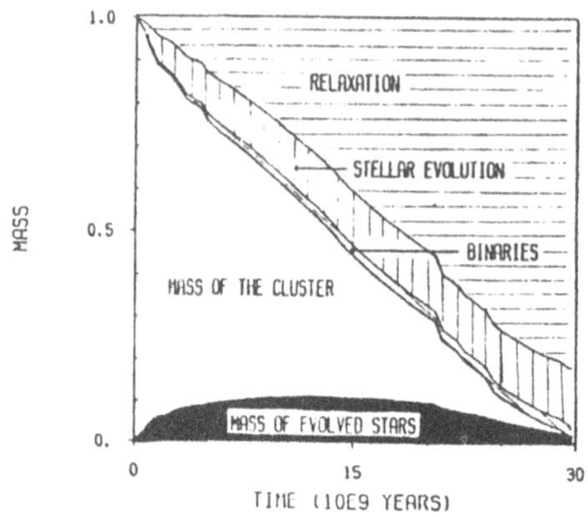

Figure 5. Balance of mass. Mass loss caused by each mechanism until a given time.

The same result was obtained by Hénon more than twenty years ago (Hénon 1961). This phenomenon has been also observed in the n-body calculations of the evolution of the open clusters. Here it is fully maintained, although the cluster mass decreases through two decades. The second process which is essential for the mass loss from the cluster, outflow of stellar envelopes, is much less efficient. Only 13 percent of the mass leaves the cluster this way. The mass of stars thrown out of the system as a result of interactions with binaries is quite negligible (2.6 percent of the initial cluster mass). The relative contribution of the dead stars systematically grows. 13 billion years after its birth the mass of the evolved stars is equal to 22 percent and reaches 85 percent at the end of the computations.

The evolution of this cluster, in which we took into account all essential sources of energy, can be compared with the evolution of globular clusters computed earlier (Stodółkiewicz 1982), where we neglected the action of binaries. In Models C1a and C1b from the quoted paper the main source of energy was the steady contraction of the inner core. The mass loss in all models and the mass distribution (except for the central part of the clusters) is very similar (Table 3). We see the evolution of the bulk of the cluster is very insensitive to the energy source. The energy demanded by the cluster evolution is delivered from this source, which can operate in the actual conditions. This is a reason that we were bold to calculate the cluster evolution without the special method to examine the structure of the central deep core.

TABLE 3. Comparison between Models C1a, C1b[x/] and presented in this paper. Total mass and radii of spheres containing 20, 40, 60 and 80 percent of cluster mass at the age of 20 billion years.

Model	r_{20}	r_{40}	r_{60}	r_{80}	M
C1a	.105	.225	.366	.541	.356
C1b	.100	.233	.387	.537	.305
This paper	.133	.240	.354	.544	.306

[x/]see text

The results presented here concern the globular cluster of initial parameters given above and cannot be directly applied to the other clusters.

REFERENCES

Aarseth, S.J. 1980, IAU Symp. 85, 325.
Angeletti, L., and Giannone, P. 1977, Astron. Astrophys., 58, 363.
Angeletti, L., and Giannone, P. 1980, Astron. Astrophys., 85, 113.
Alcock, Ch., and Paczyński, B. 1978, Astrophys. Journ., 223, 244.
Antonov, Y.A. 1962, Bull.Univ.Leningrad, 7, 135.
Bahcal, J.N., and Ostriker, J.P. 1975, Nature, 256, 23.
Bartelli, G., Bolton, A., Chiosi, C., and Nasi, E. 1979, Astron. Astrophys. Suppl., 36, 429.
Gingold, R.A., and Monaghan, J.J. 1980, M.N.R.A.S., 191, 897.
Gurevich, L.E., and Levin, B.Yu. 1950, Astron. Żurn., 27, 273.
Fabian, A.C., Pringle, J.E., and Rees, M. 1975, M.N.R.A.S., 172, 15P.
Heggie, D.C. 1975, M.N.R.A.S., 173, 729.
Hénon, M. 1961, Ann. d'Astrophys., 24, 369.
Hénon, M. 1971, Astrophys. Sp. Sci., 14, 151.
Hénon, M. 1975, IAU Symp., 69, 133.
Hills, J.S. 1975, Astron. Journ., 82, 626.
King, I. 1966, Astron. Journ., 71, 64.
Lynden-Bell, D., and Wood, R. 1968, M.N.R.A.S., 138, 495.
McMillan, S.L.W., and Lightman, A.P. 1983, Cambridge, Mass., Preprint 1941.

Mikkola, S. 1984, M.N.R.A.S., 206, 115.
Ostriker, J.P., Spitzer, L., and Chevalier, R.A. 1975, Astrophys. Journ. /Lett./, 176, L51.
Press, W.H., and Teukolsky, S.A. 1977, Astrophys. Journ., 213, 183.
Silk, J., and Arons, J. 1975, Astrophys. Journ. /Lett./, 200, L131.
Spitzer, L. 1969, Astrophys. Journ. /Lett./, 158, L139.
Spitzer, L. 1975, IAU Symp., 69, 3.
Spitzer, L., and Mathieu, L.D. 1980, Astrophys. Journ., 241, 618.
Stodółkiewicz, J.S. 1982, Acta Astron., 32, 63.
Stodółkiewicz, J.S. 1983, Publ. Astron. Inst. Czechoslovac Acad. Sci., 56, 103.

DISCUSSION

SPITZER: What is the radial distribution of the formation rate for tidal-capture binaries?

STODÓŁKIEWICZ: Most of these binaries is formed in the deep core, 85 percent of them was born inside the sphere containing the innermost 3 promille of the cluster stars.

LUPTON: How does the mass function evolve in your calculations?

STODÓŁKIEWICZ: The rate of escape of massive stars is smaller than - of the others. But the rate of evaporation of all stars with masses lower than about the mean stellar mass in the cluster is practically not dependent on their mass.

MONTE CARLO SIMULATIONS OF THE 2+1 DIMENSIONAL FOKKER-PLANCK EQUATION: SPHERICAL STAR CLUSTERS CONTAINING MASSIVE, CENTRAL BLACK HOLES

Stuart L. Shapiro
Cornell University
Center for Radiophysics and Space Research
Space Sciences Building
Ithaca, New York 14853

ABSTRACT. The dynamical behavior of a relaxed star cluster containing a massive, central black hole poses a challenging problem for the theorist and intriguing possibilities for the observer. The historical development of the subject is sketched and the salient features of the physical solution and its observational consequences are summarized.

The full dynamical problem of a relaxed, self-gravitating, large N-body system containing a massive central black hole has all the necessary ingredients to excite the most dispassionate many-body, computational physicist: it is a time-dependent, multidimensional, nonlinear problem which must be solved over widely disparate length and time scales simultaneously. The problem has been tackled at various levels of approximation over the years. A new 2+1 dimensional Monte Carlo simulation code has been developed in appreciable generality to solve the time-dependent Fokker-Planck equation in E-J space for this problem. The code incorporates such features as (1) a particle "cloning and renormalization" scheme to provide a statistically reliable population of test particles in low density regions of phase space and (2) a time-step "adjustment" algorithm to ensure integration on local relaxation timescales without having to follow typical particles on orbital trajectories. However, critical regions in phase space (e.g. disruption "loss-cone" trajectories) can still be followed on orbital timescales. Numerical results obtained with this Monte Carlo scheme for the dynamical structure and evolution of globular star clusters and dense galactic nuclei containing massive black holes are reviewed.

Recent dynamical integrations of the Einstein field equations for spherical, collisionless (Vlasov) systems in General Relativity suggest a possible origin for the supermassive black holes believed to power quasars and active galactic nuclei. This scenario is discussed briefly.

J. Goodman and P. Hut (eds.), Dynamics of Star Clusters, 373–413.

I. INTRODUCTION

Several years ago we embarked on a major program at Cornell to set up and solve on the computer the Fokker-Planck equation in 2+1 dimensional phase space. The purpose of this exploration was to study the structure and evolution of large N-body, self-gravitating, spherical stellar systems, like globular clusters and dense galactic nuclei. We were particularly intrigued by the possibility of massive central black holes residing in such systems and this aspect of the problem served as a focus for much of our work.

As it has developed, the study of the Fokker-Planck equation has become just one, albeit essential, component of a much broader effort at Cornell in large-scale computational astrophysics and relativity. Our current emphasis is on solving the general Boltzmann equation,

$$\frac{Df}{Dt} = \left(\frac{\partial f}{\partial t}\right)_{coll} , \tag{1}$$

in many different physical regimes where gravity provides the dominant long-range interaction.

Our motivation for tackling the Boltzmann equation is both physical and computational. Physically, this equation encompasses a huge range of interesting, many-body phenomena. Computationally, the Boltzmann equation provides a nontrivial, multidimensional arena to explore nonlinear dynamics on the computer. We shall elaborate on these points below with reference to the specific star cluster problem that is the subject of this review.

Some of our recent computational work revolving around eqn. (1) is summarized in Table 1. This table is a vivid illustration of the fact that an enormous and diverse range of gravitation physics is embodied in eqn. (1). This review will focus on the Fokker-Planck regime of eqn. (1). In this limit, the relaxation timescale of the system under investigation, t_r, greatly exceeds the dynamical or crossing timescale, t_d, but is shorter than characteristic integration timescale of interest, t. In stellar dynamics, t is usually the age of the stellar system and, typically, $t \sim H^{-1} \sim 10^{10}$ yrs, where H is Hubble's constant. We will, however, also provide a "sneak preview" of some recent advances we have made in solving the collisionless Boltzmann (i.e. Vlasov) equation for spherical systems in full General Relativity. For such cases we set the collision terms $(\partial f/\partial t)_{coll}$ equal to zero in eqn. (1).

As we shall shortly describe, it is the Fokker-Planck regime which is relevant to addressing the question, "What is the dynamical influence of a massive black hole in a relaxed stellar system?". However, we shall point out later that the relativistic Vlasov regime, may, following the Fokker-Planck epoch, hold the clue to answering the question, "What is the origin of the supermassive black holes believed to reside in quasars and active galactic nuclei (AGNs)?".

Most of the large-scale numerical work at Cornell is currently performed on Floating Point Systems (FPS) 190L or 164 Array Processors, hosted by an IBM 3081 mainframe computer. [For a discussion of the merits of this machine, and an overview of some of the computational work summarized in Table 1 and performed on these processors see Farouki, Shapiro and Teukolsky (1983). For a discussion of the more recent work on the Vlasov equation in General Relativity, see Shapiro and Teukolsky (1985 a,b,c).

TABLE 1

CORNELL BOLTZMANN EQUATION COMPUTATIONS

REGIME:	Vlasov (Collisionless)		Fokker–Planck (Secularly Collisional)	Fluid (Collision-Dominated)	
TIME-SCALE ORDERING:	$t_r \gg t \gg t_d$		$t \gg t_r \gg t_d$	$t \gg t_d \gg t_r$	
APPLICATIONS:	Galaxy and Galaxy Cluster Dynamics		Globular Cluster and Dense Galactic Nuclei Dynamics	Hydrodynamics	
GRAVITY	Newtonian	General Relativity	Newtonian	Newtonian	General Relativity
SPATIAL SYMMETRY	Arbitrary	Spherical	Spherical	Axisymmetric	Spherical
COMPUTATIONAL DIMENSIONS	2K+1 (K=3)	2K+1 (K=1)	2K+1 (K=1)	K+1 (K=2)	K+1 (K=1)
GOVERNING NONLINEAR EQUATIONS	ODE's	ODE's + PDE's	PDE's	PDE's	PDE's
TECHNIQUE	Particle Simulations	Particle Simulations + Finite Differencing	Monte Carlo Particle Simulations	Finite Differencing	Finite Differencing

II. 3-FOLD MOTIVATION FOR TACKLING THE PROBLEM

Studying the dynamical structure and evolution of a relaxed stellar system (e.g. a globular cluster or dense galactic nucleus) containing a massive black hole is a pursuit with a 3-fold motivation, as we shall now discuss.

a) Theoretical Motivation

The formation of a supermassive black hole is one of at least two plausible outcomes of stellar core collapse. Both possibilities were enumerated by Spitzer (1975) almost a decade ago.

In the "point-mass scenario", homological core collapse (i.e. the 'gravothermal catastrophe') terminates with the formation of hard binaries near the cluster center. Further gravitational encounters between these binaries and neighboring single stars and binaries cause the binaries to become more tightly bound at the expense of the ambient core. In fact, some stars may be ejected from the core altogether by this binary "heating" mechanism. This scenario has been the focus of considerable attention of late; it is the subject of most of the theoretical papers in this volume.

In the alternative "finite-size star scenario", core collapse ultimately leads to star collisions and coalescence in the core and to the run-away build up of one (or more) supermassive star. This short-lived supermassive star ultimately collapses to form a supermassive black hole at the cluster center. Not surprisingly, the details of the all-important collision-coalescence phase of this scenario are subtle and yet to be fully resolved. [See Lightman and Shapiro (1978) for a discussion and references.] However, as we shall discuss below, given the presence of a massive, central black hole, the dynamical fate of the cluster which evolves under its influence has been worked out, at least for suitably approximate circumstances. So for at least one of the two theoretical scenarios - that involving a supermassive black hole - the answer to the question, "What lies beyond core collapse?" can be concretely answered.

b) Observational Motivation

The presence of a supermassive black hole has been invoked often to explain diverse forms of 'anomalous' activity observed at the centers of various stellar systems. Table 2 offers a glimpse of this trend.

c) Computational Motivation

A massive black hole in a relaxed star cluster provides a unique cosmic setting in which to solve the 2+1-dimensional Fokker-Planck equation. The problem is highly nonlinear, multidimensional and is characterized by widely varying length scales. These attributes are particularly appealing to a computational physicist. Thus, for all its deceptive simplicity ("merely" Newton's laws of motion in a weak gravitational field for N point-masses), the problem of a massive black hole in a relaxed stellar system poses an exciting and difficult computational challenge.

In recent years, "computational field theory on a space-time lattice" has become a booming enterprise. This activity is pursued by many-body theorists in several branches of physics, including solid-state, particle, plasma, astrophysics and relativity. The Fokker-Planck problem discussed in this paper falls into this broad category of computational problems: the 'field' here is Newtonian gravity and the 'space' is E-J phase space.

Progress achieved in solving a nonlinear problem in physics often transcends the specific discipline in which the problem was originally formulated and the specific application initially under investigation. It was with this notion in mind that the Cornell group initiated its study of the massive black hole problem.

TABLE 2 OBSERVATIONAL BASIS FOR SUPERMASSIVE BLACK HOLES IN STELLAR SYSTEMS

Phenomenon	Examples	References
optical light 'cusps'	NGC 6624, 6681 and 7078 (M15) M87	Djorgovski and King (1984), Young et al. 1978
star count 'excesses'	NGC 1851	Bahcall et al. (1977)
velocity dispersion rise	M87	Sargent et al. (1978)
X-ray sources	X-ray globular clusters; AGNs and quasars	Grindlay (1977); Gursky and Schwartz (1977)
broad emission lines and rapid time variability	AGNs and quasars; (e.g. NGC 4151)	Strittmatter and Williams (1976); Ulrich et al. (1984)
energetic activity: high luminosities and jets	AGNs and quasars	Rees (1984)

III. HISTORY OF THE PROBLEM

Given the many motivating factors outlined in the previous section for considering the influence of a massive black hole in a star cluster, it is not surprising that an enormous amount of work has been done on this problem by a very large number of researchers over the years. Indeed, virtually every dynamicist has thought about this issue at one time or another and a good many have published on the subject!

With this in mind, it is amusing to trace the historical development of the topic. A light-hearted version of such a review is given in Box 1. Please note that this capsule summary of some of the highlights in the formulation and analysis of the problem is both highly personalized and very incomplete. Other authors may rightfully have an entirely different perspective on the development of the problem. It is also hoped the reader will excuse this author for dramatizing the Cornell group's contributions during this brief stroll down Memory Lane. They are mentioned mainly to show the extent of the group's interest and involvement in the subject and not to attach undue significance to our work.

IV. FORMULATION OF THE IDEALIZED PROBLEM

In its simplest version, the problem focuses on the steady-state distribution and tidal disruption rate of bound stars near a central black hole M embedded in a static cluster. In this version, one makes the following basic assumptions (Bahcall and Wolf 1976; Lightman and Shapiro 1977):

1. A massive hole, M, is situated at the center of a spherical cluster containing a very large number, N, of stars in the core.
2. All stars have equal mass $m \ll M$ and radius r.
3. The mass, M, is significantly less than the mass of the cluster core but much larger than the total mass of (bound) stars inside the accretion or 'cusp' radius r_a, where

$$r_a \sim \frac{GM}{\langle v^2 \rangle} \sim 4 \times 10^{-2} \text{ pc} \left(\frac{\langle v^2 \rangle}{100 \text{ km}^2/\text{s}^2}\right)^{-1} M_3 \quad , \tag{2}$$

Here $\langle v^2 \rangle$ is the mean squared velocity dispersion in the core and $M_3 \equiv M/10^3 M_\odot$.

4. Conditions in the central core of the cluster remain constant in time, so that a steady-state can be achieved near the black hole. The ambient (unbound) core stars far outside r_a satisfy an isothermal distribution.
5. A star is tidally disrupted and removed whenever it passes within a distance r_D of M, where

Box 1

THE DYNAMICAL ANALYSIS OF MASSIVE BLACK HOLES IN STAR CLUSTERS: SOME "GOLDEN MOMENTS"

AUTHOR (S)	CONTRIBUTION (S)	REMARKS
EINSTEIN (1939)	Shows impossibility of 'Schwarzschild singularities' in static spherically symmetric collisionless clusters of circularly orbiting particles	Relativistic stellar dynamics is born!
ZEL'DOVICH and PODURETS (1965) IPSER (1969) FACKERELL, IPSER and THORNE (1969)	Suggest how 'gravothermal catastrophe' and collisions might drive a cluster relativistically unstable, leading to catastrophic collapse and supermassive BH formation	Clusters of little BH's can produce a big BH
SPITZER (1971, 1975) LYNDEN-BELL (1967, 1969) COLGATE (1967) SANDERS (1970)	Suggest, alternatively, how 'gravothermal catastrophe' might drive star collisions and coalescence, leading to buildup of supermassive stars and possibly, supermassive BH's	Analysis of late core collapse forces abandonment of point-mass approx.
WYLLER (1970) WOLFE and BURBIDGE (1970) TRURAN and CAMERON (1972)	Suggest that a supermassive BH in a star cluster might be uncovered by enhanced star density around it	The stage is set!
PEEBLES (1972) (a, b)	Insightful formulation of BH in glob cluster problem: (1) delineates role of tidal disruption of stars by BH– inward $\mathcal{F}$, outward $\mathcal{E}$ (2) suggests possibility of observable light 'cusps' (3) suggests nonthermal power-law 'cusp' distribution function: $f(E) \propto \lvert E \rvert^{p}$ $\Rightarrow n(r) \propto r^{-(p+3/2)}$	Argues for $p = 3/4$ whereas $p = 1/4$ is more appropriate

Box 1 (cont.)

AUTHOR(S)	CONTRIBUTION(S)	REMARKS
BAHCALL and OSTRIKER (1975) SILK and ARONS (1975)	Suggest supermassive BH ($M/M_\odot \sim 10^2 - 10^3$) as the source of X-rays in globular clusters	The search is on!
BAHCALL and WOLF (1976, 1977)	Careful 1D FP calculation of f(E) and n(r): $p = 1/4$; formulate detectability criteria; estimate rms radial offset of BH from center	"The cusp unveiled"
ZEL'DOVICH and NOVIKOV (1971) HILLS (1975)	Stellar capture rate by BH's in galaxies is estimated	The BH appetite is assessed
FRANK and REES (1976) LIGHTMAN and SHAPIRO (1976 a,b)	Stellar capture rate by BH's in glob clusters is estimated; approximate 2D FP determination of f(E,J), n(r), $\mathcal{F}$ and $\mathcal{E}$, including scaling	
SHAPIRO (1977)	Homological model for time-dependent evolution of cluster with BH: core collapse halted and reversed; eventual cluster dissolution in Galactic tidal field.	'Born again' clusters
MARCHANT and SHAPIRO (1978) COHN and KULSRUD (1978)	Careful 2D FP calculation of f(E,J), n(r) $\mathcal{F}$ and $\mathcal{E}$	Back-of-envelopes replaced by large-scale computations
GRINDLAY (1981, 1983) GRINDLAY et. al. (1984)	Precise (1") positions of 8 cluster X-ray sources are measured by the Einstein Observatory: their radial offset argues against massive ($M/M_\odot \geq 3$) BH's	"The saddest words of Mice and Men..."
LIGHTMAN, PRESS and ODENWALD (1978) FALL and MALKAN (1978) LIGHTMAN (1982) COHN and HUT (1984)	Preliminary analysis of cluster data for post-collapse behavior	Dynamical model builders take note!

Box 1 (cont.)

AUTHOR (S)	CONTRIBUTION (S)	REMARKS
ZEL'DOVICH and NOVIKOV (1971) HILLS (1975) REES (1977, 1978) YOUNG, SHIELDS and WHEELER (1977) FRANK (1978) McMILLAN, LIGHTMAN and COHN (1982) DUNCAN and SHAPIRO (1983)	Application of theory to dense galactic nuclei: models for quasars and AGN's	"BH's are where you find them" (Peebles 1972a)

Other Contributors: Caveats and Controversies

see, e.g., IPSER (1978)
BISNOVATYI – KOGAN, CHURAYEV and KOLOSOV (1982)
DOKUCHAYEV and OZERNOI (1977)
LIN and TREMAINE (1980)
NORMAN and SILK (1982)

$$r_D \sim r\,(M/m)^{1/3} \sim 2 \times 10 \;\; \text{pc} \left(\frac{r}{R_\odot}\right) \left(\frac{m}{M_\odot}\right)^{-1/3} M_3^{\;1/3} , \qquad (3)$$

Immediate consequences of the assumption of large N are the following familiar results:

1a. Only a negligible fraction of the stars in the cluster are in binary systems (Spitzer and Hart 1971).

1b. The predominant relaxation process between stars is via repeated, two-body, small-angle scattering in the hole's r^{-1} "Coulomb" potential.

1c. The dynamical timescale t_d is significantly shorter than the relaxation timescale t_r everywhere. Thus, by Jean's Theorem, the distribution function depends only on E (energy) and J (magnitude of angular momentum): $f = f(E,J)$.

The goal, then, is to calculate $f(E,J)$. Given $f(E,J)$ we can then determine the steady-state density profile $n(r)$ and rms velocity profile $v_{rms}(r)$ of bound stars in the cusp. In addition, we want to determine the rate of tidal disruption of stars by the hole, F, which in steady-state is precisely equal to the net inward drift rate of stars from the isothermal core into the cusp. Finally, we need to determine the corresponding cluster heating rate, $\mathcal{E}$, due to the destruction of bound stars at r_D by the hole.

V. APPROXIMATE SOLUTION OF THE IDEALIZED PROBLEM

a) Simple Scaling Argument

Following Shapiro and Lightman (1976) we present here a simple scaling argument for the form of the stellar distribution inside the cusp. The derivation focuses on bound stars well inside r_a but far outside r_D. As Peebles (1972 a,b) originally pointed out, these stars cannot maintain thermal equilibrium with the ambient core stars, due to disruption at r_D. On the other hand, stars deep in the cusp but far outside r_D should presumably be moving nearly isotropically, so it is reasonable to look for a scale-free, isotropic, power-law distribution function of the form

$$f(E) \propto |E|^p \quad . \tag{4}$$

Now from eqn. (4) and the assumptions listed in Section IV it immediately follows that the mean energy per unit mass of star at r is given by

$$E \sim - GM/r \quad , \tag{5}$$

the stellar velocity dispersion is given by

$$\langle v^2 \rangle \sim GM/r \quad , \tag{6}$$

and the stellar density profile is given by

$$n(r) \propto r^{-(p + 3/2)} \quad , \tag{7}$$

so the problem of determining the stellar distribution reduces to finding p.

Focus, then, on stars in a typical spherical shell between r and 2r in the cusp. The net inward flux of stars is then

$$\mathcal{F} \sim \frac{n(r)\ r^3}{t_{\mathcal{F}}(r)} = \text{constant, independent of r} \quad , \tag{8}$$

where $t_{\mathcal{F}}$ is the net diffusion timescale for inward star transport in the cusp. The net outward flux of energy is

$$\mathcal{E} \sim \frac{n(r)\, r^3\, E}{t_{\mathcal{E}}(r)} = \text{constant, independent of } r \quad , \tag{9}$$

where $t_{\mathcal{E}}$ is the net diffusion timescale for outward energy transport. In steady-state $\mathcal{F}$ and $\mathcal{E}$ are constant, independent of radius. Now eqns. (8) and (9) immediately relate $t_{\mathcal{E}}$ and $t_{\mathcal{F}}$:

$$t_{\mathcal{E}} \propto t_{\mathcal{F}}\, r^{-1} \ . \tag{10}$$

Thus $t_{\mathcal{E}}$ is shorter than $t_{\mathcal{F}}$ because for each star moving inward from 2r to r there is almost simultaneously a corresponding star moving outward from r to 2r, so that the net star flux is small. However, at r_D there are no outgoing stars so the net diffusion timescales for star and energy transport are comparable there. This fact determines the proportionality constant in eqn. (2):

$$t_{\mathcal{E}}(r) \sim (r_D/r)\; t_{\mathcal{F}}(r) \tag{11}$$

Now no quantity can be transported in a relaxed cluster on a timescale shorter than the local relaxation timescale,

$$t_r \sim \frac{\langle v^2\rangle^{3/2}}{G^2 m^2 n^2} \ . \tag{12}$$

So setting $t_{\mathcal{E}}(r)$ ($< t_{\mathcal{F}}(r)$ for $r > r_D$) equal to t_r everywhere and using eqns. (5), (6) and (12) in eqn. (9) yields

$$n(r) \propto r^{-7/4} \ , \qquad r_a \gg r \gg r_D \tag{13}$$

or, from eqn. (6),

$$p = 1/4 \ . \tag{14}$$

This result was first derived by Bahcall and Wolf (1976) (hereafter BW) from a detailed integration of the 1-dimensional Fokker-Planck equation for f(E). The density profile in a cluster containing a black hole is shown schematically in Fig. 1.

b) Some Immediate Observational Consequences

(1) Black Hole Mass Limits. Upon deriving the bound star distribution function, BW showed how it could be applied observationally to set an upper limit to the mass of any central black hole in a star cluster. To detect a massive black hole it is necessary to resolve the cusp region $r \lesssim r_a$. If the cluster is located at distance d from the earth, with the cusp subtending an angle θ_a, we have from eqn. (2)

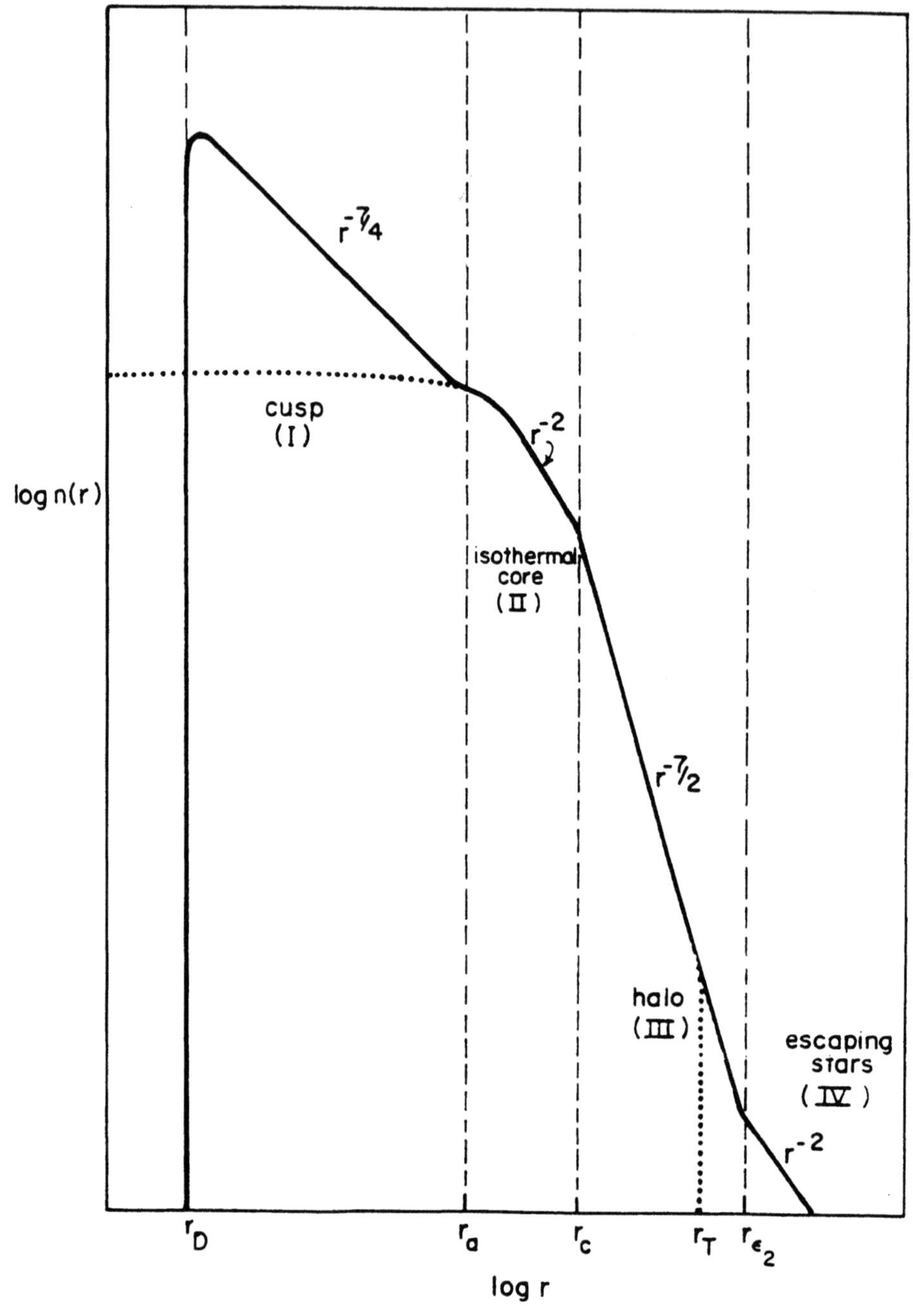

Figure 1: The stellar density n(r) as a function of radius r in an isolated spherical cluster containing a massive, central black hole. In the absence of a black hole, the isothermal core extends from r=0 to the core radius at $r=r_c$ and there is no cusp (dotted line in I). In the presence of the galactic tidal field, the density falls sharply at the galactic tidal radius r_T (dotted line in III). From Shapiro and Lightman (1976).

$$\theta_a \sim 1'' \left(\frac{d}{10\ \text{kpc}}\right)^{-1} \left(\frac{\langle v^2\rangle}{100\ \text{km}^2/\text{s}^2}\right)^{-1} M_3 \quad , \tag{15}$$

which must be compared with the "seeing" disk (i.e. resolution limit) θ_s of an optical telescope:

$$\theta_s \sim 1'' \text{ for (optimal) ground based observations,}$$

$$\sim 0.1'' \text{ for the Space Telescope} \quad . \tag{16}$$

Now the conservative criterion established by BW for believing that a black hole is present in a cluster (i.e. the number of stars within a projected angular radius θ_s from the cluster center is at least F times the unperturbed value for the core, where F ~ 3-10) yields for the minimum detectable black hole mass the value

$$M_{(\text{detectable})} \sim 5 \times 10^3\ M_\odot \left(\frac{F}{10}\right)^{4/7} \left(\frac{\theta_c}{5''}\right)^{3/7} \left(\frac{\langle v^2\rangle}{100\ \text{km}^2/\text{s}^2}\right) \tag{17}$$

where θ_c is the angular radius of the cluster core.

Evidently, ground based observations are only sensitive to black hole masses greater than ~ $5 \times 10^3\ M_\odot$ while future ST observations can detect somewhat smaller masses [see the article by Bahcall in this Proceedings for further discussions of planned ST observations of central cusps in clusters and the associated complications]. Application of eqn. (17) to ground based optical observations of the central density profiles in several X-ray globular clusters indicated central black hole masses (if present) less than ~ $10^4\ M_\odot$. This conclusion is generally consistent with the results of Bahcall et al. (1975) for NGC 7078 (M15), of Bahcall (1976) for NGC 6624 and of Bahcall and Hausman (1976) for NGC 6440 and 6441.

(2) Cusp Profiles. In principle, the observation of an optical light (projected) surface density profile increasing like

$$\sigma\ (\text{stars pc}^{-2}) \sim r^{-(p+1/2)} \sim r^{-3/4} \tag{18}$$

toward the center of a relaxed cluster would provide strong evidence for the presence of a central black hole. In practice, however, this observational test is almost impossible to apply. The difficulty stems from the fact that, observationally, it is not easy to distinguish between a black hole induced cusp and, say, an isothermal profile ($\sigma \sim r^{-1}$) predicted by several theories for the post-collapse halo of a cluster [see Inagaki and Lynden-Bell 1983, Heggie 1983, Goodman 1983, and papers in this Proceedings for theoretical discussions of post-collapse clusters without black holes; see Djorgovski and King 1984 and papers by King and Bahcall in this Proceedings for discussions of the observational difficulties].

A more distinguishing characteristic of cusps around black holes is the velocity dispersion profile [eqn. (6)]. The Keplerean rise in $\langle v^2 \rangle$ with decreasing r discriminates between black hole cusps and isothermal halos, where $\langle v^2 \rangle \sim$ constant. Unfortunately, velocity profiles are more difficult to measure than light profiles.

(3) Black Hole Cluster Locations. BW and Lightman (1976) showed that in a relaxed cluster core the projected radial offset, R_X, from the core center of a heavy mass M_X satisfies

$$\langle R_X \rangle \sim 0.7 \, r_c \, q^{-1/2} \quad , \quad q \gg 1 \tag{19}$$

where $q = M_X/m$ is the ratio of the heavy mass to the mean stellar mass m in the core and r_c is the core radius. This result has been applied statistically by Grindlay et al. (1984) to eight X-ray globular clusters for which precise (1") positions of the X-ray sources have been measured with the Einstein X-ray Observatory. From the rather large offset of these sources from their cluster centers, the X-ray source masses were determined to be in the range $0.9 - 2M_\odot$. One thus concludes rather definitively that the observed globular cluster X-ray sources are not supermassive black holes [see Grindlay, this volume, for further discussion]. Note that this conclusion by no means rules out the possibility that a massive black hole exists at the centers of these clusters - only that the observed X-ray sources are not massive black holes.

c) The Disruption "Loss-Cone"

Any discussion of the distribution of stars around a black hole which, like the one in Section V.a, presumes an isotropic stellar velocity profile everywhere, $f = f(E)$, cannot be entirely correct. For consider, say, those highly eccentric stars which possess sufficiently low J that, at pericenter, they wander inside r_D. These stars are disrupted inside r_D and are immediately removed from the system. Clearly, then, the stellar distribution function must depend on J as well as E, as originally emphasized by Frank and Rees (1976) and Lightman and Shapiro (1977).

To appreciate the two-dimensional (anisotropic) character of the problem, consider the distribution of bound stars in velocity space as depicted in Fig. 2. Focus on stars of a given energy E at radius r_E. All stars of energy E move with the same speed $v_E = (GM/r_E)^{1/2}$ at r_E; only the directions of their velocity vectors vary. If their instantaneous positions in velocity space are marked by the location of the tips of their velocity vectors, then these stars will be found on the surface of a sphere in velocity space, as shown. At this radius stars with maximal angular momentum $J_{max}(E)$ move in a circular orbit about the hole. Stars with critical angular momentum $J_{min}(E)$ just graze the tidal disruption radius r_D at pericenter. As is clear from the figure,

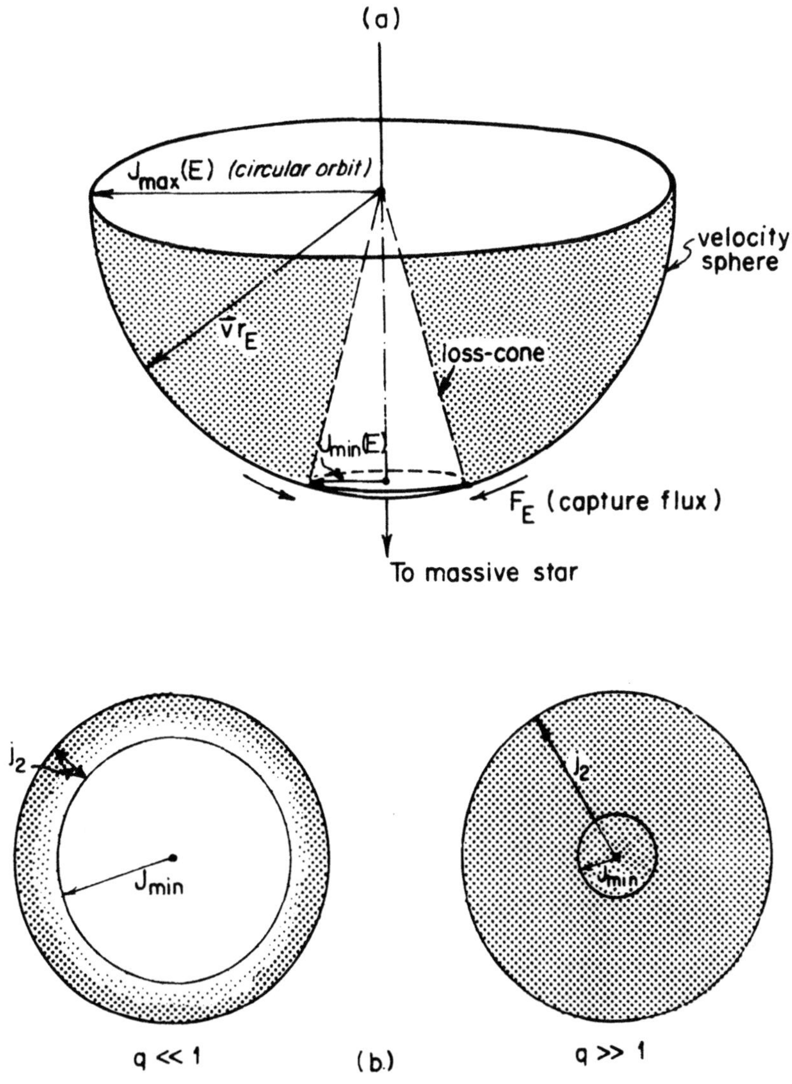

Figure 2: (a) Velocity distribution for ingoing stars with fixed energy E, radius $r_E \equiv GM/2|E|$, and speed $v = (2|E|)^{1/2}$. Stars with angular momentum J in the range $J_{min}(E) \ll J \leq J_{max}(E)$ are distributed nearly uniformly on the velocity sphere, where $J_{min}(E) \equiv [2(E + GM/r_t)]^{1/2} r_t$ and $J_{max}(E) \equiv GM/(2|E|)^{1/2}$. Stars scattered into the loss cone with $J \leq J_{min}(E)$ may be removed from the system in an orbital period. (b) The velocity sphere viewed from below. The quantity j_2 represents the dispersion in ΔJ suffered by a star in one orbital period due to stellar encounters; $q(E) \equiv j_2^2/J_{min}^2(E)$. The phase space density of stars outside the loss cone falls rapidly with J as $J \to J_{min}$ when $q \ll 1$ (the "diffusion" limit), but the density remains nearly uniform when $q \gg 1$ (the "pinhole" limit). In each dynamical time only stars in the ring $J_{min} < J \lesssim J_{min} + j_2$ may enter the loss cone. From Lightman and Shapiro (1977).

the region $J < J_{min}(E)$ maps out a "loss-cone" at the south pole in velocity space: stars inside this cone at pericenter are destroyed within one orbital period.

Far from the disruption loss-cone, those stars with $J_{max} \geq J \gg J_{min}$ are barely affected by the cone and distribute themselves nearly isotropically on the velocity sphere. However, near and inside the loss-cone, the stellar distribution is depleted. This depletion in velocity (or J) space drives a net flux of stars per unit energy, F_E, into the loss-cone from the ambient quasi-isotropic region. It is this J-space depletion and corresponding differential loss-cone flux that have been ignored in the naive 1-dimensional analysis described above.

Now consider how the stars move on the velocity sphere. In each period, stars experience a small rms change in angular momentum, j_2, due to gravitational encounters with their neighbors. This change causes them to move slightly on the velocity sphere, much like bees buzzing around a hive. What happens to those stars near the loss-cone depends critically on the ratio

$$q \equiv j_2^2/J_{min}^2(E) \tag{20}$$

which, in turn, depends on E. There are two extreme possibilities: (1) $q \ll 1$, in which case the change in J is sufficiently small that stars enter the loss-cone via two-dimensional diffusion. In this "empty loss-cone" case, the distribution function falls rapidly to zero at $J=J_{min}$ because any star found well inside the loss-cone would be destroyed in one period, long before it could be scattered out of the loss-cone; (2) $q \gg 1$, in which case the change in J is larger than the loss-cone opening so that most stars which reside inside the cone at apocenter manage to scatter out by the time they reach pericenter and are not disrupted. The distribution function remains nearly isotropic even inside the loss-cone in this "full loss-cone" case.

There exists a critical energy $E = E_{crit}$ defined by

$$q\,(E_{crit}) = 1 \tag{21}$$

Stars with energy E_{crit} typically reside at a radius

$$r_{crit} \sim GM/\left|E_{crit}\right| \tag{22}$$

from the black hole. This energy marks the transition between the "full" and "empty" loss-cone regimes.

Any attempt to analyze rigorously the two-dimensional character of the stellar distribution function $f(E,J)$ around a massive black hole must deal carefully with the disruption loss-cone region in phase space. In particular, numerical routines constructed to determine $f(E,J)$ must be able to handle stars reliably in the two opposite extreme regimes, $q \ll 1$ and $q \gg 1$ as well as handle stars in the critical transition regime in between. This realization was one of the key factors governing the design of the 2+1 Monte Carlo simulation code described below.

d) The 2+1 Fokker-Planck Equation

The 2+1 Fokker-Planck equation for $F(E,J)$ is given by (Lightman and Shapiro 1977; hereafter LS)

$$P(E)\frac{\partial f}{\partial t} = -\frac{\partial}{\partial E}[f\varepsilon_1] + \frac{1}{2}\frac{\partial^2}{\partial E^2}[f(\varepsilon_1^2+\varepsilon_2^2)] - \frac{1}{J}\frac{\partial}{\partial J}[fJj_1]$$

$$+ \frac{1}{2J}\frac{\partial^2}{\partial J^2}[fJ(j_1^2+j_2^2)] + \frac{1}{J}\frac{\partial^2}{\partial E\partial J}[fJ(\varepsilon_1 j_1+\zeta^2)] , \tag{23}$$

which must be solved in steady-state for the idealized problem formulated in Section IV. In eqn. (23), ε_1 and j_1 are the mean changes in energy and angular momentum per orbital period P, ε_2 and j_2 are the dispersions about these means, and ζ^2 is the correlation between ε_2 and j_2 for a star with fixed energy and angular momentum. These orbital perturbations are obtained directly from the locally defined velocity diffusion coefficients by integrating over orbits as shown in LS and in Shapiro and Marchant (1978).

Equation (23) is a 2+1 dimensional, integro-partial differential equation. Its nonlinear character is "hidden" by the fact that the orbital perturbations appearing in the equation are themselves integrals over f.

Consistent with the assumptions listed in Section IV, eqn. (23) must be integrated subject to the following boundary conditions:

i) The distribution of unbound stars is isotropic and isothermal: i.e. f satisfies

$$f = (2\pi v_o^2)^{-3/2}\, n_o \exp(-E/v_o^2) \quad , \quad E > 0 \tag{24}$$

where v_o is the constant line-of-sight velocity dispersion in the isothermal core and n_o is the core density.

ii) The distribution function vanishes for $J > J_{max}$, i.e. $f = 0$ for $J > J_{max}$.

iii) Stars are consumed by the black hole if and only if they lie within the loss cone $[J \leq J_{min}(E)]$ when they are at pericenter; that is, stars are destroyed when they are physically within the disruption radius, $f = 0$, $r < r_D$.

Note that (iii) is the precise statement of the loss-cone boundary condition. It is not appropriate to replace (iii) by "simplifications" such as $f = 0$ for $J < J_{min}$, which is not true in general (see Section V.c).

In addition, eqn. (23) is subject to the initial (steady-state) condition

$$\frac{\partial f}{\partial t} = 0 \tag{25}$$

e) Approximate Solution of Equation (23) and Scaling

Before discussing detailed Monte Carlo simulations of eqn. (23) for f(E,J) it is useful to summarize the results of a crude analytic analysis of the two-dimensional problem by LS [see also Frank and Rees 1976]. Indeed, the availability of the analytic analysis and the scaling behavior which it predicted proved crucial to the construction of a reliable simulation code to solve the problem more accurately.

(1) Bound Star Distribution (E < 0). LS found that the distribution could be written approximately in the power-law form of eqns. (4), (6) and (7), with

$$p \approx 1/4 + \text{logarithmic correction terms} \quad .$$

The correction terms are small except near r_D. Hence the solution of BW remained valid in the outer (observable) regions of the cusp.

(2) Critical Radius. The critical radius defined by eqn. (22) was found to satisfy

$$r_{crit}/r_a \approx 0.1 \; n_*^{-4/9} \; v_*^{32/9} \; M_3^{-20/27} \tag{26}$$

where

$$n_* \equiv n_o/(5 \times 10^4 \text{ pc}^{-3}) \quad \text{and} \quad v_* \equiv v_o/(10 \text{ km s}^{-1}) \tag{27}$$

and where $m = M_\odot$ and $r = R_\odot$. Thus one obtains the important result that r_{crit} resides inside the cusp for typical globular clusters with massive black holes (i.e. $r_{crit}/r_a < 1$) but outside the cusp for typical dense galactic nuclei (i.e. $r_{crit}/r_a > 1$ assuming $n_o \sim 10^7$ pc^{-3}, $v_o \sim 500$ km/s, $M \gtrsim 10^6 M_\odot$).

(3) Loss-Cone Disruption Rate: Bound Stars (E < 0). The maximum possible star flux that a relaxed system can accommodate is, on dimensional grounds,

$$F_{max}(r) \sim \frac{n(r)\, r^3}{t_r(r)} \tag{28}$$

where t_r is the local relaxation timescale. LS found that the differential loss-cone disruption rate F_E (stars/time/specific energy) peaked sharply at E_{crit} and that the total disruption rate satisfied

$$\mathcal{F} \equiv \int_{-M/r_D}^{0} F_E \, dE \sim \mathcal{F}_{max}(r_{crit})/\ln(r_{crit}/r_D) \tag{29}$$

$$\sim 10^{-7} \, n_*^{14/9} \, v_*^{-49/9} \, M_3^{61/27} \, yr^{-1} \quad , \tag{30}$$

where we have neglected logarithmic variations in eqn. (29). The corresponding core heating rate due to disruption was found to be

$$\mathcal{E} \equiv \int_{-M/r_D}^{0} F_E \left|E\right| dE \sim \mathcal{F}\left(\frac{GM}{r_{crit}}\right) \ln(r_{crit}/r_D) \tag{31}$$

(4) Loss-Cone Disruption Rate: Unbound Stars (E>0). Unbound stars can also enter the loss-cone at pericenter and be disrupted. Their rate of disruption was found to be

$$\mathcal{F}^u \approx \mathcal{F}\,(r_{crit}/r_a)^{5/4} \quad . \tag{32}$$

Accordingly, the bound star disruption rate dominates the unbound rate for typical globular clusters but the reverse is true for dense galactic nuclei. A similar result applies to the relative heating rates, although for unbound stars, is negative and corresponds to core "cooling":

$$\mathcal{E}^u \approx v_o^2 \, \mathcal{F}^u \quad . \tag{33}$$

f) Time-Dependent, Homological Core Evolution.

Suppose we remove the assumption of steady-state. How will a relaxed cluster core respond with time to the presence of a central black hole? This question motivated Shapiro (1977) to construct a simple homological model for the evolution of a globular cluster core with a central hole. The equations were those employed by Ambartsumian (1938) and Spitzer (1940) in their simple "evaporation model" for core collapse, modified by terms accounting for the effects of a central hole. The key effects considered were the tidal disruption of bound stars by the hole (eqn. 28) and the associated heating of the core which accompanies this process (eqn. 31).

The resulting evolution of the cluster core radius with time is shown in Fig. 3 for various initial black hole masses. In all cases, the hole eventually manages to halt and reverse core collapse. At late times the core expands asymptotically to infinity according to

$$R_c \propto t^{2/3} \quad , \quad t \to \infty \quad . \tag{34}$$

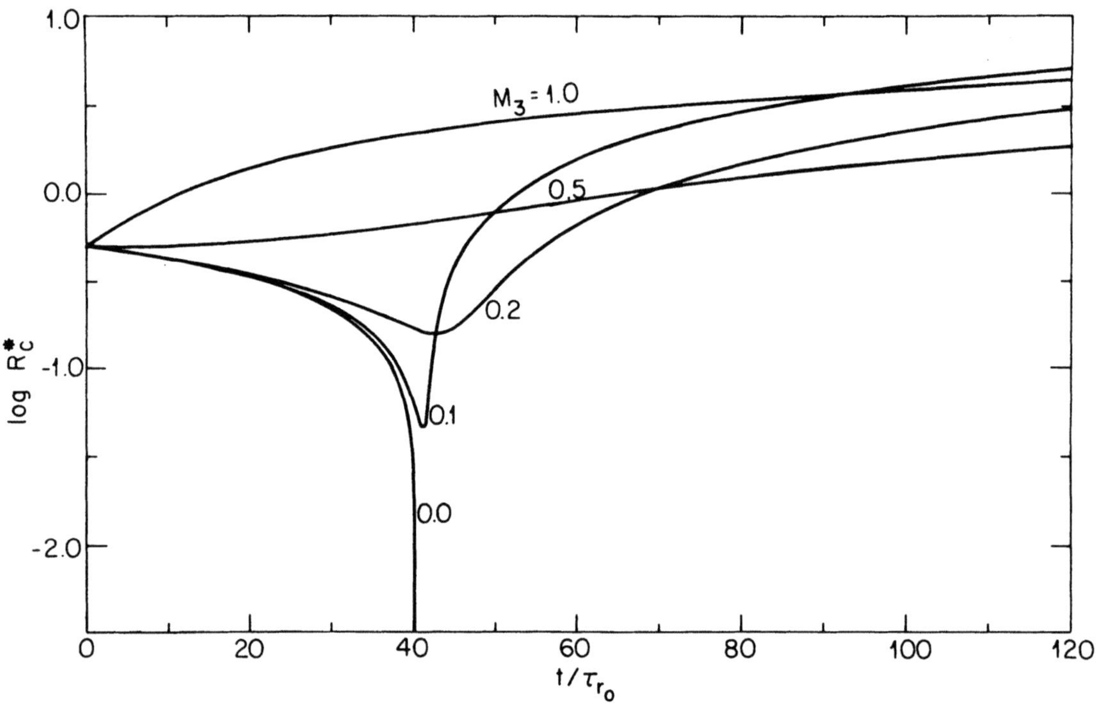

Figure 3: The core radius R_c^* as a function of time t (in units of the initial relaxation time τ_{r_0} = 6.2 × 10^6 yr) for a cluster with an initial core density $n_{c_0}^*$ = 0.2, radius R_c^* = 0.5, and black hole mass M_3 between 0 and 1. Core parameters are expressed in the following non-dimensional units: $n_c^* = n_c/(5 \times 10^4\ pc^{-3})$, $R_c^* = R_c/(1\ pc)$, $M_3 = M/(10^3 M_\odot)$. Massive central black holes invariably halt and reverse core collapse. From Shapiro (1977).

This behavior is thus quite analogous to that found by Hénon (1961; 1975) for cluster cores heated by central binaries. In spite of the crudeness of the homological model, it does suggest that core re-expansion may be a generic feature of post-collapse evolution, whatever the central energy source. It also points to the dissolution of a globular cluster in the Galactic tidal field as a plausible final outcome (Wielen 1971; Shapiro 1977).

VI. THE MONTE CARLO APPROACH

a) Key Numerical Difficulty: The Problem of Multiple Lengthscales

Simulating the full 2D problem rigorously by means of a Monte Carlo scheme immediately poses some "technical" difficulties. The key difficulties are associated with the vast dynamical range spanned by the parameters describing a relaxed cluster with a massive, central hole. This large dynamical range is in turn due to the existence of multiple, widely disparate length scales characterizing such a system. In this respect, a star cluster containing a massive black hole is not unlike many other multidimensional, nonlinear, many-body systems whose dynamical behavior on large scales requires detailed knowledge of behavior on small scales, and vice versa.

Specifically, consider the idealized problem posed in Section IV and focus on the bound stars ($E < 0$). They occupy the region $r_D \lesssim r \lesssim r_a$, which spans over six decades in radial coordinate space and a corresponding six decades in $|E| \sim GM/r$ phase space. The associated number density of stars in the cusp, $N(E) \propto |E|^{-9/4}$ thus decreases by over 14 decades through the cusp! This fact poses the following numerical challenge: <u>how do we achieve statistical reliability at high $|E|$ deep in the cusp with only a finite number of stars at low $|E|$ at the outer edge of cusp?</u>

Consider next the position of bound stars in J-space. They occupy the region $J_{min}(E) \leq J \leq J_{max}(E)$, where the ratio J_{min}/J_{max} varies between $(r_D/r_a)^{1/2} \simeq 10^{-3} \lesssim J_{min}/J_{max} \lesssim 1$. The small value of this parameter, which describes the "opening angle" of the disruption loss-cone, throughout most of the cusp poses the problem: <u>how do we achieve accuracy near the relatively small loss-cone to enforce the all-important tidal disruption boundary condition?</u>

Now consider the time variable. We first note that in a relaxed Fokker-Planck system, it is not so much dynamical or orbital time, t_d which matters but "only" the ratio of t_d to the local relaxation time, t_r. But even so, this ratio scales like $t_d/t_r \sim |E|^{-5/4}$ in the cusp and thus varies between $10^{-11} \lesssim t_d/t_r \lesssim 10^{-4}$. This vast variation poses the problem: <u>how do we follow stars for many orbital periods - some for many more than others, depending upon $|E|$ - before they relax?</u>

Finally note that only the bound stars entered into the above considerations. For general problems where the unbound star distribution is not fixed, a whole new set of lengthscales (e.g. core radius, Galactic tidal radius, etc.) enter the discussion and increase the dynamical range of the system even further!

b) Confronting the Difficulty

Below we summarize some of the tricks we have incorporated in our 2+1 Monte Carlo simulation scheme to handle the problem of multiple lengthscales. As these tricks, as well as the problems they are designed to overcome, are quite general in character, it may prove useful to incorporate them in other simulation codes.

Before discussing them we first note that the highly nonlinear character of the Fokker-Planck equation is treated by the familiar pseudo-linearization device of following a 'test' star distribution which relaxes by scattering not against itself directly, but against a fixed 'field' star background distribution (cf. Chandrasekhar 1942; Spitzer 1962). Of course, the 'field' star distribution and the orbital perturbations it induces are recomputed at regular time intervals to match the evolving 'test' star distribution, thereby achieving self-consistency (see Box 2).

Box 2

MONTE-CARLO SIMULATIONS IN (E,J) SPACE

'FIELD' STARS vs 'TEST' STARS

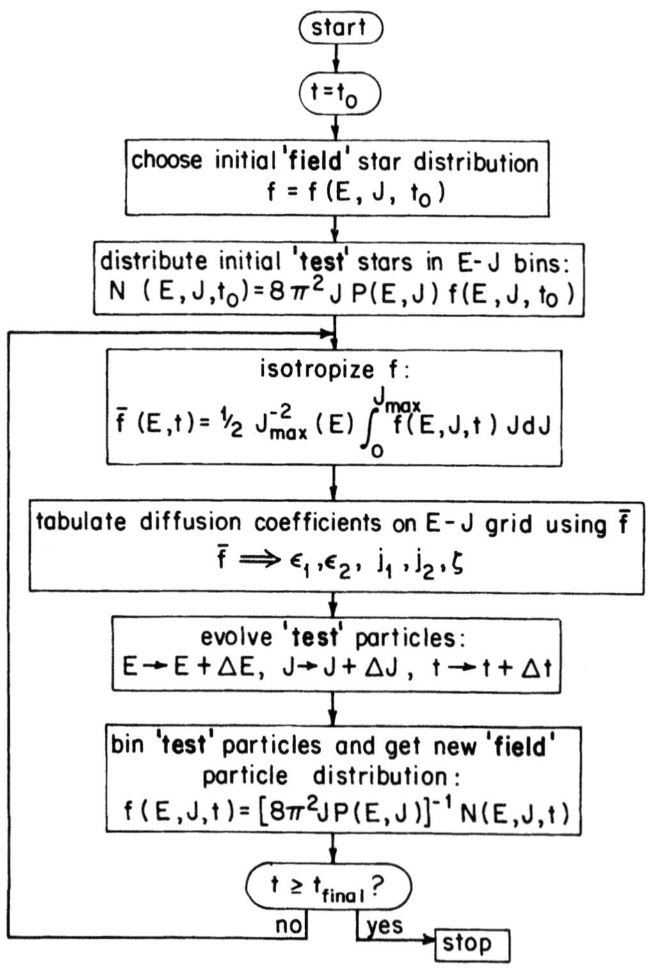

We employ one approximation in our application of the method: the orbital perturbation coefficients, which control the motions of 'test' stars in phase space, are computed from the J-averaged (isotropized) field-star distribution. This approximation is reasonable whenever the field-star distribution does not vary greatly from isotropy, as is the case in the black hole problem.

Box 3

PARTICLE 'CREATION - ANNIHILATION' SCHEME:
"DYNAMICAL RENORMALIZATION"

'Test - star' Energy - level diagram:

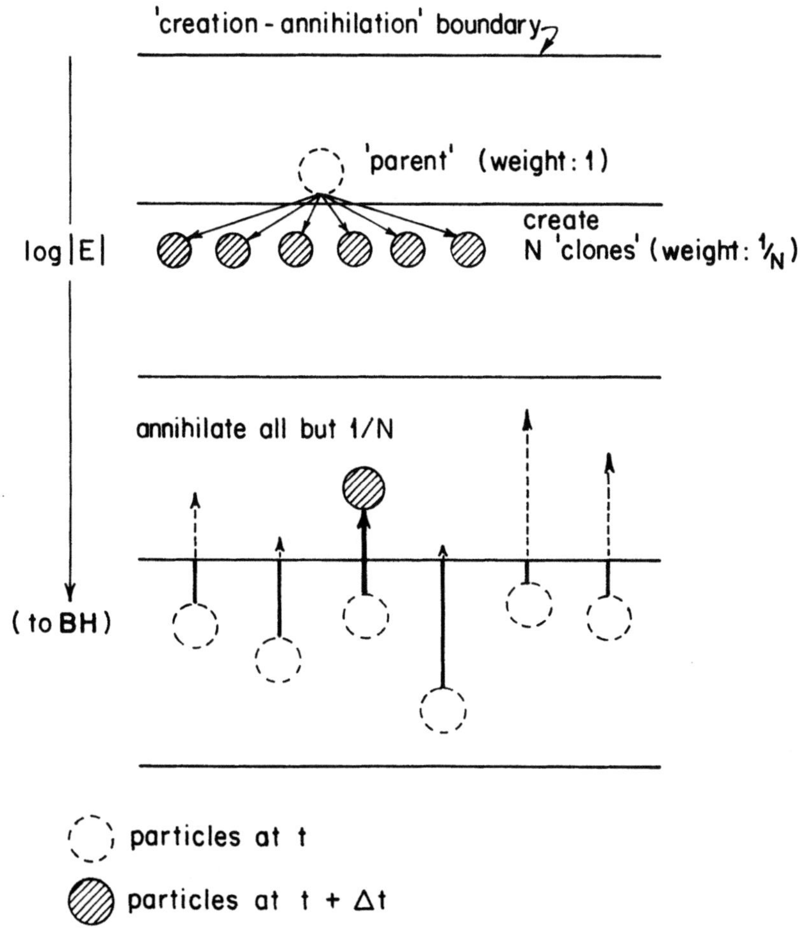

"Put particles where you need them"

1) Particle 'Creation-Annihilation' Scheme. To provide sufficient numbers of test stars and thereby guarantee statistical reliability in low populated regions of phase space (e.g. high $|E|$ regions in the cusp) we construct a particle 'creation-annihilation' scheme, illustrated schematically in Box 3. Specifically, at predeterminated boundaries in E-space, stars diffusing toward higher $|E|$ generate "clones" which may, themselves, subsequently diffuse toward the black hole independently of their "parent". To prevent artificial numerical contamination, these clones are prevented from crossing back across the energy boundary below which they formed, and, instead, are removed from the system if they attempt to do so. Since all test stars are dynamically coupled and interact only with field stars, there is no nonlinear feedback associated with this procedure.

Our method for accurately surveying many decades in phase space by particle 'cloning' has much in common with the Renormalization Group applied elsewhere in many-body physics. Accordingly, we sometimes refer to our procedure as 'Dynamical Renormalization'.

2) 'Time-step Adjustment' Scheme. To reconcile the very short dynamical timescales t_d associated with loss-cone consumption, with the rather long timescales associated with 2-body relaxation, $t_r \gg t_d$, we employ a 'time-step adjustment' algorithm (Box 4). That is, the size of an individual test particle time-step is determined by the particles' value of E and J. For typical stars, it is chosen to be some fraction of the local relaxation timescale and consists of many orbital periods (the number increasing as $|E|^{5/4}$ in the cusp). However, for stars on critical trajectories (e.g. near the disruption loss-cone), the step size is chosen to be an orbital period so that the stars' position can be examined each time they reach pericenter. The ability to chose whether or not to follow a star on an orbital timescale gives maximum flexibility and efficiency to any Fokker-Planck simulation scheme.

VII. MONTE CARLO SIMULATIONS OF BLACK HOLES IN GLOBULAR STAR CLUSTERS

(a) Simulations of the Idealized Problem

Figure 4 shows the nondimensional, isotropized distribution function and differential tidal disruption rate for bound stars obtained by Shapiro and Marchant (1978) using the E-J Monte Carlo simulation scheme. The isotropized profile is quite comparable to the 1D BW profile except that deep in cusp near the disruption region ($x_D \equiv M/(2r_D v_o^2) = 10^4$ in the case illustrated) considerable depletion is evident. A nearly identical isotropized profile to the Monte Carlo one shown here was obtained by Cohn and Kulsrud (1978), who integrated the 2D Fokker-Planck equation by finite-difference techniques.

Box 4

'TIME STEP ADJUSTMENT' ALGORITHM

$n = n(E, J)$ = number of orbital periods per Monte-Carlo time-step; guarantees $\Delta E/E$ and $\Delta J/J$ are $\ll 1$.

a. "Typical" bound stars : $J \lesssim J_{max}(E)$

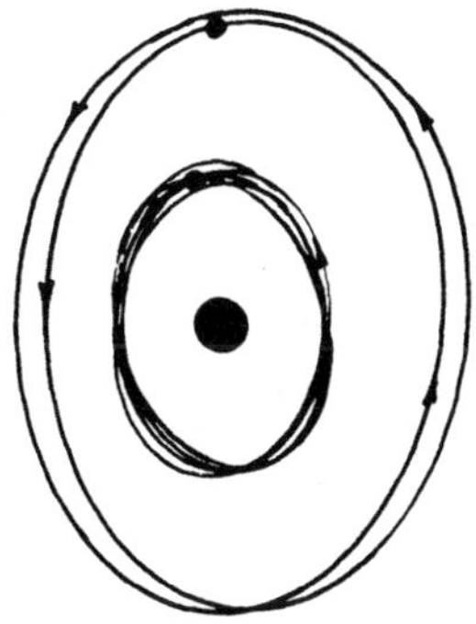

$$n \propto \frac{t_r}{t_d} \propto |E|^{5/4} \propto r^{-5/4}$$

● tidal disruption sphere

b. Stars near 'loss-cone': $J \ll J_{max}(E)$

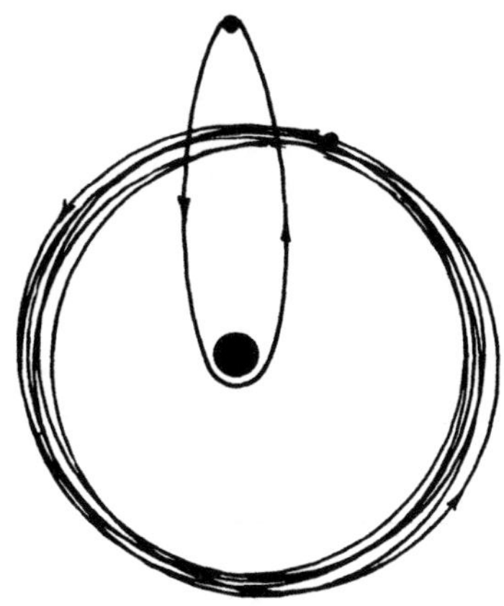

$$n \propto \left(\frac{J}{J_{max}}\right)^2$$

∴ For typical stars each step constitues many periods

BUT

stars on critical trajectories (i.e. "loss-cone" orbits) can be followed carefully on orbital timescales.

The region of maximum tidal disruption, as measured by $|E|F_E$, is strongly peaked near E_{crit}, as predicted by LS (see Section V.e). Indeed, most qualitative aspects of the solution discussed by LS, including the scaling, are verified by the numerical simulation.

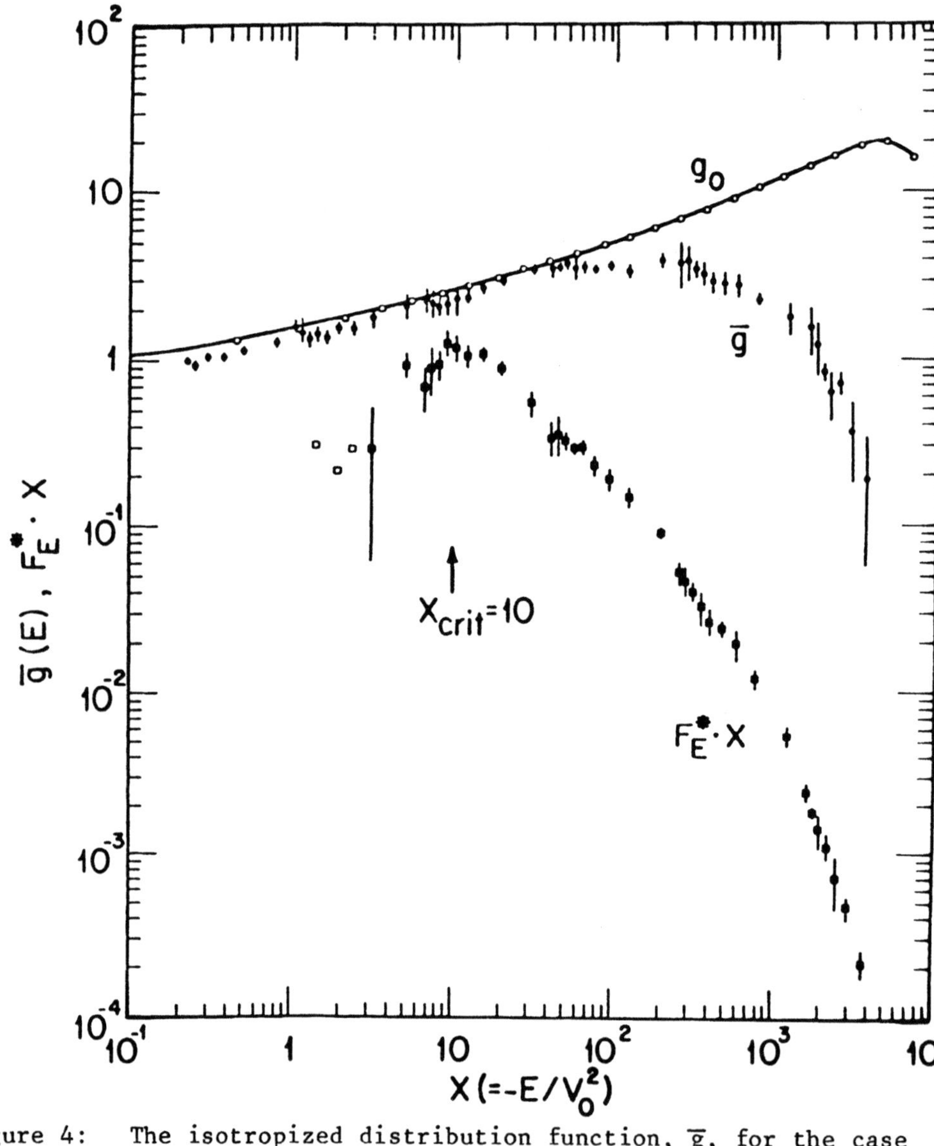

Figure 4: The isotropized distribution function, $\bar{g}$, for the case $x_{crit} = 10$ and $x_D = 10^4$, is plotted as a function of dimensionless energy $x \equiv -E/v_o^2$ (filled circles). The function $\bar{g}$ is normalized to unity at $x = 0$. The differential consumption rate, $F_E^* x$, is also plotted (filled squares); and those data points which had rather large error bars are distinguished (open squares). The function $F_E^* x$ peaks sharply near x_{crit}, indicating that the black hole principally consumes stars of energy ~ E_{crit}. The distribution g_o is the (one-dimensional) BW solution (solid line). From Shapiro and Marchant (1978).

(b) Simulations of Black Holes in Realistic Clusters

The Monte Carlo scheme was used by Marchant and Shapiro (1979) to determine the steady-state distribution and tidal disruption rate of stars around a massive black hole at the center of a realistic King-model stellar cluster. Here, the self-gravity of the stars is taken into account. The density and surface density profiles calculated for different assumed black hole masses are plotted in Figs. 5 and 6, respectively. For these computations the total cluster mass is $M_{cluster} = 3.6 \times 10^5\ M_\odot$ and the core mass is $M_{core} = 2 \times 10^4\ M_\odot$. It is clear from the figures that self-gravity becomes important whenever the black hole mass is sufficiently large: $M/M_{core} \gtrsim 0.1$. In this limit the cusp does not display any distinctive structure near r_a. Thus if such high-mass black holes reside in globular clusters, their main observable signature might be the lack of a resolvable cluster core.

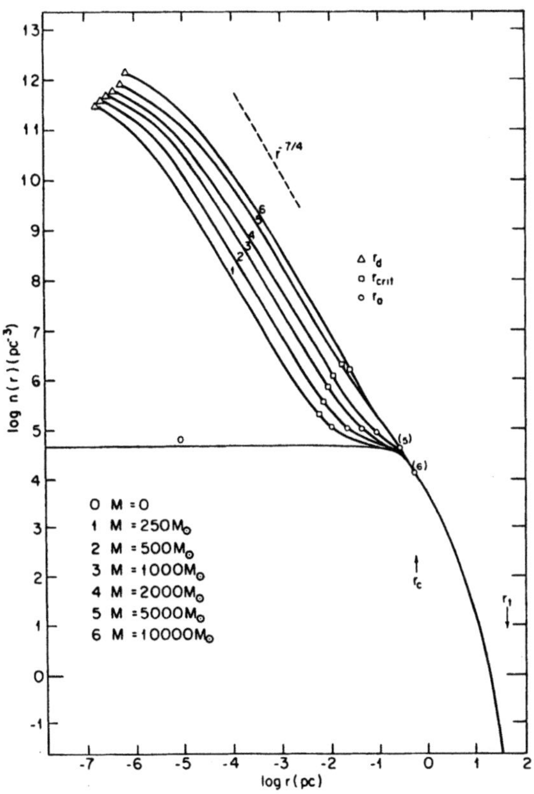

Figure 5: Density of stars n is plotted as a function of cluster radius r for clusters with black holes with different assumed masses M. The ambient King-model cluster parameters are $N = 3.6 \times 10^5$ stars, $v_o = 10$ km s^{-1}, and $\phi_s(0) = -8v_o^2$. The cluster tidal radius r_t, the core radius r_{crit}, and the tidal disruption radius r_d, are shown for all cases. All the black-hole cases show cusp profiles similar to the $r^{-7/4}$ power law. The effects of self-gravity near r_a are evident for the very massive black-hole cases. From Marchant and Shapiro (1979).

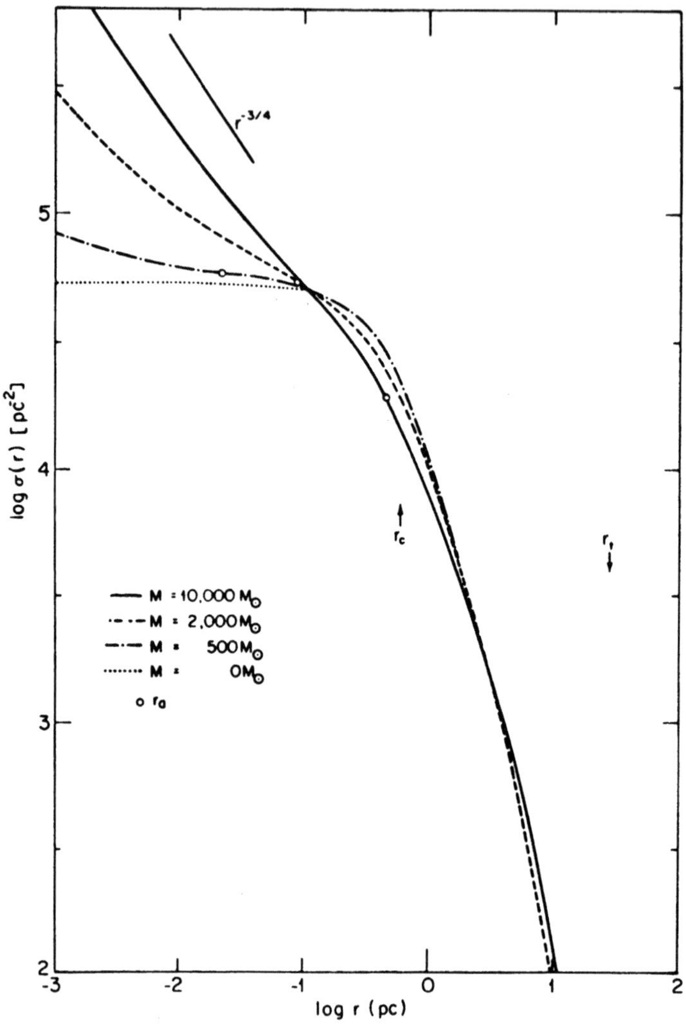

Figure 6: Surface density of stars σ as a function of the projected cluster radius r for some of the clusters shown in Figure 5. Note that the case $M = 10^4\ M_\odot$ shows no shoulder in its surface-density profile. In all cases, the distinctive $r^{-3/4}$ dependence of σ is evident, but only well inside the accretion radius. From Marchant and Shapiro (1979).

c) 2+1 Cluster Evolution Simulations

By far the most significant application of the 2+1 Monte Carlo scheme is determining the detailed, time-dependent, dynamical evolution of a realistic star cluster containing a massive, central black hole. Such calculations were performed by Marchant and Shapiro (1980) and Duncan and Shapiro (1982).

To check how well the simulation scheme could track Fokker-Planck evolution, the dynamical evolution of a cluster without a central hole was considered first. Initial data for this case consisted of a Plummer model (n = 5 polytrope) allowing for a finite Galactic tidal cut-off radius r_t. In general, the familiar features of the 'gravo-thermal catastrophe' were revealed by our integrations. Specifically, the simulations of the early and intermediate phases of core collapse, $0 \leq t/t_{rh} \leq 13.4$, agreed well with earlier 2+1 Fokker-Planck simulations of Spitzer and his colleagues [see paper by Spitzer in these Proceedings for a review and references]. Here t_{rh} is the initial half-mass relaxation time defined by Spitzer. Our simulations of the late phases, $13.4 \leq t/t_{rh} \leq 14.7$, agreed well with the Fokker-Planck integrations of Cohn (1979, 1980). In particular, the homologous nature of advanced core collapse as predicted by Lynden-Bell and Eggleton (1980) was verified. Not surprisingly, it was found to be relatively independent of r_t.

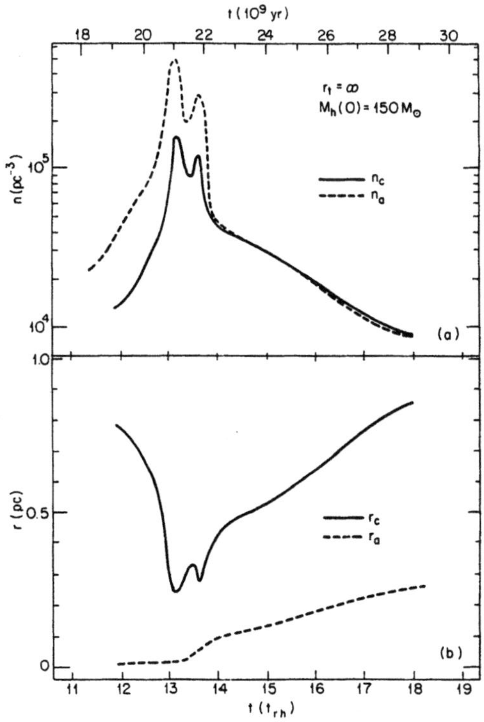

Figure 7: (a) The density at r_a, $n_a \equiv n(r_a)$ (dashed line), and the core density n_c as functions of time. (b) The core radius R_c (solid curve) and the accretion radius r_a as functions of time. From Duncan and Shapiro (1982).

Confident that our scheme could handle evolution accurately, we next considered the effect of a central black hole. The initial data again consisted of a Plummer model with a finite tidal radius, but now we inserted a black hole, with initial mass in the range $0 \leq M_i/M_\odot \leq 2000$ at the cluster center at an arbitrary time t_i (which we varied) during the cluster's evolution. We found that in all cases core collapse is eventually reversed by the heat flux from stellar disruption by the hole. The system attains a quasi-stationary, expanding state, by which time the hole (if assumed to swallow most of the disrupted stellar debris) has grown to several thousand solar masses. This expanding state appears to be roughly independent of (i) the initial hole mass, (ii) the time during core evolution at which the hole is introduced and, (iii) the value of r_t.

Figure 7 illustrates the variation with time of characteristic central densities and radii during the evolution. Core collapse, followed by re-expansion, are clearly evident. The corresponding growth of the black hole is shown in Fig. 8. Typically, core bounce occurs when the hole grows to $M \approx 0.1\ M_{core}$.

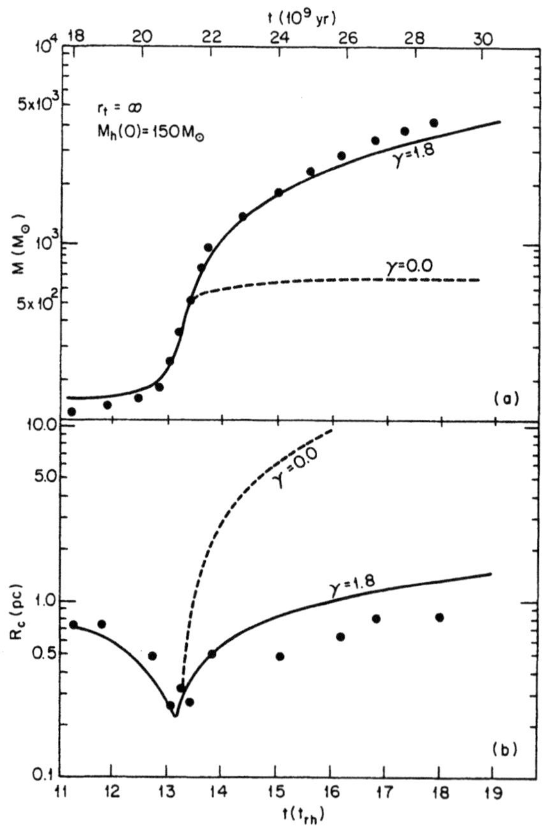

Figure 8: (a) The growth of the black hole mass M_h with time for the case shown in Fig. 7. The filled circles are "data" from a Monte Carlo simulation, while the smooth curves represent homological solutions characterized by one free parameter (γ). (b) The time evolution of the core radius. From Duncan and Shapiro (1982).

VIII. MONTE CARLO SIMULATIONS OF BLACK HOLES IN DENSE GALACTIC NUCLEI

Dense galactic nuclei may provide more hospitable environments than globular clusters for forming massive black holes. Indeed, many models of AGN's and quasars now favor the presence of a supermassive black hole ($M \gtrsim 10^6$–10^9 $M_\odot$) to power these energetic sources (see Table 2). It was therefore natural for Duncan and Shapiro (1983) to extend their 2+1 Monte Carlo simulation scheme to study the dynamical evolution of a dense galactic nucleus with a massive, central hole.

As initial data they considered King model clusters of solar-type stars. The core parameter for these models were in the range expected for AGN's and quasars,

$$\begin{aligned} 2 \times 10^7 &\leq M_{core}/M_\odot < 3 \times 10^8 \\ 300 &\leq v_o \text{ (km/s)} < 10^3 \\ 10^6 &\leq n_o \text{ (stars pc}^{-3}) < 10^7 \quad , \end{aligned} \tag{35}$$

while the initial central black hole mass was varied between

$$2 \times 10^6 < M/M_\odot < 5 \times 10^4 \quad . \tag{36}$$

In addition to tidal disruption by the black hole, stars in dense galactic nuclei can be destroyed by star-star collisions. These collisions were incorporated in the Monte Carlo simulations, where it was assumed that the stars disrupted immediately upon impact and that <u>all</u> the gaseous debris liberated during collisions and tidal disruptions is consumed by the central hole. Accordingly, the hole grows at a rate given by

$$\dot{M} = m\,(F_{coll} + F_{tide}) \quad . \tag{37}$$

This accretion by the hole results in a radiation luminosity given by

$$L = 7 \times 10^{45} \left(\frac{\varepsilon}{0.1}\right) \left(\frac{\dot{M}}{M_\odot \text{ yr}^{-1}}\right) \text{ erg s}^{-1} \tag{38}$$

where ε is the assumed conversion efficiency of rest mass into radiation.

The key results of this study were the following:

1. Star destructions result predominantly from collisions when the rms velocity v_o exceeds the escape velocity $v_{esc} \approx 620$ km/s from a star. Destructions result predominantly from tidal disruptions when the reverse is true.
2. The maximum stellar destruction rate achieved in these systems is found to be

$$m\, F_{tide}(\max) \sim \frac{M_{core}}{t_r} \lesssim 10^{-2}\ M_\odot\ yr^{-1} \tag{39}$$

for tidal disruption and

$$m\, F_{coll}(\max) \sim \frac{M_{core}}{t_{coll}} \sim M_\odot\ yr^{-1} \tag{40}$$

for star collisions. Here t_r and t_{coll} are the core relaxation and collision timescales, respectively, and the numerical values were those found for the Monte Carlo simulations (they are largely determined by the assumed initial conditions).

Equations (37)-(40) thus suggest that quasars ($L \sim 10^{45}$-10^{48} erg s^{-1}) may be dense galactic nuclei in which physical collisions are occurring, while AGN's ($L \sim 10^{42}$-10^{45} erg s^{-1}) may be nuclei where collisions are unimportant and tidal disruptions dominate. Similar conclusions have been reached by other investigations (see references in Box 1).

3. For late times, $\dot{M}$, hence L, decays like

$$L \sim \dot{M} \sim t^{-\alpha} \tag{41}$$

where α depends on the power-law stellar profile outside the core and is, typically, in the range $\alpha \sim 0.8 - 1$.

Simulation results for a candidate "quasar" model are shown in Figs. 9 and 10. In Fig. 9 the evolution of the density profile is shown. For this case the initial core relaxation time is $t_{r,i} = 4.5 \times 10^9$ yr while the collision time is $t_{coll,i} = 2 \times 10^8$ yr; evidently $t_{coll,i} \ll t_{r,i}$ so collisions dominate. In this case the bound stars in the cusp are quickly destroyed by collisions, leaving only the unbound stars with a characteristic $n(r) \propto r^{-1/2}$ profile. In Fig. 10 the black hole consumption rate is plotted vs time. This rate remains constant for $t \lesssim t_{coll,i}$ while the core is being consumed, after which it decays

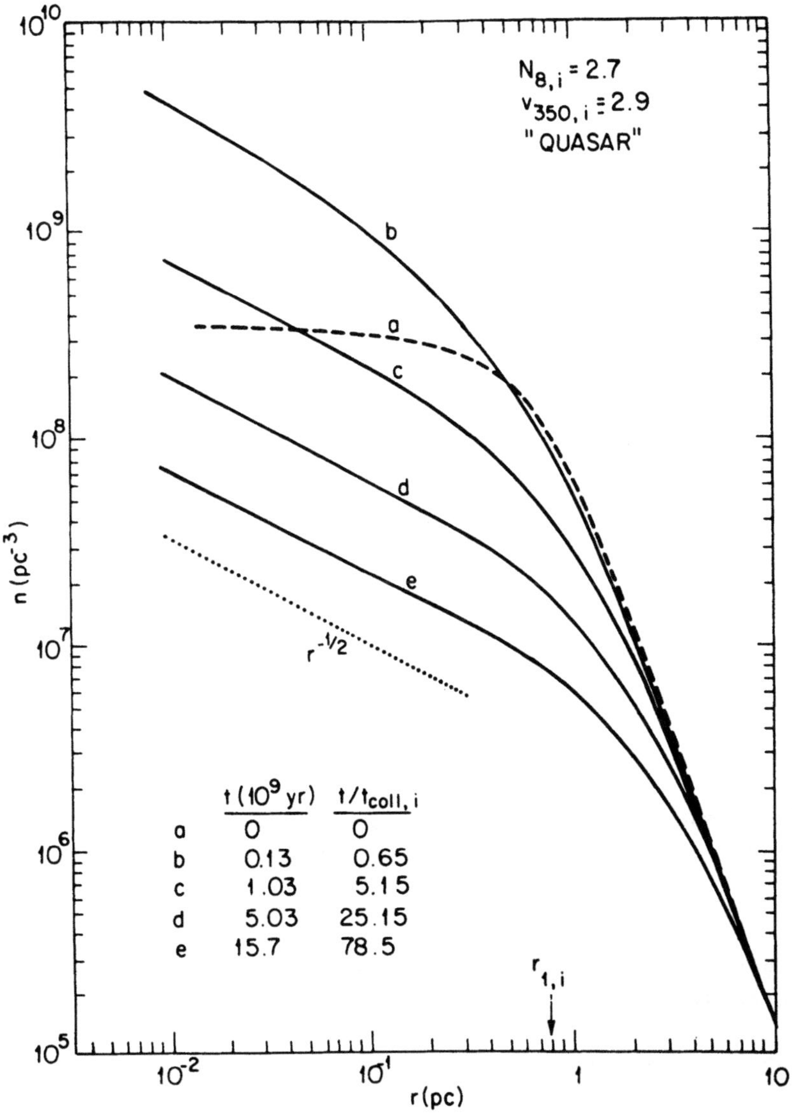

Figure 9: The stellar density profile at five successive times for the "quasar" model simulation. The dashed curve shows the initial profile, while the lowest solid curve is the profile at the end of the simulation. The initial cluster has a core radius $r_{1,i}$, star number $N_{8,i}$ and line-of-sight velocity dispersion $v_{350,i}$. From Duncan and Shapiro (1983).

like t^{-1} as the hole eats its way out into the halo. From an initial mass of $2 \times 10^6\ M_\odot$, the black hole grows to $3 \times 10^9\ M_\odot$ by the end of the simulation. The maximum luminosity achieved during the cluster's evolution is about $L \sim 10^{46}$ erg s^{-1} (for $\varepsilon = 0.1$), which is quite respectable for a typical quasar.

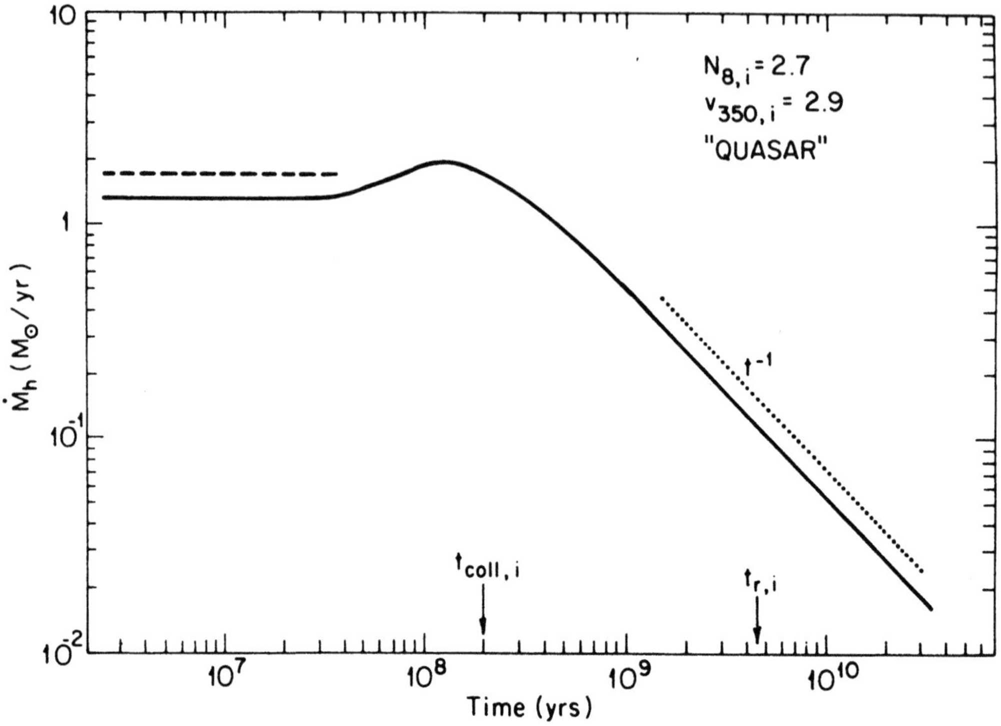

Figure 10: The black hole growth rate $\dot{M}_h$ ($\propto$ L) as a function of time for the case shown in Fig. 9. The dashed curve is the rate predicted by simple analytic arguments. From Duncan and Shapiro (1983).

IX. THE BIRTH OF SUPERMASSIVE BLACK HOLES VIA THE COLLAPSE OF DENSE STELLAR SYSTEMS

An analysis of massive black holes in star clusters would not be complete without some discussion of their origin. A possible scenario for the formation of massive holes in globular clusters via star collisions and coalescence has been proposed by Spitzer (1975) and sketched by Lightman and Shapiro (1978) (see Section II.a). Such a mechanism cannot produce black holes much more massive than ~ 100 $M_\odot$, but we have seen that black holes in clusters can grow considerably by the consumption of tidally disrupted stars (Section VII.c). This scenario has to be investigated in much greater detail before its likelihood can be properly assessed.

Several mechanisms have been suggested for the formation of supermassive black holes in dense galactic nuclei (see Box 1). One of the most appealing is the scenario first put forward by Zel'dovich and

Podurets (1965). They argued that the combined effects of secular core collapse (i.e. the 'gravothermal catastrophe', which they referred to as 'stellar evaporation') and star-star collisions would drive a cluster to states of ever higher central velocity and redshift. At sufficiently high redshift, the cluster would become relativistically unstable, at which point it would undergo catastrophic collapse on a dynamical timescale to a supermassive black hole. They had in mind initial Newtonian star clusters composed of stellar mass black holes which would thus ultimately collapse to form a single, supermassive black hole.

Some recent theoretical developments suggest that this proposal by Zel'dovich and Podurets (1965) ought to be regarded quite seriously. For the first time, it has been possible to integrate numerically the full Einstein field equations for an arbitrary spherical, collisionless gas in General Relativity (Shapiro and Teukolsky 1985 a,b). These integrations enable one to follow on the computer the evolution on dynamical timescales of relativistic star clusters, even during epochs characterized by total gravitational collapse and the formation of supermassive black holes. The formation and growth of the black hole can be followed accurately without the appearance of numerical or physical singularities.

The original speculation of Zel'dovich and Podurets (1965) that star clusters become relativistically unstable at sufficiently high central redshift ($z_c \gtrsim 0.5$) has been demonstrated rigorously in perturbation theory (see, e.g., Ipser 1969, 1980; Fackerell, Ipser and Thorne 1969 and references therein). The recent numerical integrations described above provide further verification of this relativistic instability and, more significantly, follow its nonlinear growth and the ultimate fate of unstable clusters. In particular, the integrations show quite generally that clusters of sufficiently high central redshift do undergo catastrophic collapse to a black hole on a dynamical timescale. Moreover, they reveal that, while the core may encompass only a small fraction of the cluster mass at the center initially, the hole ultimately grows to entrap virtually the entire cluster in a few mean orbital periods.

The fully relativistic, Vlasov simulations of spherical star clusters described above provide fresh support for the Zel'dovich and Podurets supermassive black hole scenario. They have motivated Shapiro and Teukolsky (1985c) to reconsider this suggestion in greater detail, in light of our current knowledge of dense galactic nuclei, the gravothermal catastrophe, etc. The result is summarized in Fig. 11. Illustrated are the evolutionary tracks of initially Newtonian, isothermal cores of dense stellar systems composed of neutron stars ($m = 1.4\ M_\odot$). At any instant, the cores are defined by two parameters: N_c, the total number of core stars and z_c, the central redshift in the core ($z_c \propto \Phi_c \propto v_c^2$, where Φ_c is the central potential and v_c is the core velocity dispersion, also shown). Each track is composed of two distinct segments, corresponding to two different epochs: an early, low-redshift ($z < z_{coll}$) "point-mass" epoch during which the core undergoes secular collapse via the 'gravothermal catastrophe' and a later,

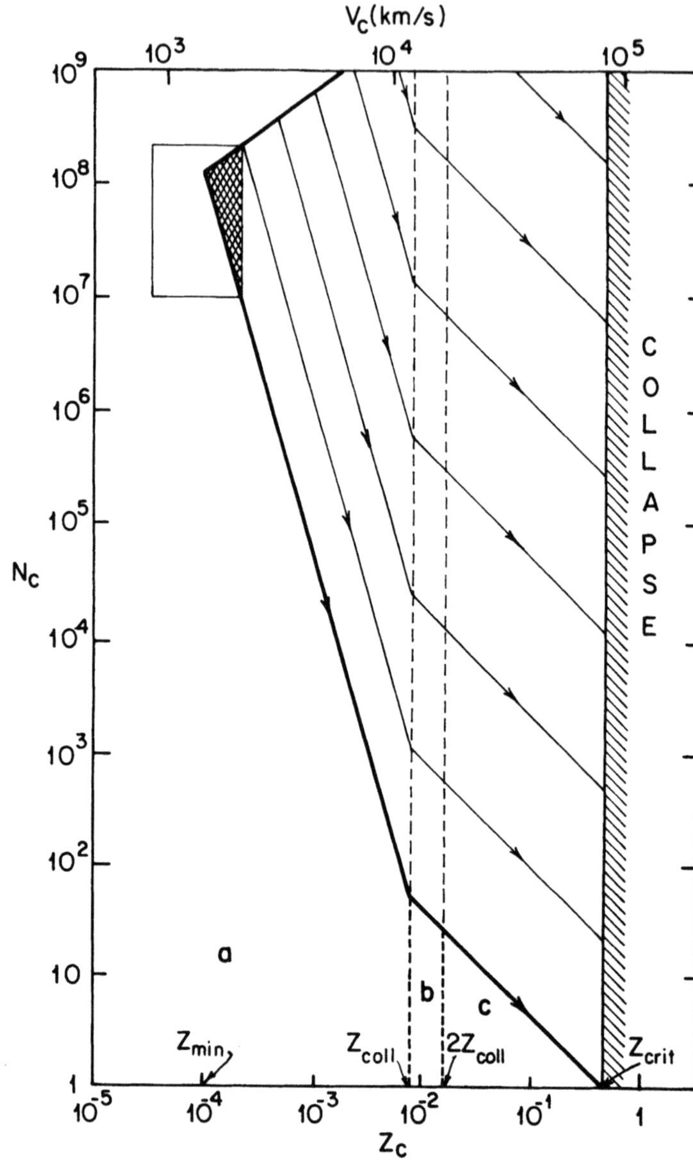

Figure 11: Dynamical evolution tracks in the N_c - z_c plane of dense cluster cores consisting of 1.4 $M_\odot$ neutron stars. During phase (a), the "gravothermal catastrophe", the core evolves by two-body scattering to central redshift $z_{coll} \sim 8 \times 10^{-3}$. During phase (b) the neutron stars collide, coalesce and collapse to form ~ 2.8 $M_\odot$ black holes. During phase (c) the remnant black holes collide and coalesce, driving the core to central redshift $z_{crit} \sim 0.5$. At this point the entire cluster becomes relativistically unstable and undergoes catastrophic collapse to a supermassive black hole in a few mean orbital periods. The range of plausible core parameters for dense galactic nuclei lie in the box shown in the upper left of the figure. The hatched region shows the fraction of these capable of evolving to a supermassive black hole in a Hubble time. From Shapiro and Teukolsky (1985c).

high-redshift ($z \geq z_{coll}$) "finite-radius" epoch during which star-star collisions and coalescences dominate the evolution. During both epochs, the number of core stars decrease while the central redshift increases. Ultimately, the core achieves relativistic central velocities and redshift ($z = z_{crit} \sim 0.5$) at which point it undergoes total gravitational collapse to a black hole. The hole quickly grows outward and eventually swallows up most of the ambient cluster in a few mean dynamical timescales.

Only those clusters which can evolve to a relativistic state in a Hubble time are relevant for our investigation. Also, only those clusters which have more than one star remaining in their core by $z = z_{crit}$ can actually reach this state. These two constraints confine the tracks to be within the heavy lines in the figure. A similar figure can be drawn for clusters consisting initially of white dwarfs or stellar-mass black holes.

Consider the consequences of this scenario. First, there is minimum central redshift ($z = z_{min} \sim 10^{-4}$) or a minimum central velocity dispersion ($v_c \sim 10^3$ km s^{-1}) below which a cluster of neutron stars cannot evolve to a relativistic state in a Hubble time. Although this minimum core velocity is large, it is not unreasonably large for conditions expected in dense galactic nuclei. Indeed, consider the range of plausible parameters calculated for dense galactic nuclei following the collision-coalescence epoch which the normal stars in the nucleus are expected to undergo (Colgate 1967; Sanders 1970). Very likely, this initial collision-coalescence epoch will convert the dense, but otherwise normal galactic nucleus into a cluster of stellar-mass compact stars - black holes or neutron stars (Begelman and Rees 1978). The parameter range calculated by Colgate (1967) and Sanders (1970) for such systems is indicated by the box in the upper left-hand corner of Figure 11. What is interesting is that (1) this box intersects a small, but finite domain occupied by those dense cores capable of evolving to supermassive black holes in a Hubble time and (2) the domain of intersection occurs in the range $10^7 \lesssim N_c \lesssim 10^8$, indicating that, following total gravitational collapse, the clusters will yield supermassive black holes with masses in the range $10^7 \lesssim M/M_\odot \lesssim 10^8$. This is roughly the mass range of black holes which can generate AGN and quasar luminosities via gas accretion near the Eddington limit! It is thus within the mass range frequently cited in black hole models of such systems (cf. Section VIII).

The supermassive black hole generation mechanism in dense clusters discussed here is attractive and uncomplicated. It leads to black holes of the "right size" to explain quasars and AGNs. Whether or not the picture is also correct will depend on the results of more detailed, future studies, both theoretical and observational.

This work has been supported in part by National Science Foundation Grant AST 81-16370 at Cornell University.

REFERENCES

Ambartsumian, V.A. 1938, Ann. Lenningrad State Univ. No. 22, (Astr. Series, Issue 4)
Bahcall, J.N., Bahcall, N.A. and Weistrop, D. 1975, Ap. J. (Lett.) 16, L159.
Bahcall, J.N. and Ostriker, J.P. 1975, Nature 256, 23.
Bahcall, J.N. and Wolf, R.A. (BW) 1976, Ap. J. 209, 214.
Bahcall, J.N. and Wolf, R.A. 1977, Ap. J. 216, 883.
Bahcall, N.A. 1976, Ap. J. (Lett.), 204, L83.
Bahcall, N.A. and Hausman, M.A. 1976, Ap. J. (Lett.) 207, L181.
Bahcall, N.A., Lasker, B.M and Wamsteker, W. 1977, Ap. J. (Lett.) 213, L105.
Begelman, M.C. and Rees, M.J. 1978, MNRAS 185, 847.
Bisnovatyi-Kogan, G.S., Churayev, R.S. and Kolosov, B.I. 1982, Astron. Astrophys. 113. 179.
Chandrasekhar, S. 1942, Principles of Stellar Dynamics (Chicago: University of Chicago Press).
Cohn, H. 1979, Ap. J. 234, 1036.
Cohn, H. 1980, Ap. J. 242, 765.
Cohn, H. and Hut, P. 1983, preprint.
Cohn, H. and Kulsrud, R.M. 1978, Ap. J. 226, 1087.
Colgate, S.A. 1967, Ap. J. 150, 163.
Djorgovski, S. and King, J.R. 1984, Ap. J. (Lett.) 277, L49.
Dokuchayev, V.I. and Ozernoi, L.M. 1977, JETP 73, 1587.
Duncan, M.J. and Shapiro, S.L. 1982, Ap. J. 253, 921 (Paper IV).
Duncan, M.J. and Shapiro, S.L. 1983, Ap. J. 268, 565.
Einstein, A. 1939, Ann. Math. 40, 922.
Fackerell, E.D., Ipser, J.R. and Thorne, K.S. 1969, Comments Astrophys. Space Phys. 1, 134.
Fall, S.M. and Malkam, M.A. 1978, MNRAS 185, 899.
Farouki, R.T., Shapiro, S.L. and Teukolsky, S.A. 1983, in IEEE Computer 16, 73.
Frank, J. 1978, MNRAS 184, 87.
Frank, J. and Rees, M.J. 1976, MNRAS 176, 633.
Goodman, J. 1983, preprint.
Grindlay, J.E. 1977 in Highlights of Astronomy Vol. 4, 111.
Grindlay, J.E. 1981, in X-ray Astronomy with the Einstein Satellite ed. R. Giacconi (Reidel, Dordrecht) p. 79.
Grindlay, J.E. 1983, Advances in Space Exploration Vol. 2, No. 9, p. 133.
Grindlay, J.E., Hertz, P., Steiner, J.E., Murray, S.S. and Lightman, A.P. 1984, Ap. J. (Lett.) 282, L13.
Gursky, H. and Schwartz, D.A. 1977 in Ann. Rev. Astron. Astrophys. 15, 541.
Heggie, D.C. 1983, preprint.
Hénon, M. 1961, Ann. d'Ap. 24, 369.
Hénon, M. 1975 in Dynamics of Stellar Systems (IAU Symposium No. 69), ed. by A. Hayli (Reidel, Dordrecht) p. 133.
Hills, J.G. 1975, Nature 254, 295.
Inagaki, S. and Lynden-Bell, D. 1983, MNRAS 205, 913.

Ipser, J.R. 1969, Ap. J. 158, 17.
Ipser, J.R. 1978, Ap. J. 222, 976.
Ipser, J.R. 1980, Ap. J. 238, 1101.
Lightman, A.P. 1976, private communication in Bahcall and Wolf (1976).
Lightman, A.P. 1982, Ap. J. (Lett.) 263, L19.
Lightman, A.P., Press, W.H. and Odenwald, S. 1978, Ap. J. 219, 629.
Lightman, A.P. and Shapiro, S.L. (LS) 1977, Ap. J. 211, 244.
Lightman, A.P. and Shapiro, S.L. 1978, Rev. Mod. Phys. 50, 437.
Lin, D.N.C. and Tremaine, S. 1980, Ap. J. 242, 789.
Lynden-Bell, D. 1967, MNRAS 136, 101.
Lynden-Bell, D. 1969, Nature 223, 690.
Lynden-Bell, D. and Eggleton, P.P. 1980, MNRAS 138, 495.
Marchant, A.B. and Shapiro, S.L. 1979, Ap. J. 234, 317 (Paper II).
Marchant, A.B. and Shapiro, S.L. 1980, Ap. J. 239, 685 (Paper III).
McMillan, S.L., Lightman, A.P. and Cohn, H. 1981, Ap. J. 251, 436.
Norman, C. and Silk, J. 1983, Ap. J. 266, 502.
Peebles, P.J.E. 1972a, Gen. Rel. Grav. 3, 63.
Peebles, P.J.E. 1972b, Ap. J. 178, 371.
Rees, M.J. 1977, Ann. N.Y. Acad. Sci. 302, 613.
Rees, M.J. 1978, Phys. Scripta 17, 193.
Rees, M.J. 1984 in Ann. Rev. Astron. Astrophys., in press.
Sanders, R. 1970, Ap. J. 162, 791.
Sargent, W.L.W., Young, P.J., Boksenberg, A., Shortridge, K., Lynds, C.R. and Hartwick, F.D.A. 1978, Ap. J. 221, 731.
Shapiro, S.L. 1977, Ap. J. 217, 281.
Shapiro, S.L. and Lightman, A.P. 1976, Nature 262, 743.
Shapiro, S.L. and Marchant, A.B. 1978, Ap. J. 225, 603 (Paper I).
Shapiro, S.L. and Teukolsky, S.A. 1985 (a,b,c) in preparation.
Silk, J. and Arons, J. 1975, Ap. J. (Lett.) 200, L131.
Spitzer, L. 1940, MNRAS 100, 396.
Spitzer, L. 1962, Physics of Fully Ionized Gases (Wiley, New York).
Spitzer, L. 1971 in Semaine d'etude sur les noyeaux des galaxies 1970 (Pontificae Academiae Scientarum Scripta Varia No. 34).
Spitzer, L. 1975 in Dynamics of Stellar Systems (IAU Symposium No 69), ed. by A. Hayli (Reidel, Dordrecht), p. 3.
Spitzer, L. and Hart, M.H. 1971, Ap. J. 164, 399.
Strittmatter, P.A. and Williams, R.E. 1976 in Ann. Rev. Astron. Astrophys. 14, 373.
Truran, J.W. and Cameron, A.G.W. 1972, Ap. and Space Sci. 14, 179.
Ulrich, M.H., Boksenberg, A., Bromage, G.E., Clavel, J., Elvius, A., Penston, M.V., Perola, G.C., Petteni, M., Snijders, M.A.J., Tanzi, E.G. and Tarenghi, M. 1984, MNRAS 206, 221.
Wielen, R. 1971, Ap. Space Sci. 13, 300.
Wolfe, A.M. and Burbidge, G.R. 1970, Ap. J. 161, 419.
Wyller, A.A. 1970, Ap. J. 160, 443.
Young, P.J., Shields, G.A. and Wheeler, J.C. 1977, Ap. J. 212, 367.
Young, P.J., Westphal, J.A., Kristian, J., Wilson, C.P. and Landauer, F.P. 1978, Ap. J. 221, 721.
Zel'dovich, Ya. B. and Novikov, I.D. 1971, Relativistic Astrophysics, Vol. 1 (University of Chicago, Chicago).

Zel'dovich, Ya. B. and Podurets, M.A. 1965, Astron. Zh 42, 963 [English translation in Sov. Astron. - A.J. 9, 742].

DISCUSSION

KING: Another version of Peebles' remark (not due to Peebles) is, "I'll believe a black hole when I see one." More seriously, what would you expect us to observe, that would distinguish between your model and those that are stabilized by binaries?

SHAPIRO: As has been pointed out for a long time, black holes will induce distinguishing 'cusp' profiles in the cores of any cluster in which they reside. These central cusps will take a number of forms: optical light cusps, star-count surface density cusps (σ), and projected rms velocity cusps, $V_{\|}$ (averaged through the line of sight). Of these, the most frequently mentioned is σ, which acquires the characteristic Bahcall-Wolf $\sigma \sim r^{-3/4}$ shape. It will be difficult, however, to distinguish this profile observationally from asymptotic isothermal profiles, $\sigma_{isoth} \sim r^{-1}$, which may characterize post-collapse binary scenarios. However, as Marchant and I emphasized in Paper III of our series (see, e.g. Fig. 12), $V_{\|}$ may provide the *best* evidence for the presence of a massive, central black hole. This function rises rapidly ad $V_{\|} \sim r^{-1/2}$ inside the core, while, as Cohn has pointed out, it does not increase very rapidly in clusters without massive holes.

One should also pursue *statistical* studies of cumulative cluster data to distinguish between the different post-collapse evolutionary tracks that clusters follow, depending on whether they contain a massive black hole or binaries. One might ultimately produce 'H-R' diagrams in which the location of clusters reveals pre- and post-collapse evolutionary paths. This may be difficult, of course, particularly since at late times the post-collapse core expansion may be quite comparable in both scenarios. Nevertheless, this study should be pursued further.

COHN: As a follow-up the previous question, I would like to note that the behavior of the velocity dispersion profile in a cusp should give the best indication of whether a massive central black hole is present. In a cusp around a black hole, $\nu \propto r^{-0.5}$ and is thus fairly steep. In contrast, in a cusp that developes as a result of core collapse with no black hole, the velocity dispersion is very much flatter, $\nu \propto r^{-0.11}$.

CAN A MODERATELY MASSIVE BLACK HOLE REVERSE CORE COLLAPSE?

Martin J. Duncan
Astronomy Department and
Physical Sciences Division, Scarborough Campus
University of Toronto
1265 Military Trail, Scarborough, Ontario, Canada M1C 1A4

Direct N-body simulations show that the heat flux from the cusp of stars surrounding an object of mass M $\sim$ 10-100 $M_\odot$ has a profound effect on the evolution of a cluster of N stars of mass m $\sim$ 1 $M_\odot$. A cluster containing an object which is $\gtrsim$5% of its total mass expands, transferring most of the cluster binding energy to a few stars deep in the cusp within several central relaxation times. The results, together with analytic estimates, suggest that core collapse in a globular cluster will be reversed when the core mass is reduced to $\sim$10 M. As is discussed below, this result does not depend on whether or not stars are disrupted by the black hole.

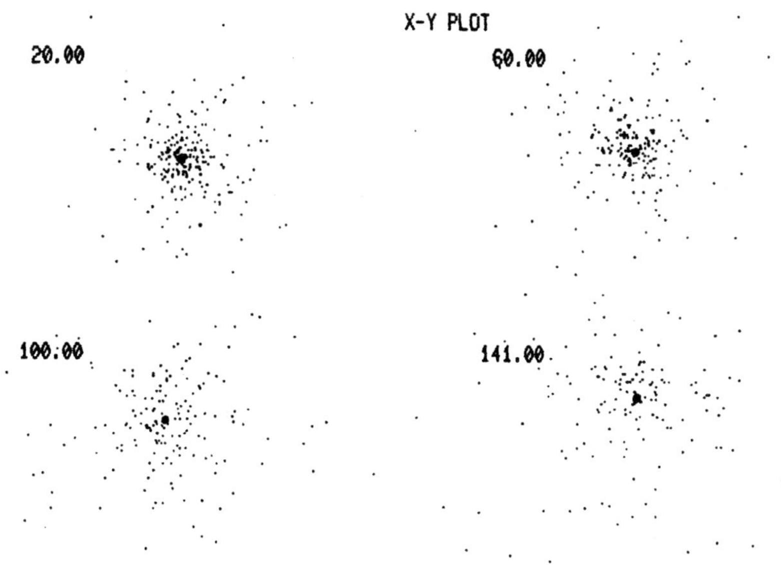

Figure 1. The evolution of a cluster of 240 stars with central object of mass = 10. Initial core mass = 60, disruption radius = 0, unit of time = core crossing time, core relaxation time $\sim$10 crossing times.

J. Goodman and P. Hut (eds.), Dynamics of Star Clusters, 415–417.

In these proceedings, Stu Shapiro has summarized Monte Carlo simulations of the dynamical effect of a very massive black hole ($\sim 10^2$-10^4 $M_\odot$) in a star cluster. In order to check the validity of the approximations used in these simulations (e.g. orbit-averaging in E-J space, small angle scattering) and to extend them to smaller M, a comprehensive series of direct N-body simulations has been completed for systems of N $\sim$200-500 stars containing a black hole of mass M $\sim$10-500. Stars are assumed to be consumed by the hole if they come within a distance r_D (so that setting $r_D = 0$ eliminates consumption). The simulations are described in detail elsewhere (Duncan 1984 a,b).

The results show that a cusp of bound stars with density profile $n(r) \propto r^{-7/4}$ develops within $\sim$ one central relaxation time, t_r, as suggested by the Fokker-Planck calculations of Bahcall and Wolf (1976). The consumption rate of stars is in rough agreement with Monte Carlo calculations (Marchant and Shapiro 1980, Duncan and Shapiro 1982). However, for cases in which $M \gtrsim mN$, the stars beyond the half-mass radius tend to have highly eccentric orbits with density profile $\propto r^{-3.5}$ (cf. Spitzer and Shapiro 1972). In all cases studies ($.05 < M/mN \lesssim 1$) the bulk of the cluster expanded on a timescale $\sim t_r$, leaving a few tightly bound stars (especially when $r_D = 0$).

When the number of cusp stars is large, Bahcall and Wolf (1976) have shown that the $r^{-7/4}$ solution implies a net outward conduction of energy through the cusp, with a "zero-flow" diffusion of stars inward on a timescale $\propto r^{-1}$. The analysis of Duncan (1984b) suggests that for cusps containing $\lesssim$10 stars, the bound stars may be approximated as black hole-star "binaries" which interact with each other and the unbound population. Stars in orbits with semi-major axes $a > r_a/3 \equiv M/v^2$ (where v is the rms velocity dispersion in the core) are soft binaries. However, those with $a < r_a/3$ are hard binaries which are formed either by the cascade inward of a fraction of the soft binaries or by three-body encounters involving the black hole and two unbound stars. A lower limit to the heating per hard binary is the heating rate due to encounters with unbound stars, $0.5\, mv^2/(t_r \ln\Lambda)$, which is a "zero-flow" result independent of r and M. Neglecting consumption, the useful lifetime of a hard binary is $\sim 10\, t_r \ln\Lambda$ after which point it ejects unbound stars from the core and the heating efficiency is reduced. Hard binaries form from unbound-unbound reactions at the rate $\sim 3(M/mN_c)^3 N_c/(t_r \ln\Lambda)$. In steady-state, then, the lower limit to the heat input is

$$\dot{E} = (30/\ln\Lambda)(M/mN_c)^3(0.5N_c mv^2)/t_r$$

The rate at which the core energy decreases during core collapse is $\sim 0.01\, N_c mv^2/t_r$, which suggests that $M \sim 0.1\, mN_c$ is sufficient to reverse core collapse. This simple analysis of course neglects bound-bound interactions which may lead to ejection or consumption of one of the participants, but the actual number of hard binaries is of order unity. As a result changes in stellar orbits will be highly stochastic and the "loss-cone" theory of the rate of consumption can at best be valid in a statistical sense and at worst may be unreliable. If the latter is true, it is not obvious that the black hole will increase in mass at the dramatic rate indicated by Monte Carlo calculations. It is clear that further N-body and/or Monte Carlo simulations (or a hybrid thereof) will be required to study the re-expansion phase when a moderately massive black hole is present, especially when there is a spectrum of stellar masses.

REFERENCES

Bahcall, J.N., and Wolf, R.A. 1976, Ap. J., 209, 214.
Duncan, M.J. 1984a. 'Dynamical Evolution of Star Clusters Containing a Massive, Central Black Hole. I. A Dominant Black Hole', preprint.
__________ 1984b. 'Dynamical Evolution of Star Clusters Containing a Massive, Central Black Hole. II. A Dominant Black Hole', preprint.
Duncan, M.J., and Shapiro, S.L. 1982, Ap. J., 253, 921.
Lightman, A.P., and Shapiro, S.L. 1977, Ap. J., 211, 244.
Marchant, A.B., and Shapiro, S.L. 1980, Ap. J., 239, 685.
Spitzer, L., and Shapiro, S.L. 1972, Ap. J., 173, 529.

TWO BODY CAPTURE IN LARGE N BODY SYSTEMS

M. Giersz
N. Copernicus Astronomical Center,
Polish Academy of Sciences
ul. Bartycka 18, 00-716 Warszawa, Poland

I have considered the rates of formation of binaries in the following processes: two-body tidal encounters, three-body encounters, self-hardening of soft binaries formed in three-body encounters and two-body encounters connected with gravitational wave radiation. I have considered also the cooling and the heating connected with formation of binaries and the subsequent interactions of binaries with single stars. I have found that the rate of formation of heating binaries in two-body tidal encounters differs from that calculated by Ozernoy and Dokuchaev (1982). These differences are connected first of all with the precise calculation of the dissipated energy of two stars passing near-by each other, maximum and minimum binary binding energy and also with the use of the incomplete gamma function instead of gamma function. The processes presently predominat in the globular cluster conditions are the formation of hard binaries in two-body tidal encounters between evolved and unevolved stars and cooling connected with collisions between white dwarfs and main sequence stars and between white dwarfs. In the systems of density greater than $10^9 pc^{-3}$ the binaries formed in three-body encounters (and to a smaller extent self-hardening of soft binaries formed in three-body encounters) predominate the globular cluster evolution. In the globular cluster heating by hard binaries exceedes cooling associated with the tidal interactions and collisions of stars.

In the nuclei of galaxies hard binaries (formed in two-body and three-body encounters) must be contact binaries. So these systems will be probably dominated by dissipative processes. The critical velocity at which cooling and heating are balanced, for stellar density less than $10^9 pc^{-3}$, slightly depends on the density of stars and is equal to about 20 km/s. If the density increases then the critical velocity grows to about 45 km/s at the density of 10^{14} pc^{-3} . The rate of formation of binaries connected

J. Goodman and P. Hut (eds.), Dynamics of Star Clusters, 419–420.

with gravitational wave radiation is negligibly small even in stellar systems of very great density (of the order 10^{12} pc^{-3}).

The ejection of matter during interactions between stars is important for the evolution of stellar systems. The rough calculations suggest that if 12 percent of mass of colliding stars is ejected from the stellar system, then the cooling associated with collisions between stars is balanced by the heating connected with the ejection of the matter.

I have found also the new analitical formulae for the formation rate of the heating tidal-capture binaries and for effectiveness of cooling and heating processes. The obtained results can be used in the investigation of the evolution of globular clusters and nuclei of galaxies.

REFERENCE

Ozernoy, L.M., Dokuchaev, V.I.: 1982, Astron.Astrophys.111,1

Submitted to Acta Astronomica

BLACK-HOLE REMNANTS IN GLOBULAR CLUSTERS

Richard B. Larson
Yale University Observatory
Box 6666
New Haven, CT 06511
U.S.A.

Many presentations at this meeting have discussed the phenomenon of "core collapse", and it seems agreed that a stellar system like a globular cluster will, within a few half-mass relaxation times, undergo a runaway increase in central density and achieve a nearly singular density distribution with a logarithmic gradient slightly steeper than -2. The collapse is halted by the formation of binaries when the core has shrunk to contain only a small number of stars, and the system subsequently expands gradually while maintaining a density profile approximating that of a singular isothermal sphere. Light profiles resembling the predicted nearly singular form have been found in a small number of globular clusters (Djorgovski, this meeting), but a puzzle remains in that many more clusters should have undergone core collapse, yet they show quite flat light profiles in their cores.

A plausible resolution of this discrepancy would be to assume that the cores of globular clusters contain remnants more massive than the visible stars; these remnants could then have the predicted nearly singular density distribution, while the lighter visible stars would have a flatter density profile because of approximate equipartition of energy. In this case the observed core properties of globular clusters would be determined by the properties of the central subcluster of heavy remnants. In order to see whether such a model can successfully explain the core structure of globular clusters, and to estimate what kind of population of heavy remnants would be required, some very simple two-component models of globular cluster cores have been calculated, and the predicted surface density profiles of the lighter objects have been compared with the observed light profiles of a number of well-studied globular clusters.

The heavy objects are assumed to form a collapsed core with a singular density distribution, and both heavy and light objects are assumed to have velocity dispersions that vary with radius according to $v^2 \propto r^{-n}$. Various numerical simulations of pre-collapse evolution have suggested values for the exponent n in the range $\sim 0.2 - 0.5$, while a singular isothermal sphere would have $n = 0$. By adjusting the assumed

J. Goodman and P. Hut (eds.), Dynamics of Star Clusters, 421–422.

mass ratio, it is possible to fit the observed light profiles of typical globular cluster cores with any value of n between 0 and 0.5; therefore it has been attempted to constrain the parameter n better by using the available information on velocity dispersions in globular clusters. A value for n of 0.5 has been found to allow a good fit to all of the available data.

With models of this type, the core light profiles of typical globular clusters are well reproduced if the assumed mass of the heavy remnants is about 3 times the mass of the visible stars, or about 2.5 $M_\odot$. Thus the heavy remnants are probably mostly black holes, although neutron stars and binaries will contribute. The anomalous cusped light profile of M15 can be fitted closely with a heavy-remnant mass that is somewhat smaller, about 1.7 $M_\odot$. The fraction of the cluster mass that is required to be in heavy remnants varies from about 1 to 5 percent, and is typically about 3 percent. There is some evidence that this fraction correlates inversely with the slope of the main-sequence luminosity function, as would be expected if these differences reflect differences in the initial mass function.

It is not implausible that 3 percent of the mass of a typical globular cluster would be in the form of remnants whose representative mass is $\sim 2.5\ M_\odot$. In the absence of any firm knowledge about the final evolution of massive stars it is not possible to say exactly what initial population of massive stars this corresponds to, but if it is assumed that the initial mass function is like that of the solar neighborhood, then a remnant population with the required properties would be produced if stars leave remnants whose mass is either 0.6 $M_\odot$ or 15 percent of the initial stellar mass, whichever is larger. These assumptions would also imply a large mass spread for the remnants, in which case the two most massive ones would almost certainly form a dominant central binary of two black holes.

The best evidence for, or constraints on, a central subcluster of heavy remnants will probably come from accurate measurements of velocity dispersions near the centers of globular clusters. Some of the data presented at this meeting show unexpectedly high velocity dispersions in the cores of 47 Tuc and ω Cen, apparently indicative of a substantial amount of unseen mass. Deviations from a gaussian distribution of velocities may also be an important indicator of core structure. Gunn and Griffin found two stars with anomalously high velocities in the core of M3; such high-velocity stars are in fact predicted by models of the type discussed here, since a few stars whose orbits pass close to the center will gain unusually high velocities by interaction with the heavy remnants. Thus the presence of such stars may be a sensitive test for the presence of a condensed core of heavy remnants.

A more complete account of this work is to be published in Monthly Notices of the Royal Astronomical Society.

TIDAL STRIPPING AND DISRUPTION OF GLOBULAR CLUSTERS

T.S. van Albada and T.R. Bontekoe
Kapteyn Astronomical Institute
Postbus 800 9700 AV Groningen

Tidal stripping and disruption of globular clusters may be responsible for the absence of low density clusters in the inner region of the Galaxy. We have studied these processes by integrating orbits of stars while the cluster is moving through a spherically symmetric galactic potential with constant circular velocity (V_{cir}=220 km/s). The response of the cluster to the tidal field of the galaxy is calculated in a selfconsistent manner with a collisionless N-body code with N=5000 (van Albada 1982, van Albada and Bontekoe in preparation).

The following cluster parameters were used:
Mass: 2×10^5 M_{sun}.
Outer radius: 50 pc.
Density distribution: polytrope n=4 ($\log r_t/r_c \simeq 0.75$).
Crossing time: 2.3×10^6 yr.
This choice represents a cluster of average mass and outer radius with low central concentration.

The cluster was put on an eccentric orbit with apocenter at 5 kpc. At this distance the instantaneous tidal radius of the cluster is much greater than the initial outer radius. As the cluster approaches the galactic center the tidal surface shrinks and for pericenter distances $\lesssim$ 2 kpc it lies well inside the cluster.

During the passage of the center two phenomena occur. First, stars outside the tidal surface are removed from the cluster. (More precisely, stars with sufficient nergy to reach the tidal surface will ultimately escape.) Second, depending on the speed with which the cluster passes the galactic center, energy is added to the motion of the stars in the cluster leading to a general decrease in binding energy. The resulting expansion of the envelope of the cluster is largest for fast passages. In first approximation the shape of the distribution of binding energies, N(E), does not change, but shifts by an amount ΔE towards more positive values. The core radius of the cluster is not immediately affected since ΔE is generally small with respect to the central potential of the cluster. In first instance the net effect of a tidal shock is therefore an

J. Goodman and P. Hut (eds.), Dynamics of Star Clusters, 423–425.

increase in central concentration of the cluster. Thus, tidal effects do in general not lead to a truncation of the system.

Two examples are shown in Figure 1 (orbits) and Figure 2 (time evolution of fractional mass shells). These results show that clusters with small central concentration can easily be disrupted totally within a Hubble time if their pericenter distances are small.

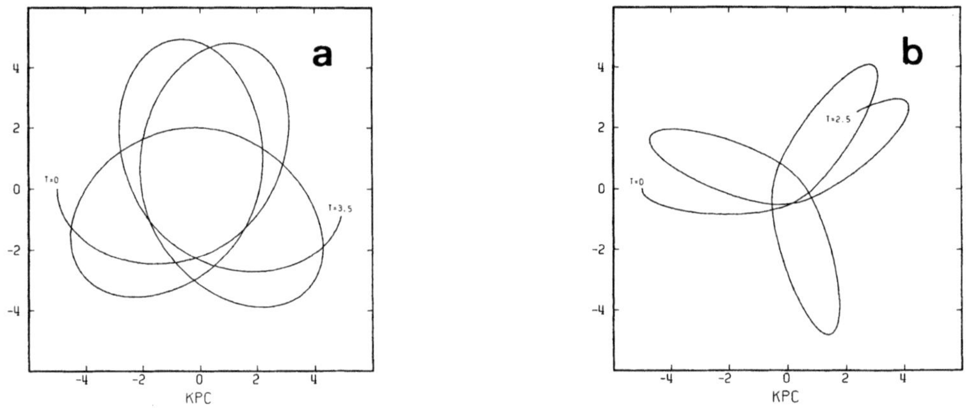

Figure 1. Orbits of two clusters in a galactic potential with constant circular velocity of 220 km/s. Apocenter in both cases is at 5 kpc. Pericenter lies respectively at 2 kpc (a) and 0.5 kpc (b).

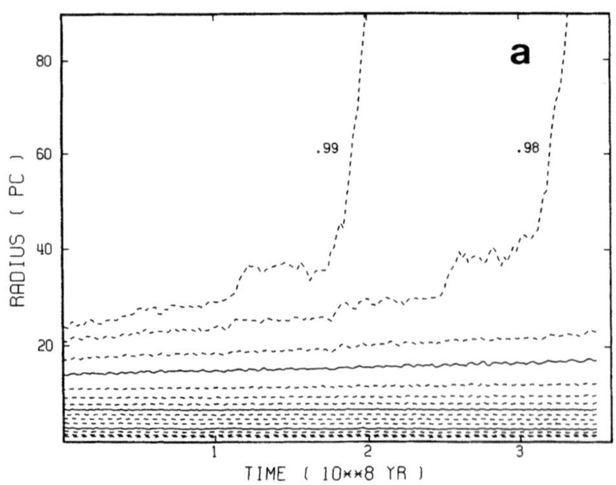

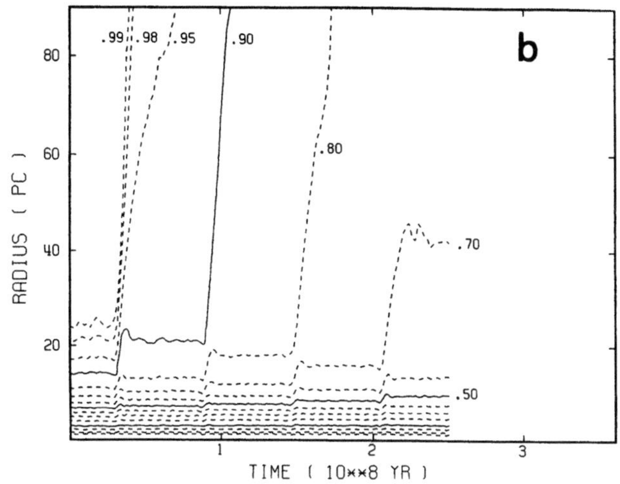

Figure 2. Time evolution of radii of shells containing given fraction of total mass for clusters whose orbits are shown in Figure 1. Note sudden expansion at pericenter passage in Figure 2b. In both cases the structure of the inner region is hardly affected.

Detailed calculations of stripping and disruption of clusters, which include tidal effects by the disk component and internal evolution, are necessary to determine the evolution of properties of globular cluster systems. The X-ray sources in the bulge of our Galaxy may be associated with remnants of disrupted clusters (Watson et al. 1981, Grindlay 1984). It is also conceivable that the spheroidal components of galaxies consist entirely of disrupted clusters (Fall and Rees 1977).

REFERENCES

Fall, S.M. and Rees, M.J., 1977, M.N.R.A.S., 181, 37P.
Grindlay, J.E., 1984, preprint.
van Albada, T.S., 1982, M.N.R.A.S., 201, 939.
Watson, M.G., Willingale, R., Grindlay, J.E., and Hertz, P., 1981, Ap.J., 250, 142.

THE STRUCTURE AND INTERNAL KINEMATICS OF OPEN CLUSTERS

Robert D. Mathieu
Center for Astrophysics
Cambridge, Massachusetts

I. INTRODUCTION

Open clusters provide an opportunity for observational study of several important facets of stellar dynamics which cannot easily be addressed in globular clusters for either fundamental or practical reasons. In particular, open clusters are case studies for:

1) The dynamics of small-N stellar systems. This aspect of open clusters has been excellently reviewed by King (1980), to which the reader is referred. The work of Retterer (1979) and Goodman (this volume) suggests that any differences in the velocity distributions of small-N systems as compared to large-N systems due to the increased importance of close stellar encounters will appear primarily in the high-energy tails of the distributions. This will be reflected in the structure of the cluster halos, where quality data are difficult to obtain. In order to guide observational study, more theoretical work is needed regarding the expected nature and magnitude of any observable deviations from large-N theory.

2) The effects of a large stellar-mass spectrum. While globular clusters probably also have a spectrum of masses spanning an order of magnitude or more, the extension of the mass spectrum to higher masses in open clusters permits the dynamical consequences of a mass spectrum to be more easily studied. In addition, many studies have pointed out the significance of stellar mass loss in massive stars for the evolution of clusters (e.g., Terlevich 1983). However, as yet the theory regarding the effects of such mass loss has not provided unique observational predictions which might act as critical tests, so this important issue will not be discussed further here.

3) A spectrum of cluster ages. Open clusters span the complete spectrum of age from birth to dissolution. Except for the classic work of Wielen (1971 and this volume) in which he analyses the observed distribution of cluster ages, this feature of open clusters has been little exploited. Yet by mapping appropriate diagnostics with cluster

J. Goodman and P. Hut (eds.), Dynamics of Star Clusters, 427–448.

age open clusters have the potential to provide empirical measures of the fundamental dynamical time-scales, e.g. the relaxation time, equipartition time and mixing time. Also, study of the very youngest clusters provides appropriate initial conditions for evolutionary models as well as insight into star- and cluster-formation processes.

4) The effect of the external environment. Open cluster halos are molded by the galactic tidal field. In addition, open clusters are subject to impulsive encounters with clouds in the interstellar medium. The effects of these environmental influences can be examined in numerous ways, including detailed analyses of the structure of cluster halos and observations of the variation with radius of anisotropy in the cluster velocity distributions.

Finally, open clusters have the advantage that many lie within 1 kpc of the sun. Thus, in principle, for any given magnitude limit stars of lower mass can be studied in open clusters than in globular clusters. In practice membership difficulties in open clusters reduce this advantage. The proximity of open clusters also allows excellent spatial resolution; every star in an open cluster can be precisely placed in two dimensions. There is no need to assume a particular mass function in order to study integrated light distributions. In addition, in the nearest open clusters the dynamically active binaries are resolvable, providing the only opportunity to test the extensive theory concerning their role as an energy source for clusters.

Unfortunately, while open clusters provide exciting opportunities for stellar dynamical research, they are, from a practical point of view, difficult systems to study. The inherent low-number statistics are often a severe limitation. The low total masses imply small internal motions (usually less than 1 km/sec in one dimension) which are a technical challenge to measure. And, perhaps most frustrating, without a great deal of work one is not even certain which stars are cluster members due to the extensive background contamination from the galactic disk. Fortunately, because of the narrowly confined distribution of open cluster members in velocity space they can be isolated through precise velocity measurements (usually via proper motions, but the radial-velocity techniques of Gieseking (1981) have also been effective). Such membership studies are extensive projects, although technical advances are rapidly lowering this threshold. In addition, proper-motion studies are significantly limited by the areal coverage and magnitude limits of first epoch plates. Nonetheless, essentially every study of open cluster dynamics in the last decade has begun with a comprehensive membership study, and the fruitful results should provide the motivation for more extensive membership studies in the near future as more recent generation plates become ripe for measurement.

2. THE STRUCTURE OF OPEN CLUSTERS

Historically the study of cluster structure has preceded the study of internal motions, and so we begin the review here. Even casual visual inspection of a deep photograph of a young cluster reveals that the more luminous (and thus more massive) stars are more centrally concentrated than the fainter cluster members. It has long been realized that a reasonable explanation for this mass segregation lies in the equipartition of energy among the cluster members, where the more massive stars give up energy to the lighter stars and fall toward the center of the potential well. One primary goal of recent work on open cluster structure has been to study more precisely to what extent the mass segregation can in fact be ascribed to energy equipartition and to determine whether the equipartition is complete among all stellar species.

One of the first clusters analysed from this perspective was the Hyades cluster (Pels, Oort and Pels-Kluyver 1975). In Fig. 1a are shown the stellar surface-density distributions (filled circles) for two stellar mass groups with average masses of 1.36 $M_{\odot}$ (Group 1) and 0.61 $M_{\odot}$ (Group 2). The mass segregation is evident in the higher central concentration of the Group 1 stars. Pels _et al._ point out that there is also mass segregation within the Group 1 stars, as noted earlier by van Bueren (1952) who showed that the A stars and red giants are all located very near the cluster center. This is also evident in the cumulative radial distributions (Fig. 1a) where the stars have been more finely divided in stellar mass. It is interesting that the lowest two mass groups are not markedly different in their spatial distribution, despite having the same ratio in average mass as the two intermediate mass groups. This will be further discussed below.

Pels _et al._ fit to the data dynamical models based on a truncated Maxwellian velocity distribution; the best-fit models are indicated by the open symbols. The fit of the models to the data is reasonable. More significantly, the best-fit models have velocity distributions with moduli of precision related as the inverse square root of the average stellar mass of each group, as would be expected for a cluster in equipartition. [1]

Quite a few proper-motion membership studies of open clusters have been done in the last 15 years. One of the best such studies examined the rich intermediate-age (2×10^8 yrs) cluster M11 (McNamara, Pratt and Sanders 1977). The structure of this cluster has been studied in detail

----

[1] Over the last decade Upgren and Weis have obtained precise photometry for roughly 800 stars with V < 16 in the Hyades field, including essentially every star on anyone's list of proper-motion candidates for Hyades membership. This work is now completed and will hopefully in the near future permit more extensive studies of the Hyades structure and luminosity function.

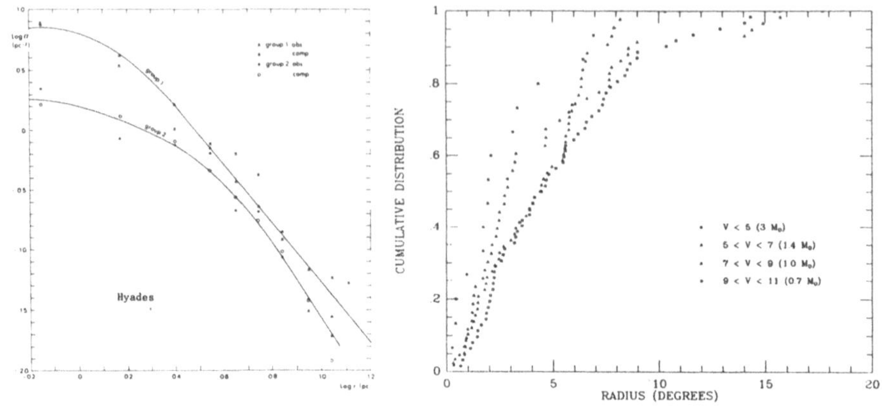

a)

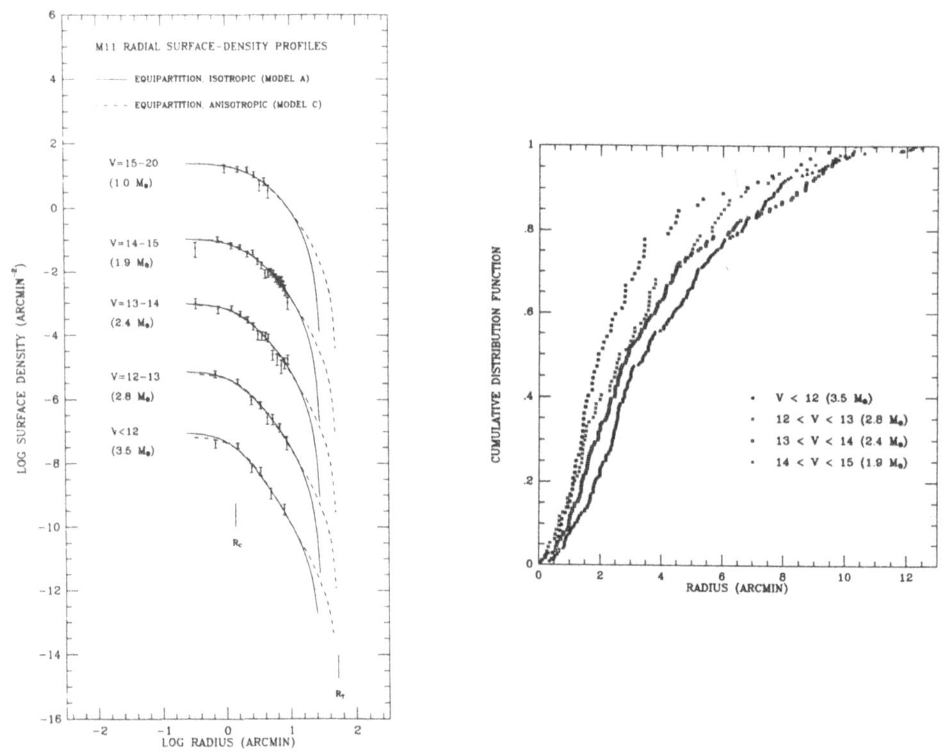

b)

Fig. 1. Surface-density and cumulative radial distributions as a function of stellar mass. The assorted dynamical models shown are described in the text. a) Surface density taken from Pels et al. 1975, cumulative distribution from data in Oort 1979, b) taken from Mathieu 1984, c) and d) taken from Mathieu 1983b and e) taken from van Leeuwen 1983.

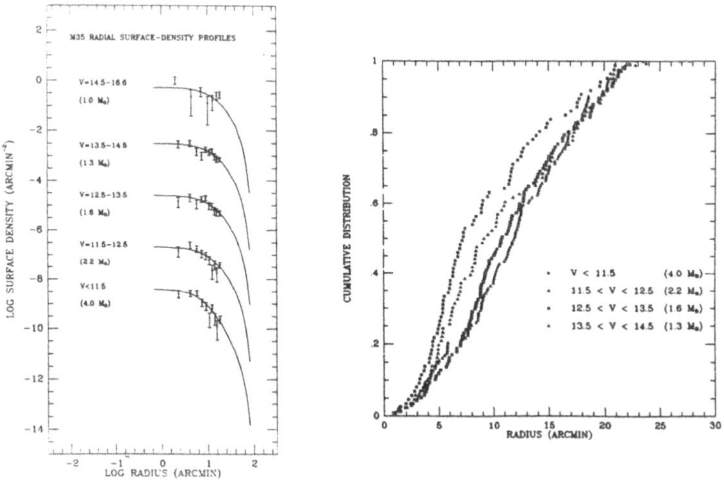
M35 RADIAL SURFACE-DENSITY PROFILES
V=14.5-16.6
(1.0 M☉)
V=13.5-14.5
(1.3 M☉)
V=12.5-13.5
(1.6 M☉)
V=11.5-12.5
(2.2 M☉)
V<11.5
(4.0 M☉)
LOG SURFACE DENSITY (ARCMIN^{-2})
LOG RADIUS (ARCMIN)
CUMULATIVE DISTRIBUTION
RADIUS (ARCMIN)
V < 11.5 (4.0 M☉)
11.5 < V < 12.5 (2.2 M☉)
12.5 < V < 13.5 (1.6 M☉)
13.5 < V < 14.5 (1.3 M☉)

c)

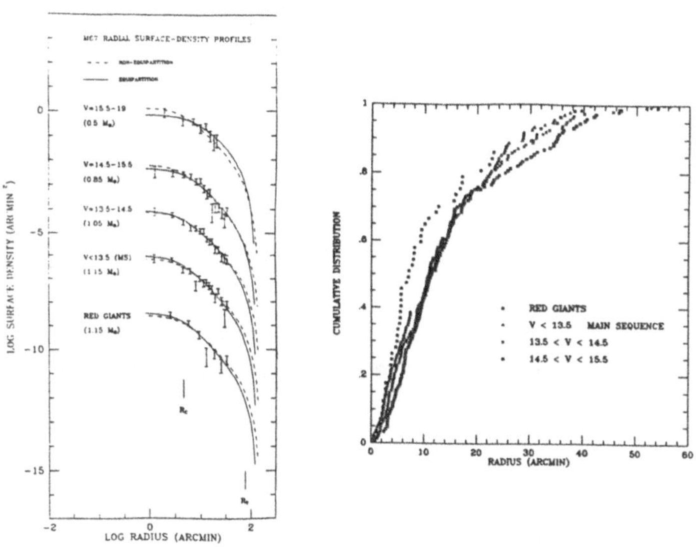
M67 RADIAL SURFACE-DENSITY PROFILES
V=15.5-19
(0.5 M☉)
V=14.5-15.5
(0.85 M☉)
V=13.5-14.5
(1.05 M☉)
V<13.5 (MS)
(1.15 M☉)
RED GIANTS
(1.15 M☉)
LOG SURFACE DENSITY (ARCMIN^{-2})
LOG RADIUS (ARCMIN)
CUMULATIVE DISTRIBUTION
RADIUS (ARCMIN)
RED GIANTS
V < 13.5 MAIN SEQUENCE
13.5 < V < 14.5
14.5 < V < 15.5

d)

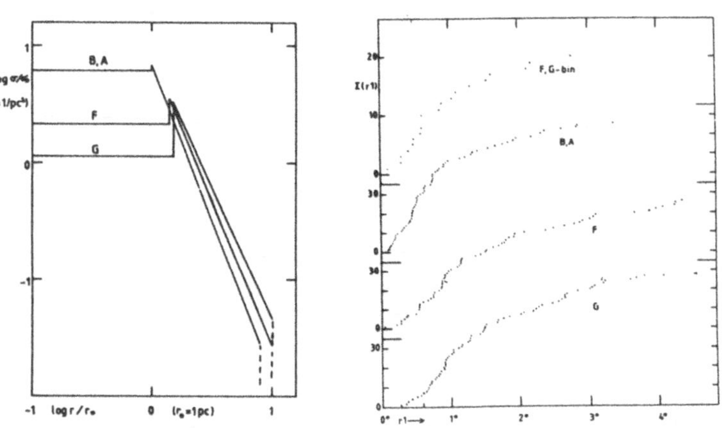
B, A
F
G
log r / r₀
F, G-bin
B,A
F
G

e)

by Mathieu (1984) who also included star counts done on deep 4m plates in order to study the distribution of stars down to 0.7 $M_\odot$. The stellar surface-density distributions and the cumulative radial distributions (the latter for the proper-motion members only) are given as a function of stellar mass in Fig. 1b. Again mass segregation is evident; in fact the degree of central concentration is rather sensitive to stellar mass. The central concentration decreases continuously with lower stellar mass down to the lower mass limit of the data, although the star-count data cannot resolve the relative distributions of stars with masses between 0.7 and 1.9 $M_\odot$. In order to study in detail the cluster dynamics, multi-mass isotropic King models were fit to the data with varying assumptions concerning the dependence of the velocity distributions on stellar mass. Models without any energy equipartition were first considered (i.e. the velocity distributions were independent of stellar mass); no acceptable fits to the data were found. Next, models in complete equipartition were fit; such models provided excellent fits, as shown in Fig. 1b. Finally, models were fit in which only the two most massive stellar components were in equipartition while the lower three components were forced to have the same velocity dispersion. (The motivation here comes from Monte Carlo simulations of three-component models with mass ratios 6.25:2.5:1 (Spitzer and Shull 1975). In these models the most massive component quickly concentrated in the core, ejecting rapidly lower mass stars into the outer regions of the cluster where relaxation times are relatively long. Thus the spatial distribution of both low-mass components were very similar despite their large mass ratio.) These models were unable to fit the data well, although the best fit case could only be rejected at the 90% confidence level. The M11 data were thus able to distinguish between possible models and strongly suggest that, at least for stars with masses greater than roughly 1 $M_\odot$ within 7 core radii, the cluster is in a state of complete energy equipartition.

In Figs. 1c and 1d are shown the stellar surface-density and cumulative radial distributions of M35 and M67 (taken from Mathieu (1983b), using the proper-motion membership studies of Cudworth (1971; M35) and Sanders (1977; M67) as well as star counts done by the author in M35 and taken from van den Bergh and Sher (1960) in M67). The young cluster M35 (2×10^7 yr) appears structurally very similar to M11, with mass segregation again being present among all stars with masses greater than 1 $M_\odot$. Interestingly, the oldest cluster studied, M67 (5×10^9 yr), has the least evident mass segregation. (The apparent central concentration of the red giants is only significant at the 85% confidence level.) While the stellar mass range of the M67 proper-motion study is small, it still spans a ratio in stellar mass larger than that of the two most massive components in M11, where the segregation is readily apparent. Formally the data cannot distinguish between equipartition and non-equipartition models. However, the lowest mass stellar component (derived from star counts) does appear less centrally concentrated and models in complete equipartition provide somewhat better fits to the data than models without equipartition. (See further discussion of this point after the question of Bahcall.)

Finally, the stellar surface-density and cumulative radial distributions for the Pleiades, taken from van Leeuwen (1983), are shown in Fig. 1e. Interestingly, while the B-A stars are notably centrally concentrated, there is no statistically significant distinction between the spatial distributions of the F (1.2 $M_\odot$) and G (0.8 $M_\odot$) stars. This lack of evident mass segregation among the lower-mass stars is similar to the spatial distribution found in the Hyades and may be a general phenomena.

It is not obvious what is the expected degree of mass segregation among the lowest mass stars in a cluster. If a cluster is subject to the galactic tidal field, then in equipartition the velocity distributions of the lowest mass stars will be the most severely truncated. This will result in a lesser degree of mass segregation among these stars, but the magnitude of the effect is dependent on the severity of the tidal truncation. As already noted, Spitzer and Shull (1975) also found a lesser degree of mass segregation among lower mass stars in their simulations of isolated clusters. Finally, Wielen (private communication) finds in his N-body simulations that mass segregation becomes much less evident below the average stellar mass; however, Terlevich (1983) finds in her models that mass segregation remains among even stars of the lowest stellar masses. In all cases the precise nature of the mass function is important in that the degree of mass segregation is dependent on which stars play the dominant roles in relaxation processes and in determining the cluster potential.

Clearly we need to turn to the clusters themselves to find the answers. Unfortunately we immediately encounter one of the primary deficiencies in the data available for open clusters. In most cases the magnitude limits of open cluster membership studies exclude low-mass stars, in particular those with masses comparable to or less than the average stellar mass in the cluster. In order to be able to determine the extent of mass segregation among the lowest mass stars and thus obtain some insight into the physical processes determining their velocity distributions, we are badly in need of membership studies that extend to 0.5 $M_\odot$ (M_V = 9) or lower.

Returning again to the Pleiades (Fig. 1e), van Leeuwen has presented the cumulative radial distribution of binaries (spectral types F-G) in the Pleiades. (These binaries were selected photometrically and are thus biased toward systems of comparable mass components.) The binaries are centrally concentrated relative to "single" stars of similar spectral type. Similar results have been found in the Hyades (Mathieu, Stefanik and Latham 1985) and in several other clusters (Abt 1980), although only in the Pleiades are the relative central concentrations so distinct. Mathieu (1983b) has also found a pronounced central concentration of the blue stragglers in M67, which coincides with the central concentration of the spectroscopic binaries in the cluster (Mathieu and Latham, 1985a). All of these results point to the expected conclusion that binaries, being more massive, tend to be more centrally concentrated.

Leaving the subject of energy equipartition we next address the structure of cluster halos and processes which mold the high energy tails of the velocity distributions. Here we are hampered by the lack of proper-motion membership studies far from cluster centers, due primarily to limitations in coverage of the early epoch plates. The Hyades are in fact the only cluster which has been studied beyond the theoretical tidal radius (Pels _et al._ 1975). The findings in this one case are indeed interesting; consider again Fig. 1a. The theoretical tidal limit is 9.5 pc. Yet there are 9 stars beyond this limit ($10 < R < 14$ pc) which appear to be members; also, the outermost region was incompletely studied so several additional stars would be expected. Pels _et al._ suggest that these stars are in the process of escaping the cluster; in particular they suggest that these stars have escape energy but have not yet found the hole in the equipotential surfaces. (See also the paper by Seitzer in this volume.) On the other hand, it remains possible that these stars are members of the Hyades moving group rather that the cluster itself. More studies beyond cluster tidal radii are needed both to examine the process of stellar escape and to determine the frequency and nature of moving groups.

Both theory and numerical simulations suggest that in the presence of the galactic tidal field the outer regions of clusters should be flattened with axial ratios of 2:1.4:1 in the directions toward the galactic center, in the galactic plane perpendicular to the galactic center and perpendicular to the galactic disk, respectively (e.g., Wielen 1975). Interestingly, after a reanalysis of the Hyades in which photometric information was included in the membership selection, Oort (1979) found a significant flattening of the cluster outside a 4 pc radius in the expected orientation. However, the degree of flattening was significantly greater than predicted by theory. Marginal evidence for flattening along the galactic plane has also been found in the Pleiades (van Leeuwen 1983). NGC 3532 shows strong flattening (Gieseking 1981), but roughly orthogonal to the galactic plane. No significant flattening has been found in the clusters studied by Mathieu (1983b); however only in the case of M67 do the data extend beyond 5 pc radius. In summary, there may be evidence for the flattening of some clusters along the galactic plane; whether the degree of flattening is consistent with the tidal theory is not clear and better data than available at present will be required before a definitive comparison of the tidal theory and cluster halo structure is possible.

It is clear from these few studies of cluster halos that 1) the findings have been divergent from simple theoretical expectations and thus the subject merits further study and 2) the data available are entirely insufficient. As already mentioned, proper-motion studies are presently severely limited by the coverage of first epoch plates. This limitation can best be solved by the construction of an appropriate plate collection and the passage of time. Van Leeuwen is presently creating such a collection which will be of value for membership studies in 10-15 years. Until then, those working with proper motions should also examine existing collections for fortuitous coverage of regions around clusters. The radial-velocity membership techniques of Gieseking (1981) may provide an alternative, short time-scale method. In any case, the subject of cluster halos remains poorly understood, both observationally and theoretically, and we should direct our efforts toward a marked improvement in membership studies at a tidal radius and beyond.

3. THE INTERNAL KINEMATICS OF OPEN CLUSTERS

The dynamical models fit to the stellar spatial distributions in open clusters make very definite predictions concerning the internal kinematics of these clusters. Technical advances now permit the measurement of the internal motions of open clusters and within the next half-decade a wealth of data should become available to permit a detailed confrontation of observation and theory. Thus the review given here is, in many ways, only a preliminary report.

3.1 Radial Velocities

Radial-velocity measurements with single-measurement precisions of 0.5 km/sec for stars with $V < 15$ have become possible at several observatories around the world. Mathieu and Latham (1985b; see also Mathieu 1983b) have recently completed a kinematic study of the clusters M11 and M67. In M11 30 giants were observed, essentially a complete sample of the cluster giants. In M67 28 giants/subgiants were observed, selected from the cluster core. In addition, 25 giants in NGC 2506 have been observed, selected from the proper-motion study of Chiu and van Altena (1981). Spectra were obtained with the echelle spectrographs (10 km/sec resolution) behind the FLWO 1.5m and the MMT. The radial velocities were measured with cross-correlation techniques (see Tonry and Davis 1979). Single-measurement precisions were 0.6 km/sec, with 5-10 observations per star obtained over 3 years. (In the following, detected spectroscopic binaries will not be considered; see Mathieu 1983b for a complete discussion.)

The apparent velocity distributions are shown in Fig. 2, both in order to emphasize the degree to which the distributions are resolved and to present an essential difficulty in the use of radial velocities to study open cluster kinematics. The single-measurement precision is comparable to the bin size in Fig. 2. Thus in M11 and NGC 2506 the velocity distributions are well resolved. In M67 the cluster velocity dispersion is comparable to the measurement precision; indeed the narrowness of the M67 distribution is indicative of the resolution achieved for the other clusters.

Attention need be drawn to the highest velocity star in NGC 2506 and the lowest velocity star in M11. The former is a proper-motion probable member, but is 4σ off the cluster mean as determined by the remainder of the stars. The latter appears to be a binary on photometric grounds. However, neither star has yet been detected as a binary spectroscopically. Therein lies what may be the fundamental limitation in the use of radial velocities to study the internal kinematics of open clusters. The internal motions of open clusters are sufficiently small as to be comparable to the orbital velocities of long-period binaries. (For example, the primary of a 100 yr period binary can easily be observed to have a radial velocity due to orbital motion of 0.5-1.0 km/sec relative to the binary center-of-mass radial velocity, which is of course what one wishes to measure in order to study cluster kinematics.) However, the periods of such binaries are sufficiently long that their binary nature is not easily detectable. Since such long-period binaries will tend to overpopulate the tails of the observed cluster velocity distributions, measurements of the velocity dispersion will be inflated.

The biases due to binaries are discussed in detail in Mathieu (1983b), where it is shown that the root mean square is a poor measure of the cluster velocity dispersion in the presence of binaries. Instead, the inter-quartile-range (a statistic less sensitive to the extremes of a distribution) was chosen as a statistic by which to measure the true velocity dispersion. This statistic was calibrated against velocity dispersion with Monte Carlo techniques. This of course necessitates modeling the significant characteristics of a cluster binary population. Fortunately, several detailed studies suggest that cluster binary populations are similar to that in the field (Mathieu,

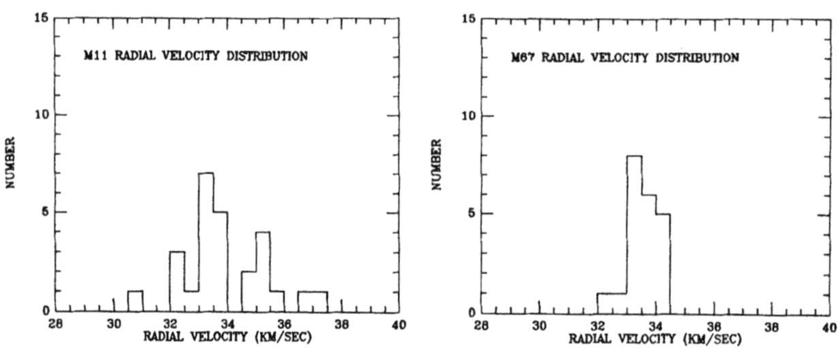

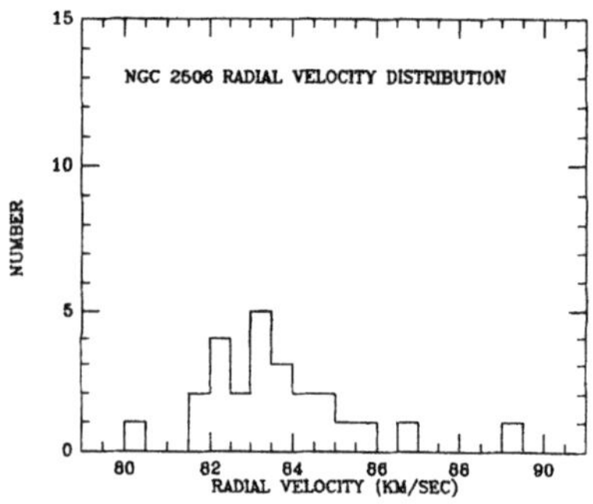

Fig. 2. Observed radial-velocity distributions in three open clusters.

Stefanik and Latham 1985 (Hyades); Brosche and Hoffman 1979 (Pleiades); Bettis 1977 (Pleiades, Praesepe, Hyades)). In addition, Mayor and Mermilliod (1983) have found little variation in the spectroscopic-binary frequency among the Pleiades, Praesepe and Coma clusters. Nonetheless this analysis technique is susceptible to systematic errors if binary populations do vary from cluster to cluster.

The radial-velocity dispersions for M11 and M67 are given in Table I, both with and without the correction for binaries.[2] As expected, correction for the presence of binaries leads to systematically lower values for the velocity dispersion, but with larger errors since valuable information in the tail of the distribution is lost when using the inter-quartile-range.

Table I

Comparison of Theoretical and Observed Radial-Velocity Dispersions

	$\langle V_R^2 \rangle^{1/2}$ (km/sec)		
	Theoretical	Observed	
		with binary correction	w/o binary correction
M11	1.09	1.21 ± 0.35	1.43 ± 0.20
M67	0.49	0.25 ± 0.18	0.48 ± 0.09

Also presented in Table I are the theoretically predicted velocity dispersions as derived from the best-fit dynamical models. Following Abt and Levy (1976) an additional 50% of the observed cluster mass has been assumed to be present in binary secondaries. Also, in the case of M11, an additional component of unseen low-mass cluster members has been included in the models, assuming a cluster mass function like the field initial mass function. This component produces only a 9% increase in the predicted velocity dispersion over that derived from models assuming no unseen mass.

The agreement between theory and observation is reasonably good. Thus the observed cluster structure and internal kinematics do appear properly related by the stellar dynamical theory. Such agreement is in fact not common. In clusters of galaxies, individual galaxies, some globular clusters and the disk of our own galaxy the particle motions

----

[2]The data are presently being reanalysed and these numbers should be considered preliminary.

are notably larger than expected from the observed distribution of matter. The conclusion usually drawn is that there exists unseen "dark" matter in the system. That as much dark matter as observable matter seems to be present in the disk of our galaxy (Bahcall 1984) is particularly relevant here in that open clusters form in the galactic disk. It is thus of interest to ask what constraints the above results place on the presence of dark matter in open clusters. The errors in these observations are such that it is difficult to clearly discriminate the presence of unseen matter at the factor 2 level unless that matter is centrally concentrated, significantly contributing to the potential where the internal motions were measured. In other words, assuming energy equipartition, the observations can only detect dark matter made up of relatively massive particles, greater than a few solar masses in M11 or roughly one solar mass in M67. Assuming such particles, the M67 observations argue against unseen matter being present in an amount comparable to the observed stars; the M11 observations cannot rule out the presence of such a component of massive particles. As a case of particular interest, we note that if M67 is taken to be comprised of 50% (by mass) white dwarfs of 0.6 $M_{\odot}$, than the expected velocity dispersion is 0.6 km/sec, only marginally distinguishable with the data at hand from models without white dwarfs.

Gieseking (1981) has successfully measured the internal motions in NGC 3532, obtaining relative radial velocities with Fehrenbach objective-prism spectra of all stars of spectral types between B3 and F5 in a 4 square degree field. Typical single-measurement errors are on the order of 8 km/sec; with approximately 30 measurements per star Gieseking computes typical errors of 1.5 km/sec. For all cluster members Gieseking measures a velocity dispersion of 1.49$\pm$0.29 km/sec. He also suggests the presence of a cluster halo population with a velocity dispersion 2-3 times larger than that of the cluster stars. Finally, since NGC 3532 is markedly elliptical in shape, Gieseking analysed his data for evidence of rotation, finding none but placing an upper limit on systematic motions of 0.5 km/sec.

Looking beyond this completed work, a great deal of high-precision radial-velocity work in open clusters is in progress at several observatories and will soon reach fruition. McClure and Harris at the Dominion Astrophysical Observatory are completing an extensive study of NGC 188. The similarity of NGC 188 and M67 should provide for a very interesting detailed comparison of the two clusters. Mermilliod, Mayor and collaborators have in progress an extensive radial-velocity study of the giants in over 100 clusters, using the CORAVELs at Haute-Provence and La Silla. Several clusters included in their survey have sufficient numbers of giants to permit detailed kinematic studies. In addition they have observed dwarfs in the Pleiades (100 stars), Praesepe (100), Coma Ber (35) and α Persei (50). When completed these data will provide a superb basis for both dynamical analyses and studies of binary populations in clusters. Finally, at the Center for Astrophysics (Mathieu, Latham, Stefanik and Gordon) detailed studies of several clusters are in progress, as mentioned elsewhere in this review.

3.2) Proper Motions

In many ways proper-motion measurement is an inherently more powerful technique for the study of internal motions in open clusters than radial velocities. Proper motions provide velocity information in two orthogonal directions, allowing direct observation of anisotropy in clusters. They are not limited to late stars, as are most high-precision radial-velocity techniques, and can thus in principle study the cluster velocity distribution over the complete stellar mass range. Finally, except for occasional blended wide binaries, they are not affected by the presence of binaries in the cluster. (Note that this is only true for proper-motions measured over long time baselines. Proper motions determined from high-precision measurements over short time baselines have the same difficulties with binaries as do radial velocities.) The essential difficulty in the use of proper motions is simply precision. Measurement precision of 0.02 arcsec century^{-1} (achievable but not common) correspond to 0.5 km/sec at 500 pc and degrades linearly with distance. Thus, at the moment, for most clusters radial-velocity precisions are superior. However, the limitations of proper motions are purely technical and can thus in principle be overcome.

Jones' studies in the Pleiades (Jones 1970, σ = 0.2 km/sec; σ is the measurement precision) and Praesepe (Jones 1971, σ = 0.3 km/sec) first demonstrated the potential of high-precision proper-motion work. Since then several studies of kinematic precision have been done by McNamara and Sanders (1977, M11, σ = 1.8 km/sec, $\langle V^2\rangle^{1/2}$ = 1.7$\pm$0.4 km/sec; 1978, M67, σ = 2.9 km/sec, $\langle V^2\rangle^{1/2}$ < 1.5 km/sec; 1983, NGC 6494, σ = 0.8 km/sec, $\langle V^2\rangle^{1/2}$ = 0.8$\pm$0.1 km/sec). ($\langle V^2\rangle^{1/2}$ is the one-dimensional velocity dispersion.) In all of these studies the dynamical masses determined from the virial theorem were in reasonable agreement with the observed cluster mass in the form of stars, although it should be noted that the use of the virial theorem in itself introduces large uncertainties in relating cluster mass and observed velocity dispersion. Recently, Vasilevskis et al. (1979) published a proper-motion study of the inner region of the Pleiades with a measurement precision of 0.1 km/sec. These data have been analysed in detail by van Leeuwen (1980, 1983), who is presently expanding the proper-motion sample to fainter stars and regions more distant from the core. Van Leeuwen (1983) finds a significantly larger dynamical cluster mass than did Jones (1970), but argues that this mass is consistent with the stellar mass content of the Pleiades if one assumes a mass function similar to that in the field.

Certainly the most interesting finding of this proper-motion work has been the detection of anisotropy in both the Pleiades and Praesepe. In Fig. 3a we plot tangential vs. radial velocity dispersions in the Pleiades as a function of cluster radius, showing both the results of Jones (1970) and van Leeuwen (1983). (The difference in the absolute measures of the cluster velocity dispersion in these two studies is not

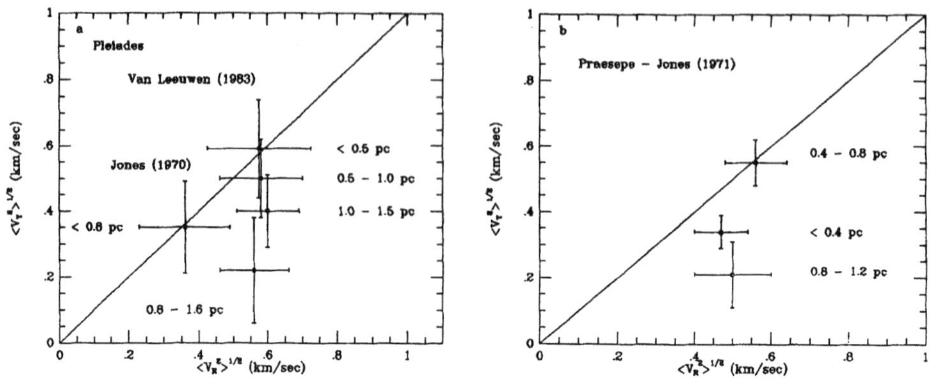

Fig. 3.a. Radial vs. tangential velocity dispersions in the Pleiades as a function of radius.
b. The same for Praesepe.

indicative of significant disagreements in the proper motions, but rather differences in models of the cluster structure used to remove the relative secular parallax [van Leeuwen, private communication]). Both studies show the velocity distribution in the central region to be isotropic. However, beyond a radius of roughly 0.5 pc (on the order of the core radius) the velocity distribution becomes anisotropic; while the radial velocity dispersion is isothermal with radius, the tangential velocity dispersion decreases. Van Leeuwen (1980) suggests that the tangential velocity dispersion decreases as 1/R beyond 0.5 pc. Thus the anisotropy in the Pleiades is present much closer to the cluster center than has been found for globular clusters (see review of Lupton in this volume and Cudworth (1976, 1979)). The results of Jones (1971) for Praesepe are shown in Fig. 3b. The degree of anisotropy in both clusters is similar.

Terlevich and Van Leeuwen are presently making detailed comparisons of her 1000-body simulations (Terlevich 1983) and the Pleiades data. One particularly interesting finding is that the rate of decrease of the tangential velocity dispersion with radius in the Pleiades appears to be less than that found in simulations of isolated clusters, but very similar to that found in simulations of clusters located in the galactic tidal field. They suggest that this result is indicative of tidal heating by the galactic field in both the simulations and the Pleiades. Such results argue eloquently for the need for additional accurate proper-motion studies outside of the cluster cores.

One of the most fundamental observable quantities for stellar dynamics is the behavior of the velocity distribution as a function of stellar mass. Unfortunately, very little data exist and, remarkably,

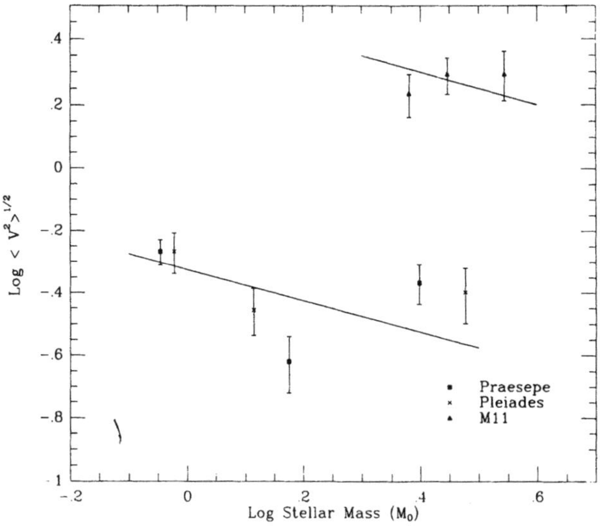

Fig. 4. The dependence of observed velocity dispersion on stellar mass. The solid line represents $\langle V^2 \rangle^{1/2} \propto M_*^{1/2}$.

the state of our knowledge has been little changed for well over a decade. The high-precision data available are, again, those of Jones' studies of the Pleiades and Praesepe and the work of McNamara and Sanders in M11, although in the latter case the data do not span a large range in stellar mass. The one-dimensional velocity dispersions as a function of stellar mass are shown in Fig. 4 for each cluster. No clear conclusions can be drawn from the data; neither the Pleiades data nor the M11 data can clearly distinguish between the $\langle V^2 \rangle^{1/2} \propto M_*^{1/2}$ relation (indicated by the solid lines in Fig. 4) or complete independence of the velocity distribution with stellar mass. The Praesepe data show an odd dependence of velocity dispersion on stellar mass, which is somewhat mimicked in the Pleiades data (although the latter are consistent with a less unusual dependence).

Hopefully the state of the observations will improve markedly in the next few years. Van Leeuwen is presently in the process of analysing his Pleiades data as a function of stellar mass, covering essentially the same stellar mass range as Jones' work. In addition, a program has been begun at the Center for Astrophysics to obtain high-precision radial velocities of Pleiades members with $10.5 < V < 16.5$ (roughly 1.0 $M_\odot$ to 0.4 $M_\odot$). The combined data set will span an order of magnitude in stellar mass and should provide a clearer picture of the dependence of the stellar velocity distribution on mass in the Pleiades. Finally, McNamara is in the process of obtaining precise proper motions in M35 which should be of kinematic accuracy and will span a factor 4 in stellar mass. The addition of another well-studied cluster will hopefully shed light on the generality of the Praesepe-like distribution.

4. BINARIES IN OPEN CLUSTERS

The final subject of this review has little to do with the structure and internal kinematics of open clusters but has been of central importance at this symposium. One of the more surprising findings in the study of clusters has been the lack of any detected binaries in globular clusters, other than those suggested by indirect evidence (e.g., x-ray sources or cataclysmic variables). However no such deficiency exists in open clusters; these systems have long been known to be rich in binaries. Indeed, several studies have shown that the binary populations in the Hyades, Pleiades and Praesepe are indistinguishable from that in the field where binaries are the dominant species (Mathieu, Stefanik and Latham 1985, Bettis 1977, Brosche and Hoffman 1979).

Using the simple formula of Heggie (1975) for binaries with 1 $M_{\odot}$ components, one finds that the boundary in binary period between hard and soft binaries in the Hyades is roughly 10^4-10^5 yr. In the Pleiades this boundary lies at 10^3-10^4 yr. Binaries in these period ranges are the most dynamically active and are significant in determining cluster evolution. The essential point here is that in open clusters many such binaries exist from the moment of cluster formation; assuming a field-like binary population (Abt and Levy 1976) roughly 10% of an initial cluster population are binaries with energies within an order of magnitude of the mean cluster kinetic energy. Furthermore, these binaries will rapidly concentrate toward the cluster core on an equipartition time scale. Thus binaries act as energy sources more or less continually throughout the lifetimes of open clusters.

The subject of binary encounters has to date been based entirely on analytic theory or numerical experiment with little observational constraint. It is natural to ask what observable consequences of such encounters might be expected in the nature of the binary population. Hut, in this symposium, has already pointed out that the spatial distribution of stars in a relaxed population of soft binaries would be indistinguishable from a random distribution. The theory thus provides little room for detailed observational confirmation with regard to soft binaries; nonetheless the distribution of wide pairs should be studied lest the observations contradict the theory. In addition, a comparison of the very youngest clusters with intermediate age clusters might provide evidence for the evolution of an initial soft binary population.

Study of the hard binary population should be more productive. A 100-body equal-mass numerical simulation run by Aarseth (1975) (see also Gianone, this volume) of a cluster with an 11% population of slightly hard binaries is instructive. After 24 crossing times, approximately two thirds of the binaries had had encounters which increased their binding energy (and significantly affected the cluster evolution). Nonetheless, none of the binaries suffered period changes of even an order of magnitude. Clearly, then, one cannot hope to detect any modification in the binary period distribution. However, exchange reactions are a common form of encounter, with massive single stars replacing lower mass binary components. Thus one might expect to detect a deficiency at the hard/soft boundary among binaries with lower-mass primaries, particularly in the cluster core. (Whether the corollary, an excess of binaries among massive stars, would be observable would depend on the population of invisible stellar remnants.) It is intriguing that in the Hyades there are several binaries with approximately solar-mass primaries and with periods near the hard/soft boundary, roughly in agreement with the number expected given a binary population similar to that in the field. However, three of these binaries are among the most distant cluster members found in Van Bueren's (1952) original survey; in fact they lie beyond the theoretical tidal radius. Have these binaries survived because they reside in the low-density halo? Or perhaps they are escaping the cluster as a result of encounters?

More study of wide binaries in other open clusters is very much needed. Clusters within 200 pc have dynamically active binaries with separations of $\gtrsim 1''$. Thus there are 3-4 relatively rich clusters, with membership studies, which can easily be studied. Indeed, as with our study of the Hyades, much or all of the necessary data may reside in the literature or plate stacks. The results of analyses of the binary populations in these clusters, combined with the Hyades study, will markedly improve statistics and should allow significant conclusions to be drawn with regard to the observable consequences of binary activity.

5. OUTSTANDING PROBLEMS AND NEW DIRECTIONS

Many of the essential problems in open cluster dynamics have received mention already, but two points merit repeated emphasis here. First, all stellar systems have a spectrum of masses to some degree and yet to date we have very little direct observational evidence of the dependence of velocity distributions on stellar mass. This knowledge is fundamental to any theoretical understanding of cluster dynamics and merits far more attention than it has been given to date. Hopefully the work in the Pleiades and M35 will provide a significant step forward in the near future.

Second, core collapse phenomena have been a primary focus of recent research in cluster dynamics, and with a substantial degree of success as has been evident at this symposium. However we should not entirely abandon the halos of clusters, where the remarkable success of the lowered Maxwellian velocity distribution may have lulled us into believing that we fully understand their dynamics, despite the fact that these outer regions are neither relaxed nor resident in a square-well potential. That the Hyades, the only open cluster whose outermost regions have been studied in detail, does not smoothly fit into the theoretical picture should motivate this direction of study. From an observational perspective the essential problem will be membership determination in that appropriate first-epoch plate material is rare.

Perhaps efficient combination of photometry and high-precision radial velocities will be a reasonable alternative in the short run.

These problems are classical ones in that they have been the subject of study since the first research into cluster dynamics; this of course makes them no less important. In the near future, however, strides will be taken in new directions as well, in particular in the study of very young clusters with ages of less than a relaxation time, less than an equipartition time, indeed less than a crossing time. The most basic questions need to be answered. What is the initial mass function? What is the initial stellar spatial distribution, in particular as a function of stellar mass? Is mass segregation the result of stellar dynamical processes or do clusters form with such structure? Burki (1978) has taken a first look at the very youngest clusters and suggests that the very massive stars form in the outer regions of clusters. However, Herbst and Miller (1982) find a marginally significant central concentration of the more massive stars in NGC 3293 (nuclear age of 6×10^6 yr). If we indeed find that the very youngest clusters are not mass segregated, then we have the means to directly measure the equipartition time-scale. If the very youngest clusters are mass segregated, we will have found an important insight into the ways stars form. Using similar reasoning it is clearly of interest to determine whether anisotropy exists in clusters younger than the Pleiades. This is a far more difficult observation, but is crucial in interpreting the observations of older clusters with regard to cluster evolution processes and time-scales.

Expanding our horizons a bit further, the dynamical principles developed for clusters can provide important insights into the star-formation process itself. Mathieu (1983a), Elmegreen (1983) and Lada, Margulis and Dearborn (1984) have applied simple stellar dynamics to forming clusters in order to argue for very high star-formation efficiencies. However, the detailed theoretical questions concerning the dynamical interaction of forming clusters with their parent molecular clouds remain unanswered. The observational side is equally untouched and promising; the stellar kinematics in star-forming regions are largely unknown. How do the stellar and gas kinematics compare? What implications does this have for magnetic fields in molecular clouds? How do the stellar kinematics of regions disturbed by shocks differ from quiescent regions? How many young clusters are in fact unbound? These questions have only begun to be answered; indeed many more still remain to be asked.

I would like to thank D. Latham, F. van Leeuwen and P. Hut for critical readings of the manuscript and my colleagues who generously provided materials for its preparation.

REFERENCES

Aarseth, S. J.: 1975, Dynamics of Stellar Systems (ed. A. Hayli), p. 57 (Dordrecht: Reidel)
Abt, H. A.: 1980, Ap. J. **241**, 275.
Abt, H. A. and Levy, S. G.: 1976, Ap. J. Supp., **30**, 273.
Bahcall, J.: 1984, Ap. J. **276**, 169.
Bettis, C.: 1977, Ap. J. **214**, 106.
Brosche, P. and Hoffmann, M.: 1979, Astrop. and Space Science, **63**, 467.
Burki, G.: 1978, Astron. and Ap. **62**, 159.
Chiu, G. and van Altena, W.: 1981, Ap. J., **243**, 827.
Cudworth, K. M.: 1971, A. J., **76**, 475.
________.: 1976, A.J., **81**, 519.
________.: 1979, A. J., **84**, 1312.
Elmegreen, B. G.: 1983, M.N.R.A.S., **203**, 1011.
Gieseking, F. J.: 1981, Astron. and Ap., **99**, 155.
Heggie, D. C.: 1975, Dynamics of Stellar Systems (ed. A. Hayli), p. 73, (Dordrecht: Reidel).
Herbst, W. and Miller, D. P.: A.J., **87**, 1478.
King, I. R.: 1980, Star Clusters (ed. J. Hesser), p. 139, (Dordrecht:Reidel).
Jones, B. J.: 1970, A.J. **75**, 563.
________.: 1971, A.J. **76**, 470.
Lada, C. J., Margulis, M. and Dearborn, D.: 1984, Ap. J., in press.
Mathieu, R. D.: 1983a, Ap. J. Letters, **267**, L97.
________.: 1983b, Ph.D. dissertation, Univ. Cal. Berkeley.
________.: 1984, Ap. J., **284**, in press.
Mathieu, R. D. and Latham, D. W.: 1985a, in preparation.
________.: 1985b, in preparation.
Mathieu, R. D., Stefanik, R. P. and Latham, D. W.: 1985, in preparation.
Mayor, M. and Mermilliod, J.C.: 1983, Les Etoiles Binaires dans le Diagramme H. R. (eds. A. Florsch, C. Jaschek and M. Jaschek), p. 45, (Observatoire de Strasbourg).
McNamara, B. J., Pratt, N. M. and Sanders, W. L.: 1977, Astron. and Ap. Supp., **27**, 117.
McNamara, B. J. and Sanders, W. L.: 1977, Astron. and Ap., **54**, 569.
________.: 1978, Astron. and Ap., **62**, 259.
________.: 1983, Astron. and Ap., **118**, 361.
Oort, J. H.: 1979, Astron. and Ap., **78**, 312.
Pels, G., Oort, J. H. and Pels-Kluyver, H. A. 1975, Astron. and Ap., **43**, 243.
Retterer, J. M.: 1979, A. J. **84**, 370.
Sanders, W. L.: 1977, Astron. and Ap. Supp., **27**, 89.
Spitzer, L., Jr. and Shull, J. M.: 1975, Ap. J. **201**, 773.
Terlevich, E.: 1983, Ph.D. dissertation, University of Cambridge.
Tonry, J. and Davis, M.: 1979, A. J., **84**, 1511.
Van Bueren, H. G.: 1952, Bull. Astron. Inst. Neth., **11**, 385.
van den Bergh, S. and Sher, D.: 1960, Pub. David Dunlap Obs., **2**, 203.
van Leeuwen, F.: 1980, Star Clusters (ed. J. Hesser), p. 157, (Dordrecht: Reidel).
________.: 1983, Ph.D. dissertation, Sterrewacht Leiden.

Vasilevskis, S., van Leeuwen, F., Nicholson, W. and Murray, C. A.: 1979, Astron. and Ap. Suppl, **37**, 333.
Weilen, R.: 1971, Astron. and Ap., **13**, 309.
__________.: 1975, Dynamics of Stellar Systems (ed. A. Hayli), p. 119, (Dordrecht: Reidel).

DISCUSSION

BAHCALL: You gave a clear discussion of the mass segregation observed among the various clusters. However, there is one puzzling fact: the oldest cluster you discussed, M67, has the smallest amount of mass segregation, markedly less than for the youngest clusters. How do you understand this fact?

MATHIEU: The lowest mass component (derived from star counts) does appear to be somewhat less centrally concentrated and as a result equipartition models do seem to fit the data better than non-equipartition models. The data cannot distinguish the two models at a statistically significant level, however. So, you are right that the mass segregation in M67 is far less evident than in the younger clusters. The model which best fits the data has a shallower potential than for M11 and thus a more severely truncated velocity distribution, which lessens mass segregation. Also, in M67 the most massive stars are dominant contributors to its potential, which also lessens mass segregation among these stars. Thus, the lack of marked mass segregation is not necessarily inconsistent with equipartition in the cluster. Still, it is interesting that it is the oldest cluster which is the least mass segregated. I look forward to seeing whether studies presently being done on other equally old clusters tend to find similar cluster structure.

APPLEGATE: What mass functions do you find in the multimass models that you showed?

MATHIEU: The mass functions of M11 and M35 are not distinguishable from that found in the field by Miller and Scalo. However, in neither cluster do the data extend to masses significantly below 1 $M_\odot$, where the deficiencies in low mass stars that have been observed in several clusters first appear. In M67 the proper-motion membership data is consistent with star-count data of Van den Bergh and Sher, but again the proper-motion data do not extend faint enough to provide an independent confirmation of their observed deficiency. I should note that observational biases due to mass segregation cannot be invoked to explain away their results. Van Leeuwen argues that the luminosity function in the Pleiades is consistent with the field down to several tenths of a solar mass, while in the Hyades, Oort suggests that indeed there is a deficiency of the lowest mass stars beginning at $M_V = 9$.

LARSON: I would like to second Dr. Mathieu's final remark about the importance of young open clusters for the study of star formation. One of the most interesting questions concerns the stellar initial mass function (IMF) and whether it is the same everywhere or varies with location. There is some folklore to the effect that some young open clusters have a deficiency of low mass stars which is difficult to account for in terms of dynamical evolution, and there are even some hints that the IMF may vary with position within individual young clusters. As was pointed out several years ago by Freeman, if the IMF in some clusters or regions contains a much higher proportion of massive stars than a "standard" IMF, the evolution of the system can be radically altered. Effects of mass loss from massive stars will be of very major importance, and the remnants left by the massive stars can become an important or even dominant constituent of the system. The best place to look for possible variations in the IMF is in very young clusters where effects of dynamical evolution have not yet become important.

KING: The spatial distribution of the Hyades offers one additional opportunity. Because of the convergent effect in the proper motions, this is then one cluster that we can study in three dimensions rather than two.

APPLEGATE: I would like to reinforce Larson's comment on the importance of stellar evolution in clusters with flat mass functions. I find that the early evolution of clusters with IMF's flatter than $(N(M) \propto m^{-2}$ can be completely dominated by mass loss; collisional evolution plays a very minor role. IMF's this flat are well within the range reported by Freeman for the young LMC clusters.

GOODMAN: Do you see any evidence in the open cluster radial-velocity data for motions in the atmospheres of giants, such as was found by Gunn and Griffin (1979)?

MATHIEU: No. The internal errors of our open cluster giant observations are essentially the same as our errors for all other observations, including for example the dwarfs in the Hyades.

DYNAMICS OF OPEN STAR CLUSTERS

Roland Wielen
Institut für Astronomie und Astrophysik
Technische Universität Berlin, Germany

ABSTRACT

The dynamical evolution of open star clusters has been studied successfully by numerical N-body simulations. We compare in detail the theoretically predicted lifetimes with those derived from the observed age distribution of open clusters. Rare but efficient encounters between clusters and giant molecular clouds are probably responsible for the sudden disruption of many open clusters. Massive black holes from the galactic corona would, on the average, affect only old open clusters.

1. INTRODUCTION

Our theoretical knowledge of the dynamical evolution of open star clusters stems largely from numerical N-body simulations. In such simulations, the orbits of all the stars in the cluster are followed numerically by integrating the equations of motions of the complete N-body problem. In addition to the mutual gravitational interactions between all the stars in the cluster, other important effects, such as the stationary tidal field of the Galaxy, the impulsive tidal fields of passing interstellar clouds, the mass loss of evolving stars etc., can be easily implemented into the numerical experiments in a very realistic manner. With modern computers and the available sophisticated integration schemes, it is not difficult to follow the evolution of a star cluster with an initial membership of N = 1000 stars over its total lifetime, i.e. until the cluster is nearly completely dissolved. For the application of the results of the numerical N-body simulations to real astronomical objects, it is, of course, important to know the correct physical input data, such as initial conditions, mass spectrum of the stars, masses and radii of passing interstellar clouds etc. . This limitation is, however, also valid for any other theoretical treatment of the problem.

J. Goodman and P. Hut (eds.), Dynamics of Star Clusters, 449–462.

The results of the N-body simulations for open star clusters have been reviewed in the last decade by Wielen (1974, 1975), Aarseth and Lecar (1975) and Aarseth (1984). E.Terlevich (1980, 1983) has carried out an important series of new N-body simulations for open clusters. Her results confirm most of the conclusions drawn in the earlier reviews from less extensive material. Instead of repeating the reviews listed above, we shall concentrate in the present paper on some new results, mainly concerning the effects of passing massive objects on the lifetimes of open clusters.

In contrast to the numerical N-body simulations, statistical theories of stellar dynamics are difficult to apply to the internal dynamics of open clusters. This is due to the relatively small number of stars in an open cluster, typically between N = 100 and N = 1000. Firstly, statistical theories assume by their very nature 'large' N, and it is uncertain wether a typical open cluster fulfills already this requirement. Secondly, in open clusters, the crossing time T_{cr} and the internal relaxation time T_{rel} are of same order of magnitude, thereby hampering a conventional statistical treatment as used for globular star clusters with their much higher values of N. King (1980) has emphasized these difficulties in his review of the dynamics of open clusters. A statistical treatment is, however, rather suitable for studying the effects of passing interstellar clouds etc. on open cluster (Section 3), since these external effects are basically independent of the number N of stars in the cluster.

2. CLUSTER LIFETIMES DERIVED FROM THE OBSERVED AGE DISTRIBUTION

A comparison of the total lifetimes of clusters derived from the observed age distribution with theoretically predicted disintegration times seems to be the most powerful observational test of dynamical theories of open clusters (Wielen 1971, 1974, 1975). The age distribution of open clusters has been recently rediscussed by van den Bergh (1981), Janes and Adler (1982) and Lynga (1982a, b, 1983). Especially the observed age distribution derived by Lynga (1982b) from a very extensive material is in perfect agreement with that derived earlier by Wielen which led to the following conclusions about the 'observed' total lifetimes of open clusters (Wielen 1971): 50% of newly born clusters disintegrate within $2 \cdot 10^8$ years (median lifetime of open clusters), 10% have a total lifetime longer than $5 \cdot 10^8$ years, and only 2% live longer than $1 \cdot 10^9$ years. Hence the typical lifetime of an open cluster is short, but there exists a wide spread in individual lifetimes.

According to van den Bergh and McClure (1980), Janes and Adler (1982) and Lynga (1982a, b, 1983), open clusters in the outer parts of the Galaxy ($R > R_0$) have longer lifetimes than those in the inner parts ($R < R_0$).

3. ENCOUNTERS BETWEEN CLUSTERS AND MASSIVE OBJECTS

Encounters of open clusters with passing massive objects can reduce, or even determine, the lifetimes of the clusters. Spitzer (1958) discussed the effect of interstellar HI clouds on open clusters. We shall derive here the effects of more massive objects, such as giant molecular clouds (GMC) or massive black holes (BH). GMCs certainly exist in our Galaxy, and van den Bergh and McClure (1980) proposed already that GMCs may strongly affect the lifetimes of open clusters. Massive black holes may constitute the 'dark matter' in the galactic corona. Ostriker (1983) has pointed out that the existence of such black holes would nicely explain the observed age-velocity relation of disk stars (Wielen 1977). Lacey (1984) has used this idea to calculate the typical mass of such coronal black holes.

3.1 General Theory

We shall use basically the formalism developed by Spitzer (1958) for deriving the disruption time of clusters due to passing objects. In the impulsive approximation, the increase ΔE_c in the total energy of a cluster (index c), due to the tidal field of a passing object (index n), is given by

$$\Delta E_c = (4\alpha^2/3)\ (Gm_n/(p^2V))^2\ m_c r_c^2 \quad , \tag{1}$$

where m_c is the mass of the cluster, r_c the median radius of the cluster (containing half of m_c in projection), α the ratio between the root-mean-square cluster radius and r_c, G the gravitational constant, and m_n the mass of the object (GMC, BH) which moves relative to the cluster center on a straight line with constant velocity V and shortest distance p (impact parameter). Eq.1 overestimates ΔE_c, if $p < r_n$ (GMC) or if $p < r_c$. In these cases, $p < \max(r_n, r_c)$, we calculate ΔE_c by replacing p in Eq.1 by

$$p_0 = \max(r_n, r_c) \quad . \tag{2}$$

This approximation is sufficient for our purpose. For a Plummer model, the total energy E_c of the cluster is given by

$$E_c = -(3\pi/64)\ Gm_c^2/r_c \quad . \tag{3}$$

We ask now for that impact parameter p_1 for which ΔE_c is already equal to $-E_c$ after just one passage of an object. From Eqs.1 and 3, we get

$$p_1 = 4(\alpha/3)^{1/2}(G/\pi)^{1/4} m_n^{1/2} V^{-1/2} (m_c/r_c^3)^{-1/4} \quad . \tag{4}$$

3.1.1 The Case $p_1 > p_0$

If $p_1 > p_0$, then a single passage of an object with $p < p_1$ is able to disrupt the cluster immediately. The average number Z of such encounters with $p < p_1$ in a period Δt is

$$Z = \pi p_1^2 \nu_n V \Delta t \qquad , \qquad (5)$$

where ν_n is the number density of the objects. We get an estimate for the time T_1 after which a cluster is disrupted by a single encounter with a massive object, if we set $Z = 1$ and $\Delta t = T_1$ in Eq.5 and insert p_1 from Eq.4:

$$T_1 = (3/(16\alpha))\pi^{-1/2}(Gm_c/r_c^3)^{1/2}/(G\rho_{an}) \qquad , \qquad (6)$$

where

$$\rho_{an} = m_n \nu_n \qquad (7)$$

is the overall mass density of the objects (GMCs or BHs). It is remarkable that T_1 does not depend on the individual mass m_n of the passing objects but only on ρ_{an}. Furthermore, T_1 is independent of the velocity V of the passing objects.

A short calculation shows that all the distant encounters with $p > p_1$ increase the total energy of the cluster just by $-E_c$ within the period T_1, if we neglect the increase of αr_c due to successive encounters and the adiabaticity of very distant encounters. Since the distant encounters ($p > p_1$) dissolve the cluster within the same period T_1 as the close encounters ($p < p_1$), the total dissolution time scale $T_{n,1}$ due to the combined effect of close and distant encounters is approximately given by

$$T_{n,1} = T_1/2 \qquad (8)$$

with T_1 according to Eq.6. This shortening of the time scale by a factor of two has been experimentally confirmed by Bahcall et al. (1984) for the disruption of wide binaries.

The dissolution of clusters due to close encounters ($p < p_1$) with massive objects introduces a strong accidentalness into the lifetimes of open clusters. Even the effect of distant encounters ($p > p_1$) is governed by a few encounters with $p \sim p_1$ because of the strong decrease of the energy transfer with increasing p. Hence the lifetime of an individual cluster can significantly deviate from its expectation value, $T_{n,1}$, given by Eq.8.

3.1.2 The Case $p_1 < p_0$

If $p_1 < p_0$, then a number of successive encounters are necessary for disrupting the cluster. This condition has been implicitely assumed by Spitzer (1958). The rate of change of E_c is given by Eq.1 multiplied by the number of collisions per unit time and per interval of p, and integrated over p:

$$\dot{E}_c = (4\alpha^2/3)\, G^2 m_n^2 \nu_n V^{-1} m_c r_c^2 \left(\int_0^{p_0} 2\pi p p_0^{-4} dp + \int_{p_0}^{\infty} 2\pi p p^{-4} dp \right)$$

$$= (8\pi\alpha^2/3)\, G^2 m_n \rho_{an} V^{-1} m_c r_c^2 p_0^{-2} \quad . \qquad (9)$$

If we neglect the increase in αr_c due to successive encounters, we obtain an estimate for the dissolution time $T_{n,0}$ of the cluster from $-E_c/\dot{E}_c$, using Eqs.3 and 9:

$$T_{n,0} = (9/(512\alpha^2))\, G^{-1} m_n^{-1} \rho_{an}^{-1} V m_c r_c^{-3} p_0^2 \quad , \qquad (10)$$

where p_0 is either r_n or r_c according to Eq.2.

If we had taken into account the increase in αr_c due to successive encounters by using Eq.1 and integrating over t until $E_c = 0$ (Spitzer 1958), we would have found $T_{n,0}/3$ for the dissolution time scale instead of $T_{n,0}$. We prefer, in accordance with Spitzer and Chevalier (1973), to use $T_{n,0}$ as a more conservative estimate. Equation 9 overestimates probably the real increase of E_c, because the outermost stars in a cluster gain energy so rapidly that they often escape before sharing their energy gain with the remaining cluster by encounters with other stars in the cluster. Furthermore, $T_{n,0}$ according to Eq.10 equals then $T_{n,1}$ in the limit $p_0 = p_1$. The disruption time t_d, derived by Spitzer (1958), is twice as long as $T_{n,0}/3$, because he neglected in that paper encounters with $p < p_0$.

Why is $T_{n,0}$ not also valid for $p_1 > p_0$? The reason is the following: For $p_1 > p_0$, each encounter with $p < p_1$ has an 'overkill' capacity, i.e. ΔE_c is larger than $-E_c$. The overshooting parts of these energy gains must not be used in calculating the average energy change $\dot{E}_c$. In order to take into account the upper limit $-E_c$ for ΔE_c, we have to replace p_0 by p_1 in Eqs.9 and 10 for $p_1 > p_0$. By replacing p_0 by p_1 in Eq.10 for $T_{n,0}$, we recover in fact $T_{n,1}$ derived in Section 3.1.1 (Eq.8).

3.1.3 Other Aspects

If a cluster is exposed to an ensemble of passing objects with different values of m_n, r_n and V, some encounters belong to the case $p_1 > p_0$, while others correspond to $p_1 < p_0$. For simplicity, we shall use then either $T_{n,1}$ or $T_{n,0}$, according to the majority of the efficient encounters. In all of our applications, we assume furthermore $\alpha = 1$.

In calculating the disruption times $T_{n,1}$ and $T_{n,0}$, we have neglected the effect of the stationary tidal field of the Galaxy. Due to this galactic tidal field, the real dissolution times will be shorter than given here. The effect is, however, small for typical open clusters for which the median radius r_c is much smaller than the tidal radius (Bouvier 1971).

As pointed out by Spitzer (1958), slow or distant encounters may not be impulsive. The impulsive approximation is valid only for encounters with

$$p < p_{ad} = V\, T_{cr} \quad , \qquad (11)$$

i.e. the internal crossing time T_{cr} of the cluster should be longer than the effective duration of the encounter, p/V. Due to the strong decrease of ΔE_c with increasing p, our earlier results are essentially correct as long as

$$\max(p_0, p_1) < p_{ad} \qquad (12)$$

holds.

The general theory for encounters between open clusters and massive objects has many similarities to the problems of the disruption of wide binaries (Bahcall, Hut, and Tremaine 1984) and of the dissolution of the Oort Comet Cloud (e.g. Bailey 1983). There are, however, also important differences. For example, a wide binary can be disrupted by a very close encounter between one component and a passing compact object (e.g. a star), while for open clusters, the effect of encounters with $p < r_c$ on the dissolution of the whole cluster are essentially independent of p.

3.2 Giant Molecular Clouds

Observations indicate for a typical giant molecular cloud (GMC) a mass of $m_n = 5 \cdot 10^5\ m_\odot$ and a diameter of $2r_n = 50$ pc. The contribution of GMCs to the overall mass density in the solar neighbourhood, i.e. at $R = R_0$, is rather uncertain. We shall use $\rho_{an} = 0.02\ m_\odot/pc^3$, perhaps a somewhat optimistic value. The typical velocity V of a young cluster relative to a GMC is about 10 km/s. Since r_n is larger than r_c for open clusters, p_0 is equal to r_n. From Eq.4, we obtain that the disruption of the cluster by a single passage of a GMC is possible ($p_1 > p_0 = 25$pc), if

$$m_c/r_c^3 < 250\ m_\odot/pc^3 \quad . \qquad (13)$$

Many open clusters belong to this class. The disruption time of the cluster due to the GMCs is then given by

$$T_{n,1} = 6.2 \cdot 10^8 \text{ years } ((m_c/r_c^3)/250\ m_\odot pc^{-3})^{1/2} \quad . \qquad (14)$$

For clusters which are denser than the limit set by Eq.13, the longer disruption time $T_{n,0}$ applies.

For the effect of GMCs on open clusters, the overall mass density ρ_{an} is of primary importance, because this is the only property of clouds which enters into the disruption time $T_{n,1}$. Even much smaller masses m_n of the clouds would not alter our conclusions if ρ_{an} is kept fixed and if $p_1 > p_0$ still holds. For example, we can allow molecular clouds with $m_n = 2 \cdot 10^4\ m_\odot$ and $2r_n = 10$ pc, because even then many open clusters will belong to the case $p_1 > p_0 = r_n = 5$ pc ($p_1 > 5$ pc for $m_c/r_c^3 < 250\ m_\odot/pc^3$).

The effect of encounters between open clusters and GMCs depend on the overall density ρ_{an} of GMCs at the position of the cluster. Hence older clusters with higher distances z from the galactic plane suffer less from GMCs. Furthermore, the average space velocity of older clusters is higher, thereby decreasing p_1 and hence shifting some of these clusters from the case $p_1 > p_0$ to $p_1 < p_0$.

The adiabatic limit for p is given by

$$p < p_{ad} = 100 \text{ pc } (V/10 \text{ kms}^{-1})\ (T_{cr}/10^7 \text{ years}) \tag{15}$$

with the internal crossing time of the cluster (using a Plummer model)

$$T_{cr} = (32/(3\pi))^{3/2}(Gm_c/r_c^3)^{-1/2}$$
$$= 5.9 \cdot 10^6 \text{ years } (250\ m_\odot pc^{-3}/(m_c/r_c^3))^{1/2} \quad . \tag{16}$$

Hence the impulsive approximation is not valid anymore for encounters between GMCs and open clusters with $m_c/r_c^3 > 600\ m_\odot/pc^3$, because then $p_0 = r_n > p_{ad}$ holds.

3.3 Massive Coronal Black Holes

The local density of the invisible galactic corona is $\rho_{an} = 0.006\ m_\odot/pc^3$, according to the galactic mass model constructed by Caldwell and Ostriker (1981). Lacey (1984) derives the quantity $\nu_n m_n^2 = \rho_{an} m_n = 2 \cdot 10^4\ m_\odot^2 pc^{-3}$ for the postulated coronal black holes. Using ρ_{an} from Ostriker and Caldwell, we find for the typical mass of a coronal black hole $m_n = 3 \cdot 10^6\ m_\odot$. The relative velocity between a cluster and a black hole should be typically of the order of $V = 250$ km/s. For black holes, we have $r_n \sim 0$ and hence always $p_0 = r_c$. The limiting impact parameter p_1 for the coronal black holes is given by

$$p_1 = 12 \text{ pc } (250\ m_\odot pc^{-3}/(m_c/r_c^3))^{1/4} \quad . \tag{17}$$

Hence, most open clusters belong to the case $p_1 > p_0 = r_c$. For these open clusters, the disruption time of the cluster due to coronal black holes is given by

$$T_{n,1} = 2.1 \cdot 10^9 \text{ years } ((m_c/r_c^3)/250\ m_\odot pc^{-3})^{1/2} \quad . \tag{18}$$

The rather long time scale $T_{n,1}$ shows that coronal black holes affect, on the average, only old open clusters. The encounters between open clusters and coronal black holes are nearly always impulsive, because the adiabatic limit p_{ad} (Eqs.11 and 15) is typically of the order of 1 kpc, i.e. $p_1 << p_{ad}$.

4. LIFETIMES OF OPEN CLUSTERS

We shall now discuss the theoretically predicted dissolution times of open clusters and compare these predictions with the 'observed' lifetimes derived from the age distribution (Section 2).

First, we consider a 'typical' open star cluster. We assume for such a typical cluster at birth N = 500 stars (with a realistic spectrum of stellar masses) and a total initial mass of $m_c = 250\ m_\odot$ (see Lynga (1983) and Bruch and Sanders (1983)). In Figure 1, we illustrate the total lifetime T of such a cluster as a function of the median radius r_c of the whole cluster (r_c encircles $m_c/2$ or N/2 in projection). For $r_c < 0.1$ pc, the cluster is essentially isolated and evaporates due to internal relaxation (encounters between cluster stars) on a time scale shorter than $2 \cdot 10^8$ years. For $r_c > 0.1$ pc, the stationary tidal field of the Galaxy enhances the dissolution with respect to isolated systems significantly. In the range 0.5 pc $< r_c <$ 3 pc, the total lifetime, measured in years, depends then only weakly on r_c. The full curve and the symbols in Figure 1 represent results from N-body simulations which include the galactic tidal field. There is beautiful agreement between the N-body simulations of Terlevich (1983, e.g. model IV) and Wielen (1975, e.g. model FG3). The presence of initial binaries (Aarseth 1980) and the mass loss of evolving stars seem to affect the lifetimes of clusters only marginally.

The tidal radius ξ_L of a cluster (King 1962) in the direction towards the galactic center is given by

$$\xi_L = (\ Gm_c/(4A(A-B))\ )^{1/3} \quad . \qquad (19)$$

The tidal limits in the other two directions are smaller, $\eta_L = (2/3)\xi_L$ and $\zeta_L \sim 0.5\ \xi_L$ (Wielen 1974, Fig.6). Clusters with median radii $r_c \gtrsim 0.5\ \xi_L$ are already unbound in the gravitational tidal field of the Galaxy and expand on the time scale of the galactic rotation, $T_{rot} = 2\pi/\omega_0 = 2\pi/(A-B) = 2.5 \cdot 10^8$ years.

The disruption time of an open cluster due to passing interstellar HI clouds (dashed line in Fig.1) is proportional to r_c^{-3} for a given m_c (Eq.10). It has been predicted by Wielen(1974, 1975) on the basis of Spitzer's impulsive approximation that the disruption of open clusters is significantly accelerated by passing interstellar HI clouds of the standard type only if the cluster is already either weakly bound or even unstable because of the galactic tidal field (i.e. $r_c > 3$ pc

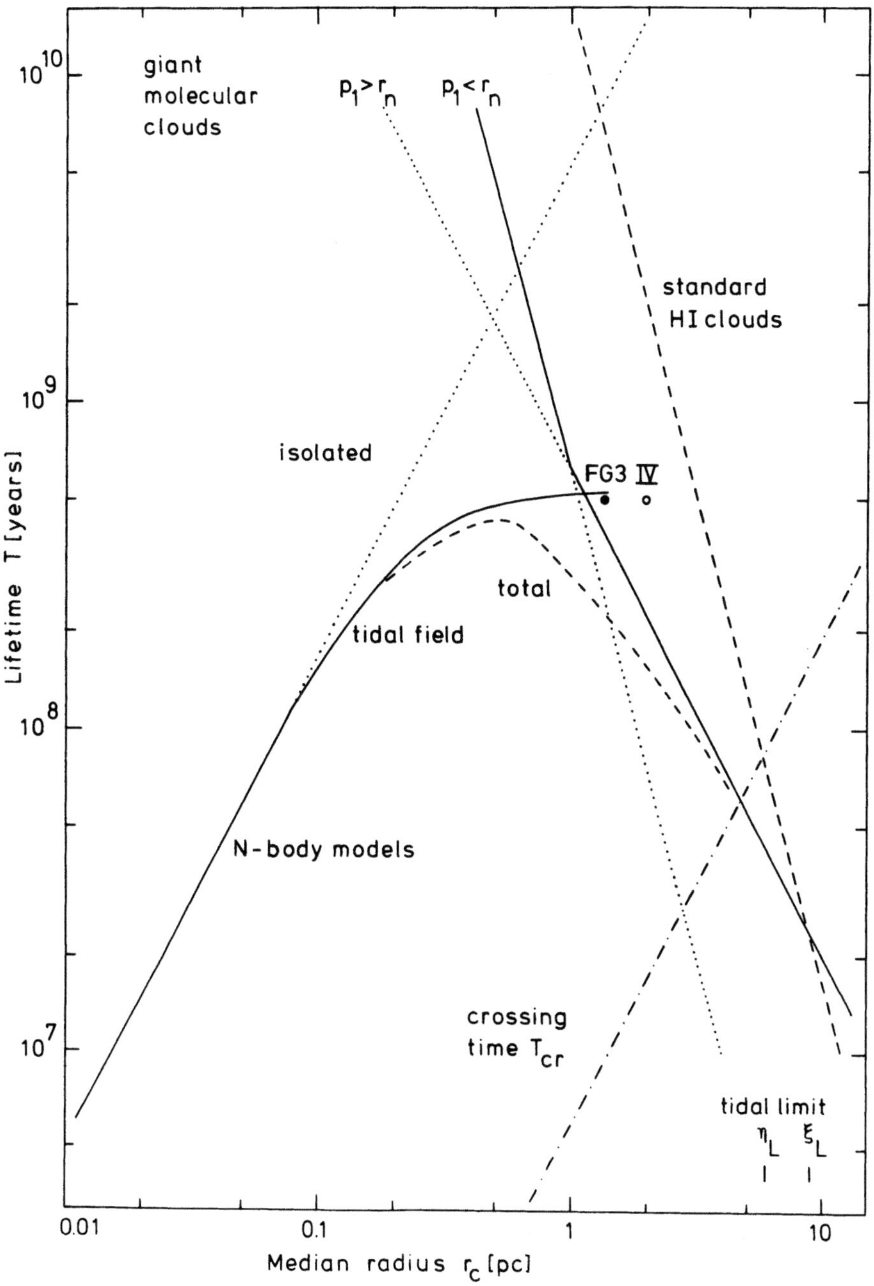

Figure 1. Total lifetime T of an open star cluster with an initial total mass of $m_c = 250\ m_\odot$ and N = 500 stars as a function of the median radius r_c of the cluster. Details are explained in the text.

in Fig.1). This has now been nicely confirmed by direct N-body simulations: Terlevich (1983) found no differences in the lifetimes of cluster models with and without HI clouds for $r_c \sim 2$ pc.

If there were no GMCs or BHs, we would derive from Fig.1 for a typical open cluster with $m_c = 250\ m_\odot$ and $r_c \sim 1$ pc a predicted total lifetime of $T = 5 \cdot 10^8$ years. This is higher than the observed typical lifetime of $2 \cdot 10^8$ years. The N-body results for N = 500 stars and $m_c = 250\ m_\odot$ can be scaled to other values of N and m_c (keeping $m_c/N = 0.5\ m_\odot$ constant) by shifting the full curve in Fig.1 first in radius r_c by a factor of $(N/500)^{1/3}$ (to the right for N > 500) and then in lifetime T by a factor of (N/500) (upwards for N > 500). This scaling law (Wielen 1974) is theoretically predicted, but is also confirmed by N-body simulations ($N \leq 500$: Wielen (1974), N = 1000 : Terlevich (1983)). In order to explain the observed typical lifetime, we would require $m_c = 100\ m_\odot$ and $r_c \sim 1$ pc as typical initial values for an open cluster. The observed large spread in cluster lifetimes (10^8 to 10^{10} years) can be explained mainly by different initial masses (50 to 5000 $m_\odot$) of the clusters.

We shall now consider the lifetimes of open cluster including the effect of the passages of massive objects, using the results derived in Section 3. The disruption time of a cluster of $m_c = 250\ m_\odot$ due to passing giant molecular clouds are given by the full lines in Fig.1. The disruption time $T_{n,1}$ (Eqs.8 and 14) applies for $p_1 > r_n$, i.e. for $r_c > 1$ pc (Eq.13), while $T_{n,0}$ (Eq.10) is correct for $p_1 < r_n$ or $r_c < 1$ pc. $T_{n,0}$ decreases with r_c^{-3} (as for HI clouds), while $T_{n,1}$ falls off more slowly with $r_c^{-3/2}$. In order to derive the lifetime T_{total} due to the combined effect of GMCs and internal relaxation including the galactic tidal field, we add the corresponding escape rates,

$$T_{total}^{-1} = T_{GMC}^{-1} + T_{N\text{-body}}^{-1} \quad . \tag{20}$$

The result is indicated by the dashed curve labelled 'total' in Fig.1. Terlevich (1983) has carried out a few N-body simulations in which a single encounter of the cluster with a molecular cloud occurs. Using $p < p_1$, she found indeed a sudden disruption of the cluster due to the encounter. From these experiments, it is not possible, however, to derive directly a mean disruption time of a cluster for comparison with our theoretical results.

Using the results shown in Fig.1 after including the effect of GMCs (dashed curve), we now predict for a typical open cluster ($m_c = 250\ m_\odot$, $r_c = 1$ pc) a shorter total lifetime, namely $T \doteq 3 \cdot 10^8$ years, in rough agreement with the observational value of $2 \cdot 10^8$ years. For clusters with larger median radii r_c, the average lifetime is even more strongly limited by the GMCs. The typical lifetime of open clusters is so short that they spend most of their life close to the galactic plane and mainly in the GMC layer. The decrease of the overall density ρ_{an} of GMCs with the galactocentric radius R may explain the observed increase of cluster lifetimes with increasing R (Section 2), because $T_{n,1}$ and $T_{n,0}$ are proportional to $1/\rho_{an}$. Another explanation of the R-dependence of the cluster lifetimes could be a systematic change of the typical initial cluster masses and radii with R, due to variations in the conditions for cluster formation (e.g. Jeans masses).

If one or a few encounters with GMCs are largely responsible for the disruption of an open cluster, then the stochastic occurence of such encounters produces already a wide spread in the lifetimes of individual clusters with the same initial values of m_c and r_c. The existence of very old open clusters can be explained by a combination of two effects: (1) a high initial mass m_c (otherwise the cluster dissolves too rapidly by internal relaxation even without GMCs), and (2) older clusters have been accelerated by the irregular part of the galactic gravitational field (Wielen 1977) and spent only a small fraction of their life inside the GMC layer, thereby increasing the mean T_n due to GMCs drastically.

Encounters with massive black holes of the galactic corona do not affect a typical open cluster, because the time scale $T_{n,1}$ (Eq.18) is much longer ($\sim 2\cdot10^9$ years) than the typical lifetime of a cluster with $m_c = 250\ m_\odot$ and $r_c = 1$ pc. The existence of the postulated massive black holes would, however, limit the lifetime of old open clusters severely. In order to explain the survival of an open cluster over $5\cdot10^9$ years, we would, on the average, require an initial mass of $m_c = 16000\ m_\odot$ for $r_c = 2$ pc ($\sim$ present median radius of M67 and NGC188) and of $m_c = 5000\ m_\odot$ for $r_c = 1$ pc. In deriving the numbers, we have added the escape rates (1) due to passing black holes and (2) due to internal relaxation and galactic tidal field, using the scaling law for reaching higher values of N as described earlier. Such high values for the initial masses of open clusters are rare today. They may have been more frequent in the past, however. Furthermore, due to the stochasticity of encounters of clusters with black holes, some of the old clusters may have avoided, by chance, efficient encounters with black holes up to now. Because very old open clusters are extremely rare, all these explanations seem to be reasonable. Therefore, the existence of massive coronal black holes cannot be ruled out from the point of view of open cluster dynamics. The black holes may even help to explain the scarcity of very old open clusters.

If an open cluster is suddenly disrupted by a single encounter with a massive object, most probably with a GMC, the system after the disruption will show many characteristics of a moving group. The internal velocity dispersion after the encounter will be typically of the order of 1 km/s. A moving group produced by such a sudden disruption of an old open cluster would look old from the ages of the stars, while it is actually young from a dynamical point of view, because the stars have spent most of their time in the bound cluster. In that case, a moving group composed of old stars would not be in contradiction to the existence of the irregular gravitational field in the Galaxy deduced from the age-velocity relation of nearby disk stars (Wielen 1977).

We conclude that encounters of open clusters with giant molecular clouds can probably help to understand better the dynamics, especially the lifetimes, of open clusters.

ACKNOWLEDGEMENTS

I would like to thank Professor John Bahcall and his colleagues for their kind hospitality and for many interesting discussions during my stay at the Institute for Advanced Study at Princeton.

REFERENCES

Aarseth,S.J. : 1980, in IAU Symposium No.85, Star Clusters, Ed.J.E.Hesser, D.Reidel Publ.Comp.,Dordrecht, p.325.
Aarseth,S.J. : 1984, in Multiple Time Scales, Eds. J.U.Brackbill and B.I.Cohen, Academic Press, New York (in press).
Aarseth,S.J., Lecar,M. : 1975, Ann.Rev.Astron.Astrophys. 13, 1.
Bahcall,J.N., Hut,P., Tremaine,S. : 1984, Astrophys.J. (in press).
Bailey,M.E. : 1983, Mon.Not.R.Astron.Soc. 204, 603.
Bouvier,P. : 1971, Astron.Astrophys. 14, 341.
Bruch,A., Sanders,W.L. : 1983, Astron.Astrophys. 121, 237.
Caldwell,J.A.R., Ostriker,J.P. : 1981, Astrophys.J. 251, 61.
Janes,K., Adler,D. : 1982, Astrophys.J.Suppl. 49, 425.
King,I. : 1962, Astron.J. 67, 471.
King,I.R. : 1980, IAU Symposium No.85, p.139.
Lacey,C.G. : 1984, in Formation and Evolution of Galaxies and Large Structures in the Universe, Eds. J.Audouze and J.Tran Thanh Van, D.Reidel Publ.Comp.,Dordrecht, p.351.
Lynga,G. : 1982a, Astron.Astrophys. 109, 213.
Lynga,G. : 1982b, Reports from the Observatory of Lund No.18, p.68.
Lynga,G. : 1983, in Star Clusters and Associations and their Relation to the Evolution of the Galaxy, Eds. J.Ruprecht and J.Palous, Publ.Astron.Inst.Czech.Acad.Sci. No.56, p.292.
Ostriker,J.P. : 1983, private communication.
Spitzer,L. : 1958, Astrophys.J. 127, 17.
Spitzer,L. Chevalier,R.A. : 1973, Astrophys.J. 183, 565.
Terlevich,E. : 1980, IAU Symposium No.85, p.165.
Terlevich,E. : 1983, Ph.D. Dissertation, Univ.Cambridge, England.
van den Bergh,S. : 1981, Publ.Astron.Soc.Pacific 93, 712.
van den Bergh,S., McClure,R.D. : 1980, Astron.Astrophys. 88, 360.
Wielen,R. : 1971, Astron.Astrophys. 13, 309.
Wielen,R. : 1974, in Proceedings of the First European Astronomical Meeting, Vol.2, Stars and the Milky Way System, Ed. L.N.Mavridis, Springer-Verlag, Berlin, p.326.
Wielen,R. : 1975, in IAU Symposium No.69, Dynamics of Stellar Systems, Ed. A.Hayli, D.Reidel Publ.Comp.,Dordrecht, p.119.
Wielen,R. : 1977, Astron.Astrophys. 60, 263.

DISCUSSION

TREMAINE: To estimate cluster disruption times one can replace the cluster by a hypothetical binary star with the same mass and rms velocity, and then use existing formulae for the half-life of binary stars (e.g. Bahcall, Hut & Tremaine, submitted to Ap.J.Lett.).

FALL: If open clusters are disrupted explosively by single encounters with giant molecular clouds, then one might expect, by analogy with a Poisson process, that the distribution of cluster ages should decline exponentially at large ages. It appears from your earlier work, however, that the distribution declines as a power of age. Are these results consistent, or is the expected distribution modified significantly by the finite velocity dispersion of the clouds and other factors?

WIELEN: If all the open clusters would have the same initial values for the total mass and the median radius, you would indeed expect, in a first approximation, an exponential decline of the age distribution for the case that close encounters with massive objects are the most important source of cluster disruption. In reality, you have a wide spread in initial cluster masses and radii. This produces already a more complex age distribution.

GRINDLAY: May I comment simply that the importance of giant molecular clouds (GMCs) for cluster disruption as you have discussed, may also be very important for globular clusters as I discussed in my talk on Tuesday. Globulars with the unfortunate initial conditions of having prograde orbits with inclinations near the plane (within $\sim \pm 10^{o}$) and apogalacticon distances $R \sim 4 - 6$ kpc will be disrupted within a few orbital periods by the GMC ring, if it exists, between 4-6 kpc. These disruptive encounters may also give rise to the high velocity moving groups of halo-type stars (e.g. Groombridge 1830, as discussed by Oort in his 1965 paper in the Stars and Stellar Systems Vol. V) should these also prove to be real groups. In this way yet another parallel may be drawn with the processes you discussed for open clusters.

SEMENZATO: You have an idea of the order of magnitude of the velocities of the stars of the cluster after it gets disrupted by a massive cloud. If they are high enough the disruption of the clusters might be one of the mechanisms responsible for ejecting stars out of the galactic plane and account for the early type high velocity stars which are observed at high galactic latitudes.

WIELEN: The final velocities of stars after the disruption of an open cluster by even a close encounter with a giant molecular cloud are insufficient for ejecting stars out of the galactic disk. The internal velocity dispersion of the disrupted system is typically 1 km/s, while the change in the <u>total</u> space velocity can reach about 20 km/s.

MATHIEU: You have discussed the effect of encounters with giant molecular clouds. It seems fairly certain that open star clusters also form in GMCs, and in the cases of NGC 2244 and NGC 2264, for example, its parent GMC is not entirely disrupted by the massive stars in its cluster. What is the dynamical effect of the parent cloud on the young cluster?

WIELEN: The dynamical effects of the parent cloud on an open cluster are difficult to predict theoretically, depending on many highly uncertain input data. From observations it is clear that (at least some) open clusters survive this period, since most of the clusters older than a few 10^7 years are found to be not anymore connected with molecular clouds.

TOOMRE: Please explain again what you have against those hypothetical massive black holes from the halo blasting apart the open clusters. You seem to say that the "hits" from 10^6 $m_\odot$ objects would be far too rare to do their needed job in 10^8 years - even though they would engage in some real overkill if they did happen to come close. But what about cutting up those very massive objects into, say, 100 black holes of 10^4 $m_\odot$ apiece. Wouldn't they thus suffice for this task?

WIELEN: I have nothing against those black holes. A higher number of less massive black holes would have roughly the same effect on the dissolution of open clusters. I used 10^6 $m_\odot$ for the mass of a black hole in the galactic corona, because a value like this was proposed by Lacey, Ostriker and Schmidt for explaining the observed heating of the galactic stellar disk.

KING: In small-N clusters about half the escapers are relatively high-velocity ones, from a single close encounter. Won't those stars avoid being retained in the corona?

WIELEN: According to my numerical experiments, even escapers with positive energies are often retained for a few galactic rotation periods, e.g. for $5 \cdot 10^8$ or 10^9 years, if no external forces help to remove them rapidly from the cluster vicinity.

LARSON: I didn't understand your remark that the equality of the scale heights of old disk stars and open clusters of the same age is consistent with acceleration of the stars by 10^6 $m_\odot$ black holes. In this picture the older field stars have larger scale heights because of acceleration, but the larger scale heights of old open clusters can't be explained this way because, as you stated earlier, an encounter between a clusters and a 10^6 $m_\odot$ black hole would destroy the cluster, so you would just end up with fewer clusters but not with a larger scale height for the surviving clusters.

WIELEN: Clusters are disrupted by the tidal field of passing objects, while the center-of-mass of a cluster is accelerated by the total force provided by the objects. Due to the much stronger decrease of the effect of the tidal field with increasing impact parameter, the more frequent distant encounters accelerate all the clusters, while only some of the clusters are destroyed by the rare close encounters.

THE DYNAMICAL EVOLUTION OF YOUNG OPEN CLUSTERS

Michael Margulis and Charles J. Lada
University of Arizona, Steward Observatory
Tucson, Arizona 85721 U.S.A.
David Dearborn
Lawrence Livermore National Laboratory
P. O. Box 808, Livermore, California 94550 U.S.A.

Using numerical N-body calculations we have simulated the dynamical evolution of young clusters as they emerge from molecular clouds. Starting with initially virialized systems of stars and gas we follow the evolution of these systems from the time immediately after the stars have formed in a cloud until a time long after all the residual star-forming gas has been dispersed. In the models stellar systems were composed of 50, and in some cases 100, stars and these stars were represented as point masses. The stellar mass function followed a power law with an index of -2.5 and ranged over two decades in mass (Scalo 1978). Gas in the models was represented as an extra term in the gravitational potential function governing stellar motions, and was set to follow a density distribution corresponding to a spherically symmetric Plummer potential function (Plummer 1911). Starting with these initial conditions, stellar motions were then integrated and evolution of each stellar system was followed as gas was dispersed from the vicinity of the stars as a function of time.

From these simulations we have found that whether a cloud produces a cluster depends primarily on two parameters: the star formation efficiency (i.e. fraction of total system mass which is stellar) at the time of gas removal and the duration of that gas removal from the system (Hills 1980, Mathieu 1983, Elmegreen 1983, Wilking and Lada 1983). Clusters which remain bound after gas removal undergo significant expansion and may lose a large fraction of their stars. Figure 1 below displays the fraction of stars which remain in a bound system after gas removal versus the initial star formation efficiency. Triangles and circles represent actual models run. For models 1, 2, and 3, gas removal was instantaneous, while for models 4 to 7 gas was removed over three initial system cross times, and for models 8 and 9 gas was removed over four initial system crossing times.

Comparison of the initial and final states of our models with observations of molecular clouds and young open clusters allowed us to place constraints on the initial mass densities, radii, and velocity

J. Goodman and P. Hut (eds.), Dynamics of Star Clusters, 463–465.

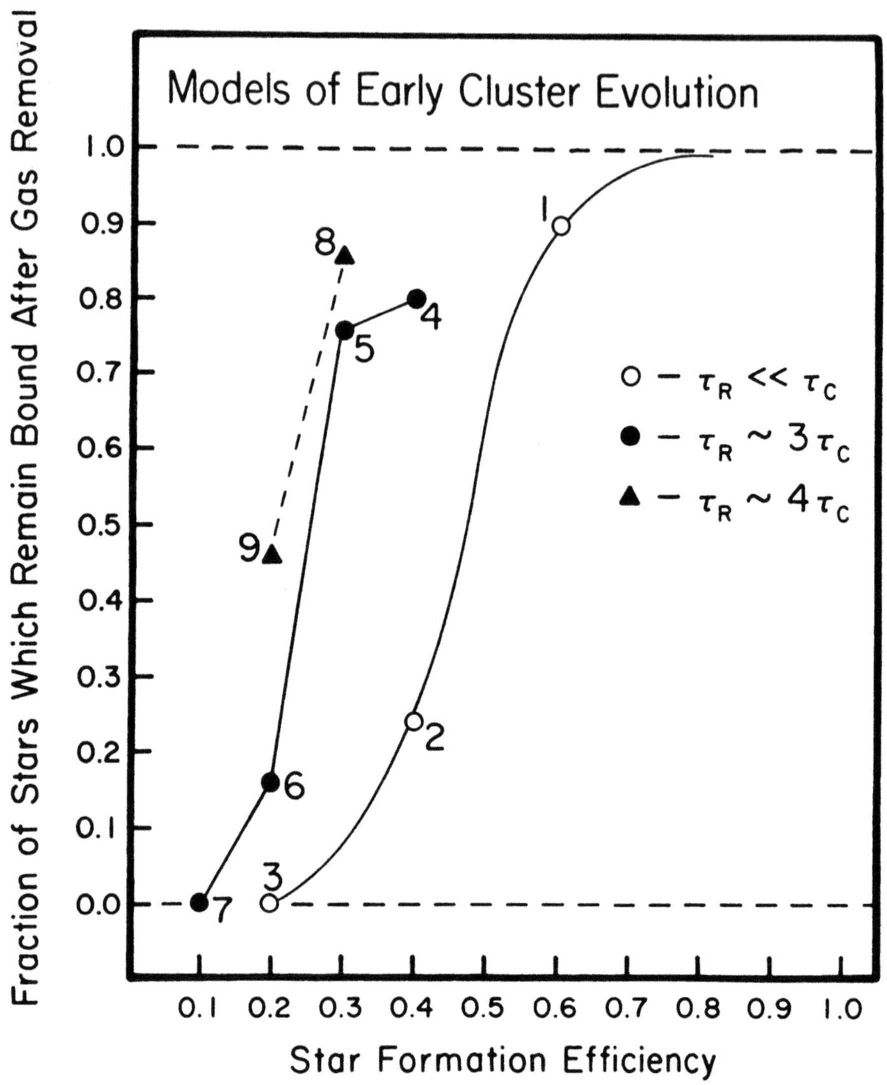

Figure 1. The fraction of stars which remain bound after gas removal as a function of star formation efficiency. The different symbols correspond to different gas removal times.

dispersions of clouds likely to produce open clusters. In particular, we find that in order to produce final clusters similar to the Pleiades, protocluster star formation must occur in the relatively dense (hydrogen number density > 10^3 - 10^5 cm^{-3}), small (size ~ 0.1 - 1.0 pc) condensations found in molecular clouds. In addition, star formation efficiencies in excess of 30% and gas removal times of at least one million years are required in order for these clumps to evolve into such clusters. In fact, even more stringent requirements on cluster forming regions may exist since the tidal shear due to the parent molecular clouds on emerging protoclusters may be significant.

In light of a global star formation efficiency in our galaxy of only 0.1% - 5.0% (Blaauw 1964, Duerr, Imhoff, and Lada 1982), it is not yet clear whether these very high efficiencies, which are required to form clusters, can be achieved in typical molecular clouds. There is, however, one cloud where such high efficiencies have been observed. This is the Ophiuchus cloud, with a star formation efficiency of at least 25% (Lada and Wilking 1984) and relatively quiescent gas dynamics. In this cloud, stellar winds from recently formed stars might be capable of slowly dispersing the gas on time scales of three million years or more (Lada, Margulis, and Dearborn 1984).

Although it may be difficult to achieve high efficiencies in clouds intrinsically, initial conditions in the dynamical structure of clouds may help to enhance the star formation efficiencies there. If the size scale of turbulence in clouds is much less than the Jean length, or if non-gravitational forces are important in the dynamical equilibrium of the gas, stars may form with little or no kinetic energy with respect to the center of mass of the cloud. In this case a newly formed protocluster of stars will quickly collapse and reach a virialized state in about a free fall time, well before significant gas dispersal is likely to occur. Although the subsequent dynamical evolution of the system will then proceed in a way similar to that in the models described above, the star formation efficiency at the time of gas removal will be significantly enhanced over the initial value. For example, calculations concerning such a collapse in constant density clouds indiate that the star formation efficiency in such clouds could be enhanced from 20% to 46% (Lada *et al*. 1984). Since the star formation efficiency is a pivotal parameter governing the likelihood that a particular cloud will produce a cluster, knowledge of these dynamical factors may ultimately be very important in the study of cluster formation.

REFERENCES

Blaauw, A. 1964, *Ann. Rev. Astr. Ap.*, **2**, 213.

Duerr, R., Imhoff, C. L., and Lada, C. J. 1982, *Ap. J.*, **261**, 135.

Elmegreen, B. G. 1983, *M.N.R.A.S.*, **203**, 1011.

Hills, J. G. 1980, *Ap. J.*, **235**, 986.

Lada, C. J., Margulis, M., and Dearborn, D. 1984, *Ap. J.*, **285**, in press.

Lada, C. J., and Wilking, B. A. 1984, *Ap. J.*, in press.

Mathieu, R. D. 1983, *Ap. J. (Letters)*, **267**, 97.

Plummer, H. C. 1911, *M.N.R.A.S.*, **71**, 460.

Scalo, J. M. 1978, in *Protostars and Planets*, ed. T. Gehrels (Tucson: University of Arizona Press), p. 265.

Wilking, B. A., and Lada, C. J. 1983, *Ap. J.*, **274**, 698.

THE DENSE STELLAR CORES OF GIANT HII REGIONS

A.F.J. Moffat [a,b]
Dépt. de Physique, Univ. de Montréal, Canada
W. Seggewiss [a]
Univ.-Sternwarte Bonn, F.R. Germany
M.M. Shara [b]
Space Telescope Science Inst., Baltimore, USA

ABSTRACT. We explore the bright central diffuse objects of the closest visible giant HII regions 30 Dor (LMC) and NGC 3603 (Galaxy), with the aid of CCD imagery in 1" seeing. Both central objects are interpreted as dense cluster cores composed of normal stars with masses ranging up to $\approx 100\ M_\odot$. Each object appears to contain at least one close massive binary. Being significantly denser, the core of NGC 3603 seems to be dynamically more evolved than the core of 30 Dor.

1. CCD IMAGERY OF 30 DOR AND NGC 3603

The closest visible giant HII regions with total mass $> 10^5\ M_\odot$ are 30 Doradus in the LMC and NGC 3603 in the Galaxy. Both of the accompanying broadband blue CCD images are shown on the same angular scale (Fig.1); a linear scale of 2 pc is shown for each. At the center of 30 Dor is a fuzzy, luminous object known as R136, which has recently been argued to be or to contain a supermassive star (Feitzinger et al. 1980,Astron.Ap.84,50; Cassinelli et al. 1981,Science 212,1497; Ebbets and Conti 1982,Ap.J.263,108; Savage et al. 1983,Ap.J.273,597; Chu et al. 1984,Ap.J. in press). Our data do not support this hypothesis; they favor the alternative interpretation that both central objects represent dense cluster cores, composed of normal stars ranging up to $\approx 100 M_\odot$ (Moffat, Seggewiss and Shara 1984, preprint, = MSS). Similar conclusions have been reached by Melnick (1983,ESO Messenger 32,11), Moffat and Seggewiss (1983,Astron.Ap.125,83) = MS, and Walborn (1984,IAU Symp. 108,p.243).

[a]Visiting Astronomer, European Southern Observatory, Chile
[b]Visiting Astronomer, Cerro Tololo Inter-American Obs., Chile

J. Goodman and P. Hut (eds.), Dynamics of Star Clusters, 467–469.

2. THE WOLF-RAYET POPULATION: MASS SEGREGATION ?

Many if not all extragalactic giant HII regions exhibit weak, broad emission lines, indicating the presence of WR stars (Rosa 1983,Highlights of Astronomy 6,625). 30 Dor and NGC 3603 are no exception, as the accompanying net HeII 4686 Å emission images show (Fig.2). While the ~ 15 WR stars in 30 Dor (only ~ half of these are seen in Fig.2) are spread out much like the overall light distribution (R136a itself even contains several WR stars), the ~ 3 WR stars in NGC 3603 are located only in its core. Thus, since WR stars probably represent the advanced evolutionary stage for the most massive stars, it appears that NGC 3603 may have succeeded in concentrating its more massive members in its core, i.e. it may be relaxed (see below). Furthermore, both central WR components reveal diluted, low-amplitude, periodic radial velocity variations (cf. MS and MSS for 30 Dor, and Moffat and Niemela 1984,Ap.J. in press, for NGC 3603), showing that (hard) binaries are present in the cores.

3. RADIAL SURFACE BRIGHTNESS DISTRIBUTION

The overall surface brightness distribution over two orders of magnitude in radius nicely fits a King profile for either case (MSS), with formal core radii:

r_c = 0.26 pc (1"02) for 30 Dor, and

r_c = 0.026 pc (0"76) for NGC 3603.

Since these values are very close to the seeing radius (HWHM ~ 0"7), the true core radii could be considerably smaller.

We now compare best estimates of the central star density and relaxation times of the core regions of each cluster, defined by a radius r_i = 1"5 (~ 2 seeing radii):

	r_i(pc)	ρ_c($M_\odot pc^{-3}$)	t_{rh}(yr)
30 Dor	0.38	$\sim 10^5$	$\sim 20\cdot10^6$
NGC 3603	0.051	$\sim 10^7$	$\sim 10^6$

Since each region contains early O-stars, the evolutionary ages are 2-3·10^6 years. Thus, while the core region of 30 Dor is still far from being relaxed, the core of NGC 3603 is well relaxed (as suspected above) and may be entering a core collapse stage.

4. CONCLUSIONS

NGC 3603 seems to be dynamically more evolved than 30 Dor, with more of its massive stars concentrated in its core, whereas the 30 Dor cluster shows no apparent differences in stellar content between its core and its halo. The fate of the collapse stage for NGC 3603 may depend on the presence of the central hard binary.

The 30 Dor cluster is possibly the youngest among the populous clusters in the LMC. Although NGC 3603 verges on being a populous cluster in the LMC sense, it is a rare type of object in the Galaxy and indeed may not survive for long in its present state.

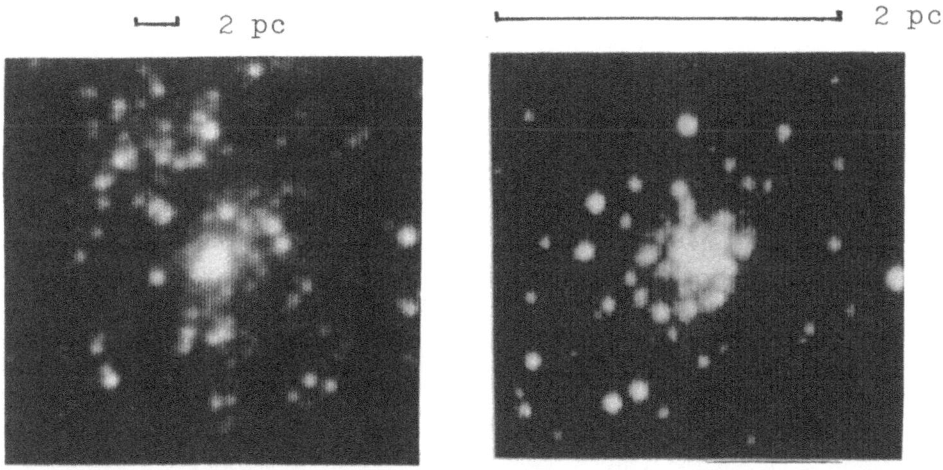

Fig.1: Broadband blue images of the central 80" of 30 Dor (left) and NGC 3603 (right).

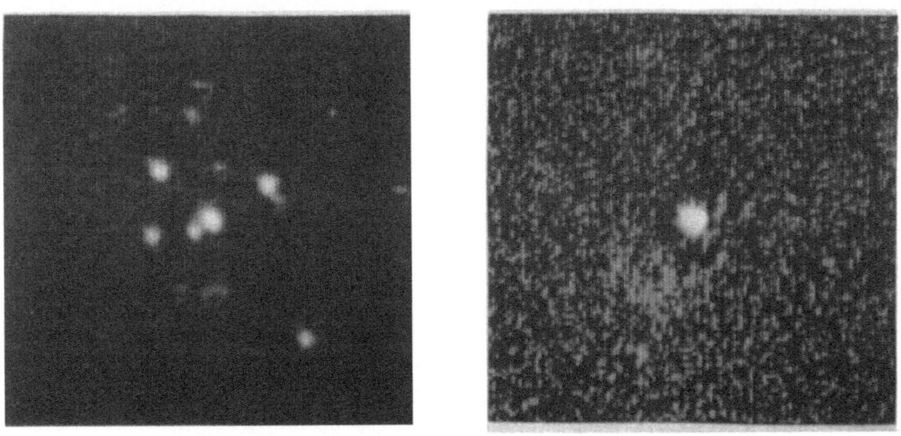

Fig.2: Net HeII 4686 Å CCD images of the central 80" of 30 Dor (left) and NGC 3603 (right).

N-BODY SIMULATIONS OF REALISTIC OPEN CLUSTERS

Elena Terlevich
Royal Greenwich Observatory
Herstmonceux Castle
UK

ABSTRACT. N-body simulations of dynamical evolution of open clusters have been computed with the purpose of comparing them with observations. Special effort has been put into reproducing conditions present in galactic clusters. Most of the models contain 1000 bodies with masses following a power-law mass function of slope α =-2.75 and mean mass 0.5$M_\odot$. Neutron stars or white dwarfs (depending on the initial stellar mass) are generated by instantaneous changes in individual masses, when stars reach the end of their main sequence life. Close approaches between particles are treated by a two-body regularization technique that allows to follow binary evolution in detail. Two types of tidal perturbation are considered: a smooth linearized galactic tidal field is simulated assuming that the clusters move in a circular orbit at 10kpc from the galactic centre; transient shocks are simulated by encounters with extended interstellar clouds of different mass-spectrum, densities and space concentration. It is found that the combined action of evolutionary mass loss and binaries (when the cluster has a realistic mass function) is enough to arrest the core collapse. Tidal heating shapes the halo of the cluster. There is good agreement with the observed density and velocity distribution of open clusters and with reported changes in their mass function.

1. INTRODUCTION

Assuming that the observed age distribution of open clusters is not biased by selection effects (Wielen 1971; Lyngå 1982), that the rate of formation in the galaxy has been constant for the last few billion years and that the lifetime of the clusters depends on individual parameters but not on the time of formation, the total lifetime of galactic clusters can be deduced from their observed age distribution (Oort 1958).

The purpose of this work is to put limits to the conditions that determine the total lifetime of clusters: total initial mass, stellar mass distribution, external forces, evolutionary mass loss and internal dynamical evolution; using new observational and theoretical

J. Goodman and P. Hut (eds.), Dynamics of Star Clusters, 471–475.

parameters. Both the asymmetry involved in these effects and the small number of stars concerned, make open clusters ideal targets for N-body solutions. The models described in this paper have been produced using Aarseth's N-body code in which the equations of motion are regularized in the case of close encounters between pairs of particles (Aarseth 1985). To produce realistic simulations of open clusters, the following effects have been included in addition to direct N-body interactions: mass function of the cluster components, mass loss due to stellar evolution, galactic tidal field and transient tidal shocks produced by passing interstellar clouds. In order to compare the results with observations, careful consideration has been given to the selection of the initial parameters which I describe in the following section. A detailed analysis of the initial conditions and the method in general is given elsewhere (Terlevich 1983).

2. THE MODELS

The equations of motion used correspond to the direct N-body solution (exact newtonian potential) plus tidal terms assuming that the cluster follows a direct plane circular orbit with uniform velocity at 10kpc from the galactic centre, in an axially symmetric tidal field (Wielen 1965; Hayli 1967). For close two-body encounters the equations of motion are regularized according to a Kustaanheimo-Stiefel scheme. The critical distance for regularization is given by the force between two approaching particles relative to the perturbing force by the rest.

The initial mass function chosen is assumed to follow a power law. The values adopted for the slope α , the mean mass <M> and the mass range are taken from published observations of open clusters (Taff 1974; Tarrab 1982; van Altena 1966). Table I shows the relevant parameters for selected models.

Table I- Model parameters

MODEL	N	<R> pc	α f(M)	R_t pc	$T_{1/2}$ 10 yr	T_{cr} 10 yr
I	250	2	2.35	7	174	10.8
IV	500	2	2.35	9	252	7.6
V	1000	2	2.35	11	420	5.4
VI	1000	2	2.75	11	460	5.5
VII	1000	3	2.75	11	365	9.9
IX(1)	1000	2	2.75	11	460	5.4
XII(2)	1000	2	2.75	11	770	5.4
XIII(3)	1000	2	2.75	11	30	5.4
XV	1000	2	1.10	27	47	1.4

Notes to Table I-(1) Includes encounters with Spitzer's standard clouds. (2) Initial mass segregation: more massive stars outside; Q=0.5. (3) Identical to Model VI until T=100 million years when it suffers a catastrophic encounter with one GMC.

All but model XV (with $\langle M \rangle = 7M_\odot$) have a $\langle M \rangle$ of $0.5M_\odot$; $\langle R \rangle$, sometimes called the virial radius, has the meaning of a mass weighted harmonic radius averaged over time; R_t is the tidal radius (King 1962). Stars at a distance of $2R_t$ are considered escapers. $T_{1/2}$ is the time when half of the initial number of stars have escaped from the cluster. T_{cr} is the crossing time defined as $2\langle R \rangle/(r.m.s.)v$.

The initial coordinates follow a random spherical distribution with density proportional to $1/r^2$. The initial velocities are isotropic. To simulate violent relaxation through an initial contraction of the cluster (Henon 1967) the parameter Q relating kinematic and potential energy is chosen as 0.25 for all the models but one; for a virialized system Q=0.5.

In the range of masses that is relevant for open clusters (0.3 to $15M_\odot$) processes of mass loss like supernova events or formation of planetary nebulae can be assumed to occur on time scales which are very short compared with the characteristic dynamical times for the cluster. The pattern used for instantaneous mass loss is as follows: at the end of its mass dependent main sequence life, a star suffers a single mass loss event. If its mass is lower than $6M_\odot$ it becomes a $0.7M_\odot$ white dwarf. Otherwise it becomes a supernova with a $1.5M_\odot$ remnant to which high velocity is given (four times the r.m.s. velocity of the stars in the cluster) to account for the recoil effect.

Standard HI clouds (Spitzer 1958) are simulated as polytrope spheres of index 5, with masses between 50 and 500 $M_\odot$ and half mass radii of 3.6 to 7.7 pc. Five clouds are confined inside a boundary of radius R =28pc to give a number density of 5.4×10^{-5} pc^{-3}. A cloud reaching the boundary is replaced by another at a random position over the sphere, using a "cloud rise" and "cloud set" procedure to avoid sudden changes in the force. The three dimensional velocity of the clouds is taken at random from a maxwellian distribution with mean value $\langle V \rangle$=10km/sec and dispersion σ=6.25km/sec. The mean field inside the boundary is subtracted when including the force due to the individual clouds. Models XIII suffer a single encounter with one Giant Molecular Cloud of $1\text{-}5\times10^4 M_\odot$ and 10pc radius, approaching the cluster from a distance of 75pc at a speed of 5km/sec (Solomon, private communication).

3. DISCUSSION

Mass loss from the stars affects the cluster in various ways. The most obvious one, though it seems to be only marginal, is the effect on the cluster lifetime. Model V, losing more mass up to half-life due to its initial mass function, disrupts quicker; model XII, for which the energy increment due to mass loss is smaller (because the more massive stars are originally in the outer regions) lives longer. It is evident from the analysis of model XV that if clusters originally containing around a thousand stars are going to survive as we see them today, they cannot have formed with a mass function much flatter than the Salpeter one.

More striking is the effect of stellar evolution on the density distribution and the time evolution of the mass function, through a combined action with binary activity. The components of dynamically formed binaries, tend to be the most massive stars in the cluster and therefore mass loss occurs frequently among them. This causes the pair to lose binding energy inducing frequent change of components. As a result, binaries attain a larger cross section and remain "active" for longer periods (avoiding the final hardening process, Heggie 1972) in scattering light stars towards the halo. Consequence of this process are: preferential depletion of light stars, arrest of the core collapse and enhanced mass segregation. The contraction of the core can be analysed by plotting the time evolution of the radii containing 5, 10, 50 and 90% of the cluster mass (Figure 1). Figure 2 shows the density profile for four mass groups in model VI, where the different behaviour of light and heavy stars in the core and halo are apparent. In order to by-pass the small number problem, five consecutive time steps have been combined in figure 2.

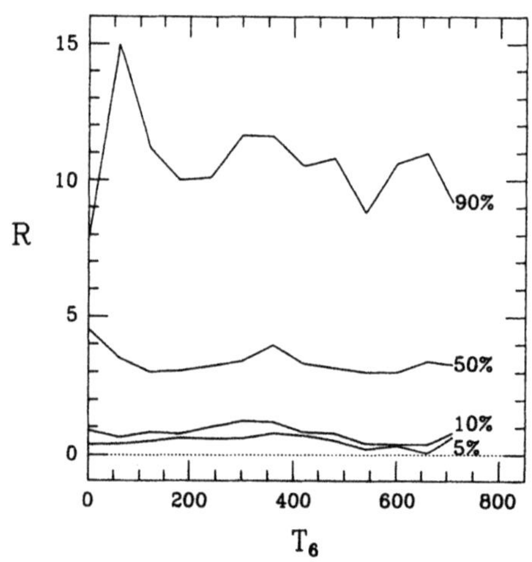

Figure 1-Evolution of the radii for 5, 10, 50 and 90% of the cluster total mass.

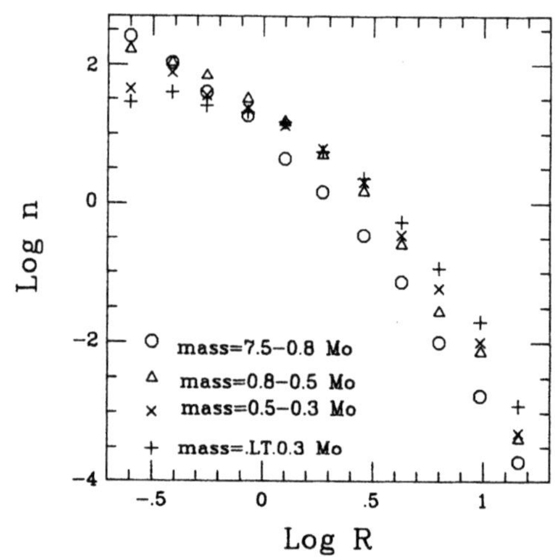

Figure 2-Density profiles for Model VI at time=580 million years.

The tidal field has two effects on the cluster. The equipotential surfaces around a non-isolated cluster are non spherical. There is a compression in the direction perpendicular to the galactic plane (Z). Figure 3 shows the projection on the (X.,Z) plane of model VI at half life, where this flattening is most notorious. X is the direction towards the galactic centre. The second effect is a "heating" of the halo by the galactic tidal field and by tidal shocks. Figure 4 shows the tangential velocity dispersion as a function of distance to the cluster centre, for three models with tidal heating. The difference in slope with the one obtained for isolated models (Aarseth 1974) is notorious. In spite of the difficulties inherent in velocity distribution studies of haloes in open clusters, a similar flat slope seems to be observed (van Leeuwen 1983). A more detailed comparison with observations is to be published elsewhere (Terlevich and van Leeuwen, in preparation).

An inspection of Table I shows that the lifetime of open clusters is not affected by encounters with HI clouds, but an eventual encounter with a giant molecular cloud is catastrophic (Wielen, this symposium).

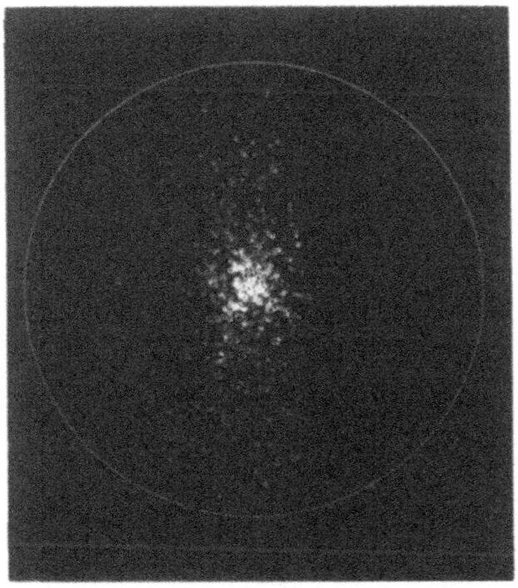

Figure 3-Tidal flattening;(x,z) projection of model VI at half life.R=2R_t .

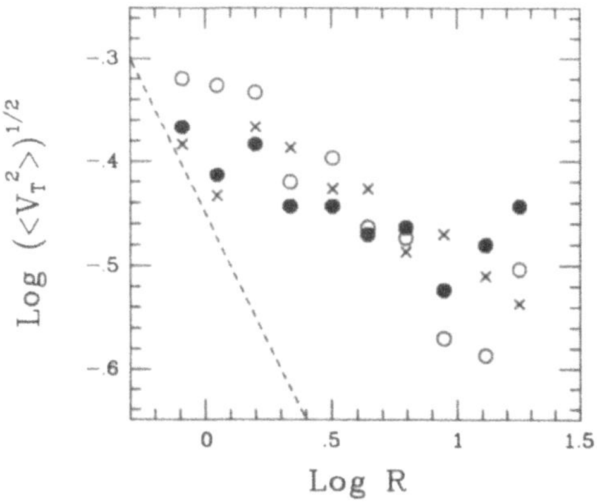

Figure 4-Tangential velocity disp. as a function of radius for models with tidal heating. O:model VI;●:model VII; X:model IX;straight line for isolated models (Aarseth 1974)

It is a pleasure to thank the IAU and the Cambridge Philosophical Society for financial support to attend this Symposium.

REFERENCES

Aarseth, S.J. 1974, A.and A., 35, 237

Aarseth, S.J. 1985, to appear in Multiple Time Scales by J.U. Brackbill and B.I. Cohen, Academic Press, N.Y.

Altena, W.F. van 1966, A.J. 71, 482.

Hayli, A. 1967, Bull.Astr. (3) 2, 67.

Heggie, D.C. 1972, Ph.D. Thesis, Cambridge University.

Henon, M. 1967, Mem.Soc.Roy.Sci.Liege, Series 5 15, 243.

King, I.R. 1962, A.J. 67, 471

Leeuwen, F.van 1983, Ph.D. Thesis, Leiden.

Lyngå, G. 1982, A.and A. 109, 213.

Oort, J.K. 1958, Stellar Populations (ed.D.J.K.O'Connell), Vatic. 5, 63

Spitzer, L. 1958, Ap.J. 127, 17.

Taff, L.G. 1974, A.J. 79, 1280.

Tarrab, I. 1982, A.and A. 109, 285.

Terlevich, E. 1983, Ph.D. Thesis, Cambridge University.

Wielen, R. 1965, Ph.D. Thesis, Heidelberg.

Wielen, R. 1971, A.and A. 13, 309.

A PROPER MOTION STUDY OF THE PLEIADES CLUSTER

Floor van Leeuwen

Royal Greenwich Observatory,
Herstmonceux Castle,
Hailsham, East Sussex, BN27 1RP,
United Kingdom

Abstract. The first results of an extensive proper motion study of the Pleiades cluster are presented. A total of 166 exposures covering a 3 by 3 degrees area are now incorporated. The accuracies of the centennial proper motions range from 0″.015 in the central region to 0″.2 in the outermost region.

Introduction. A good segregation of cluster members in the central 2.5x2.5 degrees region of the Pleiades down to m(pg)=16 has been obtained by Hertzsprung(1947). The internal motions in the cluster centre have since been measured successfully by Jones(1971) and Vasilevskis et al (1979). Three papers dealt with an astrometric detection of members in the outer parts of the cluster: Trumpler(1921), Arthyukhina and Kalinina(1971) and Van Leeuwen(1983) (vL hereafter). The astrometric selection criteria obtained in these papers are not sufficient to rely on for a membership selection. In general, improvements should come from an extention of the field and magnitude range for the internal motions, improvement of the accuracies of internal motions, and an extention and improvement of the membership segregation for stars that occupy the cluster halo.

For the length of astronomical photography, the Pleiades has been a popular subject to picture. The plate material existing at different observatories is therefore likely to be sufficient to improve significantly each of the above mentioned studies. This would allow us to derive a reliable picture of the distribution of masses, energy and angular momentum in the cluster. Moreover, the detection of late type members by means of proper motions has become very important for the study of the pre-main sequence evolution of these stars (see Van Leeuwen 1983, Stauffer et al 1984).

The present paper is an intermediate report on an extensive study of the Pleiades cluster. It provides in the first section a review of the plate material and reduction techniques used. The second section describes the star catalogue and the updating mechanism used in

J. Goodman and P. Hut (eds.), Dynamics of Star Clusters, 477–480.

collecting data. Section three provides some of the results obtained and section four gives a brief outline of further extensions to the catalogue.

The data. In the most recent catalogue the following data have been incorporated: 161 exposures taken with the Leiden Astrograph (39".3/mm) and 5 exposures taken with the Paris Astrograph (60".0/mm). The Leiden plates cover 1.5x1.5 degrees, the Paris plates 2.5 by 2.5 degrees. The epochs of the plates range from 1898 to 1979. Positions of stars on these plates have been measured with the Leiden Observatory ASTROSCAN automatic measuring device (see Swaans 1981, vL).

The Leiden plates cover the 3x3 degrees field with 9 plate centres, in a four-fold overlap. The first problem that had to be solved was the connection of these non-central overlapping plates. A similar problem has been treated by Eichhorn et al(1970) using multiple plate solutions, involving large matrix manipulations. Systematic errors in the positions thus obtained were claimed by Vasilevskis et al(1979). Here too, we made as a first step a multiple plate solution, using only stars within 7.5 cm from each plate centre. A transformation from rectangular to spherical coordinates is necessary, for which a determination of the position of the projection point was needed. The iterative solution to this position was described by vL. The overlap reduction has been satisfactory and allowed me to perform a reduction of each individual plate, including those exposed with the Paris Astrograph, to the composite Leiden field without significant loss of accuracy. A detailed comparison with the positions presented by Eichhorn et al(1970) is foreseen for the near future.

In addition to an astrometric reduction, also photometric reductions were performed. The density measurements by ASTROSCAN in squares of 20 by 20 micron over the whole stellar image alowed me to obtain from Leiden Astrograph plates magnitudes with an accuracy better than 0.05. As calibrating sequence I used the B magnitudes of the VBLUW photometry presented by vL. Thus, most of the mean magnitudes in the catalogue have an internal accuracy better than 0.01.

The catalogue. The catalogue consists of two files. The first contains information on all exposures incorporated in the catalogue. The second file contains all the star records, each of which contains a reference number for the star and three information matrices (results of least squares solutions by means of Householder transformations). These matrices contain all the information on position, proper motion and magnitude needed for updating the catalogue. An interpretation program is used to represent the catalogue in positions, proper motions and magnitudes, giving for each of these the up to date accuracies. Incorporating new data into the catalogue involves updating with Householder transformations the information matrices, which represent all the previously incorporated data. The updating process is automatic, but allows for manual interaction in order to prevent erratic data to enter the catalogue. A star is only accepted as a catalogue entry if at least five independent data points are available, so that checks on errors are possible. Stars that do not fulfil this requirement yet are stored on a temporary data file in the form of the observation equations.

Some results. The present catalogue contains information on some 1850 different stars, out of which 193 are likely to be cluster members. The accuracies of their centennial proper motions range from 0".015 to 0".1 for 95 percent of these stars. It allows one to study the internal proper motion dispersions up to 2.5 pc (1.2 degrees) from the cluster centre, for stars down to m(pg)=14.5 (mass range 4.0 to 0.6 Mo). A preliminary investigation into these dispersions for stars brighter than m(pg)=11.5 confirms the results obtained by Jones(1971) and Van Leeuwen(1980) out to a radius of 1.5pc. Beyond that radius further interpretation in radial and transverse proper motion dispersions becomes virtually impossible due to the uncertainty in the parallactic influences on the proper motions. Beyond 2 pc these influences cause dispersions greater than the internal dispersions and introduce too much noise too allow further interpretation. The stars fainter than m(pg)=11.5 show projected on the cluster centre a proper motion dispersion of at most 0".06/cent., which is only half the value observed for the brighter stars. Both groups, however, have entirely different spatial density distributions, which do not allow any straight forward interpretation. At larger radii, these fainter stars show virtually only effects of parallactic motion and measuring inaccuracy, and do not allow for a dynamic interpretation.

Five new probable members and one possible member have been detected. In addition, six stars from the list of probable and possible members in the outer regions of the cluster by G.Pels (see vL) have been confirmed : P52, 57, 58, 60, 71 and 142. One of the probable new members is a star from the catalogue by Hertzsprung(1947), Hz 5. This star is situated very close to the edge of the field measured by Hertzsprung. Search maps for these stars have been prepared and are to be presented for publication to the supplements of one of the main journals soon.

Future work. The measurements of more than 500 exposures are ready and waiting to be added to the existing catalogue. They cover mainly the outlying region of the cluster. An incorporation of the original measurements used by Eichhorn et al(1970) is also foreseen. They will not add much weight to the proper motions, but allow for a check on the accuracy of the positional system and the origin of the magnitude error claimed by Vasilevskis et al (1979). In addition, I hope to measure some other old and deep exposures in order to increase the magnitude range of the catalogue to m(pg)=16 or 16.5. This would allow a study of the internal proper motion dispersion over a still wider mass range, and may in addition add some certainty about the membership in this spectral region of the cluster. Any communication on the existence of old plates (within a radius of 5 degrees from the cluster centre and preferably with scales not worse than 60" per mm) would be most appreciated.

Acknowledgement. I would like to thank Profs. S.Vasilevskis and A.Blaauw for initiating this project, and R.S.LePoole and A.A.Schoenmaker for their assistence during the measuring of the plates. In addition, I enjoyed discussions with C.A.Murray on the techniques employed in and the contents of the present paper.

References.

Arthyukhina, N.M. and Kalinina, E.P., 1970: Trudy Gos.Astron.Sternberga, 39, 111

Eichhorn, H., Googe, W.D., Lukac, C.F., Murphy, J.K., 1970: Mem.Royal Astron.Society, 73, Part 1, 125

Hertzsprung, E., 1947: Ann.Sterrew.Leiden, Vol XIX, part one

Jones, B.F., 1970: Astron.J., 75, 563

Stauffer, J.R., Hartmann, L., Soderblom, D.R. and Burnham, N., 1984: Ap.J., 280, 999

Swaans, L.W.J.G., 1981: Ph.D. thesis, Leiden

Trumpler, R., 1921: Lick Obs.Bull. 333, 110

Van Leeuwen, F., 1980: in "Star Clusters", IAU Symp.85, p157; ed. J.E. Hesser, Reidel Publ., Dordrecht, Netherlands

Van Leeuwen, F., 1983: Ph.D. thesis, Leiden

Vasilevskis, S., Van Leeuwen, F., Nicholson, W. and Murray, C.A., 1979: Astron.Astroph.Suppl., 37, 333

SPACE TELESCOPE OBSERVATIONS OF GLOBULAR CLUSTERS

John N. Bahcall
Institute for Advanced Study
Princeton, New Jersey

ABSTRACT

Six proposed Space Telescope programs involving globular clusters are described. The projects appropriate for galactic clusters are: the detection of white dwarfs, the study of the faint end of the Population II luminosity function, the measurement of mass segregation, and the search for a cusp in the density distribution caused by core collapse or by a massive black hole. The two programs that involve extragalactic globular clusters are: the determination of the luminosity function of clusters around different galaxies and the measurement of tidal radii of clusters surrounding elliptical galaxies.

I. INTRODUCTION

All fields of astronomy will benefit from observations with Space Telescope and the study of globular clusters is no exception. It is easy to be enthusiastic in general about the potentialities of the ST Observatory, but it is more difficult to decide which observations will provide the largest scientific return. I will give my preliminary guesses as to what are some of the most appropriate lines of research for ST observations of globular clusters, but I will look forward to hearing from you - and finally from the observations themselves - as to what really are the most important areas of investigation.

II. CHARACTERISTICS OF THE OBSERVATORY AND THE INSTRUMENTS

I will only review briefly the characteristics of the observatory and the instruments that are most relevant for the present discussion. There are comprehensive discussions in the literature of the telescope and its instruments (see, e.g., O'Dell 1980, Leckrone 1980, Bahcall and Spitzer 1982, and the articles in the Patras session on the Space Telescope Observatory, available from the STScI). As most of you know, the ST mirror is 2.4 m (94 inches), which is not large by ground-based

J. Goodman and P. Hut (eds.), Dynamics of Star Clusters, 481–498.

standards. The optical design is the Ritchey-Chretien variation of the Cassegrain configuration. The wavelength coverage is broad, from about 1150 angstroms to about 1 micron. The optics is expected to be stupendous. At 6330 angstroms, the full width of a stellar image at half of maximum intensity will be 0.06 arcseconds and 70% of the total image energy will be enclosed within 0.1 arcseconds radius. The actual performance of the mirror may exceed even these wonderful specifications.

There are six instruments in the initial configuration for the Space Telescope Observatory. They are: (1) a Wide Field/Planetary Camera; (2) a Faint Object Camera (supplied by the European Space Agency); (3) a High Speed Photometer; (4) a Faint Object Spectograph; (5) a High Resolution Spectograph; and (6) the Fine Guidance System (which will be used for astrometry). Except for the absence of an infra-red sensitive detector, the initial set of instruments is comparable to the best available instrument complement at a ground-based observatory. For each of the instruments, a number of different modes and options are available. Anyone wishing to know more details should consult one of the review articles cited above.

The characteristics of the instruments determine in large part which projects are most promising for study in the first several years of ST operations. I will therefore limit myself to discussing programs that can be carried out with the existing selection of instruments.

Much of our discussion will center on the use of the two sets of cameras. I will summarize some of the principal characteristics of the cameras. The Wide Field/Planetary Camera (WF/PC) can be used in the following two modes (among others): (1) Field 2.'6 x 2.'6 (f/12.9) and pixel size of 0."1 or (2) Field 1.'1 x 1.'1 (f/30) and pixel size 0."04. The Faint Object Camera (FOC) has many options. The two modes that we consider most often have: (1) Field 0.'4 x 0.'4 (f/48) and pixel size of 0."04 or (2) Field 0.'2 x 0.'2 (f/=96) and pixel size of 0."02. The cameras can reach a visual apparent magnitude V = 28 mag in about 30 minutes, with the WF/PC being faster for $\lambda > 5000$ Å and the FOC being faster for $\lambda < 5000$ Å. The cameras will often be used in a novel way: they can simultaneously take exposures on fields whose centers are 6.6 arcminutes apart. This will be especially valuable for extended objects like globular clusters; one can study at the same time a field close to the core and one in the cluster halo.

What are the characteristics of the Space Telescope Observatory that most affect its usefulness for particular kinds of research? The good characteristics are well known: high spatial resolution, broad wavelength coverage, accurate pointing and stability requirements, and high time resolution. The bad characteristics should be equally well known, for they are critical in deciding what should not be attempted. In my opinion, the most limiting features of the ST Observatory are: (1) the telescope is photon starved for many observations (because the aperture is not large); (2) only relatively small fields can be covered in any one exposure (this is not primarily a survey instrument); and (3) ST time will be vastly oversubscribed. This latter characteristic is of special importance. There are only 24 x 365 hours in a year. With an operating efficiency for observations of about 1/3, this means that there are only about 3000 hours available per year for the thousands of astronomers that want to use the telescope. There obviously will not be very many long programs, at least by ground-based standards where a lucky astronomer may hope to get 30 or more nights a year.

III. PREDICTING STAR DENSITIES FOR ST

In order to design some of the Space Telescope projects appropriately, it is necessary to calculate the expected number of stars per unit area that correspond to a total surface brightness that is measured from the ground. This calculation can be carried out conveniently using the conversion function, F:

$$F(\leq M_V) \equiv \frac{\int_{-3}^{M_V} \Phi\ (M_V)\,dM_V}{\int_{-3}^{\infty} \Phi\ (M_V)L/L_\odot dM_V} \ . \qquad (1)$$

This function, with different numerical values, was first used by Baum and Schwarzschild (1955) in a different context and, for purposes of predicting ST observations, by I. King (1984). I adopt for the purposes of this talk the Population II luminosity function, $\Phi(M_V)$, used by Bahcall and Soneira (1984). At the bright end, this luminosity function is the same as the 47 Tuc luminosity function of Da Costa (1982). The adopted luminosity function, when combined with the Bahcall-Soneira Galaxy model, also fits the observed counts of spheroid field stars. The available counts are mainly sensitive to the region $M_V = +3$ to $M_V = +8$. The luminosity function used here includes both an estimate for the effect of the horizontal branch and the Wielen dip, near $M_V = 6$. The luminosity function and the contribution to the luminosity from different absolute magnitudes (the integrand of the integral in the denominator of equation 1) are shown in Figure 1.

Table 1 gives some numerical values for the conversion function, $F(\leq M_V)$. The number of stars expected per unit of area to a given absolute magnitude can be calculated from the observed surface brightness, σ (in stars per square arc second or stars per square arc minute), the estimated distance modulus $(m - M_V)$, and the conversion function F. One has:

$$N(\leq M_V) = F(\leq M_V) \times 10^{-0.4[\sigma - (m - M)_V - 4.7]} . \qquad (2)$$

For several of the projects that we will discuss below, it is important to know at what star density the effects of crowding will become significant. A reasonable criterion can be established in analogy with ground-based experience, which is summarized in the classic article of King, Hedemann, Hodge, and White (1968). For purposes of this discussion, we can anticipate that crowding will become significant when the area on a detector that is covered by star images reaches about 3%.

This criterion suggests that the interpretation of star densities of more than about 3 per square arc second in the visual wavelengths will be complicated by crowding effects. It is possible that the optical properties of the mirror will be considerably better in the ultraviolet, although we will not know the ultraviolet characteristics until they are measured when the telescope is operating in space.

For deep exposures, the condition for crowding to be significant can be expressed, using Table 1 and equation (2), as

$$10^{-04[\sigma_V(r) - (m - M)_V - 4.7]} \leq 3/F(\leq M_V),$$

when σ is expressed in magnitudes per square arc second.

IV. SOME IMPORTANT GALACTIC ST PROGRAMS

I will now discuss four illustrative galactic programs that seem to me to be both particularly important and especially appropriate for ST observations.

A. The Detection of White Dwarfs

The study of white dwarfs in globular clusters is a "made-for-ST" project! Even the youngest of the white dwarfs are expected to have $M_V = +10$. Therefore their apparent magnitudes are $V \approx 24$ or fainter in the clusters - difficult to detect with ground-based telescopes. On the other hand, the apparent visual magnitude range between 24 and 28 is convenient for observations with both cameras on the ST. Moreover, calculations suggest (see below) that a number of white dwarfs should be visible on a single FOC frame (f/48) of the inner regions of nearby globular clusters.

The study of the number of white dwarfs in globular clusters will be important for the theory of stellar evolution since it will be possible to check the general expectation that stars with masses less than a critical mass (variously estimated between $3M_\odot$ and $10M_\odot$) end up as white dwarfs. Moreover, it will be possible to study the cooling of white dwarfs in a much simpler context than can be done with disk stars: all of the objects are at the same distance and the only variation in the time of the formation of the white dwarfs is caused by the stellar evolution time of the parent. This study is especially important since there is a well known discrepancy between the theory of white dwarf cooling and the observations of disk white dwarfs.

There are two physical effects that must be taken into account in order to predict the expected number of white dwarfs in globular clusters. (1) The total number of white dwarfs that are created depends upon the initial mass function of the globular cluster. (2) The white dwarfs formed by the more massive stars will already have cooled down to fainter magnitudes. One can express, symbolically, the expected number of white dwarfs per unit area, N_{WD}, in terms of the total number of stars, N, given in equation (2). One has,

$$N_{WD} = G_{cool}(\leq M_V) \times R \times [F(M_{V,turnoff}) - F(M_{V,critical})]$$

$$\times 10^{-0.4[\sigma - (m - M)_V - 4.7]} . \quad (3)$$

The function G_{cool} represents the fraction of the white dwarfs that have not cooled down below M_V. R is the average amount by which the initial mass function exceeds the Population II luminosity function used to calculate the conversion function, F. Assuming a Salpeter (1955) initial mass function, a critical mass for the production of white dwarfs of $5M_{\odot}$, and a turnoff mass of $0.7M_{\odot}$, I find: $R \approx 6$.

Only those white dwarfs that have formed relatively recently, from masses just above the turnoff mass, will still be bright enough to be detected with ST. I have estimated G by using the fact (see VandenBerg 1983) that the turnoff luminosity is approximately proportional to the inverse age of the cluster and to the fourth power of the turnoff mass. This implies that the ratio of the range of masses from which observable white dwarfs have come to the present day turnoff mass is about

$$\frac{t_{cool}}{4t_{age}},$$

where t_{cool} is the time for white dwarfs to cool down to the limit of detectability and t_{age} is the age of the cluster. Carrying out the integrals over a Salpeter initial mass function, I find:

$$G_{cool}(\leq M_V) = \frac{0.35 \times t_{cool}}{t_{age}} . \quad (4)$$

The quantity G is very small and greatly reduces the expected number of detectable white dwarfs. For example, one might expect to be able to study white dwarfs down to absolute visual magnitudes of about 13.5 (or with a major effect perhaps to 14.5) in the nearer clusters. Using the white dwarf cooling times of Lamb and van Horn (1975) scaled to $0.7M_{\odot}$ as given in equation (7) of Green (1980), I find that

$$\frac{t_{cool}}{t_{age}} \approx 0.1 \ (0.2). \quad (5)$$

Thus G is only of order 0.035 for $M_V \leq 13.5$ and rises to 0.07 for $M_V \leq 14.5$. For the more distant clusters, we may be able to reach to $M_V = 12.5$, in which case G decreases to 0.015. More generally, $G(\leq M_V) \cong 0.035 \times 10^{0.3(M_V - 13)}$.

I have estimated for the high surface brightness cluster 47 Tuc the number of white dwarfs that may be visible in a deep exposure (1/2 hour) with the FOC (f/48 mode) to U = 27 mag and V = 27 mag. For 47 Tuc, I took (Peterson and King 1975) a distance modulus of 13.5, a central surface brightness in the visual of 5.55 magnitudes per square arc minute, and a core radius of 0.7 pc. In order to avoid crowding effects, I assumed that the part of the frame that was closest to the center was at a radial distance of three core radii. I also averaged the predicted surface density over the surface of the FOC detector, assuming that the cluster is described by a King profile.

The net result of these calculations is that only of order 15 white dwarfs are expected to be visible on the FOC frame of 47 Tuc. This is about 1% of the total number of stars that are expected to be detected in the FOC exposures discussed above.

I have made a similar set of calculations for M4 which has a much lower central surface brightness (about 9.15 mag per square arc minute), and is closer than 47 Tuc (2.1 kpc compared to 4.9 kpc). In this case, one can center the exposure on the core of the cluster. Assuming the same limiting magnitudes of U = 27 and V = 27, I find that only of order 5 white dwarfs are expected in a single FOC (f/48) frame of M4.

B. The Low Mass Stars

The detection of the faint end of the globular cluster luminosity function is of great theoretical interest because of the light it may shed on the dynamics of globular clusters and on theories of star formation. Are the low mass stars concentrated on the outside of the clusters? Have they been stripped from the outermost regions? What is the spectrum of stellar masses that were formed? Since all of the stars are at the same (known) distance as the parent globular cluster, some aspects of this study will be simpler than for the corresponding investigations with disk stars. Moreover, it is possible that the study of the faint end of the cluster luminosity function will provide some clue as to whether or not the so far unobserved mass in the disk could be in the form of low mass stars.

I have calculated the expected number of stars that will be observed in an exposure to V = 27 mag if the Pop II luminosity function shown in Figure 1 is appropriate to the clusters. In order to verify that stars are indeed of low mass, it will be necessary to take also a deep exposure in I with the wide field camera. For 47 Tuc, the total number of stars is about 10^4 in a frame situated as described above. For a central exposure on M4, the corresponding number is also about 10^4. About one-fourth of these stars may have masses between $0.1M_\odot$ and $0.3M_\odot$, if the stars were originally present and dynamical effects have not depleted the core of low mass stars.

C. Mass Segregation

The effects of mass segregation can be studied with ST observations over a wide range of masses, from the turnoff mass ($\approx 0.7M_\odot$) to the mass corresponding to the faintest limiting magnitude ($\approx 0.15M_\odot$ for the nearest clusters).[1] This range corresponds to a ratio of masses of more

[1]In the absence of an empirical mass to visual light relation for low mass Population II stars, I have interpolated between the low-metallicity and normal metallicity parameters at a given effective temperature in the theoretical study of VandenBerg et al. (1983). I normalized the results to the empirical mass versus visual luminosity relation for normal metallicity disk stars.

than four. Observations with the cameras should yield star surface densities over a very large range in radii also, from $\approx 0 \times r_{core}$ to more than $25 \times r_{core}$. These data can be acquired as a natural by-product of the programs to detect white dwarfs and low mass stars. It will be particularly useful to take simultaneous exposures with the WF/PC and the FOC, one close to the cluster center and one quite far away (6.6 arc minutes or of order 20 core radii for some typical cases).

Multi-mass models for clusters should become the norm, not the exception, for both analytic studies and for many-body simulations. The beautiful data of Da Costa (1982) already show that mass segregation is an observationally accessible and important phenomenon in 47 Tuc. In addition, various theoretical studies that were discussed at this conference have confirmed the importance of mass segregation. This topic will become even more lively when ST observations are available over a large range of masses and radial distances for many different clusters.

D. Core Collapse/Massive Black Hole

As we have heard at this conference, the dynamical evolution of globular clusters is expected to lead to a strong density enhancement in their centers. The expected density profile has been calculated in two limiting cases: 1) the cluster undergoes "core collapse" (see, e.g., the references in Heggie's talk at this conference, in the recent discussion by Cohn and Hut 1984 and in Lightman and Shapiro 1978); and 2) the core contains a massive black hole (see the informative and clear summary of this subject in S. Shapiro's talk at this conference). The predictions that are made from these two scenarios are similar. I shall discuss the expectations for a density cusp using the language appropriate to a massive black hole because the details of the predictions have been worked out for this case - including the effects of mass segregation and the interactions with unbound stars.

In the presence of a massive black hole, the stellar distribution function in the central region of the cluster has the form (Bahcall and Wolf 1976, hereafter Paper I)

$$f(E) \propto E^{1/4}; \qquad n(r) \propto r^{-7/4}. \tag{6a}$$

The square, R, of the ratio of the velocity dispersion of the bound stars to the unbound stars is (equation 94 of Paper I):

$$R = \frac{4}{11}\left(\frac{r_h}{r}\right), \tag{6b}$$

where

$$r_h = \frac{GM_{BH}}{(\text{LOS velocity dispersion})^2}$$

is the characteristic radius over which the black hole perturbs the star distribution (Peebles 1972). Equation (6) is correct when:

1) all the stars have the same mass; 2) the boundary conditions are ignored; 3) large angle collisions are neglected; 4) the loss cone is ignored; and 5) binaries are unimportant. The Boltzmann equation can be solved numerically (Paper I and Bahcall and Wolf 1977, hereafter Paper II) when all of these effects are included.[2] The only effect that alters significantly the density distribution (but see below) is the mass spectrum. If, for example, there are two classes of stars with masses M_1 and M_2, then the distribution functions of the bound stars are of the form (Paper II):

$$f_i(E) \propto E^{p_i(E)} \; ; \qquad n_i(r) \propto r^{-(3/2 + p)} . \tag{7a}$$

The cusp exponents satisfy a symmetry relation:

$$\frac{1}{M_1} p_{M_1} = \frac{1}{M_2} p_{M_2} . \tag{7b}$$

The numerical solutions show that the larger mass always has (an approximately constant) cusp coefficient $p_{\text{more massive}} = 0.31$. For $M_1 = 4M_2$, corresponding to an extreme range of masses for globular clusters, $p_1 \approx 0.31$ and $p_2 \approx 0.078$.

The sensitivity of observational tests to the presence of a massive black hole scales approximately as

$$\frac{M_{BH}}{(\text{LOS velocity dispersion})^{3.5} r_{\text{core}}} .$$

Ground-based measurements of the line-of-sight velocity dispersions are needed for many of the nearby clusters in order to predict what one might be able to see with Space Telescope and to interpret future ST observations.

For illustrative purposes, I have chosen parameters that are appropriate to 47 Tuc (see Illingworth 1976), a core radius of 0.7 pc and a line of sight velocity dispersion of 10.5 km s^{-1}. This cluster is not a particularly likely candidate for harboring a black hole, but at least Illingworth has measured the relevant structural parameters.

Figure 2a compares the surface brightness of equal mass stars in the presence of a black hole with mass $M_{BH} = 5 \times 10^3 M_\odot$ with the distribution expected in the absence of a black hole. [The results shown here and in Figure 2b and 2c are based upon equation 88 of Paper II.]

I show in Figure 2b the projected number densities of stars that would be observed near the center of 47 Tuc if it contains a massive black hole (the dark circles) of projected mass $5 \times 10^3 M_\odot$. The open triangles represent a King model in the absence of any massive singularity at the center. The error bars are estimates of what may be obtainable with the Space Telescope using star counts (here assumed to

[2]Lightman and Shapiro 1977 have given an elegant discussion of the effect of the loss cone.

be a maximum of 10 stars in the inner one arc second). One can infer from Figure 2b that, even in unfavorable cases such as 47 Tuc it may still be possible to detect black holes with masses as large as $5 \times 10^3 M_\odot$ and, perhaps in more favorable cases, (with smaller velocity dispersions and core radii, see scaling relation quoted above) $10^3 M_\odot$.

A comparison of Figures 2a and 2b suggests that surface brightness measurements may be a more powerful diagnostic than star counts. This inference is correct provided that the star counts cannot go very deep in the central region because of crowding effects (see the detailed comparison of star counts versus surface brightness measurements in section VId of Paper II) and provided one can prevent the bright stars from dominating the light. One can bias against the bright red giants by using ultraviolet exposures.

Figure 2c compares the distributions predicted for two stellar populations when $M_1 = 4M_2[n_1(r) \propto r^{-1.81}; n_2(r) \propto r^{-1.58}]$. It does not seem feasible to detect even this substantial difference in the density laws of the two stellar populations within the core of a globular cluster. Unfortunately, this conclusion implies that it will not be easy to distinguish observationally between core collapse and massive black hole solutions, since the difference between two such dynamical explanations is probably only of order 0.3 in the cusp exponent. The effect of a loss cone is even more difficult to detect since it changes the cusp exponent by less than 0.02 (Bahcall and Wolf 1977), an order of magnitude less than the unobservable difference shown in Figure 2c.

I conclude that it may be possible in favorable cases to detect, by measurements of the surface brightness or by star counts, a density enhancement of order $10^3 M_\odot$, but it will probably not be feasible to determine many theoretically interesting details of this density distribution.

Complementary information can be obtained by measuring the central velocity dispersion and line profiles (see Vb of Paper I).

V. TWO IMPORTANT EXTRAGALACTIC ST PROGRAMS

I will discuss next two important ST programs that can be carried out on extragalactic globular clusters.

A. The Luminosity Function of Globular Clusters

The measurement of the luminosity function of globular clusters is another "made-for-ST" project! Ground-based observations suggest (see Hanes 1977 and Harris and Racine 1979) that the relative number of globular clusters per unit of luminosity is a universal (perhaps Gaussian) function, which may have a peak near $M_V = -7.3$. If correct, this conjecture would imply that the globular cluster luminosity function of Virgo galaxies would have a peak near V = 23.5 mag, implying that the declining part of the luminosity function would be just beyond the ability of ground-based telescopes to carry out reliable observations.

If the luminosity function does indeed have a characteristic shape that permits a well-defined luminosity calibration, then it will be possible to carry out an "all ST, Population II" determination of the Hubble constant. Direct parallax measurements of RR Lyrae field stars in the Galaxy can be carried out with the astrometric capability of the fine guidance system and will allow a direct calibration of the intrinsic luminosities of the RR Lyraes and hence the distances and luminosities of the galactic globular clusters.

How many globular clusters are expected in a single WF field centered near a prominent galaxy in the Virgo cluster? I have calculated the number of clusters that are expected to be detected in frames that begin 2 arc minutes ($\approx$ 9 kpc, in order to avoid confusion) from particular galaxies of interest. The results are shown in Table 2 and indicate that of order 50 to 100 globular clusters should be detectable for the more luminous galaxies. The principal uncertainty in these predictions - assuming that the form of the luminosity function is exactly Guassian - is the uncertainty in the plate limit of the ground-based observations (which is known in some of the best cases to only about $\pm$ 0.5 mag; see, e.g., Harris and Petrie 1978).

Observations should be carried out on a variety of galaxies to determine to what extent the shape of the globular clusters luminosity function is dependent on galaxy morphology or luminosity. In order to provide some indication of what is required in order to relate the Virgo observations directly to the globular cluster population of the Galaxy, I have calculated the number of globular clusters that would be visible on a WF frame if the Galaxy with its 180 globular clusters (Harris and Racine 1979) were moved to Virgo. I find that only 5 clusters would be visible on a frame that began 2 arc minutes (9 kpc) from the Galaxy's center. If the Galaxy happened to be seen edge-on so that observations as close as 1 effective spheroid radius (2.7 kpc) were possible, the number of clusters would be about 16 on a WF frame.

The main background in these measurements will be field galaxies. Figure 3 shows a simulation of what the WF/PC is expected to see in a typical deep exposure at high galactic latitude. About 50 galaxies are expected in each of the four quadrants of the WF/PC. These galaxies must be separated from the program globular clusters by their colors and by the lack of concentration to the center of the program galaxy (as determined by the number density from serendipidous exposures in similar areas of the sky).

B. Tidal Radii of Clusters Around Elliptical Galaxies

K. Freeman (1979) pointed out the importance of measuring with ST the tidal radii of globular clusters around distant elliptical galaxies. The mass of the galaxy, M_{galaxy}, contained within a given distance D is proportional to the cube of the cluster tidal radius, i.e.,

$$M_{galaxy} \approx (D/r_{tidal})^3 M_{cluster}.$$

The expected tidal radii are about 1 arc second at the distance of the Virgo cluster for a globular cluster located 10 kpc from the galactic center. Thus one should be able to measure the tidal radii of clusters all the way out to the Virgo cluster and perhaps a little beyond. This measurement will be one of the few reliable ways of determining the potential field and mass of elliptical galaxies. The recent demonstration by Elson and Freeman (1984) that the mass for the LMC cluster NGC 1835 that is inferred from measurements of the velocity dispersion and from the tidal radius is the same to within a factor of two strengthens confidence in the method.

VI. SUMMARY

There are many important globular cluster programs that can be done with Space Telescope. I have mentioned only a few that are of special interest to me. Undoubtedly, you can think of a number of other projects that ought to be carried out (see the excellent review by Freeman 1979 for different suggestions). The limitation will not be the lack of good ideas, but the small amount of observing time that is available.

I am grateful to C. Alcock, K. Freeman, I. King, and B. Paczynski for valuable discussions. This research was supported by NASA #8-32902.

TABLE 1. The Conversion Function, F. The function $F(M_V)$ converts surface brightness per unit area to star counts per unit area using a Population II luminosity function (from Bahcall and Soneira 1984).

M_V	F	M_V	F
-2.0	3.3 E-5	7.0	3.0 E-1
-1.0	6.0 E-4	8.0	3.9 E-1
-0.5	1.2 E-3	9.0	5.0 E-1
0.0	2.1 E-3	10.0	6.6 E-1
0.5	3.8 E-3	11.0	9.0 E-1
1.0	9.7 E-3	12.0	1.2
2.0	1.4 E-2	13.0	1.7
3.0	1.9 E-2	13.5	1.9
4.0	4.0 E-2	14.0	2.2
5.0	1.1 E-1	15.0	2.6
6.0	2.1 E-1	16.5	3.2

TABLE 2. The Expected Number of Globular Clusters in a WF/PC Field to V = 27 mag for Virgo Galaxies. The exposures reach to within 2 arcmin (9 kpc) of the galaxy centers.

Galaxy	Type	Number of Clusters
NGC 4486	E0	$10^{2.5}$
NGC 4374	E1	$10^{1.9}$
NGC 4472	E2	$10^{2.1}$
NGC 4526	S0	$10^{2.0}$

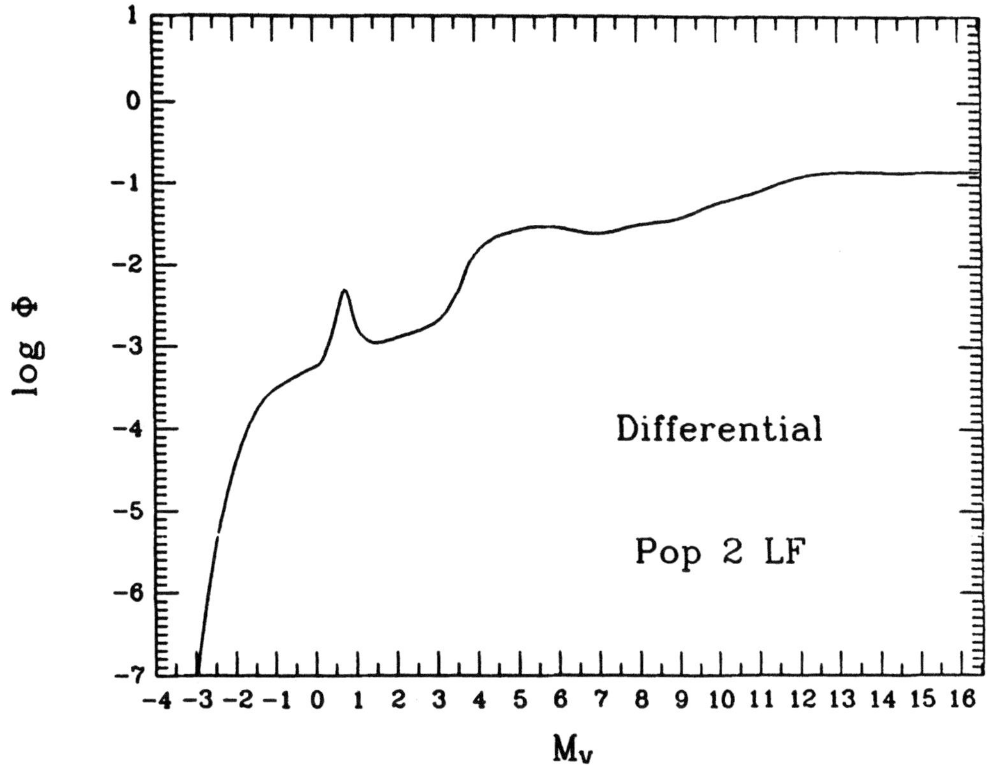

Figure 1a. The Population II luminosity function used by Bahcall and Soneira (1984). The bright end of the luminosity function is the same as the 47 Tuc luminosity function used by Da Costa (1982). The luminosity function fits the spheroid field counts in the observable region between M_V = +3 and M_V = +8.

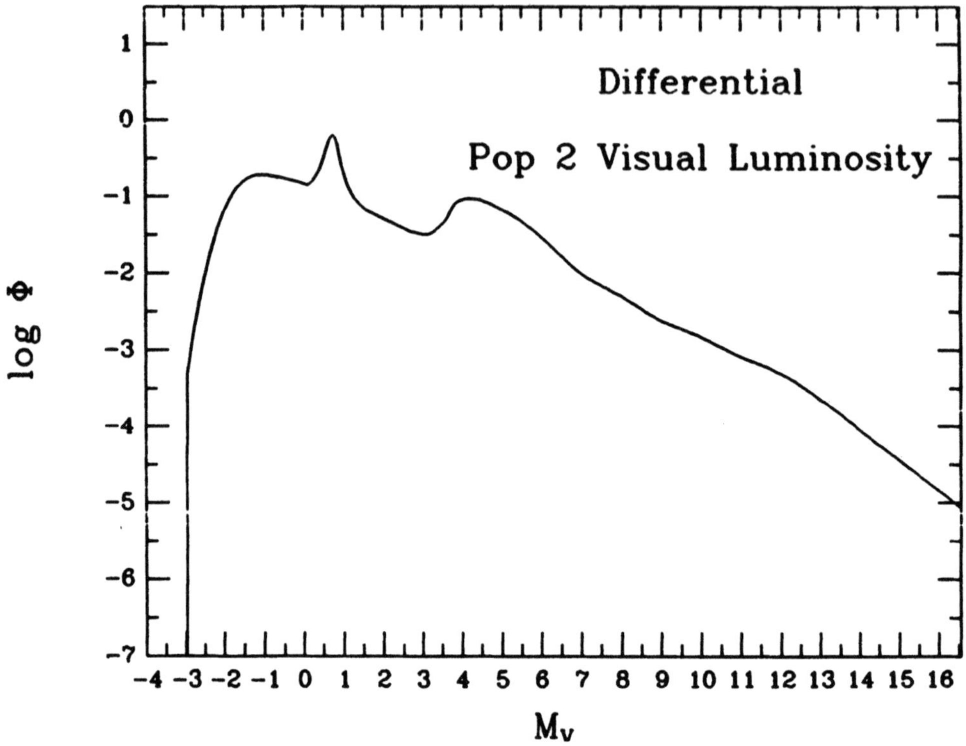

Figure 1b. The contribution to the luminosity from different absolute magnitudes for the luminosity function shown in Figure 1a.

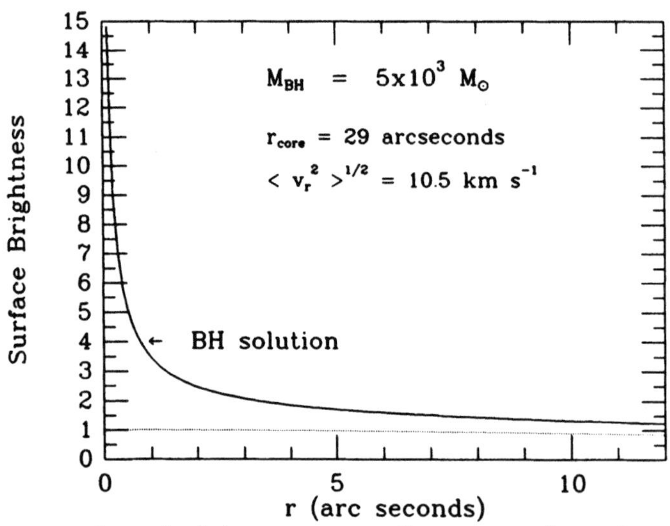

Figure 2a. The surface brightness as a function of projected distance from the center of a globular cluster in the presence of a massive black hole, $M_{BH} = 5 \times 10^3 M_\odot$. The stars are assumed to all have equal masses. The parameters for the cluster are taken from Illingworth's (1976) study of 47 Tuc. The dotted line is a King model in the absence of a central singularity.

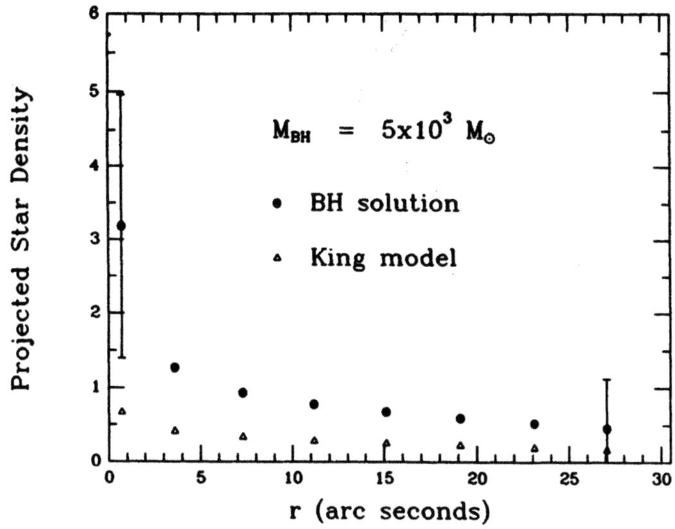

Figure 2b. Star counts as a function of projected distance. .Otherwise, the same as for figure 2a.

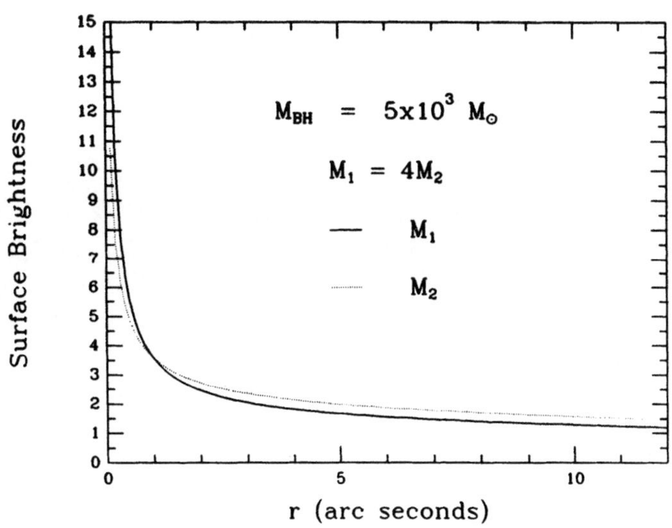

Figure 2c. Surface brightness as a function of radius for stars with two different masses, $M_1 = 4M_2$. Otherwise, the same as Figure 2a.

Figure 3. Simulated WF/PC V-band picture with Space Telescope. The simulation [J. Bahcall and R. Soneira, 1979, for the STWG] refers to one quadrant of a WF/PC field at a high galactic latitude. A Hubble constant of $H_0 = 50$ km $s^{-1}$$^{-1}$ with $q_0 = 0.1$ and Tinoley x = 2 evolution is assumed. The exposure reaches to V = 28 mag with a signal/noise ratio of 3.
There are 6 stars and 51 galaxies in this field which is 1.3 arc min. on a side.

REFERENCES

Bahcall, J. N. and Soneira, R. M. 1980, Ap. J. Suppl., 44, 73.
Bahcall, J. N. and Soneira, R. M. 1984, Ap. J. Suppl., 55, 67.
Bahcall, J. N. and Spitzer, L. 1982, Sci. Am, 247, 40.
Bahcall, J. N. and Wolf, R. A. 1976, Ap. J., 209, 214 (Paper I).
Bahcall, J. N. and Wolf, R. A. 1977, Ap. J., 216, 883 (Paper II).
Baum, W. A. and Schwarzschild, M. 1955, A. J., 60, 247.
Cohn, H. and Hut, P. 1984, Ap. J., 277, L45.
Da Costa, G. S. 1982, A. J., 87, 990.
Da Costa, G. S. and Freeman, K. C. 1976, Ap. J., 206, 128.
Elson, R. A. W. and Freeman, K. C. 1984 (to be published).
Freeman, K. C. 1979 in Scientific Research with the Space Telescope, IAU Colloquium 54, NASA CP-2111, ed. M. S. Longair and J. W. Warner.
Green, R. F. 1980, Ap. J., 238, 685.
Hanes, D. A. 1977, Mem. R. Astron. Soc., 84, 45.
Harris, W. E. and Petrie, P. L. 1978, Ap. J., 223, 88.
Harris, W. E. and Racine, R. 1979, Ann. Rev., 17, 241.
Harris, W. E. and Petrie, P. L. 1978, Ap. J., 223, 88.
Illingworth, G. 1976, Ap. J., 204, 73.
King, I. R. 1984 (unpublished).
King, I. R., Hedemann, E., Hodge, S. M., and White, R. E. 1968, A. J., 73, 456.
Lamb, D. Q. and Van Horn, H. M. 1975, Ap. J., 200, 306.
Leckrone, D. S. 1980, P. A. S. P., 92, 5
Lightman, A. P. and Shapiro, S. L. 1977, Ap. J., 211, 244.
Lightman, A. P. and Shapiro, S. L. 1978, RMP, 50, 437.
O'Dell, C. R. 1980, Annual Reviews Monograph: Telescopes for the 1980's, ed. Burbidge, G. R. and Hewitt, A., Palo Alto, CA: Annual Reviews, Inc., 129.
Peebles, P. J. E. 1972, Ap. J., 178, 371.
Peterson, C. J. and King, I. R. 1975, A. J., 80, 427.
Salpeter, E. E. 1955, Ap. J., 121, 161.
VandenBerg, D. A. 1983, Ap. J. Suppl., 51, 29.
VandenBerg, D. A., Hartwick, F. D. A., Dawson, P. and Alexander, D. R. 1983, Ap. J., 266, 747.

DISCUSSION

COHN: I would like to note that the difference between the slopes of the radial density profiles for components of different mass is much larger in the case of simple core collapse (with no black hole) than in the case of a black hole cusp. This may provide another observational means of distinguishing between these two alternatives.

BAHCALL: Good point. But I am not optimistic.

KING: With regard to the brightness distribution at the center of 47 Tucanae, I believe that published data of Djorgovski and me (Ap.J. Letters 277, 49, 1984) already exclude a central brightness peak.

BAHCALL: It's a matter of resolution. You can always have an unresolved peak if the central mass concentration is not too large. In any event, I only used 47 Tuc as an illustration, not a likely case.

FALL: There are good dynamical reasons for thinking that the luminosity function of globular clusters should differ from one galaxy to another. Given that, is it not more likely that we will learn from ST observations about the presence or absences of such variations rather than obtaining an independent distance-scale?

BAHCALL: You could be right. We will see. I wouldn't want to bet on what we will learn until the observations are made and the analysis is complete.

WHAT NEXT? PRIORITIES IN THEORY AND OBSERVATIONS

Summary of the panel discussion.

Panel Members: I. R. King, L. Spitzer (chair), A. Toomre, S. D. Tremaine, T. S. van Albada, S. D. M. White

---

Lyman Spitzer, Jr.
Princeton University Observatory
Princeton, NJ 08544 USA

The papers presented at this symposium show that the theory of a spherical cluster composed of mass points has made great progress; the major processes occurring seem to be well understood in principle. In particular, evaporation and mass stratification account for the early contraction of a cluster core, while the gravothermal instability is responsible for the final collapse. Formation of binaries by three-body encounters can stop the collapse, though the detailed nature of the ensuing expansion is still not fully explored. While further interesting research remains to be done on the evolution of a spherical aggregation of mass points, the broad outlines of the subject seem clear.

There are two main areas where pioneering research seems needed in the future years. The first of these areas concerns the finite size of the stars. While tidal capture seems reasonably well understood, the somewhat closer and almost as numerous direct physical collisions pose many problems. How does a composite object formed from such a collision evolve with time? Its evolution as an isolated object and its interaction with single stars, binaries and other objects all need further study. Such objects may be stepping stones in the complex path to the formation of neutron stars and black holes, and may affect core dynamics in ways not yet analyzed.

A second area that calls for new research includes the orgin and history of actual clusters. Some properties of clusters, such as the observed chemical gradient of stars as a function of distance from the cluster center, may provide clues to the early history of these systems. The possible early disruption of clusters, resulting partly from extensive mass loss by the more massive stars, and partly from interactions with the Galaxy, needs further analysis. While many traces of initial condition and early history have been obliterated by later evolution, some may still remain. Their study may clarify our views of the early history of the Galaxy.

J. Goodman and P. Hut (eds.), Dynamics of Star Clusters, 499–509.

Alar Toomre
Department of Mathematics
Massachusetts Institute of Technology
Cambridge, MA 02139 USA

As an outsider, I have enjoyed this Symposium very much. It has certainly taught me again that the globular clusters -- though, with my profuse apologies to Bob Mathieu, probably not their poor open cousins found in our present-day disk -- constitute a fascinating subject, so teasingly simple at first sight, but so rich with dynamical phenomena and paradoxes. An this setting at the Institute for Advanced Study has also proved suitably glorious, especially now that the organizers have kindly brought back the Sun after all that rain.

Perhaps more interesting than such praise, though, may be two aspects of this Symposium that leave me at least mildly disappointed. One was simply the fact that all the hard and clever work by several of you from this past decade, carrying the dynamics soundly beyond the core collapse that loomed large as recently as the 1974 meeting in Besancon, has not been rewarded by anything remotely as memorable as, say, the formation of a supermassive central object or even any rapid dissolution of the cluster itself. In that sense, your recent solid consensus that binaries form and take over the heating is certainly a pity. It means that the answer to Haldan Cohn and Piet Hut's cheerful query "Is there life after core collapse?" is merely "Yes, but it is pretty dull."

The other thing that disappointed me, a bit more seriously, is that during the past few days I have heard scarcely a word of speculation or of reasoned puzzlement directed at the grand unsolved questions of how the globulars came into being in the first place, or how they ended up where they are now. When we next gather to discuss clusters, I hope we will all agree that at the 1984 meeting in Princeton we were silly to shy so completely from these vague, fuzzy, untidy, and yet ultimately very vital issues. Given all that we know already about the puzzling correlations of metallicities with the present kinematics and whereabouts of globular clusters -- and also about giant molecular clouds, young LMC globulars, some remarkably metal-rich globulars claimed in Cen A, giant starbursts detected in other peculiar galaxies especially by infrared techniques, and yes, even about the mergers of galaxies -- I cannot help feeling that the time is ripe for some bright astrophysicists to pull it all together into a cohesive dynamical picture that afterwards, like core collapse, will be declared only too obvious!

For what it is worth, my own guess remains that, like today's scrawny open clusters, all globulars were basically manufactured in the relative peace, quiet, and high densities found in the <u>disks</u> of galaxies, and that they were evicted from such nurseries only subsequently during the mechanical violence of early mergers that helped to create the spheroids or bulges. I would not grieve desperately if such conjectures were proved wrong. But I would be sorry indeed if they and other plausible stories like, say, the old Dicke-Peebles theme emphasizing Jeans masses did not receive a close and critical scrutiny at nice future gatherings like this one.

Tjeerd S. van Albada
Kapteyn Astronomical Institute
P.O. Box 800
Groningen, The Netherlands

I would like to raise a few questions related to the origin of globular clusters, which should enter sooner or later theories describing their evolution.

First there is the question whether or not the present structure of globular clusters still contains information regarding the formation phase. One might think that internal relaxation and gravitational shocks have obliterated all memory of the initial conditions. But there is at least one piece of information which suggests otherwise: the absence of clusters with high central and mean densities in the outer parts of the Galaxy; there is a corresponding lack of low density clusters in the inner parts of the Galaxy. On the whole, mean and central densities of clusters are correlated with the background density of the population II component, indicating that the present central densitites of globular clusters still bear some relation to their central densities after formation. The absence of low density clusters in the inner region of the Galaxy can probably also be attributed to disruption by gravitational shocks, but it is hard to understand the absence of high density clusters in the outer regions other than in terms of primeval conditions. (If it is true that entire clusters have been disrupted by shocks it would be important to include energy input by shocks in studies of core collapse.)

Next one may ask whether there are other differences between clusters that may affect their evolution. In other words, is it proper to work with some standard cluster representing the 'mean,' or is there such a wide range in individual properties that there is also a large variety of evolutionary 'path?' Take, for example, the number of primeval binaries. The blue stragglers in M3 may be an indication that this cluster is exceptionally rich in binaries. If so, this has probably a pronounced effect on its dynamical evolution. Can this effect be predicted, and observed? Further, there are probably differences in the mass spectra, which will also effect the evolution.

These considerations show that, from the theoretical point of view, there is probably not a unique path of dynamical evolution. This implies that globular cluster evolution is not such a 'clean' problem. A full description will involve several physical processes, leading to a large number of parameters. Yet, the overall similarity of globular clusters shows that different conditions may still lead to similar global properties.

Scott D. Tremaine
Department of Physics
Massachusetts Institute of Technology
Cambridge, MA 02139 USA

For the first time we seem to have a coherent picture of core collapse. Three features of this picture have particularly impressed me:

1. As Jeremy Goodman phrased it, core collapse seems to be "robust." Every model presented here, no matter how crude, has exhibited core collapse. N-body computations, whether with few or many bodies, chosen with a single mass or from a spectrum of masses; Monte Carlo calculations by Hénon, by Spitzer, or by Stodółkiewicz; self-gravitating gas spheres in insulating boxes; numerical solutions of the Boltzmann equation with or without strong encounters and with or without anisotropic velocity distributions -- all of them collapse in a qualitatively similar fashion. The pre-collapse evolution of a cluster appears to be a calculation which is almost impossible to do wrong. Even McMillan and Lightman's complicated and ingenious hybrid code undergoes a standard core collapse, which implies that either their code or core collapse is surprisingly reliable, though I am not sure which. The re-expansion of the core after collapse appears to be a more subtle process, which is rather more difficult to get right, but I am confident that this process as well will be understood in the next few years.

2. Core collapse is surprisingly unspectacular. Here we have a cluster of 10^5 stars, whose only goal in life is to increase its entropy, and all it does is to spit out a few garden-variety binary stars of which hundreds are seen already in any small volume of the galactic disk. A few years ago many astronomers believed that core collapse represented the death of globular clusters; now it seems to be only the sort of mild upheaval which one might associate with puberty.

3. The study of globular cluster evolution is now similar to the study of stellar evolution in that almost all of the important physical processes are known and relatively well understood. Dr. Stodółkiewicz's code, in particular, is strongly reminiscent of a stellar evolution code, and just as in the study of stellar evolution we can look forward to ever more accurate evolution tracks as the cross sections and transport coefficients are improved. The great advantage over stellar evolution is that we don't have to treat convection. The disadvantage is that we don't have a Hertzsprung-Russell diagram (yet); there is still no accepted observational criterion even for distinguishing pre-collapse from post-collapse clusters.

Progress on other theoretical topics has been much less rapid. We need more work on the evolution of the outer parts of clusters, by evaporation and by external gravitational shocks from giant molecular clouds and the galactic disk. Our understanding of even the classical problem of stripping by the Galactic tidal field is still rudimentary, although some progress was reported here by Seitzer, and a series of beautiful papers on tidal stripping were written by Hénon around 1970. Since we now know that clusters survive core collapse, they must ultimately die from processes like evaporation, tidal stripping and external shocks. We need to know the lifetimes of clusters and what they leave behind when they die. It is quite possible that hundreds or thousands of globular clusters have already died, and a proper understanding of their fate may resolve many puzzles in galactic astronomy (such as the galactic bulge X-ray sources, to name just one).

Simon D. M. White
Steward Observatory
University of Arizona
Tucson, AZ 85721 USA

I have been impressed by the extent to which the discussion at this meeting has centered on the evolution of the inner core of globular clusters. The investigation of core collapse and its posterity raises many fascinating and fundamental questions in theoretical N-body dynamics, and, as we have seen, the last few years have brought remarkable progress in understanding the relevant processes. Nevertheless, I think that we should be wary of concentrating too much on a region which is extremely difficult to observe from the ground, and may well not give up its secrets even to Space Telescope. Excellent kinematic and metallicity data are now being acquired from stars from the core right out to the tidal radius in several clusters. It seems to me that we may learn more about the present dynamical state of clusters, as well, perhaps, as about their origin and evolution, by studying the bulk of the cluster which we can see, rather than the innermost core which we cannot. The very pretty analogies between stellar evolution and cluster evolution which Piet Hut likes to draw should not blind us to the fact that the stellar distribution of many clusters may be affected more by external than by internal influences. The data shown by Freeman, Lupton, Mayor, Seitzer and others provide much material for theoreticians to digest, and complex modelling of the type discussed by Stodółkiewicz, Seitzer, Terlevich and Wielen is needed to decide which processes determine the present structure of clusters.

A question which has been barely touched on here is that of the origin of star clusters. Stellar dynamics alone will clearly be insufficient for an understanding of the processes involved, but it would be valuable to know the extent to which cluster evolution has erased any memory of initial conditions from the present structure. The problem of cluster formation is as ill understood and as messy as the related problems of star and galaxy formation but it is clearly of major importance. I believe the investigation of young clusters and of cluster-forming regions both in our own Galaxy and in the Magellanic Clouds offers us some hope of understanding how, where and why clusters form, as well as of showing us the structure of clusters as they are born. The relationship between these objects and Galactic globular clusters is unclear, but it would become much clearer if we understood how the properties of a cluster depend on the medium from which it condenses. We should perhaps also remember that most stars may form in loose associations and in open clusters, rather than in rich globular systems. For these reasons I was happy to hear Bob Mathieu discuss the properties of young open clusters, and I would have welcomed some discussion of the dynamics of cluster-forming regions as inferred from recent infrared and millimetre wave observations. While it is difficult to observe very young clusters and to provide a coherent theoretical framework in which to interpret such observations, I believe the possible payoff in increased understanding of the cosmogony of stars and clusters more than justifies the risk of prospecting in such a mire of complexities: in the words of an old saying from the north of England "where there's muck there's brass."

Ivan R. King
Department of Astronomy
University of California
Berkeley, CA 94720 USA

This meeting has brought theoreticians and observers together, but it has seemed to me that the communication of each group has been among its own members, without a lot of dialogue between the two. As I said in my own talk, I am interested in answers and believe that they come from interpretation, which is so often the result of juxtaposing observation with theory. I wish we could do more of this.

Clearly, the topic that has had the most attention is core collapse. There seems to be general agreement that it must occur, and also fairly good agreement about how it proceeds. (There is of course some disagreement about gravothermal oscillations, but they are each of such short duration that I don't think they are likely to be observed.)

What I have been most interested in is knowing how a cluster should behave after core collapse. Here I thought that the answer was becoming clear: The collapsed core (presumably stabilized by binaries) becomes isothermal with only a little re-expansion, and its center is for practical observational purposes still singular. Further expansion is very slow, so the core keeps this singular isothermal profile thereafter. Tidal-capture binaries are more numerous than 3-body-formed binaries, but they don't matter much, because they're too hard. But now I'm not so sure. I've heard it said that perhaps tidal binaries do have an important effect and collapsed cores re-expand completely. I wish we had a clear statement on this.

What I do know is that several clusters have what look like collapsed cores -- but not many. What's more, there don't seem to be intermediate cases; either a core is collapsed or it's normal. But there are things about this that bother me. Qualitatively it fits with the picture of cores remaining in a permanent post-collapse state, but quantitatively there aren't enough such clusters. There are other cores with short relaxation times, but they haven't collapsed. Or maybe they've collapsed and re-expanded -- but in that case why don't we see other intermediate types that are still in process of re-expanding?

I don't think that either the theories or the observations are good enough to answer these questions. I've said where I think the theories are lacking (or perhaps where my understanding of them is lacking), but there's one more area that ought to be mentioned: mass mixtures and segregation. At the same time let me say that I doubt that this is a serious problem, because otherwise real clusters wouldn't look so much like single-mass models. (I think that this is because a lot of the mass of a cluster is in white dwarfs, which have nearly the same individual mass as the red giants.)

On the observational side, I think that we clearly need better studies of clusters that have collapsed cores and of other high-concentration ones that don't. A lot of this can be done with carefully directed ground-based observations, and it very much needs to be done before Space Telescope flies. But there is no doubt that ST will have a major impact on this problem, because it's a problem that very much needs high-resolution star counts. I hope that a meeting 10 years from now will have all of this settled and can indeed treat cluster evolution not as a speculation but as a science.

DISCUSSION

SEVERNE: I would like to suggest that a fundamental problem remains in connection with the Fokker-Planck equation itself, which lies at the basis of so much of the work on globular clusters. The encounter term in the Fokker-Planck equation is derived by perturbing rectilinear trajectories. This approximation is essentially inconsistent for self-gravitating systems: it leads to mean free paths, orders of magnitude larger than in the systems studied and gives rise to the oldest unresolved divergence in statistical mechanics. It now appears possible to study encounters by perturbing more realistic (quasi-) periodic trajectories. My poster contained a suggestion in this direction; may I plead that more work be done on this basic problem.

HEGGIE: Concerning the disagreements on gaseous calculations between myself and Drs. Bettwieser and Sugimoto, Dr. Sugimoto has made the interesting suggestion that there may be an analogy with numerical difficulties encountered many years ago in the computation of helium shell burning in stars. On his advice I intend to check the time-step control in my code.

I wish also to express my agreement with those panel members who urged the development of satisfactory models for the outer parts of clusters. I suggest that those with dynamical codes should produce a sequence of analytic fits to their models at different stages of dynamical evolution, and present these in a form suitable for use by observers. But I would discourage attempts to build elaborate multi-component cluster models from lowered Maxwellians in equipartition, since there is little dynamical justification for such distributions even in a relaxed core, let alone in the rest of a cluster.

SHAPIRO: Would the panel be willing to anticipate the results of their posed question on the origin of clusters by answering the query "do you believe in miracles?"? For if you don't, how do you explain the curious fact, first pointed out by Spitzer 10 years ago, that roughly half of all clusters have sufficiently short relaxation timescales to have undergone collapse already while the rest have not. Why half and not something closer to unity or to zero? This fact is telling us something - perhaps that there were once many more clusters than we can now observe but have since disrupted.

TOOMRE: Shall we have a poll on whether we believe in miracles or not (laughter)? You may well be right, globular clusters lead a dangerous life and we are after all dealing only with the survivors.

SPITZER: The possibility of cluster disruption is a very fascinating one, and has to be taken very seriously. It is not easy to find the small residue that a cluster would leave, but it would be an interesting problem observationally.

VAN ALBADA: I think there is some evidence for moving groups amoung population II stars.

TOOMRE: Let me turn Shapiro's "miracle"-comment around. It has now become rather disreputable to claim that the solar system has been formed by miracle, even though we know of only one example yet. In contrast, we are dealing here with over a hundred miracles around our galaxy alone, so we have a certain safety of numbers, folks!

WHITE: I am not quite sure what the "miracle" is either. One thing which impressed me from all these calculations of core collapse, is how little it seems to do to the cluster. They do not look that much different after core collapse than they did before. It is not clear to me that by changing the parameters of the cluster you could not make many of the observed clusters have their core collapsed and yet have that fact not show up in the observations. Therefore I do not see any need for them to dissipate completely as a result of collapse-like processes. There may very well have been very many more clusters at the time of the formation of the galaxy, but various people have put forward theories that explain why the ones we see are the ones that are left now.

KING: I just do not see the mechanism for destroying a cluster. Core collapse binds it more strongly. In particular the calculations that Haldan Cohn showed us, indicate that a cluster does not come very quickly out of its collapsed state but rather reexpands very slowly. I don't see it at all as a way of getting rid of clusters. I think globular clusters are with us, to stay.

SPITZER: I think the disruption has to be done by some other, external, agent.

KING: Yes.

LARSON: I was glad to hear some of the panelists, especially Alar Toomre, emphasize the problem of the origin of globular clusters. In addition to their size, mass and chemical content, another important characteristic that globular clusters carry around from the time of formation (perhaps with some modification) is the stellar mass function. I've become increasingly impressed with the importance of the Initial Mass Function in controlling the structure and evolution of clusters, expecially if the IMF contained substantial numbers of massive stars that evolved to leave white dwarf, neutron star, or black hole remnants. The new data on velocity dispersions now becoming available suggest that at least some clusters contain substantial amounts of mass in dark form, presumably remnants, and such remnant populations would constitute another important tracer of formation conditions.

SPITZER: Thank you for pointing out this important aspect of the mass function.

DJORGOVSKI: I would like to iterate Ivan King's remark that there seems to be a discrepancy between the number of detected post-collapse clusters and the number of clusters which should have done it, but show no signs of it. Post-collapse cores are apparently easy to detect, and there are no "intermediate" cases presently known.

My second remark concerns another queer duck, M 79. That cluster looks like nothing that anybody ever predicted, with its strange bend in the surface brightness profile. It cannot be fitted with either a King model, or a power law. According to the error-bars, the bend is highly significant. Unless Nature is playing a particularly devious statistical fluctuation on us, I think that we may be dealing with something new. Please remember that there was only one M15, until somebody tried to look for more.

INAGAKI: I think there are two important problems. (1) Why are only a few binaries observed in globular clusters though theories predict that a lot of binaries are formed by tidal dissipational encounters? (2) It is necessary to construct realistic models of globular clusters including a mass spectrum, and to take account of deviations from equipartition.

GOODMAN: I'd like to make a few scattered points. First, a plea to the observers: I would like to know how far out the globular cluster system extends in our own galaxy and in nearby bright spirals such as M31. Second, I would like to point out to Drs. King and Djorgovski that it *is* easy to hide post-collapse phenomena in clusters: it is only necessary to have a component of heavy, non-luminous stars that collapse and puff up the luminous component. If the heavier stars do shine, one might see a shoulder in the luminosity profile such as Dr. Djorgovski noted for M79. Thirdly, I'd like to call everyone's attention to the calculations of Dr. Stodolkiewicz, in which 98% of the cluster mass evaporated. I do not understand which of the many physical effects he included is principally responsible for the evaporation, although tidal truncation clearly plays an important role. His results may have very important implications for the possibility of cluster disruption.

Finally, a plea to the purest theoreticians: I think there needs to be more work on the connections between short-mean-free-path gaseous systems and long-mean-free-path stellar systems. I do not understand why gaseous models work so well and agree in such detail with stellar-dynamical calculations.

GRINDLAY: Let me add several comments: (1) As I mentioned before, I urge this from dynamicists among us, to seriously attack the problem of cluster disruption by Giant Molecular Clouds. Since a simple application of Spitzer's (1958) impulse approximation theory suggests disruption *will* occur, specifically: what will such a dissolving cluster look like? What is the expected timescale for dissolution of the final $\sim$100 stars? I would hope that by the time of the next symposium on stellar dynamics of clusters these problems will have been treated. (2) We have discussed the question of dark matter in globulars extensively, but only in the context of dark remnants (neutron stars, white dwarfs, etc.) which we now know *must* be in cluster cores. What about the questions of dark matter in globular cluster halos - such as the axions suggested by Peebles? What about the question of "low mass" dark matter, such as proposed by Bahcall for the galactic plane, as a constituent throughout globular clusters? Again, it seems these questions must be addressed for a complete discussion (at the next such symposium) of the internal dynamics of globulars. (3) Finally, partially in response to many of the comments just made on such questions of cluster origin, evolution and relationships (dynamical , chemical, etc.) to their parent galaxies - let me mention that we have proposed an I.A.U. Symposium on the "Large Scale Properties and Structure of Globular Cluster Systems" to be held at Harvard in August, 1985. This would also serve to commemorate the centennial anniversary of Harlow Shapley's birth. I hope this conference will take place and that you all can participate.

COHN: The results that I presented here, as Dr. King has noted, indicated that three-body binary heating does not cause cores to expand back to their pre-collapse size. However, Dr. Stoldolkiewicz's additional effects do appear to indiate an expansion of the cluster cores to their pre-collapse size. Thus, I would like to speculate that a post-collapse cluster might look rather similar to a pre-collapse cluster, the differences in structure being subtle. This is an important issue for both theoretical and observational investigation.

FREEMAN: I was delighted at the emphasis that the panel put on cluster formation. Please remember the LMC: it has globular clusters of all ages, down to a few million years. We can in principle provide a lot of observational information about their mass functions, chemical properties and internal kinematics. So, if anyone working on the theory of cluster formation would like to know what a real forming cluster is actually doing, please ask (laughter).

SOME SUMMARY REMARKS

Jeremiah P. Ostriker
Princeton University Observatory
Princeton, New Jersey 08544 USA

I. THE HISTORICAL BACKGROUND: DEFINITION OF THE PROBLEM

Of the classical problems of physics there is one which might be picked out as simplest to state but slowest to approach solution. "What is the fate of a self-gravitating system of point masses interacting according to Newton's laws?" The frustration felt by theoreticians, who, despite the accumulation of more than 300 years of mathematical knowledge and numerical technology, have not solved the problem, even to understanding the qualitative late evolution of such systems, is compounded by the existence of observations known since Messier's time of over 100 globular star clusters. These are real and ancient systems, well approximating the idealized problem; astronomical study of their properties should have led us to an understanding of the physical situation. Galactic clusters containing smaller numbers of stars also exist in abundance. However, most of these are much younger systems and they are also subject to a variety of additional complex processes, not important in the large N, globular cluster, systems. Because of this complexity they are less well understood than globulars and will not be covered in this theoretically oriented review.

In the last decade and especially in the last few years progress in understanding large N systems on both observational and theoretical fronts has been rapid. New findings concerning this old question have accumulated from studies by diverse groups working in various parts of the world. Thus, this IAU Symposium occurs at an extremely propitious moment. Personally, this meeting has been more valuable to me scientifically than any other I can recall. It is one of the happy instances when the whole is greater than the sum of the parts. Pieces of the puzzle put together by many of the lecturers have combined to generate a clear, if not definitive, conception of the evolution of stellar clusters.

As early as 1930 Heckman and Seidentopf had argued that clusters should tend to establish an isothermal equilibrium with a Maxwellian distribution of velocities, and that the appropriate time to approach this state was, for many clusters, comparable with cosmic timescales. They further realized that the clusters would be paradoxically unable

J. Goodman and P. Hut (eds.), Dynamics of Star Clusters, 511–520.

to reach a true equilibrium since the only possible stationary state was that of an isothermal (Emden) sphere infinite in mass and radius. They exposed the basic problem but did not attempt to solve. it.

Ambartsumian (1938) and Spitzer (1940) soon showed the way that evolution must proceed. Both pointed out that the average escape velocity for a cluster star was only twice the rms random velocity, so that each relaxation time, as gravitational interactions tended to set up a Maxwellian velocity distribution, of order 1% of the cluster would be given energy sufficient to escape. Thus the cluster would evaporate. But now a new paradox presented itself. Since the escaping stars have zero or positive energy and the cluster as a whole is gravitationally bound (negative total energy), it is impossible for the cluster to evaporate totally. Furthermore, Spitzer noted, the evaporation once started "proceeds at a continually accelerating rate as the cluster contracts", and would formally lead to total evaporation in a finite time. He then proposed that evaporation "would presumably proceed until one of two alternatives occurred. Collisions between stars can become important. Or as the cluster continued to lose stars, the remaining ones might conceivably find themselves in periodic orbits".

We will return to these two possibilities. But it was clear from astronomical observations that neither of them, nor in fact the whole evaporation scenario, was a complete description of the problem. Many astronomers have pointed out that a significant fraction of all globular clusters had central relaxation times less than or comparable to 10^8 years (e.g., Ostriker, Spitzer and Chevalier, 1975; Lightman, Press and Odenwald, 1978). Does this mean that these clusters will soon totally evaporate? Two objections to this hypothesis immediately come to mind. First, it would give to our epoch a preferred status since it would imply that many of the globular clusters waited 1.5×10^{10} years until just now, to self-destruct. It could be shown that this is not statistically probable (cf. Lightman, 1982). Also, the relaxation times at the half-mass point are much longer in these same clusters and it is difficult to see why anything occurring in the inner 1% of mass fraction will very much alter the slowly changing equilibrium state of the bulk of the cluster.

But the problem is more complex. Antonov (1962) and Lynden-Bell and Wood (1968) provided a theoretical explanation for numerical findings of Henon and Spitzer in the 1960s. To wit: the central regions of concentrated clusters will tend to collapse due to conduction of heat outwards even in the absence of evaporation of stars to infinity. Both this phenomenon and evaporation are driven by the "negative specific heat" of self-gravitating systems and both proceed on the two-body time scale, but they are distinct. The "gravitational collapse", unlike evaporation, proceeds at fixed total mass and energy. Recent investigations by (cf. Spitzer, 1975, for a review of earlier work) Cohn (1979, 1980), Heggie (1979), Lynden-Bell and Eggleton (1980), Marchant and Shapiro (1980), and others established the nature of the asymptotic approach to the singular state but gave no insight into how a system could pass through this trial or what, if any, recovery was possible for its post-collapse existence. From the observations previously mentioned it seems that many real globular clusters must have in fact passed through this phase so the question was of more than theoretical interest.

Thus the development of theory and numerical modelling provided increasingly precise detail concerning the question raised by Spitzer and Ambartsumian in their early work. What happens to these systems accelerating towards infinite central density? An important clue was provided in Henon's thesis (cf. Henon, 1961, 1965, 1975) where he showed that, if clusters had central energy sources, they could survive for very long times and would in fact gradually and slowly expand with the nature of the expansion quite insensitive to the energy source. Recent work by Inagaki and Lynden-Bell (1983), Goodman (1984) and others (cf. review by Heggie in this volume) has provided much more insight into this hypothetical post-collapse state. The problem may be seen as analogous to the one solved decades earlier in the field of stellar evolution. As the hydrogen fuel is exhausted in a stellar core the central regions become degenerate and ever more massive as more and more fuel is burned. Are they due to collapse? The answer found was, no, they have a reprieve so long as other fuel sources exist. The central cores will contract until, at last, helium-burning begins abruptly and then a new equilibrium phase is possible as an expanding red giant. In the case of a stellar cluster, what are the possible energy sources? Three mechanisms have been proposed. They are associated with

a) A central black hole,
b) Central binary systems,
c) Central mass loss;

we will take each up in turn. Even if we can envision a post-collapse asymptotic expansion phase fueled by one of these energy input processes another question remains. Can a collapsing core "bounce", that is, arrest the contraction and re-expand? It was not a priori obvious that it can find a way to jump from a point on the sequence approaching collapse to one on the sequence of slowly expanding states (but see Heggie, 1984).

The possibility that stellar systems can harbor massive black holes has been attractive for some time (cf. Zeldovich and Podurets, 1965; Bahcall and Ostriker, 1975) with the current consensus of opinion that no known clusters contain black holes, but many galactic nuclei are good candidates. In this context it has been shown, first with relatively crude calculations (Shapiro, 1977) and then with more sophisticated two-dimensional Fokker-Planck numerical codes (Duncan and Shapiro, 1982) that the presence of a massive central black hole will invariably halt and reverse core collapse. Furthermore, the expansion phase satisfies the $r_{core} \propto t^{2/3}$ law pre-figured in Henon's work. The mechanism is simple. Stars are consumed by the central hole either due to tidal disruption or (for massive systems) they are "eaten whole". As we follow a prospective victim we find it diffusing downward into the cluster center and heating other stars (releasing kinetic energy) as it sinks deeper and deeper into the potential well. When, at last, it is consumed by the central hole, its kinetic energy and gravitational binding energy both disappear from the accounting, but, since its mass is added to that of the central hole, the other stars do not notice the event. Thus, for all stars except the victim, the result is a net energy input into the cluster due to the energy released in the migration of the victim to the central hole. The input is spatially distributed (since it goes as $\phi(r,t)\dot{M}_H(t)$), but nevertheless the asymptotic $t^{2/3}$ behavior results.

Binaries can be a heat source if they are "hard"; i.e., their orbital velocities are larger than the cluster rms velocity. For hard binaries, the energy input rate per binary is independent of binary separation (see, for example, Heggie, 1975), and Spitzer and Mathieu (1980) showed that primordial binaries could in fact temporarily arrest core collapse. However, interactions amongst them and between single stars and binaries will ultimately lead to the ejection of all binaries from the cluster core at which time collapse should resume; so primordial binaries will not prevent collapse.

Binaries can also be produced during the core collapse and since the rate will increase with increasing central density, one might hope that enough would be made to avert and reverse the collapse. These hopes, we shall see when we review this conference, have been realized.

The third listed mechanism, central mass loss, occurs when stars are ejected from the central cluster regions due to interactions with binaries but it can also occur due to normal stellar evolution (cf. Angeletti and Giannone, 1977, 1980; Stodolkiewicz, 1982 and Applegate, 1984). Since in either case (due to mass segregation) the mass loss will come from the central regions, the heating effect on the cluster is similar. For stars at radius r the energy input rate per unit mass is clearly $\partial\phi(r,t)/\partial t$ which, it can be shown, is just equal to $\phi(r,t)\dot{M}_{ej}(t)$. Thus the radial dependence of energy input is the same as in the black hole case, and proportional to the gravitational potential which, we have seen, does lead to the classical asymptotic expansion. But mass loss, if occurring autonomously at a rate unrelated to core density would not be expected to reverse collapse. However, the collisions mentioned in Spitzer's early work, or near collisions causing fusion and accelerated stellar evolution are density dependent so we might anticipate a chain of reactions to occur in high density stellar cluster cores (see, for example, Colgate, 1967; Lynden-Bell, 1967 or Spitzer and Saslaw, 1966 and recent reviews by Ozerony and Dokuchaev,1982; Dokuchaev and Ozernoy, 1982) which produce a mass ejection rate proportional to the square of the stellar density that could arrest and reverse core collapse. In globular clusters as opposed to galactic nuclei the total number of stars in the central regions might be small enough so that (cf. McMillan, 1983) the mass loss rate was sporadic but one would still expect the repeated expansions and contractions to statistically average out to give a net expansion rate of $r_{core} \propto t^{2/3}$.

In this review I have focused on the physical processes rather than mathematical methods. But it is comforting to know that codes such as Spitzer's and Henon's which combine collisional with dynamical time scales, direct Fokker-Planck codes such as Cohn's and Shapiro's, hybrid codes such as McMillan's and even the fluid approach of Hachisu et al. (1978); Lynden-Bell and Goodman all give similar results. There is every reason to believe that our growing understanding of the collapse, rebound and asymptotic expansion phases is not an artifact of one particular, and perhaps flawed manner of approximating the physics of this old and for so long untractable N-body problem.

II. CONFERENCE PAPERS

My review is not intended to be complete nor even to identify the most significant contributions. Rather, I will focus on those papers which, it seemed to me, contributed most to the central question defined in my opening paragraph. [References given without dates are to papers presented at the conference and included in this volume.]

Papers by Aarseth, Lightman and Jernigan showed that significant progress is being made on the computational front. My own perhaps partial view is that these techniques (which accurately treat small N systems) will be needed to understand galactic clusters, but that globular clusters and galactic nuclei can be efficiently modeled with one- or two-dimensional Fokker-Planck codes (cf. papers by Cohn and Shapiro) which can be enhanced to include large energy changes in close encounters (cf. Goodman) and even binary formation (cf. Ostriker; Cohn, Goodman and Hut). Spitzer and Heggie's reviews showed the understanding we have reached on pre- and post-core collapse behavior although the details of the post-collapse evolution are not yet agreed upon by all participants (cf. Sugimoto and Bettwieser).

To me the most exciting discussions were those which focused attention on close interactions amongst stars showing in detail how processes involving tidal capture binaries (cf. Ozernoy, Ostriker), central black holes (Shapiro), binaries made by three-body processes (Cohn, Lightman) would enable clusters to make the transition between contraction and re-expansion. Mass loss from stellar evolution can also be important. Most impressive were the calculations of Stodolkiewicz which, using rather minimal computing power, combined all of these processes and others of importance such as the galactic tidal field to produce some fascinating results. To summarize, it is now clear that evolving clusters can, by various processes, pass through the collapse phase into a slowly expanding quasi-equilibrium state.

My own view (cf. Ostriker, this conference) is that, of the various possible processes, those which follow from the two-body tidal capture (Fabian, Pringle and Rees, 1975; Press and Teukolsky, 1977) are dominant. These lead either to hard binary formation or to stellar fusion but in either case result in mass loss from the cluster center at a rate determined by the two-body capture cross section. Such processes lead to predictions that in the condensed clusters (cf. Djorgovski, Penner and King) the central regions will contain hundreds of cataclysmic variables, WUrsa Maj systems or rapidly rotating fused stars of $\sim 1.4\ M_{\odot}$. It is comforting for theoretical interpretations which make heavy use of the tidal capture process that Grindlay finds the distribution of X-ray luminosities consistent with that expected from neutron star-main sequence and degenerate dwarf-main sequence binaries formed by tidal capture.

III. FUTUROLOGY

The general subject, in both its observational and theoretical aspects, is now perhaps in a similar condition to that of stellar evolution in the 1950s. Most of the relevant physical processes have at least been outlined, if only in highly schematic form, so computations can begin the arduous pilgramage from the interesting calculation to the realistic model. Important effects include those due to:

a) Many mass components, including degenerate stars,
b) Binaries: primeval, tidally formed and those made by three-body processes,
c) Stellar evolution including angular momentum loss from binaries,
d) Tidal shocks and tidal truncation of clusters,
e) Velocity anisotropy, cluster rotation and close collisions.

To plan computations including all these processes is ambitious; it will be an impressively complex undertaking. It will be pointless to do so unless observations simultaneously improve sufficiently to enable confrontation between fact and model. As the reviews of Gunn and Bahcall suggest, we can reasonably hope for this to happen. In the next decade both ground-and space-based observations should improve dramatically. Counts, velocity distributions and a better knowledge of the populations should all emerge from the onslaught of new instrumentation.

The expected confrontation between models and observations is not likely to be smooth. We will find out that we have not included many of the relevant processes nor made many of the pertinent observations. One can expect that the increasing rigor of our discipline will have various useful spinoffs. Out of a better understanding of globular cluster evolution should come information concerning the structure and evolution of our galaxy. It should also help us in understanding the evolution of clusters of galaxies. And finally, since the cores of galaxies are, in the lowest approximation, simply very rich globular clusters one can hope that the modeling of galactic nuclei will also graduate from the now almost theological state of simple views and simple modeling to a correspondingly more rigorous condition.

REFERENCES

Ambartsumian, V. A. 1938, Ann. Len. State Univ., #22, Issue 4, p. 19.
Angeletti, L. and Giannone, P. 1977, Astron. & Astroph., 58, 363.
________ 1980, Astron. & Astroph.,
Antonov, Y. A. 1962, Bull. Univ. Leningrad, 7, 135.
Applegate, J. 1984, preprint.
Bahcall, J. N., and Ostriker, J. P. 1975, Nature, 256, 23.
Cohn, H. 1979, Ap. J., 234, 1036.
________ 1980, Ap. J., 242, 765.
Colgate, S. A. 1967, Ap. J., 150, 163.
Dokuchaev, V. I. and Ozernoy, L. M. 1982, Astron. Astrophys., 11, 16.
Duncan, M. J. and Shapiro, S. L. 1982, Ap. J., 253, 921.

Fabian, A. C., Pringle, J. E. and Rees, M. 1975, M.N.R.A.S., 172, 15P.
Goodman, J. 1984, Ap. J., 280, 298.
Hachisu, I., Nakada, Y., Nomoto, K. and Sugimoto, D. 1978, Prog. Theor. Phys. (Japan), 60, 393.
Heckman, O. and Seidentopf, H.1930, Zs. f. Ap., 1, 67.
Heggie, D. C. 1975, M.N.R.A.S., 173, 729.
__________ 1979, M.N.R.A.S., 188, 525.
__________ 1984, M.N.R.A.S., 206, 179.
Henon, M. 1961, Ann. d. Ap., 24, 369.
________ 1965, Ann. d. Ap., 28, 62.
Henon, M. 1975, IAU Symposium #69, Dynamics of Stellar Systems, ed. A. Hayli (Dordrecht: Reidel) p. 133.
Inagaki, S. and Lynden-Bell, C. 1983, M.N.R.A.S., 205, 913.
Lightman, A. P. 1982, Ap. J. (Lett.), 263, L19.
Lightman, A. P., Press, W. H. and Odenwald, S. F. 1978, Ap. J., 219, 629.
Lynden-Bell, D. 1967, M.N.R.A.S., 136, 101.
Lynden-Bell, D. and Wood, R. 1968, M.N.R.A.S., 138, 495.
Marchant, A. B. and Shapiro, S. L. 1980, Ap. J., 239, 685.
Lynden-Bell, D. and Eggleton, P. 1980, M.N.R.A.S., 191, 483.
McMillan, S. L. W. 1983, Ph.D. Harvard University.
Ostriker, J. P., Spitzer, L. and Chevalier, R. A. 1975, Ap. J. (Lett.), 176, L51.
Ozernoy, L. M. and Dokuchaev, V. I. 1982, Astron. Astrophys., 111, 1.
Press, W. H. and Teukolsky, S. A. 1977, Ap. J., 213, 183.
Shapiro, S. L. 1977, Ap. J., 217, 281.
Spitzer, L. 1940, M.N.R.A.S., 100, 397.
__________ 1975, IAU Symposium #69, Dynamics of Stellar Systems, ed. A. Hayli (Dordrecht: Reidel), p. 3.
Spitzer, L. and Mathieu, R. D. 1980, Ap. J., 241, 618.
Spitzer, L. and Saslaw, W. 1966, Ap. J., 143, 400.
Stodolkiewicz, J. S. 1982, Acta Astronomica, 32, 63.
Zeldovich, Y. A. and Podurets, M. A. 1965, Astron. Zh., 42, 963 (English translation in Sov. Ast. A.J., 9, 742).

DISCUSSION

SPITZER: I would like to add a historical note, about the first evidence of the gravothermal instability. I think that one should pay tribute to Dick Larson's work with his gas-dynamical code, which showed the first evidence that something else was happening beyond evaporation.

OSTRIKER: My apologies (some laughter).

GOODMAN: I'd like to take issue with that (lots of laughter). Not that Dr. Larson's work was not extremely important to get this subject going, but in fact Michel Hénon in his thesis, which was published in *Annals d'Astrophysique* 24, 369 (1961), first considered the time-dependent evolution of a stellar system with a finite central density, using the Fokker-Planck equation. He could not follow the evolution for very long, on his ancient IBM machine, but his numerical calculations showed that the central density rapidly increased. From there, he jumped to the conclusion that one could only hope to find a self-similar solution for a finite-mass system in a post-collapse, singular state. Thus, I believe that his was really the first numerical, time-dependent evidence that core collapse actually occurs.

SPITZER: Thank you for pointing out this important early work by Hénon.

OSTRIKER: A comment on a topic I did not touch on: I want to point out that Haldan Cohn, and perhaps others, are unwillingly supporting gravothermal oscillations if they say that a post-core-collapse cluster can look the same as a pre-core-collapse cluster, because the equations are all reversible, and there is nothing to prevent it from collapsing again.

VAN LEEUWEN: I am here as an observer of open clusters, and you have not heard much about my field during this conference. Therefore I would like to make a few comments. First, despite all modern instrumentation, we heavily depend, and will depend in the future, on what people have done fifty or eighty years ago. We simply need the old plates in order to measure proper motions in order to identify cluster members, since without knowing which stars are members we cannot do anything like what observers of globular clusters do. That leads me to one question and one request. The question is: if anyone knows of old deep plates in the field of the Pleiades, within five degrees around the centre, I would be happy to add them to my catalogue of proper motions. The request is: if you are erecting a new telescope which is good in the photometric sense and has a reasonable field, be careful that you take first-epoch plates of open clusters. These will be the plates that people, later on, will depend heavily on in the study of the dynamics of these clusters. The old plates we have available at the moment are generally not deeper than 14th or 15th magnitude. We need old plates, and the old plates can be the present plates, for the future, exposed to 20th or 21st magnitude.

TREMAINE: One comment about the origin of globular clusters. If one ever does arrive at a theory of the origin of globular clusters, one obvious thing that has to be explained, which I don't think has been mentioned, is the following. Both in our galaxy and in other galaxies, there is a very sharp cut-off in the mass and the luminosities of the globulars, corresponding to masses of slightly over one million solar masses. As Mike Fall and other people have mentioned, there is no known reason why a cluster of, say, three or five million solar masses could not survive for a Hubble time. Somehow one has to explain why the formation process produces this very sharp upper limit.

TOOMRE (seeing the many hands raised): Aha, theories are coming (laughter).

SPITZER: Mike, what is the explanation (more laughter)?

FALL: Martin Rees and I have recently proposed a theory for the origin of globular clusters in which roughly the observed masses and sizes arise naturally in the collapsing gas of a proto-galaxy.

SPITZER: Are there any other problems which can be solved quickly (laughter)?

COHN: I would like to amplify on the final point which Jerry Ostriker made in his very nice summation. As has been pointed out by members of the panel, globular clusters may well serve as analogs for galactic nuclei, which are necessarily more poorly observed than globulars because of their greater distance. If we can develop and verify a sound dynamical theory for clusters including all of the relevant physical effects, this should be of great aid in developing models for active galactic nuclei in which the fireworks of which Dr. Toomre is so fond are observed to occur.

WEBBINK: I think it is important over the next few years to pursue problems of the *hydrodynamics* of globular clusters: Observationally, just how little gas do clusters contain? And theoretically, how is gas so efficiently removed from clusters? In this connection, I think it is important to use the Space Telescope to attempt identifications of weak cluster X-ray sources. Some years ago (at the NATO workshop in Cambridge, England) Andy Fabian floated the idea that globular cluster X-ray sources could be single neutron stars passing through the winds of nearby giants. A very rough calculation indicates that, even appealing only to ambient gas (at the present upper limits to its density), isolated neutron stars should begin to appear near $L_x \sim 10^{30}$ erg s^{-1}. One might expect that post-core-collapse clusters would be very efficient at collecting gas and neutron stars in their cores, and so particular attention to this phenomenon would appear warranted in the case of clusters showing central cusps.

OSTRIKER: Are you referring to globular clusters collecting neutron stars from the background, from outside the clusters?

WEBBINK: No, I am referring to neutron stars formed in the clusters, from the initial mass function.

GRINDLAY: We looked at that possibility, but it was not consistent with our observations of low-luminosity X-ray sources in globular clusters.

WEBBINK: I agree.

OSTRIKER: Another point is that given the high velocities of neutron stars seen in the field around the sun, it is a bit hard to understand why there are as many in globular clusters as there are.

WEBBINK: That is true, but notwithstanding that, there seems to be an embarassing number of neutron stars in globular clusters. The question is: are they all tied up in binaries?

SEMENZATO: I would like to mention a problem which nobody seems to have touched upon: the possibility of existence of massive halos around globular clusters. There seems to have been some controversial evidence, recently, about their existence.

OSTRIKER: A comment on that: If I saw correctly the velocity dispersion vs. radius curves, presented on the first day of this symposium, I think there is strong evidence against massive halos. They would produce flat velocity dispersion curves, in analogy to flat rotation curves. If the velocity dispersion really significantly decreases in the outer parts, that would be a dramatic and definite test. But one would have to make a detailed comparison with Peebles' models.

BAHCALL: Perhaps Josh Grindlay could tell us about observations expected to be made by AXAF.

GRINDLAY: A few brief comments: As probably everybody here knows, the AXAF is the X-ray analog of Space Telescope, and is of course not nearly as far along but will appear one day, we all hope. If it does, and if it has the properties we hope it will have, it should allow these low-luminosity X-ray sources to be observed and located to sub-arcsecond precision right down to luminosities of order 10^{30} erg/s, well below where we think the peak is in the luminosity function. In a typical cluster one might have ten or twenty of these sources, and one can play the game of using the radial offsets once again to map out the potential within a single cluster, rather than applying a statistical analysis over a variety of different clusters, as we have been forced to do so far. Therefore, the possibilities for studying cluster dynamics that this will provide are really exciting.

AXAF will also allow to study the whole area which we did not touch on here at all, except for the comment we just heard, the question of diffuse gas in globular clusters. We did have significant evidence in a couple of clusters for hot diffuse X-ray emitting gas, in ω Cen, in 47 Tuc, and to a lesser extent in M22. This hot gas, detected by Einstein, was interpreted by us as being due to mass loss from individual cluster giants, the gas on its way out of the cluster being shock-heated running into the interstellar medium, given the large proper motion of the cluster. That process can be studied by AXAF in enormous detail, and has a whole range of exciting implications for densities of the interstellar medium and the galactic halo, and for mass loss rates and stellar evolution of cluster giants.

APPLEGATE: Will AXAF be able to see X-rays from M dwarfs in globular clusters? If so we have another method of studying the population of low mass stars in clusters.

GRINDLAY: I am glad you mentioned this third possibility of generation of X-ray sources in globular clusters. The answer is very definitely yes, one expects to detect pop. II M dwarfs.

ON THE DYNAMICS OF OPEN CLUSTERS

(orig.: Uch. Zap. L.G.U. No. 22, p. 19; 1938)

V. A. Ambartsumian

It has already been pointed out in the literature that due to several causes, open star clusters dissipate with time. For instance, Rosseland showed that when external stars move through a cluster, they cause a perturbation of the motion of the stars in the cluster and could transfer enough momentum to individual stars to cause their escape from the cluster's gravitational field. In this way the cluster will lose stars gradually, i.e., it will dissipate. According to Rosseland the time needed for the star cluster to dissipate following the outlined mechanism is 10^{10} years. However, as pointed out by the author of this article in the supplement to the Russian edition of Rosseland's book, there is another factor that makes the life of the open cluster even shorter: the stars in the cluster have close encounters with each other, as a result of which they exchange kinetic energy and gradually tend towards the most probable distribution, i.e., a Maxwell-Boltzmann distribution. And this, as we shall see shortly, also causes the dissipation of the cluster.

The relaxation time, i.e. the time in which the encounters of the stars in the cluster will lead to statistical equilibrium, is given approximately by the formula:

$$\tau = \frac{3\sqrt{2}}{32\pi n} \quad \frac{v^3}{G^2 m^2 \ln\left(\frac{\rho}{\rho_0}\right)} , \qquad (1)$$

where n is the number of stars per unit volume, m the stellar mass, G the gravitational constant, v the average stellar velocity in the cluster, ρ the radius of the cluster, and ρ_0 the distance at which the potential energy of two stars is equal to the average kinetic energy of stars in the cluster, i.e.

$$\rho_0 = \frac{2Gm}{v^2} . \qquad (2)$$

Formula (1) was derived for the case of stars with equal masses.

J. Goodman and P. Hut (eds.), Dynamics of Star Clusters, 521–524.

The average velocity v enters in formula (1) both explicitly and through ρ_0. To find v we shall assume that the cluster is stationary at any given moment of time. We can do that as the time necessary to change the distribution function of the stars in the cluster, considered as a system in phase space, is large compared to the time necessary for a star to cross the cluster. In the case of a stationary system consisting of particles attracted to each other according to Newton's law, using the virial theorem we can write

$$U = 2T, \tag{3}$$

where U is the absolute value of the potential energy of the system, and T is its kinetic energy.

The exact formula for U is:

$$U = \frac{1}{2} \sum_{i \neq k} \frac{Gm^2}{r_{ik}} , \tag{4}$$

where we shall assume again that all stellar masses are equal, and r_{ik} denotes the distance between the ith and the kth star. We shall replace all of the r_{ik}'s with their mean harmonic value which is apparently close to the radius of the cluster ρ. Therefore, approximately:

$$U = \frac{1}{2} \frac{GN(N-1)m^2}{\rho} ,$$

where N is the total number of stars in the cluster. For $N \gg 1$,

$$U = \frac{1}{2} \frac{GN^2m^2}{\rho} .$$

On the other hand

$$2T = Nmv^2 .$$

Therefore the virial theorem assumes the form:

$$v^2 = \frac{GNm}{2\rho} . \tag{4}$$

Comparing (4) with (2), we find that

$$\ln\left(\frac{\rho}{\rho_0}\right) = \ln\left(\frac{N}{4}\right); \tag{5}$$

substituting (4) and (5) in (1) and taking into account that

$$n = \frac{N}{\frac{4}{3}\pi\rho^3} ,$$

we find

$$\tau = \frac{2}{16\ln\frac{N}{4}} \sqrt{\frac{N\rho^3}{Gm}} . \tag{6}$$

Assuming that for a typical cluster N = 400, ρ = 2 parsecs, $m = 2\times10^{33}$g, we find the relaxation time to be $\tau \cong 4\times10^7$ years.

The result of the evolution of the distribution towards a Maxwell-Boltzmann distribution is the appearance of stars with kinetic energy larger than the escape energy for the cluster. Such stars will leave the cluster. The whole question is, what is the percentage of such stars in a cluster with the Boltzmann distribution ? If this percentage is small, then the dissipation of the cluster as a result of this process will be very slow. It is apparent that the ratio of the number of stars which can escape in a relaxation time τ to the total number of stars in the cluster is equal to:

$$\rho = \frac{\int_{\varepsilon_0}^{\infty} \varepsilon^{-\frac{\varepsilon}{\Theta}} \sqrt{\varepsilon}\, d\varepsilon}{\int_{0}^{\infty} \varepsilon^{-\frac{\varepsilon}{\Theta}} \sqrt{\varepsilon}\, d\varepsilon} , \tag{7}$$

where ε_0 is the escape energy, and Θ is equal to two thirds of the average kinetic energy, i.e.

$$\Theta = \frac{2}{3}\frac{T}{N} = \frac{1}{3}\frac{U}{N} . \tag{8}$$

On the other hand the mean value of ε_0 , i.e. the escape energy, is equal

$$\varepsilon_0 = \overline{\sum_k} \frac{Gm^2}{r_{ik}} = \frac{2U}{N} , \tag{9}$$

where the line over the sum indicates an average over i. Comparing (8) with (9) we find:

$$\varepsilon_0 = 6\Theta .$$

Substituting in (7), we obtain the approximation:

$$P \cong \frac{e^{-\varepsilon_0/\Theta}\ \Theta\sqrt{\varepsilon_0}}{\frac{1}{2}\sqrt{\pi}\ \ \Theta^{3/2}} = 2e^{-6}\sqrt{\frac{6}{\pi}} ,$$

i.e. one hundredth of the total number of stars should escape the cluster in a relaxation time. Therefore the dissipation time of the cluster should be of the order of several billion years.

This result was obtained for a cluster consisting of stars with equal mass. Therefore these numbers are applicable only to stars of the cluster with masses close to the average stellar mass in the cluster. For stars with masses two to three times smaller than the average, the escape time will be of the order of a few hundred million years. It is known that open clusters have few dwarfs. Perhaps the poverty in dwarfs shows the position of the cluster on its evolutionary path.

If we assume that the open clusters we observe are different stages in the evolution of one and the same cluster, in as much as the stars escaping from the cluster carry out positive kinetic energy, the total cluster energy

$$H = T - U \tag{10}$$

should decrease with the transition from richer to poorer clusters.

If we substitute (3) in (10), we find:

$$H = \frac{1}{2} U . \tag{11}$$

Therefore under the above assumption, U should increase. The data in the article of Orlova shows that such an increase of U with the decrease of N is not observed.

Another possible hypothesis is that all clusters were formed approximately at the same epoch (perhaps even at the epoch of the formation of the galaxy itself). Then the evolution of rich clusters with a large diameter should be slower. Among other things, these rich and large clusters should contain a higher percentage of dwarfs. It seems to the author that this result is backed by observations. E.g. the clusters h and χ Persei are rich and contain a high percentage of dwarfs at the same time. On the other hand, a number of poor clusters have hardly any dwarfs.

Hence, it becomes clear that to make further conclusions it is of great interest to determine not only the luminosity function for different clusters, but also the total energy H, which according to (11) could be determined from the absolute value of the potential energy.

MOST PROBABLE PHASE DISTRIBUTION IN SPHERICAL STAR SYSTEMS AND CONDITIONS FOR ITS EXISTENCE

(orig.: Vest. Leningrad Univ. 7, 135; 1962)

V. A. Antonov

The initial stages of the evolution of a spherical star system can be pictured in the following way. The stars have their beginning in a comparatively small region; the mechanism of the origin of initial velocities is not known exactly. From the point of view of the contemporary state of stellar astronomy, it is an acceptable view that the initial velocities have their origin in the attraction of each individual star towards the center of gravity of the system. The second possible assumption is to attribute the origin of initial velocities to forces of an explosive character. But in either case, initially, all orbits have to be almost exactly radial.

However, immediately after the formation of a spherical star system, irregular forces enter the picture. Such a process is investigated in the paper by T. A. Agekian[1], where it is shown that in the initial phase many stars will escape from the central region. We shall consider the next phase, i.e. when in the central part a Maxwellian distribution of velocities has already been established. This distribution gradually propagates towards the periphery. In reality, the transition from the center to the periphery has to be continuous, but in order to simplify the computations, we shall suppose that when the evolution stops (more precisely, slows down strongly) it is possible to distinguish the main body of the star system from the corona." In the main body a Maxwellian distribution is established (we neglect the truncation of this distribution). In the corona, the original distribution of velocities is preserved. This means that there are no orbits which entirely avoid the main body.

Boltzmann's well-known theorem[2] asserts that irregular forces can only increase the probability of the phase distribution and, consequently, its logarithm (entropy). Boltzmann's theorem is proven under the assumptions that only binary collisions of an elastic character take place and that the interactions exhibited are central forces. However, this circumstance is not that essential. The reason is[3] that close multiple encounters of stars are extremely rare phenomena, and multiple encounters at large distances are subject to the law of superposition, i.e. are equivalent to several consecutive binary passages.

On the other hand, regular forces exhibit no influence on the

J. Goodman and P. Hut (eds.), Dynamics of Star Clusters, 525–540.

entropy. From Boltzmann's theorem it immediately follows that the achievement of the maximum possible entropy is characteristic of the state of statistical equilibrium.

It is known from statistical mechanics that the state of statistical equilibrium (in the absence of rotation) has a Maxwelliam distribution as a necessary property at every point in space.

The aim of the present article is to show that this, generally speaking, is not sufficient, i.e. that a stationary state with a Maxwellian distribution of velocities is a state of statistical equilibrium only under certain conditions.

It has been remarked before that the region with a Maxwellian distribution in a real system is always bounded. Therefore, a stationary spherical star system with a Maxwellian distribution can be determined by three parameters: the radius, R, of the Maxwellian region, the stellar number density at the center, ν_c, and the root-mean-square velocity, σ. We are not, naturally, concerned with characteristics of individual stars. Escape of stars we neglect, proceeding as though the sphere at radius R were a reflecting boundary.

Thus, we have to establish under what conditions the above-described phase distribution maximizes the entropy

$$L = -\iint \Psi \ln \Psi \, dr \, dv \tag{1}$$

when the total mass, M, and total energy, H, are given. Strickly speaking, the total mass varies with time due to the fact that the system is not isolated from the surrounding world and may lose stars which reach a velocity greater than the escape velocity. But this process is comparatively slow and therefore is only of secondary importance for the configuration of a star system at any particular moment of time.[4]

Expressions for total mass and total energy

$$M = m\iint \Psi \, dr \, dv, \tag{2}$$

$$H = M \iint \Psi \left(\frac{v^2}{2} + \frac{\Phi}{2}\right) dr \, dv = K + W \ . \tag{3}$$

In formulas (1), (2) and (3) we use the following notations:

- r - radius vector
- v - velocity vector
- $\Psi(r,v)$ - phase density
- K - kinetic, and W - potential energy
- $\Phi(r)$ - potential energy per unit mass

In (3) it is necessary to divide Φ by two in view of non-additivity of potential energy. If we took not one half, but the full potential, then in the integration over all stars, each pair of gravitational interactions would be counted twice, i.e. one would get K + 2W.

The phase density of a stationary system will be designated by Ψ_o.

The condition that L be maximized by Ψ_o is equivalent to validity of the inequality

$$\delta L = \iint (\Psi_o \ln \Psi_o - \Psi \ln \Psi)\, dr\, dv < 0 \tag{4}$$

for all permissible Ψ (absolute maximum) or only those close to Ψ_o (relative maximum). We notice that the distinction between the relative and the absolute maximum corresponds in molecular physics to the distinction between metastable and fully-stable states.

In the beginning, we shall consider only the relative maximum. We shall introduce a certain classification of functions Ψ. Namely, we shall consider Ψ_1 and Ψ_2 as belonging to one class if they have the same

1) star density $\nu(r) = \int \Psi\, dv$, and hence also potential uniquely determined by the density;
2) kinetic energy K.

It is easy to see that all functions of one class are equally permissible or not permissible for a given M and H. But within each class considerable freedom is left in the choice of Ψ (e.g. different variations of velocity dispersion from point to point are possible). In each class, we choose a function Ψ to which corresponds the largest L. When such a choice is made, then validity of inequality (4) for the "worst case", Ψ, insures the validity for all the rest of the functions in the same class.

The selection process just described represents essentially an auxiliary constraint on the maximum, which we solve by Lagrange multipliers[5]

$$L + \int \lambda(r)\nu(r)dr + \mu K = \iint \left(-\Psi \ln \Psi + \lambda(r)\Psi + \frac{\mu v^2}{2} \Psi \right) dr\, dv \quad .$$

We find the variation of this expression with accuracy up to terms of second order

$$\iint \left[-(\ln \Psi + 1)\partial\Psi - \frac{(\partial\Psi)^2}{2\Psi} + \lambda(r)\delta\Psi + \frac{\mu}{2} v^2 \delta\Psi \right] dr\, dv \quad . \tag{5}$$

Equating to zero terms of first order, we obtain the solution of our auxiliary problem

$$-\ln \Psi - 1 + \lambda(r) + \frac{\mu v^2}{2} = 0$$

or

$$\Psi = (2\pi D)^{-\frac{3}{2}} \nu e^{-\frac{v^2}{2D}} \quad , \tag{6}$$

where

$$D = \frac{1}{\mu} \, , \quad \nu(r) = (2\pi D)^{\frac{3}{2}} e^{\lambda - 1} \, ,$$

It is easy to verify, that ν is indeed the stellar number density. D is the dispersion of velocities and is independent of the coordinates.

In agreement with the above, it is sufficient to verify the inequality (4) for functions of type (6), i.e. a Maxwellian distribution. Among such functions is also Ψ_0.

Substitute (6) into (1)

$$L = \int \nu\left(\frac{3}{2}\ \ln\ 2\pi D + \frac{3}{2} - \ln\nu\right) dr \quad . \tag{7}$$

Write down also total mass and energy

$$M = m\int \nu\, dr, \tag{8}$$

$$H = m\int \left(\frac{3D}{2} + \frac{\Phi}{2}\right) \nu dr \quad . \tag{9}$$

In (7) the terms containing $\frac{3}{2}$ and $\frac{3}{2}$ ln2π play no role, as they vanish in variation.

Suppose that ν is the density of an inhomogeneous but incompressible fluid. Generally speaking ν does not have to have exact spherical symmetry. In such case we can displace the elements of the fluid, without changing their density, in such a way that the fluid will assume a spherically symmetrical configuration with the density monotonically increasing towards the center. During such a displacement, L, M and K do not change. The functions under the integral signs in (7) and (8) do not contain coordinates explicitly and only the function ν varies.

On the other hand, W decreases since the fluid mass in a spherically symmetric configuration has minimal potential energy (or, what is the same, is in the state of stable equilibrium). If we compensate for the decrease in H by increasing D, L increases. In this fashion, it is possible to replace an arbitrary stellar number density by a spherically symmetrical one, while L increases. From this it can be seen that the maximum of L existing for symmetrical configurations, has to remain maximal also with respect to more general, non-symmetrical configurations. In the following, we can limit ourselves to ν depending on $r = \sqrt{(x^2+y^2+z^2)}$ only.

We shall find now the variation of entropy with an accuracy up to terms of second order

$$\delta L = \iiint\left\{\nu\left[\frac{3}{2}\ \frac{\delta D}{D} - \frac{3}{4}\left(\frac{\delta D}{D}\right)^2\right] + \frac{3}{2}\delta\nu\frac{\delta D}{D} + \frac{3}{2}\ \delta\nu\ \ln\ D \right.$$
$$\left. - (\ln\nu + 1)\ \delta\nu - \frac{(\delta\nu)^2}{2\nu}\right\} dx\ dy\ dz \quad . \tag{10}$$

Also, the variations of M and H must be equated to zero:

$$\delta M = m \iiint \delta\nu \; dx\, dy\, dz = 0 \quad , \tag{11}$$

$$\delta H = \frac{3m}{2} \iiint \left(\nu\delta D + D\delta\nu + \delta D\delta\nu\right) \; dx\, dy\, dz$$

$$+ \frac{m}{2} \iiint (2\Phi\delta\nu + \delta\Phi\delta\nu) \; dx\, dy\, dz = 0 \tag{12}$$

In deriving formula (12) we have used the reciprocity relation known from potential theory

$$\iiint \Phi \; \delta\nu \; dx\, dy\, dz = \iiint \nu\delta\Phi \; dx\, dy\, dz \quad .$$

For a stationary state ν has to satisfy a relationship which can be obtained by substitution of expression (6) - in the form

$$\Psi_0 = (2\pi D)^{-\frac{3}{2}} \nu e^{\frac{\Phi}{D}} e^{-\frac{1}{D}\left(\frac{v^2}{2} + \Phi\right)}$$

-into the Boltzmann equation.

Here, the last coefficient is a function of the energy integrals of each star separately; consequently, it cancels out after substitution into Boltzmann's equation. The remaining function of coordinates satisfies Boltzmann's equation and reduces to a constant. As a result, for stationary systems

$$\nu = c e^{-\frac{\Phi}{D}} \quad (D = \sigma^2) \quad . \tag{13}$$

We multiply the variations δM and δH by Lagrange multipliers and combine with δL:

$$\delta L - \frac{\delta H}{mD} + \left(\ln c + \frac{5}{2} - \frac{3}{2}\ln D\right)\frac{\delta M}{m} =$$

$$= \iiint \left[\frac{3\ln D}{2} - (\ln\nu + 1) - \left(\frac{3}{2} + \frac{\Phi}{D}\right)\right.$$

$$\left. + \left(\ln c + \frac{5}{2} - \frac{3}{2}\ln D\right)\right] \delta\nu \; dx\, dy\, dz \; +$$

+ Second order terms

Here the values of the Lagrange multipliers were already substituted, making the first order terms in $\delta\nu$ and δD vanish. This may be easily verified, having in mind (13). The remaining terms of second order give

$$\delta L = -\iiint \left[\frac{(\delta\nu)^2}{2\nu} + \frac{3}{4D^2}(\delta D)^2 + \frac{\delta\nu\delta\Phi}{2D}\right] dx\, dy\, dz \quad . \tag{14}$$

We shall solve (12) for δD to be substituted into (14), remembering that D and δD are independent of the coordinates.

$$\delta D = -\frac{2}{3}\frac{\iiint \Phi\delta\nu \; dx\, dy\, dz}{\iiint \nu \; dx\, dy\, dz} + \dots \tag{15}$$

The remaining terms are not needed, since they yield terms of higher order, for the substitution into (14).

$$\delta L = -2\pi \int_0^R \left[\frac{(\delta\nu)^2}{2\nu} + \frac{\delta\nu\delta\Phi}{D}\right] r^2 dr - \frac{4\pi\left(\int_0^R r^2\Phi\delta\nu dr\right)^2}{3D^2\int_0^R r^2\nu dr}. \tag{16}$$

where $\delta\nu$ must satisfy the additional condition (11), which is equivalent to

$$\int_0^R r^2\delta\nu \; dr = 0 . \tag{17}$$

Let us now introduce a variable

$$T = \frac{r^2}{4\pi mG} \cdot \frac{d\delta\Phi}{dr} ,$$

and notice that in agreement with Poisson's equation

$$4\pi mG\delta\nu = \frac{1}{r^2} \cdot \frac{d}{dr}\left(r^2 \frac{d\delta\Phi}{dr}\right) . \tag{18}$$

It follows that

$$\delta\nu = \frac{1}{r^2} \cdot \frac{dT}{dr} , \tag{18}$$

which, considering (17), gives $T(R) = T(0) = 0$, where $T(r=0)=0$ is of third order in r.

In (16) integrate by parts:

$$\int_0^R r^2 \ \delta\nu\delta\Phi \ dr = \int_0^R \delta\Phi \frac{dT}{dr} \ dr = -4\pi mG \int_0^R \frac{T^2}{r^2} dr \ ,$$

$$\int_0^R r^2\Phi\delta\nu \ dr = \int_0^R \Phi \frac{dT}{dr} \ dr = - \int_0^R T \frac{d\Phi}{dr} \ dr,$$

and (16) becomes

$$\delta L = -2\pi \int_0^R \left[\frac{1}{r^2\nu} \left(\frac{dT}{dr}\right)^2 - \frac{4\pi mG}{D} \cdot \frac{T^2}{r^2}\right] dr$$

$$- \frac{4\pi \left(\int_0^R T \frac{d\Phi}{dr} dr\right)^2}{3D^2 \int_0^R r^2\nu \ dr} \tag{19}$$

Consider the function

$$\phi \ (r) = r^2 \frac{d}{dr} \left(r \frac{d\Phi}{dr}\right) = r^2 \frac{d}{dr} (rF) = r^3 \ 4\pi mG\nu - r^2 F \ ,$$

where F(r) is the strength of the gravitational field. The asymptotic behavior of F is well known[6]

$$\lim_{r\to\infty} rF = 2D \ , \tag{20}$$

where the approach to the limit is oscillatory. Thus, $\phi(r)$ has an infinite number of zeros, which we arrange in ascending order: $r_o = 0$, r_1, r_2, ...

We will show that for $R < r_1$ the first integral in (19) and consequently the whole expression is always negative. For this purpose we first establish that $\phi(r)$ satisfies the differential equation

$$\frac{d}{dr} \left(\frac{1}{r^2\nu} \cdot \frac{d\phi}{dr}\right) + \frac{4\pi mG}{D} \cdot \frac{\phi}{r^2} = 0 \ . \tag{21}$$

We verify (21)

$$\frac{d\phi}{dr} = 4\pi mG \frac{d}{dr} (r^3 \nu) - 4\pi mGr^2\nu = 4\pi mG \left(r^3 \frac{d\nu}{dr} + 2r^2\nu\right) .$$

But from (13) easily follows

$$\frac{1}{\nu} \cdot \frac{d\nu}{dr} = - \frac{F}{D} . \tag{22}$$

Consequently,

$$\frac{1}{r^2\nu} \cdot \frac{d\phi}{dr} = - \frac{4\pi mG}{D} rF + 8\pi mG ,$$

from which we deduce the validity of (21).

Take $R = r_1$. Then ϕ may be considered as an eigenfunction of the equation

$$\frac{d}{dr}\left(\frac{1}{r^2\nu} \cdot \frac{d\phi}{dr}\right) + \frac{4\pi mG}{D} \lambda \frac{\phi}{r^2} = 0 , \tag{23}$$

It is the first eigenfunction in view of its constant sign, and the first eigenvalue $\lambda = 1$. This corresponds to the value of the minimum

$$\min \frac{\int_0^R \frac{1}{r^2\nu} \left(\frac{dT}{dr}\right)^2 dr}{\int_0^R \frac{4\pi mG}{D} \cdot \frac{T^2}{r^2} dr} , \tag{24}$$

which is equal to unity, Q.E.D.

Take further $R = r_2$. Then $\lambda = 1$ is the second eigenvalue of (23) with the same ϕ. We shall prove that (19) can assume positive values. Let T be subject to the additional constraint

$$\int_0^R T \frac{d\Phi}{dr} dr = 0 .$$

With this supplementary condition, the minimum (24) - in agreement with Courant's theorem[7] - is between the first and the second eigenvalue, and therefore is smaller than 1. The function T for which this minimum is attained makes δL positive, Q.E.D.

For a more exact investigation of expression (19) replace T by the function

$$T_0 = r^2 \frac{d}{dr}(rF) + \mu r^2 F \quad , \tag{25}$$

where the parameter μ is selected so that the boundary condition $T_0(R) = 0$ is satisfied,

$$\mu = - \frac{\left(\frac{d(rF)}{dr}\right)_{r=R}}{F(R)} \quad .$$

Integrating per parts and taking into consideration that by (25), T has a zero of third order at r=0, we obtain

$$\int_0^R \frac{1}{r^2\nu}\left(\frac{dT_0}{dr}\right)^2 dr = - \int_0^R T_0 \frac{d}{dr}\left(\frac{1}{r^2\nu} \cdot \frac{dT_0}{dr}\right) dr \quad ,$$

$$\frac{d}{dr}\left(\frac{1}{r^2\nu} \cdot \frac{dT_0}{dr}\right) + \frac{4\pi mG}{D} \cdot \frac{T_0}{r} = \mu\left\{\frac{d}{dr}\left[\frac{1}{r^2\nu} \cdot \frac{d(r^2F)}{dr}\right] + \frac{4\pi mG}{D} rF\right\} = \frac{4\pi mG}{D} \mu F \quad ,$$

$$\delta L = 2\pi \int_0^R T_0 \frac{4\pi mG}{D} \mu F \, dr - \frac{4\pi\left(\int_0^R T_0 F \, dr\right)^2}{3D^2 \int_0^R r^2 \nu dr} =$$

$$= \left(\frac{4\pi mG}{D} \mu - \frac{2 \int_0^R T_0 F \, dr}{3D^2 \int_0^R r^2\nu dr}\right) 2\pi \int_0^R T_0 F \, dr \quad . \tag{26}$$

Notice that for $R \to 0$ $\mu \to -2$, since near the center, the gravitational force may be considered proportional to the distance from the center. Consequently, for $0 < R < r_1$ we have $\mu < 0$. If we left out the last integral in (19) then in place of (26) we would get

$$\frac{4\pi mG}{D} \mu 2\pi \int_0^R T_0 F \, dr \quad ,$$

but this expression - as was already shown - is negative for $0 < R < r_1$ and consequently the integral itself

$$\int_0^R T_0 F \, dr > 0 \quad . \tag{27}$$

Let us gradually increase R, beginning with value r_1. Then for some R_k lying between r_1 and r_2, the expression in parentheses in (26) changes its sign from (-) to (+) (detailed calculation shows this). Since in the interval (r_1, R_k) we have $\mu > 0$, the inequality (27) is still valid in this interval.

From (26) it follows that for $R > R_k$ δL can assume positive values. It remains to prove that when $R=R_k$ (and hence also when $R < R_k$) δL cannot assume positive values. To show this represent an arbitrary T by

$$T = \tau T_0 + T_1 \quad , \tag{28}$$

for $R=R_k$, where T_1 satisfies the constraint

$$\int_0^{R_k} T_1 \psi \, dr = 0 \quad , \tag{29}$$

and T_0 is determined by (25). Obviously, $T_1(0)=T_1(R)=0$

Here, ψ is the first eigenfunction of equation (23) with boundary conditions $\psi(0) = \psi(R_k) = 0$. From (28) and (29) the value of τ is easily found

$$\tau = \frac{\int_0^{R_k} T \psi \, dr}{\int_0^{R_k} T_0 \psi \, dr} \quad . \tag{30}$$

The substitution $T_1 = T - \tau T_0$ in (29) shows that (29) is satisfied. The consistency of (28) will be proven if we can establish that the denominator in (30) does not vanish.

The minimum of (24) with the constraint

$$\int_0^R T \psi \, dr = 0$$

is indeed equal to the second eigenvalue of the equation (23); this value is equal to unity when $R = r_2$ and consequently is larger than 1 when $R = R_k < r_2$. It follows that if T_0 were orthogonal to ψ, then (19) with $T = T_0$ would be negative, but this contradicts the choice of R_k. We substitute (28) into (19):

1) The terms of the first and the second order in τ are

$$-2\pi\tau \int_0^R \frac{1}{r^2\nu} \cdot \frac{dT_0}{dr} \left(\tau \frac{dT_0}{dr} + 2 \frac{dT_1}{dr}\right) dr$$

$$+ 2\pi\tau \frac{4\pi mG}{D} \int_0^R \frac{T_0(\tau T_0 + 2T_1)}{r^2} dr$$

$$- \frac{4\pi\tau}{3D^2} \cdot \frac{\int_0^R (2T_1+\tau T_0)\, F\, dr \int_0^R T_0 F\, dr}{\int_0^R r^2\nu\, dr} .$$

Integration by parts as in the derivation of (26) gives

$$- \int_0^R \frac{1}{r^2\nu} \cdot \frac{dT_0}{dr} \left(\tau \frac{dT_0}{dr} + 2 \frac{dT_1}{dr}\right) dr =$$

$$= \int_0^R (\tau T_0 + 2T_1) \frac{d}{dr} \left(\frac{1}{r^2\nu} \cdot \frac{dT_0}{dr}\right) dr .$$

Consequently

$$-\int_0^R \frac{1}{r^2\nu} \cdot \frac{dT_0}{dr} \left(\tau \frac{dT_0}{dr} + 2 \frac{dT_1}{dr}\right) dr$$

$$+ \frac{4\pi mG}{D} \int_0^R \frac{T_0(\tau T_0 + 2T_1)}{r^2} dr =$$

$$= \int_0^R (\tau T_0 + 2T_1) \frac{4\pi mG}{D} \mu F\, dr .$$

Taking into account the definition of R_k, we can see that the sum of the terms of the first and second order in τ vanishes.

2) The remaining terms do not contain τ ; these are obtained from (19) by substituting T_1 for T. The resulting expression is negative, which follows by inspection of the minimum (24) in view of (29).

Proving in this manner the critical character of the value $R = R_k$, we shall now write the equation for R_k explicitly. In particular, replacing μ by its definition and replacing T_0 by the definition (25), we deduce from the vanishing of the parenthesized term in (26) for $R=R_k$ that

$$6\pi mGD(4\pi mGR\nu - F) \int_0^R r^2\nu\, dr - 4\pi mGR\nu \int_0^R r^2 F^2\, dr + + 4\pi mGF \int_0^R r^3\nu F\, dr = 0 \tag{31}$$

The integrals in equation (31) may be eliminated using

$$\int_0^R r^2\nu\, dr = \frac{R^2F(R)}{4\pi mG} \tag{32}$$

and the remaining two integrals we integrate by parts:

$$\int_0^R r^2F^2\, dr = \frac{R^3}{3}\,(F(R))^2 - \frac{2}{3}\int_0^R r^3FF'\, dr =$$

$$= \frac{R^3}{3}\,(F(R))^2 - \frac{2}{3}\int_0^R r^3F\left(4\pi mG - \frac{2F}{r}\right)dr = \frac{R^3(F(R))^2}{3}$$

$$- \frac{2}{3}\int_0^R 4\pi mGr^3\nu F\, dr + \frac{4}{3}\int_0^R r^2F^2 dr \quad ,$$

whence

$$\int_0^R r^2F^2\, dr = - R^3(F(R))^2 + 8\pi mG \int_0^R r^3\nu F\, dr \quad .$$

From (22) follows the relation

$$\int_0^R r^3\nu\, F\, dr = - D\int_0^R r^3\nu'\, dr = - D\left(R^3\nu(R) - 3\int_0^R r^2\nu\, dr\right) =$$

$$= - D\left(R^3\nu\,(R) - \frac{3}{4\pi mG}\,R^2F(R)\right) \quad . \tag{33}$$

Therefore

$$\int_0^R r^2F^2\, dr = - R^3\,(F(R))^2 - 8\pi mGDR^3\nu\,(R) + 6DR^2F(R) \quad . \tag{34}$$

Substituting the derived expressions for the integrals in (31), we have

$$\frac{3}{2}\ (4\pi mGR\nu\ -\ F)\ DR^2F\ -\ 4\pi mGR\nu\ (-R^3F^2\ -\ 8\pi mGDR^3\nu\ +\ 6DR^2F)$$

$$-\ 4\pi mGDR^3\nu F\ +\ 3DR^2F^2\ =\ 0\quad . \tag{35}$$

On the other hand, combining (13) with Poisson's equation, we obtain

$$\frac{1}{r^2}\cdot\frac{d}{dr}\left(r^2\ \frac{d\ \ln\ \nu}{dr}\right) = -\ \frac{4\pi mG}{D}\ \nu\ . \tag{36}$$

The order of equation (36) can be reduced in several different ways[6]. Thus, e.g., we introduce new variables

$$x\ =\ -\ r\ \frac{d\ \ln\ \nu}{dr}\quad \text{and}\ y\ =\ r\ \frac{dx}{dr}\quad . \tag{37}$$

Clearly, by its definition $x > 0$. We substitute x into (22) and get

$$F\ =\ \frac{Dx}{r}\quad . \tag{38}$$

Substitution of x in the left-hand side of (36) yields

$$\frac{1}{r^2}\ \frac{d}{dr}\ (rx)\ =\ \frac{4\pi mG}{D}\ \nu$$

or

$$\frac{4\pi mG}{D}\ \nu\ =\ \frac{x\ +\ y}{r^2} \tag{39}$$

Next, we take the logarithm of (39) and differentiate with respect to r, keeping in mind the transformation (37):

$$\frac{\frac{dx}{dr}+\frac{dy}{dr}}{x\ +\ y}\ -\ \frac{2}{r}\ =\ \frac{d\ \ln\ \nu}{dr}\ =\ -\ \frac{x}{r}\ ,$$

Substituting y as given by (37) and solving for $\frac{dy}{dx}$, we have

$$\frac{dy}{dx}\ =\ \frac{2x\ +\ y\ +\ x\ \ -\ xy}{y}\quad . \tag{40}$$

The initial conditions are obtained by taking r=0; then x=0, y=0 and dy/dx=2. The previously introduced variable ϕ differs from y only by the coefficient r. Therefore, the values of radius 0, r_1, r_2 .... correspond to the points of intersection of integral curve (40) with the axis of abscisses 0(0,)0, $A_1(x_1,0)$, $A_2(x_2,0)$... The integral curve passes from the point 0 from left to the right in the upper half-plane while in the lower half-plan, from A_1 to A_2, from right to the left (since for y=0 $\frac{dy}{dx}$ changes sign, passing through infinity) and further continues to turn around the point C(2,0) as around a focus.

We shall utilize now (38) and (39) in the transformation (35):

$$\frac{3}{2} D^2Rx \frac{Dy}{R} - \frac{D(x+y)}{R} \left[-D^2Rx^2 - 2D^2R(x+y) + 7D^2Rx\right] +$$

$$+ 3D^3x^2 = 0 \quad ,$$

from which after obvious algebra we get

$$y = \frac{3x - 2x^2 \pm x \sqrt{4x^2 - 44x + 73}}{8} \quad . \tag{41}$$

In the region of interest $0 < x < x_1$ the curve represented by (41) has the shape of a closed loop, passing, in particular, through the points 0 and C. Therefore, the point corresponding to R_k, i.e. the point of intersection of curves (40) and (41) necessarily exists. It must be located on the section of the integral curve between A_1 and A_2. Numerical computation gives the coordinates of the point of intersection: x=+2.031; y=-0.364. From the values of x and y we go back to ν:

$$\frac{d \ln \nu}{dr} = - \frac{x}{r} , \quad \ln \frac{\nu(r)}{\nu(0)} = - \int_0^r \frac{x}{r} dr = - \int_0^x \frac{x\, dx}{y} \quad .$$

Integration with $r=R_k$ yields

$$\frac{\nu(R_k)}{\nu(0)} = \frac{1}{709} \quad .$$

Thus, we have proven that the density in the center of the main body of a star system with Maxwellian distribution of velocities, cannot exceed the density on the boundary of the main body more than 709 times.

On the other hand, by equation (39) we have

$$R_k^2 = \frac{(x + y)D}{4\pi mG\nu} = \frac{1{,}667 \cdot 2^2 \cdot 709}{4\pi mG\nu(0)} \quad ,$$

$$R_k = \frac{9{,}70\sigma}{\sqrt{d \cdot G}} \quad ,$$

where d is mean density of the substance in the center and R_k defines the region which tends to establish within itself a Maxwellian distribution.

The fact that δ L may assume positive values for sufficiently large $R(R > R_k)$ means that Maxwellian distribution does not give a maximum of the entropy and that a phase distribution more probable (in the sense of Boltzmann's formula) than a stationary Maxwellian distribution is possible. What has to be the nature of the evolution of a system if at the initial moment of time it finds itself in such an "excessively probable" state? In the first place, the system will not be approaching the Maxwell's distribution but going further away from it, on account of Boltzmann's theorem. Secondly, when the entropy reaches a certain limit the system will become non-stationary. We shall prove this last assertion.

Let us divide the phase volume bounded in physical space by sphere $x^2+y^2+z^2 \leq R$ into two regions such that in the first region

$$\Psi \geq e^{-\frac{v^2}{2h}-1},$$

and in the second region

$$\Psi < e^{-\frac{v^2}{2h}-1} \leq \frac{1}{e}$$

where h is an arbitrary positive constant.

Now, estimate the entropy

$$L = -\iint\limits_{\text{region I}} \Psi \ln \Psi \, dr\, dv - \iint\limits_{\text{region II}} \Psi \ln \Psi \, dr\, dv \leq$$

$$\leq \iint\limits_{\text{region I}} \Psi \left(1+\frac{v^2}{2h}\right) dr\, dv + \iint\limits_{\text{region II}} \left(1+\frac{v^2}{2h}\right) e^{-\frac{v^2}{2h}-1} dr\, dv \; .$$

The substitution in the second integral was made on the basis of the fact that $x \ln x$ is a decreasing function for $x < \frac{1}{e}$.

We may strengthen the inequalities by integrating both integrals throughout the whole phase volume.

$$L \leq \frac{M}{m} + \frac{K}{hm} + \frac{5}{2e}(2\pi h)^{3/2}\frac{4}{3}\pi R^3$$

or

$$K \geq hmL - hM - \frac{5hm}{2e}(2\pi h)^{3/2}\frac{4}{3}\pi R^3 \; .$$

It is clear from these relations that when the entropy increases, beginning at some moment, the inequality $K>-H$, i.e. $2K+W>0$ will hold. In that case virial theorem shows that the system must disintegrate or else expand beyond the initial boundaries, which is what we wanted to prove.

Finally, we will show that in our problem the entropy maximum can only be relative and not absolute. For a special phase distribution we take the following. Let mass αM be uniformly distributed inside the sphere N_1 with radius ρ_1, with a uniform distribution of velocities. Let the velocities have upper limit C_1. Let the second part of the total star mass βM be distributed throughout sphere N_2 of radius ρ_2 in analogous way, with maximum velocity c_2. Let the distance between the centers of N_1 and N_2 be ρ_{12}, the spheres being non-intersecting. Outside N_1, N_2 the phase density is equal to zero. By the law of conservation of mass $\alpha + \beta = 1$.

The corresponding phase densities are equal to

$$\Psi_1 = \frac{\alpha M}{m \frac{4}{3} \pi \rho_1^3 \frac{4}{3} \pi c_1^3} , \quad \Psi_2 = \frac{\beta M}{m \frac{4}{3}\pi \rho_2^3 \frac{4}{3}\pi c_2^3} .$$

The entropy is

$$L = - \frac{\alpha M}{m} \ln\Psi_1 - \frac{\beta M}{m} \ln\Psi_2 = - \frac{M}{m} (\alpha \ln \alpha + \beta \ln \beta) + \tag{42}$$

$$+ \frac{3M}{m} (\beta \ln c_2\rho_2 + \alpha \ln c_1\rho_1).$$

The total energy, after some algebra, is seen to be

$$H = - \frac{3G\alpha^2M^2}{5\rho_1} - \frac{3G\beta^2M^2}{5\rho_2} - \frac{\alpha\beta M^2}{\rho_{12}} + \frac{3}{10} (\alpha M c_1^2 + \beta M c_2^2) . \tag{43}$$

c_2, ρ_1 and also the quantity $f = -\beta \ln (c_2 \rho_2)$ will be considered constant. On the other hand, let ρ_2 be infinitesimally small. The remaining parameter c_1 will be chosen such that (43) is always satisfied, i.e.

$$c_1 = \sqrt{\frac{1}{\alpha} \left[\frac{10}{3M} \left(H + \frac{\alpha\beta M^2}{\rho_{12}}\right) + \frac{2G\alpha^2 M}{\rho_1} - \beta c_2^2 + \frac{2Gf^2M}{\rho_2(\ln c_2\rho_2)^2}\right]} .$$

In the limit that $\rho_2 \to 0$, $\beta \to 0$, $\alpha \to 1$, we have $c_1 \to + \infty$.
In (42), all terms with the exception of the last one tend to finite limits:

$$\alpha \ln \alpha \to 0, \ \beta \ln \beta \to 0, \ \beta \ln c_2\rho_2 = - f ,$$

but the increase of c_1 has as a consequence the increase of L beyond any limit, i.e. there is no absolute maximum.

In conclusion the author expresses his gratitude to Professor K. F. Ogorodnikov for his great help in the formulation of this article.

References

1. Agekyan, T.A. Vestnik L.G.U. No. 1 (1962)
2. Carleman, T. Problemes Mathematiques dans la theorie cinetique des gaz (Uppsala: Almqvist & Wiksells, 1957)
3. Ogorodnikov, K.F. Dynamics of Stellar Systems (New York: Pergamon, 1965)
4. Ambartsumian, V.A. transl. in Appendix I in these proceedings. (1938)
5. Lavrent'ev, M.A. & Lyusternik, L.A. Kurs variatsionnovo ischisteniya (Course on variational calculus; Gostekhizdat, 1950)
6. Chandrasekhar, S. An Introduction to the Study of Stellar Structure (Dover, 1967)
7. Courant, R. & Hilbert, D. Methods of Mathematical Physics (Wiley, 1962)

STRUCTURE PARAMETERS OF GALACTIC GLOBULAR CLUSTERS

R. F. Webbink*
Department of Astronomy, University of Illinois,
1011 W. Springfield Ave., Urbana, IL 61801 and
Joint Institute for Laboratory Astrophysics, University of
Colorado and National Bureau of Standards, Boulder, CO 80309

ABSTRACT. Observed and derived structure parameters are tabulated for 154 galactic globular clusters, 7 dwarf spheroidal satellites of the Galaxy, and 6 globular clusters in the Fornax dwarf spheroidal. Observational parameters listed include equatorial coordinates, apparent level of the horizontal branch, reddening, subgiant branch color at the horizontal branch level, limiting and core angular radii, integrated magnitudes, and central surface brightnesses. Derived parameters include galactic coordinates, heliocentric and galactocentric distance, metallicity, limiting and core radii, central relaxation time scale, central mass density, central velocity dispersion, and central escape velocity.

1. INTRODUCTION

Nearly a decade has passed since the first comprehensive application of King's (1966) dynamical models of star clusters to the system of galactic globular clusters (Peterson and King 1975; Peterson 1976). The interim has seen a tremendous growth in the body of observational material available, particularly in integrated and surface photometry of individual clusters, and also in detailed color-magnitude studies of the more difficult clusters. At the same time, despite growing recognition of their shortcomings, the King models retain much of their attractiveness as tools in probing the dynamical structure of individual clusters by virtue of their great simplicity and ease of application. This paper incorporates this greatly increased body of observational data in a reanalysis of the galactic globular cluster system in terms of those models.

---

*1984-85 JILA Visiting Fellow.

J. Goodman and P. Hut (eds.), Dynamics of Star Clusters, 541–577.

For the sake of completeness, a number of recently-discovered objects are included here as possible to certain globular clusters. These include: AM 1 = E 1 = ESO 201-SC 10 (Holmberg et al. 1975; Lauberts 1976; Cannon, Hawarden, and Tritton 1978; Madore and Arp 1979), Eridanus = ESO 551-SC 01 (Cesarsky et al. 1977; Lauberts et al. 1981b), Reticulum = Sé 40/3 = ESO 118-G 31 (Sérsic 1974; Holmberg et al. 1975), AM 2 = ESO 368-SC 07 (Holmberg et al. 1978b; Madore and Arp 1979), E 3 = ESO 037-SC 01 (Lauberts 1976; Holmberg et al. 1978a; Cannon, Hawarden and Tritton 1978), ESO 093-SC?08 (Holmberg et al. 1977), AM 4 (Madore and Arp 1982), BH 176 = ESO 224-SC 08 (van den Bergh and Hagen 1975; Holmberg et al. 1977; Cannon, Hawarden and Tritton 1978), ESO 452-SC 11 (Lauberts et al. 1981a), TJ 5 (Terzan and Ju 1980), TJ 16 (Terzan, Bernard, and Ju 1978b; Terzan and Bernard 1978; Terzan and Ju 1980), TJ 15 (Terzan and Ju 1980), TJ 17 (Terzan, Bernard, and Ju 1978a; Terzan and Bernard 1978; Terzan and Ju 1980), Grindlay 1 (Grindlay and Hertz 1981), Liller 1 (Liller 1976a,b), TJ 23 (Terzan and Ju 1980), UKS 1 (Malkan, Kleinmann and Apt 1980), and Kodaira 1 (Kodaira 1983). UKS 2 = ESO 166-SC 12 = BH 66? (van den Bergh and Hagen 1975; Holmberg et al. 1977; Malkan 1981) was discussed by Malkan (1981) as a globular cluster, and is included here, although in the opinion of Holmberg et al. (1977) and this author, it appears to be an open cluster. In addition to these objects, several previously known objects have been added to the list: Ruprecht 106 = ESO 218-SC 10 (Ruprecht 1959; Holmberg et al. 1977), Terzan 3 = ESO 390-SC 06 (Terzan 1968; Holmberg et al. 1978b; Cannon, Hawarden, and Tritton 1978), Terzan 8 = ESO 398-SC 21 (Terzan 1968; Cannon, Hawarden, and Tritton 1978; Lauberts et al. 1981a), and Terzan 10 = ESO 521-SC 16 (Terzan 1968; Lauberts et al. 1981a). In this author's opinion, only the last of these four objects remains in doubt as a true globular cluster.

The Reticulum cluster noted above, NGC 1466, and NGC 1841 are sometimes discussed as members of the Large Magellanic Cloud (LMC) complex. Although they share a very similar apparent distance modulus with the LMC, they have been included here among the galactic globular clusters, NGC 1466 because of the large velocity difference from the LMC found by Cowley and Hartwick (1981) (although Freeman, Illingworth and Oemler [1983] find a much smaller difference), and Reticulum and NGC 1841 because, at angular distances of more than 10° from the center of the LMC, they can scarcely now be bound to the LMC, even though they may well share a common origin with it.

In the following section, the methods are discussed by which the observational data were obtained which form the basis of this study. These data are catalogued in Table I. The basic assumptions employed in deriving the individual cluster structure parameters are discussed in the subsequent section, and these parameters are listed in Table II.

2. OBSERVATIONAL DATA

Table I is organized in three parts: Ia, the galactic globular clusters, Ib the dwarf spheroidal satellites of the Galaxy, and Ic the globular clusters in the Fornax dwarf spheroidal. The observational methods employed are explained below in the context of the galactic globular clusters, but the other two parts of this table differ only slightly, if at all, and those differences will be noted at the end of this section.

The following discussion includes as well some estimates of the typical errors in each of the measured quantities. Where a numerical value is quoted for this error or uncertainty, it is meant to indicate the standard deviation of results obtained by that technique, but only when the source material is of good quality. A colon (:) or double colon (::) indicates source material of poor or very poor quality, or poor or very poor suitability for the technique employed, and the corresponding uncertainties in the quoted results are likely to be substantially worse. It must be emphasized, however, that this notation has been used here in a relative fashion, within the context of the technique involved. Poor results obtained by a reliable method may still be better than good results obtained with a poor one.

References in the table refer to the numbered bibliography immediately following it. Where more than one method, or more than one reference, have been employed, these are enumerated in the notes to each part of the table. These notes are identified by cluster designation (from column 1), with the relevant column numbers indicated in boldface type. Where multiple methods are used, the sets of reference numbers corresponding to different methods are set off by semicolons.

Table Ia is arranged as follows:

Cluster Designation. The observed structure parameters of the galactic globular clusters are listed in order of right ascension. Each cluster is identified in column (1) by its designation in IAU coordinate format. Where an object has not yet received an official designation, one has been assigned here using the same convention. The common name or names in contemporaneous use for each cluster are listed in columns (2) and (3).

Coordinates. The right ascension and declination for equinox 1950, rounded off to the nearest 1^s or $0\farcm1$, are listed in columns (4) and (5), respectively. These have been compiled from numerous sources, dating from 1850 up through the recent measurements by Shawl and White (1984), and carry a median net uncertainty of $2\farcs45$ in the positions of the cluster centers. Individual uncertainties are in general closely comparable to ε_V, the expected root-mean-square difference in position between the cluster visual photocenter and its dynamical center. This difference is given by the expression

$$\varepsilon^2 = \frac{\langle b^2 \rangle}{\langle b \rangle} \frac{\int r^3 \, \sigma(r) dr}{[\int r \, \sigma(r) dr]^2} \quad , \qquad (1)$$

where b is the apparent brightness of an individual star, r the angular distance from dynamical cluster center, and $\sigma(r)$ the surface brightness. For V-band observations, a good approximation is

$$\varepsilon_V = 24''.2\ c^{-1.43}\ 10^{0.5\,c\,+\,0.2(\sigma_c - V_{HB})} \quad , \qquad (2)$$

where

$$c \equiv \log(\theta_t/\theta_c) \quad , \qquad (3)$$

(see columns 15 and 18), σ_c is the central surface brightness (V magnitudes per square arcminute: see column 22), and V_{HB} the apparent level of the horizontal branch (column 6).

Attention should be called to significant discrepancies, exceeding $0^\circ.1$, discovered in published positions of several objects: NGC 1841, AM 4, Ton 2, Ter 10, and NGC 6749. These have been resolved here, and the standard errors in the positions of these clusters listed in Table I are 16" or smaller.

Horizontal Branch. Columns (6), (7), and (8) contain the apparent level of the cluster horizontal branch, the method of its determination, and the reference from which this determination was drawn, respectively. The horizontal branch is commonly used as the standard candle in establishing the globular cluster distance scale, and this convention is followed here. Its level is fixed at the blue edge of the RR Lyrae gap wherever possible, as in the earlier compilations by Harris (1976, 1980).
The methods listed in column (6) identify the observational feature from which this estimate derives. These are, in order of preference (together with their estimated errors for first-rank data):

- CM - color-magnitude diagram ($\pm 0^m.05$)
- RR - mean magnitudes of RR Lyrae variable ($\pm 0^m.1$)
- BG - magnitudes of the brightest giants ($\pm 0^m.3$)
- IR - combination of the index of richness (Kukarkin 1974), integrated apparent magnitude of the cluster, and estimated foreground reddening ($\pm 0^m.6$)
- A - measured foreground reddening plus assumed reddening law ($\pm 1^m.4$).

The quantitative calibrations of the first four of these methods are described in detail by Harris (1976, 1980). The fifth assumes the modified cosecant law specified below to infer the cluster distance modulus, and is useful only for very highly reddened clusters at extremely low galactic latitude.

Reddening. Color excesses, methods for their determination, and corresponding references may be found in columns (9), (10), and (11), respectively. A wide variety of methods have been employed, here, and considerable judgment was exercised in preferring the results of some methods over others. The first-echelon measures rely on the facts that the intrinsic colors of blue horizontal branch stars are nearly independent of metallicity, and that their downward tail in the color-magnitude diagram is essentially a bolometric effect. These measures are:

HB - two-color diagram of blue horizontal branch stars
NB - narrow-band photometry and spectroscopy of horizontal branch stars fit to model atmospheres
B - color-magnitude diagram of blue horizontal branch stars fit to a fiducial curve.

With accurate photometry, the resulting estimates of E(B-V) are probably accurate to $\pm 0^{m}.02$ or better.

The second-echelon measures rely on intrinsic colors of RR Lyrae stars. These are:

rr - Sturch's method, using RR Lyrae colors at minimum
RR - mean RR Lyrae colors

Third-echelon measures employ comparisons with intrinsic flux distributions of individual red giant stars:

Ψ - slope of the red continuum of individual giants
ΔU - correlation of ultraviolet excesses of giant stars with the color of the subgiant sequence at the level of the horizontal branch.

These and the second-echelon measures are probably not greatly inferior to the first-echelon measures.

Fourth-echelon measures consist of asymptotic reddenings of foreground field stars:

UB - Johnson UBV photometry of field stars
uy - Strömgren uvby photometry of field stars
XY - Vilnius UPXYZVS photometry of field stars.

Because these estimates are essentially statistical in nature, they are normally significantly more uncertain than higher-echelon measures. If the field encompassed in the study is sufficiently small, however, an accuracy of ± 0.15 E_{B-V} can be achieved in most cases, giving quite satisfactory results for high-latitude clusters.

Fifth-echelon measures employ the blue and red edges of the RR Lyrae instability strip:

GB - blue edge of the RR Lyrae gap
GR - red edge of the RR Lyrae gap.

These are also statistical in nature, depending heavily on having a dense distribution of stars in this part of the Hertzsprung-Russell diagram. Superior methods of estimating reddenings are available for the majority of clusters to which this method is most suited; among the clusters for which it is used here, an accuracy of $\pm 0^{m}.04$ is probably the best that can be expected. It should be noted that there is increasing empirical evidence (see, for example, Sandage, Katem, and Sandage 1981; Sandage 1981; Bingham et al. 1984) that the edges of the instability strip are not constant in color, as this method assumes.

The sixth-echelon measures are those which depend on comparisons of integrated photometry of the cluster in question with the intrinsic colors of clusters of similar spectral type and known reddening. They are:

CS - integrated optical or near-ultraviolet colors
IR - integrated infrared colors.
MI - correlation of metallicity index (Cowley, Hartwick, and Sargent 1978) with integrated color.

Because the calibration of these methods depends upon cluster reddenings derived by the above methods, they have been adopted only in the absence of a determination by one of the above methods. Nevertheless, at their best, the reddening estimates obtain from integrated color-spectral type relations appear comparable in accuracy with the second- and third-echelon measures discussed above.

Calibrations of integrated colors exist in a number of different photometric systems; it is the weighted mean of these independent determinations which is quoted in Table I. For the sake of consistency, a single calibration has been used for each photometric system. These are listed below, in decreasing order of accuracy, as determined by comparison with clusters whose reddenings were derived by one of the above methods:

ANS (λλ 1550,1800,2200,2500,3300): van Albada, de Boer, and Dickens (1981) ($\pm0\overset{m}{.}030$, n = 17)
Strömgren-Crawford (uvbyβ): Johnson and McNamara (1969) ($\pm0\overset{m}{.}036$, n = 12)
Zinn-Gunn (uvgr,39B,39N): Zinn (1980); Zinn and West (1984) ($\pm0\overset{m}{.}039$, n = 44)
Washington (CMT_1T_2): Harris and Canterna (1978) ($\pm0\overset{m}{.}040$, n = 34)
Johnson (UBV): Racine (1973); Harris (1976); Harris and Racine (1979) ($\pm0\overset{m}{.}046$, n = 48)
Kron-Mayall (PVI): Lohmann (1963) ($\pm0\overset{m}{.}048$, n = 31)
Stebbins-Whitford (UVBGRI): Kron and Guetter (1976) ($\pm0\overset{m}{.}052$, n = 39)
Vilnius (UPXYZVS): Zdanavičius (1983) ($\pm0\overset{m}{.}055$, n = 24)
DDO (35,38,41,42,45,48): Bica and Pastoriza (1983) ($\pm0\overset{m}{.}065$, n = 46).

Infrared photometric values are of rather lower precision ($\pm0\overset{m}{.}11$; Malkan 1981).

The third method of this group (MI) was calibrated using nine clusters with well-determined colors among those listed by Cowley, Hartwick and Sargent (1978). They give

$$(B-V)_0 = 0.479 + 0.0933 \cdot MI \ (\pm0\overset{m}{.}072) \quad . \qquad (4)$$

There are a number of clusters for which individual reddening studies are still lacking. For those lying at galactic latitude, $|b| \geq 10°$, use was made of the high-latitude extinction map of Burstein and Heiles (1982):

HI - neutral hydrogen column density plus galaxy counts.

Comparison with 43 clusters of known reddening indicates that these values are reasonably good ($\pm 0\overset{m}{.}060$). For clusters of lower galactic latitude, it was necessary to resort to a cosecant-like law:

CX - modified cosecant law.

The assumed form of this extinction law is intended to reflect the apparent absence of reddening at the galactic poles, and the finite scale height (assumed to be 150 pc) of obscuring matter in the galactic plane:

$$E_{B-V} = \frac{0.06}{\sin|b|}\left[1 - \exp(-R_o \sin|b|/0.15)\right] - 0.06 \qquad (5)$$

where R_o is the distance to the cluster in kpc. Where the exponential term in this expression is non-negligible, it is necessary to solve self-consistently for both reddening and distance. The accuracy of these estimates is only ±0.4 E(B-V).

It should be noted here that the values of reddening listed in Table I portray individual clusters as suffering uniform extinction. However, an examination of most of the clusters included in this table on deep-sky survey plates reveals that many clusters with E(B-V) $\gtrsim$ 0.2 show evidence of nonuniform extinction, and this is almost invariably the case among highly reddened (E(B-V) $\gtrsim$ 0.5) clusters. This variable extinction could conceivably introduce systematic errors in the mean reddening estimates of highly obscured clusters as derived from their integrated colors.

Subgiant Branch Color. Columns (12), (13) and (14) of Table I contain, respectively, the observed color of the cluster subgiant branch at its horizontal branch level, the method by which this color was determined, and the reference to the source material for that determination. Whenever possible, the subgiant branch color was estimated directly from the cluster color-magnitude diagram:

CM - color-magnitude diagram.

Comparison of the values measured here with the un-dereddened values of Sandage (1982), for 34 clusters with common color-magnitude diagrams, indicates an internal error of only $\pm 0\overset{m}{.}016$. Uncertainties in the dereddened colors, $(B-V)_{o,g}$, are therefore dominated by uncertainties in the reddening correction and systematic errors in the photometry.

In the absence of a suitable color-magnitude diagram, several indirect methods were employed. These correlate the intrinsic color of the subgiant branch at the horizontal branch level with other properties of the cluster. In order of decreasing accuracy, they are:

ΔP - period difference of RR Lyrae variables at constant amplitude and effective temperature (see Sandage, Katem, and Sandage 1981)

ΔS - cluster metallicity determined by the difference in spectral types, as derived from metallic lines and from hydrogen lines, observed in individual RR Lyrae stars (see Butler 1975)
UB - integrated UBV colors of the cluster
Q - line-blanketing index derived from integrated narrow-band photometry of the cluster (see Zinn 1980)
BV - integrated BV colors of the cluster

The correlations actually used, and their respective uncertainties, are:

$$(B-V)_{0,g} = 0.839 + 1.81\ (\Delta \log P)_s \qquad (\pm 0\overset{m}{.}041,\ n = 24) \qquad (6)$$

$$(B-V)_{0,g} = 1.038 + 0.172\ [Fe/H]_{\Delta S} \qquad (\pm 0\overset{m}{.}057,\ n = 26) \qquad (7)$$

$$(B-V)_{0,g} = 0.372 + 0.563\ (B-V)_0 + 0.513\ (U-B)_0$$

$$(\pm 0\overset{m}{.}059,\ n = 65) \qquad (8)$$

$$(B-V)_{0,g} = 0.719 + 0.874\ Q_{39} \qquad (\pm 0\overset{m}{.}060,\ n = 55) \qquad (9)$$

$$(B-V)_{0,g} = 0.133 + 0.992\ (B-V)_0 \qquad (\pm 0\overset{m}{.}065,\ n = 67)\ . \qquad (10)$$

To each of these values must be added the cluster reddening, E(B-V). In dereddening the observed integrated UBV colors of a cluster to apply equation (8), the ratio E(U-B)/E(B-V) has been calculated following Racine (1973).

Limiting and Core Radii. Columns (15), (16), and (17) contain the logarithm of the value (in arcminutes), method of its determination, and reference for the limiting radius of each cluster, with columns (18), (19), and (20) containing the corresponding quantities with respect to the core radius. Because the form of the surface brightness profile of the core of a strongly condensed cluster is only very weakly dependent on the degree of central concentration of the cluster, and likewise for the form of the surface brightness profile near its limiting radius, I have regarded these two quantites as independently determinable, and not attempt a reanalysis of all available data. Both determinations share many of the same methods, which are:

SC - star counts
SP - surface photometry
AP - concentric aperture photometry
D - correlation of apparent angular diameter with limiting radius (see, Kukarkin and Kireeva 1979)
e - eye estimate of the core radius.

At low surface densities, the ability to see stars several magnitudes fainter than those which contribute the bulk of the cluster light generally bestows superior statistical properties upon the star count method, in comparison with surface photometry or aperture photometry. It is therefore the method of choice for determining limiting radii. Unfortunately, the derived limiting radii are very sensitive to the adopted background star densities. Even in the best cases, they involve extrapolations of a factor of two or more in radius beyond the last measurable point, and the accuracies of the resulting radii appear to be no better than ± 0.10 in log θ_t. In comparison, it is indeed very surprising to find that a crude apparent angular diameter/ limiting radius correlation (method D) gives results which are not terribly inferior (± 0.17 in log θ_t, as deduced from the dispersion in cluster mean luminosity densities at fixed galactocentric distance -- see Innanen, Harris and Webbink [1983]).

In the cores of clusters, star counts provide an accurate description of the surface density distribution only in a very few, extremely open clusters. For the remainder, surface photometry is the method of choice, since it both provides more detailed information and is less sensitive to centering errors than either concentric aperture photometry or star counts. Provided that the cluster brightness profile is normal (see the remarks below), core radii derived from surface photometry appear to be accurate to ± 0.049 in log θ_c, those from star counts to ± 0.12 in log θ_c, and those from aperture photometry to ± 0.14 in log θ_c. Peterson and King (1975) determine an uncertainty of ± 0.16 for eye estimates of log θ_c, but this method requires photographic images of the cluster cores which are unsaturated.

It should be noted that the surface brightness distributions in the cores of some clusters are poorly represented by King models. Of particular interest are those which show central brightness excesses, the most famous of which is NGC 7078 (Newell and O'Neill 1978; Aurière and Cordoni 1981). Recent surface photometry studies by Djorgovski and King (1984), Kron, Hewitt, and Wasserman (1984), and Djorgovski and Penner (1985) have identified a number of other examples of this phenomenon, including NGC 4147, 6266, 6293, 6342, 6624, 6642, 6681, and 7099. The values of log θ_c (and of σ_c) listed in Table I for these clusters are, in effect, best fit values, but the deficiencies of the cluster model should be recognized here.

Integrated Magnitude and Central Surface Brightness. Columns (21), (22), (23), and (24) of Table I contain, respectively, the integrated V magnitude of the cluster, its central surface brightness (in V magnitudes per square arcminute), the method of their determination, and the source of the data or estimate for these values. The methods employed were:

- AP - concentric aperture photometry
- Σ - summation over the cluster color-magnitude diagram with corrections for faint stars and for stars lying outside the region studied
- R - fit of the upper end of the cluster luminosity function to a fiducial luminosity function
- IR - index of richness (see Kukarkin 1971, 1974).

Where the central surface brightness is enclosed in parentheses, it has been calculated from the integrated magnitude, using the King model appropriate to that cluster.

All available concentric aperture photometry of each cluster was collected and fit to a King model with core and limiting radii given by the values in columns (18) and (15), respectively. A complex weighting scheme was used which, in effect, weights most heavily those observations of greatest precision which involve the smallest extrapolation to find the integrated magnitude or central surface brightness, as the case may be. The median uncertainty in the derived integrated magnitude is $\pm0\overset{m}{.}024$, and that in the central surface brightness is $\pm0\overset{m}{.}035$, but these nominal errors do not include uncertainties in the limiting and core radii themselves.

For a number of highly obscured clusters, only infrared photometry is available. In these cases, the V magnitudes have been calculated by correcting the observed K magnitudes for the intrinsic $(V-K)_0$ color of the cluster (following Aaronson et al. 1978) appropriate to its deduced metallicity, and for reddening, with $E(V-R) \approx 2.82 \cdot E(B-V)$. The deduced integrated magnitudes in these cases are naturally of much lower precision than for other clusters with direct observations in the V band.

The remaining methods for estimating integrated magnitudes are of relatively low accuracy. Summation over color-magnitude diagrams gives values uncertain by an estimated $\pm0\overset{m}{.}6$. Fitting of the luminosity function of bright stars is of similar accuracy. This procedure, as applied here, amounted merely to estimating the number of cluster stars within two magnitudes of the brightest member (where possible), and scaling the integrated intrinsic luminosity of the comparison cluster accordingly. For this purpose, the fiducial luminosity function adopted here is that of M3, as compiled by Sandage (1954, 1957), corrected for the value of apparent distance modulus listed in Table Ia. Lastly, the calibration of Kukarkin's index of richness (IR) employed here is that due to Harris (1980). Among the highly obscured clusters for which this estimate has been adopted, the uncertainty in integrated magnitude is of order $\pm0\overset{m}{.}8$.

Cluster designations from column (1) are repeated in column (25) as a reference aid.

Table Ib differs slightly in format from Table Ia. Ellipticities of the dwarf spheroidals have been included in column (17). Because of their very low degree of central concentration, these objects are not amenable to independent determinations of core and limiting radii [columns (16) and (15), respectively], so a single method (column 18) and source (column 19) have been used for both of these quantities, as well as for their ellipticities. At the same time, they are poor subjects for concentric aperture photometry because of their very low surface brightnesses, and, not surprisingly, very little data of this sort exists. Surface photometry (SP) and other methods have therefore been widely used to measure integrated magnitudes (column 20) and central surface brightnesses (column 23), for which separate methods and references are given here (columns 21 and 24, and 22 and 25, respectively).

Table Ic parallels Table Ia exactly, and requires no further explanation.

3. DERIVED PARAMETERS

Table II is arranged in three parts, paralleling Table I, and contains positional and structural data derived from the observational data of Table I.

There now exists compelling evidence (Sandage 1982; Frenk and White 1982) that the horizontal branch level, V_{HB}, is not constant from cluster to cluster, but depends upon cluster metallicity. This phenomenon was anticipated from the early theoretical pulsation studies of Christy (1966), but the magnitude of the effect is not yet well established empirically, nor are its theoretical ramifications concerning differences in cluster ages and/or helium abundances well understood. Therefore, in fixing the cluster distance scale, a very conservative approach has been taken here, setting $M_V(HB) = +0.6$ (Harris 1976; Stothers 1983) without regard to metallicity, and adopting a ratio of total to selective absorption $R = A_V/E(B-V) = 3.2$. The distance from the Sun to the galactic center is assumed to be 8.8 kpc (Harris 1976).

The basis for calculating the internal structure parameters of individual globular clusters is the single-parameter family of King (1966) models. A mean mass-luminosity ratio $M/L_V = 1.6\ M_\odot/L_V(\odot)$ (Illingworth 1976) has been adopted throughout. It should be noted that a χ^2-test applied to Illingworth's results clearly indicates the presence of significant cluster-to-cluster differences in M/L_V, but they do not appear correlated in any obvious way with differences in cluster metallicity, core concentration, or galactocentric position. For purposes of estimating central relaxation time scales, a mean stellar mass (for those stars dominating the cluster light) of 0.65 $M_\odot$ was assumed. Finally, we note that a King model is completely specified by any three of the four quantities: (i) limiting radius, (ii) core radius, (iii) integrated magnitude, and (iv) central surface brightness. Where all four of these quantities are available independently, the limiting radius has been discarded in favor of the other three parameters, as it appears most susceptible of observational errors.

The columns of Tables IIa and IIb contain the following data:

(1) - Cluster designation in IAU coordinate format (as in column 1 of Tables Ia and Ib)
(2)-(3) - Cluster name (as in Tables Ia and Ib)
(4) - Galactic longitude (in degrees)
(5) - Galactic latitude (in degrees)
(6) - Heliocentric distance (in kpc)
(7)-(9) - Galactocentric coordinates (in kpc). The (X,Y,Z)-axes of this coordinate system are parallel to the directions $(\ell,b) = (0°,0°)$, $(90°,0°)$, and $(0°,90°)$, respectively
(10) - Galactocentric distance (in kpc)

(11) - Logarithm of the metallicity, in solar units, derived from $(B-V)_{0,g}$ (see below)
(12) - Integrated absolute magnitude of the cluster
(13) - Core radius (in pc)
(14) - Limiting radius (in pc)
(15) - Core concentration parameter, $c \equiv \log(r_t/r_c)$, as derived implicitly from the apparent core radius, central surface brightness, and integrated magnitude, for those clusters having multiple aperture photometry
(16) - Logarithm of the central relaxation time scale (in yr)
(17) - Logarithm of the central mass density (in $M_\odot$ pc^{-3})
(18) - Central velocity dispersion (in km s^{-1})
(19) - Central escape velocity (in km s^{-1})
(20) - Cluster designation (as in column 1)

With the exception of the metallicities (column 11), formulae and explanations for all of the structural parameters listed here may be found in the papers of King (1966) and Peterson and King (1975). It should be noted that the central velocity dispersion listed in column (18) is the true dispersion, which exceeds the projected dispersion by 4 percent or less for clusters with $c > 1.0$. The central escape velocity in column (19) is that calculated by treating the cluster as an isolated potential. This exceeds the escape velocity from cluster center to the potential corresponding to the cluster limiting radius by 8 percent or less for clusters with $c > 1.0$. The parameters listed in columns (18) and (19) are the same as those tabulated previously by Peterson and King (1975) and Peterson (1976).

The cluster metallicities calculated in column (11) were derived from the following correlation between dereddened subgiant colors and the high-dispersion spectroscopic metallicities on Cohen's scale, as tabulated by Zinn and West (1984):

$$[m/H] = -4.94 + 4.23\,(B-V)_{0,g} \qquad (\pm 0.28,\ n = 16) \quad . \qquad (11)$$

This correlation does not differ significantly from that obtained from a larger, but more heterogeneous, data base by Zinn and West themselves. While clearly present, it is of modest accuracy, and although values derived from this correlation are listed for the dwarf spheroidal satellites of the Galaxy in Table IIb, it is well known (see, for example, Stetson 1980, 1984; Carney 1984) that metallicities derived for these objects from line-blanketing indicators are systematically much lower than those derived here.

Table IIc, referring to the globular clusters in Fornax, differs from Tables IIa and IIb only in that columns (6)-(10) refer to the following positional parameters:

(6) - Angular distance of the cluster from the center of the Fornax dwarf spheroidal (in arcminutes)
(7) - Position angle (eastward from North) of the cluster (in degrees)

(8)-(9) - Cartesian coordinates of the cluster in the plane of the sky (in pc). The +X-axis points eastward along the major axis of Fornax, the +Y-axis northward along its minor axis.

(10) - Projected linear distance (in pc) of the cluster from the center of Fornax.

The position angle of the semi-major axis of Fornax has been taken as $\theta = 49°$ (Hodge and Smith 1974).

Although no formal error estimates have been made here for the derived cluster parameters, they are obviously no better than the data and assumptions upon which they are based. Except for the galactic coordinates (ℓ,b), the values quoted here are unlikely to be accurate to more than two decimal places, even in the very best of cases. The reader should, of course, consult the entries in Table I and their accompanying discussion in judging the reliability of the derived parameters of Table II.

ACKNOWLEDGMENTS

It is a pleasure to acknowledge the many individuals who contributed in some way to this effort. My gratitude goes to the following individuals who provided data prior to publication: M. Aaronson, G. Alcaino, B. J. Anthony-Twarog, R. D. Cannon, B. W. Carney, K. M. Cudworth, G. S. Da Costa, J. C. Forte, R. Gratton, D. A. Hanes, H. C. Harris, W. E. Harris, D. A. Hunter, G. E. Kron, M. H. Liller, M. Mendez, C. J. Peterson, C. A. Pilachowski, A. Sandage, P. Seitzer, S. J. Shawl, H. A. Smith, V. Trimble, R. E. White, and R. Zinn. I am especially indebted to Bill Harris for calling my attention to a number of studies not yet published, and to Gwendy Romey for preparing this camera-ready typescript. This work was supported in part by National Science Foundation grants AST78-12309, AST80-18859, and AST83-17916 to the University of Illinois, and by a Visiting Fellowship from the Joint Institute for Laboratory Astrophysics, University of Colorado and National Bureau of Standards.

REFERENCES

Aaronson, M., Cohen, J. G., Mould, J., and Malkan, M. 1978, Ap. J., **223**, 824.

Aurière, M., and Cordoni, J.-P. 1981, Astr. Ap., **100**, 307.

Bica, E. L. D., and Pastoriza, M. G. 1983, Ap. Space Sci., **91**, 99.

Bingham, E. A., Cacciari, C., Dickens, R. J., and Fusi-Pecci, F. 1984, M.N.R.A.S., **209**, 765.

Burstein, D., and Heiles, C. 1982, A. J., **87**, 1165.

Butler, D. 1975, Ap. J., **200**, 68.

Cannon, R. D., Hawarden, T. G., and Tritton, S. B. 1978, unpublished.

Carney, B. W. 1984, invited paper presented at the 96th Annual Scientific Meeting of the Astronomical Society of the Pacific.

Cesarsky, D. A., Laustsen, S., Lequeux, J., Schuster, H.-E., and West, R. M. 1977, Astr. Ap., **61**, L31.
Christy, R. F. 1966, Ap. J., **144**, 108.
Cowley, A. P., and Hartwick, F. D. A. 1981, A. J., **86**, 667.
Cowley, A. P., Hartwick, F. D. A., and Sargent, W. L. W. 1978, Ap. J., **220**, 453.
Djorgovski, S., and King, I. R. 1984, Ap. J. (Lett.), **277**, L49.
Djorgovski, S., and Penner, H. 1985, in Dynamics of Star Clusters, I.A.U. Symp. No. 113, eds. J. Goodman and P. Hut (Dordrecht: D. Reidel), this volume.
Freeman, K. C., Illingworth, G., and Oemler, A., Jr. 1983, Ap. J., **272**, 488.
Frenk, C. S., and White, S. D. M. 1982, M.N.R.A.S., **198**, 173.
Grindlay, J. E., and Hertz, P. 1981, Ap. J. (Lett.), **247**, L17.
Harris, H. C., and Canterna, R. 1977, A. J., **82**, 798.
Harris, W. E. 1976, A. J., **81**, 1095.
_____. 1980, in Star Clusters, I.A.U. Symp. No. 85, ed. J. E. Hesser (Dordrecht: D. Reidel), p. 81.
Harris, W. E., and Racine, R. 1979, Ann. Rev. Astr. Ap., **17**, 241.
Hodge, P. W., and Smith, D. W. 1974, Ap. J., **188**, 19.
Holmberg, E. B., Lauberts, A., Schuster, H.-E., and West, R. M. 1975, Astr. Ap. Suppl., **22**, 327.
_____. 1977, Astr. Ap. Suppl., **27**, 295.
_____. 1978a, Astr. Ap. Suppl., **31**, 15.
_____. 1978b, Astr. Ap. Suppl., **34**, 285.
Innanen, K. A., Harris, W. E., and Webbink, R. F. 1983, A. J., **88**, 338.
Johnson, S. L., and McNamara, D. H. 1969, P.A.S.P., **81**, 415.
King, I. R. 1966, A. J., **71**, 64.
Kodaira, K. 1983, I.A.U. Circ., No. 3846.
Kron, G. E., and Guetter, H. H. 1976, A. J., **81**, 817.
Kron, G. E., Hewitt, A. V., and Wasserman, L. H. 1984, P.A.S.P., **96**, 198.
Kukarkin, B. V. 1971, Astr. Zh., **48**, 113 [English transl.: Soviet Astr., **15**, 89].
_____. 1974, Globular Star Clusters (Moscow: Nauka) [English transl.: 1977, NASA TT F-16, 175].
Kukarkin, B. V., and Kireeva, N. N. 1979, Astr. Zh., **56**, 465 [English transl.: Soviet Astr., **23**, 261].
Lauberts, A. 1976, Astr. Ap., **52**, 309.
Lauberts, A., Holmberg, E. B., Schuster, H.-E., and West, R. M. 1981a, Astr. Ap. Suppl., **43**, 307.
_____. 1981b, Astr. Ap. Suppl., **46**, 311.
Liller, W. 1976a, I.A.U. Circ., No. 2929.
_____. 1976b, I.A.U. Circ., No. 2936.
Lohmann, W. 1963, Z. Ap., **57**, 288.
Madore, B. F., and Arp, H. C. 1979, Ap. J. (Lett.), **227**, L103.
Malkan, M. 1981, in Astrophysical Parameters for Globular Clusters, I.A.U. Colloq. No. 68, eds. A. G. D. Philip and D. S. Hayes (Schenectady: L. Davis Press), p. 533.
Malkan, M., Kleinmann, D. E., and Apt, J. 1980, Ap. J., **237**, 432.

Newell, E. B., and O'Neill, E. J. 1978, Ap. J. Suppl., **37**, 27.
Peterson, C. J. 1976, A. J., **81**, 617.
Peterson, C. J., and King, I. R. 1975, A. J., **80**, 427.
Racine, R. 1973, A. J., **78**, 180.
Ruprecht, J. 1959, Bull. Astr. Inst. Czech., **12**, no. 1, App. 2.
Sandage, A. R. 1954, A. J., **59**, 162.
_____. 1957, Ap. J., **125**, 422.
_____. 1981, Ap. J., **248**, 161.
_____. 1982, Ap. J., **252**, 553.
Sandage, A., Katem, B., and Sandage, M. 1981, Ap. J. Suppl., **46**, 41.
Sérsic, J. L. 1974, Ap. Space Sci., **28**, 365.
Shawl, S. J., and White, R. E. 1984, A. J., in press.
Stetson, P. B. 1980, A. J., **85**, 398.
_____. 1984, P.A.S.P., **96**, 128.
Stothers, R. B. 1983, Ap. J., **274**, 20.
Terzan, A. 1968, C. R. Acad. Sci. Paris, **267**, 1245.
_____. 1971, Astr. Ap., **12**, 477.
Terzan, A., and Bernard, A. 1978, Messenger, No. 15, p. 14.
Terzan, A., Bernard, A., and Ju, K. H. 1978a, C. R. Acad. Sci. Paris, **287**, sér. B, 157.
_____. 1978b, C. R. Acad. Sci. Paris, **287**, sér. B, 235.
Terzan, A., and Ju, K. H. 1980, Messenger, No. 20, p. 6.
van Albada, T. S., de Boer, K. S., and Dickens, R. J. 1981, M.N.R.A.S., **195**, 591.
van den Bergh, S., and Hagen, G. L. 1975, A. J., **80**, 11.
Zdanavičius, K. V. 1983, Astr. Zh., **60**, 44 [English transl.: Soviet Astr., **27**, 26].
Zinn, R. 1980, Ap. J. Suppl., **42**, 19.
Zinn, R., and West, M. J. 1984, Ap. J. Suppl., **55**, 45.

TABLE Ia. OBSERVED PARAMETERS OF GALACTIC GLOBULAR CLUSTERS

Number	Name		α_{1950}	δ_{1950}	V_{HB}			E_{B-V}		
(1)	(2)	(3)	(4)	(5)	(6)	(7)	(8)	(9)	(10)	(11)
0021-723	NGC 104	47 Tuc	00 21 53	-72 21.5	14.06	CM	136	0.04	UB	136
0050-268	NGC 288		00 50 19	-26 51.0	15.30	CM	160	0.04	HB	33
0100-711	NGC 362		01 01 32	-71 07.1	15.43	CM	91	0.04	UB	91
0310-554	NGC 1261		03 10 53	-55 24.1	16.63	CM	8	0.00	CS	*
0325+794	Pal 1		03 25 59	+79 24.3	16.76	CM	51	0.15	HI	29
0344-718	NGC 1466	SL 1	03 44 49	-71 49.7	18.8	CM	164	0.07	UB	165
0354-498	AM 1	E 1	03 53 35	-49 45.6	20.93	CM	1	0.00	HI	29
0422-213	Eri		04 22 35	-21 18.1	20.24	CM	50	0.00	HI	29
0435-590	Ret	Sé 40/3	04 35 21	-58 57.8	19.17	CM	60	0.018	UB	60
0443+313	Pal 2		04 42 54	+31 17.5	20.9	*	37	1.45:	MI	*
0444-840	NGC 1841		04 52 45	-84 04.8	18.88	RR	123	0.07	UB	165
0512-400	NGC 1851		05 12 28	-40 06.1	16.05	CM	197	0.018	uy	197
0522-245	NGC 1904	M 79	05 22 08	-24 34.2	16.20	CM	198	0.01	GB	198
0647-359	NGC 2298		06 47 13	-35 56.9	16.20	CM	3	0.15	CS	*
0734+390	NGC 2419		07 34 46	+38 59.7	20.50	CM	176	0.03	GB	176
0737-337	AM 2		07 36 53	-33 43.7	21.1:	BG	146	0.53	CX	215
0911-646	NGC 2808		09 11 04	-64 39.4	16.20	CM	88	0.22	UB	88
0921-770	E 3		09 21 31	-77 04.2	16.15:	CM	*	0.30	UB	107
0923-545	UKS 2		09 23 43	-54 30.1	17.75	BG	215	0.74	IR	148
1003+003	Pal 3	Sex C	10 02 58	+00 18.9	20.48	CM	*	0.05	UB	38
1015-461	NGC 3201		10 15 34	-46 09.7	14.75	CM	137	0.21	UB	137
1117-649	ESO 093-SC?08		11 17 34	-64 56.8	22.0::	BG	215	0.79	CX	215
1126+292	Pal 4	UMa	11 26 37	+29 15.3	20.45	CM	28	0.00	HI	29
1207+188	NGC 4147		12 07 33	+18 49.2	16.85	CM	186	0.02	Ψ	190
1223-724	NGC 4372		12 22 51	-72 22.8	15.50	CM	102	0.45	UB	102
1235-509	Rup 106		12 35 53	-50 52.6	18.5	BG	215	0.24	HI	29
1236-264	NGC 4590	M 68	12 36 49	-26 28.2	15.60	CM	85	0.03	GB	85
1256-706	NGC 4833		12 56 12	-70 36.3	15.45	CM	152	0.32	GB	152
1310+184	NGC 5024	M 53	13 10 28	+18 26.0	16.94	CM	48	0.00	Ψ	190
1313+179	NGC 5053		13 13 59	+17 57.7	16.63	CM	184	0.01	*	*
1323-472	NGC 5139	ω Cen	13 23 46	-47 13.0	14.52	CM	35	0.11	*	*
1339+286	NGC 5272	M 3	13 39 53	+28 37.8	15.68	CM	183	0.00	HB	178
1343-511	NGC 5286		13 43 16	-51 07.5	16.20	CM	97	0.21	CS	*
1353-269	AM 4		13 53 31	-26 55.6	18.2:	CM	118	0.06	HI	29
1403+287	NGC 5466		14 03 12	+28 46.1	16.60	CM	26	0.00	Ψ	190
1427-057	NGC 5634		14 26 59	-05 45.3	17.75	CM	174	0.05	CS	*
1436-263	NGC 5694		14 36 42	-26 19.4	18.4:	CM	93	0.10	CS	*
1452-820	IC 4499		14 52 10	-82 00.8	17.65	CM	42	0.24	CS	86
1500-328	NGC 5824		15 00 54	-32 52.4	18.00	CM	85	0.14	CS	*
1513+000	Pal 5	Ser	15 13 32	+00 03.9	17.35	CM	181	0.03	*	*
1514-208	NGC 5897		15 14 32	-20 49.6	16.25	CM	182	0.09	CS	*
1516+022	NGC 5904	M 5	15 16 01	+02 15.8	15.11	CM	17	0.03	rr	151
1524-505	NGC 5927		15 24 24	-50 30.0	16.70	CM	154	0.43	UB	154
1531-504	NGC 5946		15 31 49	-50 29.6	17.2	*	*	0.56	CS	*
1535-499	BH 176		15 35 28	-49 52.9	22.6::	BG	215	0.73	CX	215
1542-376	NGC 5986		15 42 47	-37 37.4	16.50	CM	97	0.25	CS	*
1608+150	Pal 14	AvdB	16 08 47	+15 05.2	20.08	CM	*	0.03	HI	29
1614-228	NGC 6093	M 80	16 14 03	-22 51.2	15.82	CM	95	0.22	uy	45
1620-720	NGC 6101		16 20 10	-72 05.2	16.76	CM	141	0.04	CS	*
1620-264	NGC 6121	M 4	16 20 31	-26 24.4	13.35	CM	135	0.36	*	31
1624-259	NGC 6144		16 24 10	-25 54.7	16.46:	CM	9	0.30	CS	*
1624-387	NGC 6139		16 24 17	-38 44.3	17.5	*	90	0.74	CS	*
1625-352	Ter 3		16 25 24	-35 14.2	18.8	BG	34	0.32	CX	215
1629-129	NGC 6171	M 107	16 29 43	-12 56.7	15.63	CM	64	0.33	rr	151
1636-283	ESO 452-SC 11		16 36 18	-28 18.0	16.66	CM	216	0.31	HI	29

TABLE Ia. OBSERVED PARAMETERS OF GALACTIC GLOBULAR CLUSTERS

$(B-V)_g$			$\log \theta_t$			$\log \theta_c$			V_t	σ_c			Number
(12)	(13)	(14)	(15)	(16)	(17)	(18)	(19)	(20)	(21)	(22)	(23)	(24)	(25)
1.03	CM	136	1.65	SC	49	-0.41	SP	81	3.83	5.55	AP	*	0021-723
0.88	CM	160	1.19	SC	167	0.22	SP	167	8.11	11.02	AP	*	0050-268
0.88	CM	91	1.01	SC	117	-0.69	*	117	6.46	6.07	AP	*	0100-711
0.89	CM	8	0.90	SC	167	-0.40	AP	167	8.28	8.70	AP	*	0310-554
1.08:	CM	51	0.70:	SC	167	-0.80:	SC	167	13.62	(12.36)	IR	132	0325+794
0.73	CM	164	0.81	D	215	-0.61	e	215	10.66:	10.23	AP	212	0344-718
0.77	CM	1	0.38	SC	161	-0.55	SC	161	15.84	14.97:	AP	1	0354-498
0.88	CM	50	0.54:	SC	217	-0.60	SC	162	14.70:	13.92	AP	217	0422-213
0.71:	CM	60	1.00:	SC	169	0.00	SC	169	12.7	(14.70)	Σ	169	0435-590
2.22	UB	*	0.50	SC	89	-1.10	SC	37	13.04	10.48:	AP	43	0443+313
0.87	BV	*	0.90	SC	123	-0.18	AP	168	10.90:	12.13	AP	*	0444-840
0.89	CM	197	0.91:	SC	117	-1.18	SP	65	7.15	5.23	AP	*	0512-400
0.83	CM	198	1.03	SC	167	-0.76	SP	130	7.81	7.38	AP	*	0522-245
0.83	CM	3	0.90:	*	133	-0.36	AP	167	9.20	9.71	AP	*	0647-359
0.73	CM	176	1.03	SC	167	-0.38	SP	130	10.32	10.99	AP	*	0734+390
			0.77	D	146	0.19	e	215	14.0	(16.03)	R	146	0737-337
1.04	CM	88	1.15	SC	117	-0.60	*	117	6.13	6.28	AP	*	0911-646
1.24:	CM	*	1.02:	SC	211	0.27	SC	211	11.35	(14.21)	R	211	0921-770
1.82:	Q	148	0.55	D	215	-0.11	e	215	13.0	(13.75)	R	215	0923-545
0.82	CM	191	0.63	SC	167	-0.33	SC	78	13.92:	14.20	AP	168	1003+003
1.00	CM	137	1.56	SC	167	0.04	*	167	6.68	9.69	AP	*	1015-461
			0.64::	D	215	-1.21:	e	215	14.5:	(11.75)	R	215	1117-649
0.86	CM	28	0.49	SC	167	-0.27	SC	167	14.20	(14.39)	Σ	96	1126+292
0.79	CM	186	0.86	SC	167	-0.89	SP	130	10.22	8.99	AP	*	1207+188
1.20	CM	102	1.5:	D	138	0.42:	AP	168	7.29:	11.54	AP	*	1223-724
			1.06	D	215	-0.02	e	215	10.9	(12.93)	R	215	1235-509
0.76	CM	85	1.47:	SC	167	-0.21	SP	130	7.74	9.62	AP	*	1236-264
1.02	CM	152	1.30	*	138	0.06	AP	167	7.03	9.54	AP	*	1256-706
0.72	CM	48	1.34	SC	167	-0.39	SP	130	7.48	8.51	AP	*	1310+184
0.70	CM	184	1.14	SC	167	0.39	SC	167	9.94	13.45	AP	*	1313+179
0.90	CM	35	1.64	SC	49	0.42	*	49	3.52:	7.85	AP	*	1323-472
0.86	CM	183	1.59	SC	167	-0.38	SP	130	5.92	7.40	AP	*	1339+286
1.00	CM	97	1.08	SC	166	-0.67	AP	166	7.18	7.05	AP	*	1343-511
0.70:	CM	118	0.60	D	215	-0.41	e	215	15.9	(15.86)	Σ	118	1353-269
0.73	CM	26	1.32:	SC	167	0.20	SC	20	8.95	12.08	AP	*	1403+287
0.80	*	*	0.88	D	133	-0.63	SP	130	9.38	8.90	AP	*	1427-057
0.82	*	*	1.18	*	96	-1.30	SP	130	9.17:	7.31	AP	*	1436-263
0.99:	CM	70	1.20	SC	166	0.10	SC	166	9.42:	12.02	AP	*	1452-820
0.84	CM	85	1.30:	SC	167	-1.26	SP	65	7.84:	6.15	AP	*	1500-328
0.86	CM	181	1.26:	SC	181	0.46	SC	167	11.75	(15.67)	Σ	96	1513+000
0.91	CM	182	1.06	SC	167	0.11	SP	130	8.64	11.48	AP	*	1514-208
0.82	CM	17	1.46	SC	167	-0.36	SP	130	5.69	7.15	AP	*	1516+022
1.44	CM	154	1.14	*	133	-0.32	AP	167	8.02	9.11	AP	*	1524-505
1.41	*	*	1.0:	D	133	-0.78	AP	168	9.48	8.77	AP	*	1531-504
			0.79::	D	215	-0.35:	e	215	14.0:	(14.47)	R	215	1535-499
1.01	CM	97	1.10	SC	166	-0.32	SC	166	7.48	8.52	AP	*	1542-376
0.88	CM	*	0.67	SC	99	-0.12	SC	99	14.68	(15.68)	Σ	52	1608+150
0.88	CM	95	0.97	SC	117	-0.98	SP	65	7.33	6.12	AP	*	1614-228
0.81	*	*	1.2:	D	133	-0.10	AP	168	9.16:	11.05	AP	*	1620-720
1.27	CM	135	1.64	SC	167	0.09	SP	130	5.76	9.00	AP	*	1620-264
1.04	CM	9	1.07	*	133	0.09	SP	130	9.01	11.61	AP	*	1624-259
1.53	*	*	1.10	SC	166	-0.65	AP	166	8.91	8.52	AP	*	1624-387
			0.88	D	*	-0.08	e	215	12.0	(13.53)	R	215	1625-352
1.29	CM	64	1.36	SC	166	-0.14	SP	130	8.10	9.91	AP	*	1629-129
1.24:	CM	216	0.54	D	215	-0.41	e	215	12.0	(11.86)	Σ	216	1636-283

TABLE Ia. (cont'd.)

Number	Name		α_{1950}	δ_{1950}	V_{HB}			E_{B-V}		
(1)	(2)	(3)	(4)	(5)	(6)	(7)	(8)	(9)	(10)	(11)
1639+365	NGC 6205	M 13	16 39 54	+36 33.2	14.95	CM	179	0.03	HB	178
1644-018	NGC 6218	M 12	16 44 38	-01 51.6	14.90	CM	172	0.21	uy	45
1645+476	NGC 6229		16 45 34	+47 36.9	18.10	CM	190	0.00	Ψ	190
1650-220	NGC 6235		16 50 25	-22 05.7	16.70	CM	139	0.38	GB	139
1654-040	NGC 6254	M 10	16 54 31	-04 01.4	14.65	CM	97	0.25	Ψ	190
1656-370	NGC 6256	Ter 12	16 56 10	-37 02.8	18.2:	CM	13	0.88:	GR	13
1657-004	Pal 15		16 57 28	-00 28.1	20.2:	CM	99	0.12	HI	29
1658-300	NGC 6266	M 62	16 58 02	-30 02.5	15.95	CM	*	0.45	*	6
1659-262	NGC 6273	M 19	16 59 32	-26 11.8	16.95:	CM	97	0.38	CS	*
1701-246	NGC 6284		17 01 25	-24 41.8	16.55	*	90	0.28	CS	*
1702-226	NGC 6287		17 02 09	-22 38.4	16.52	*	90	0.51	CS	*
1707-265	NGC 6293		17 07 04	-26 31.2	16.14	*	90	0.35	CS	*
1708-271	TJ 5		17 08 08	-27 08.0				≥0.65:	UB	157
1711-294	NGC 6304		17 11 22	-29 24.4	16.15	CM	106	0.52	CS	*
1713-280	NGC 6316		17 13 28	-28 05.1	17.8	*	90	0.52	CS	*
1714-237	NGC 6325		17 14 57	-23 42.8	17.3:	CM	85	0.86	CS	*
1715-277	TJ 16	TBJ 2	17 15 21	-27 43.4				≥0.65:	UB	157
1715-262	TJ 15		17 15 25	-26 14.8				≥0.65:	UB	157
1715-278	TJ 17	TBJ 1	17 15 33	-27 47.1				≥0.65:	UB	157
1715+432	NGC 6341	M 92	17 15 35	+43 11.4	15.10	CM	179	0.02	HB	178
1716-184	NGC 6333	M 9	17 16 16	-18 27.9	16.09	*	90	0.35	CS	*
1718-195	NGC 6342		17 18 13	-19 32.3	17.4	*	90	0.46	CS	*
1720-177	NGC 6356		17 20 40	-17 46.1	17.67	CM	187	0.30	CS	*
1720-263	NGC 6355		17 20 52	-26 18.5	17.2	*	90	0.73	CS	*
1721-484	NGC 6352		17 21 40	-48 22.6	15.15	CM	101	0.14	*	104
1724-307	Ter 2	HP 3	17 24 20	-30 45.6	19.8:	BG	79	1.31	IR	148
1725-050	NGC 6366		17 25 04	-05 02.1	15.70	CM	171	0.65	CS	86
1726-670	NGC 6362		17 26 45	-67 00.6	15.30	CM	2	0.08	*	69
1727-315	Ter 4	HP 4	17 27 24	-31 33.5	21.6::	BG	215	1.55	IR	148
1727-299	HP 1		17 27 53	-29 56.9	20.0:	BG	200	1.41	CX	215
1728-338	Gri 1		17 28 40	-33 47.9	26.2::	BG	80	3.2:	IR	80
1730-333	Lil 1		17 30 07	-33 21.3	24.4	A	215	2.91	IR	148
1731-390	NGC 6380	Ton 1	17 30 59	-39 01.9	18.0	BG	*	1.38	IR	148
1732-304	Ter 1	HP 2	17 32 34	-30 27.0	20.6:	BG	215	1.52	IR	148
1732-447	NGC 6388		17 32 38	-44 42.3	17.24	CM	10	0.31	CS	*
1733-390	Ton 2	Pis 26	17 32 43	-38 31.2	18.2	BG	34	0.91	CX	215
1735-032	NGC 6402	M 14	17 34 59	-03 13.2	17.50	CM	126	0.58	CS	*
1735-238	NGC 6401		17 35 34	-23 53.0	17.3	BG	90	0.76	CS	*
1736-536	NGC 6397		17 36 38	-53 38.9	12.90	CM	33	0.18	HB	33
1740-262	Pal 6		17 40 36	-26 12.1	19.1	BG	90	1.45	IR	148
1741-328	TJ 23		17 41 54	-32 45.1				1.73	CX	215
1742+031	NGC 6426		17 42 25	+03 11.4	18.0	IR	90	0.37	CS	*
1745-247	Ter 5		17 45 00	-24 45.8	21.7	BG	*	2.14	IR	148
1746-203	NGC 6440		17 45 54	-20 20.7	18.4:	CM	150	1.11	CS	*
1746-370	NGC 6441		17 46 49	-37 02.3	17.10	CM	105	0.36	uy	104
1748-346	NGC 6453		17 47 32	-34 35.2	17.7	*	90	0.61	CS	*
1747-312	Ter 6	HP 5	17 47 32	-31 15.7	20.8::	BG	132	1.46	CX	215
1751-241	UKS 1		17 51 24	-24 08.2	25.5	A	215	3.07	IR	148
1755-442	NGC 6496		17 55 24	-44 15.8	14.9	*	90	0.10	CS	*
1758-268	Ter 9		17 58 31	-26 50.4	20.3:	BG	215	1.71	IR	148
1759-089	NGC 6517		17 59 06	-08 57.6	18.0:	CM	85	1.09	CS	*
1800-260	Ter 10		17 59 51	-26 04.1	21.9::	BG	215	1.71	CX	215
1800-300	NGC 6522		18 00 23	-30 02.2	16.25	CM	18	0.49	UB	*
1801-003	NGC 6535		18 01 17	-00 18.0	15.85	CM	*	0.33	GB	140
1801-300	NGC 6528		18 01 37	-30 03.6	16.75	CM	213	0.62	CS	*

TABLE Ia. (cont'd.)

$(B-V)_g$			$\log \theta_t$			$\log \theta_c$			V_t	σ_c			Number
(12)	(13)	(14)	(15)	(16)	(17)	(18)	(19)	(20)	(21)	(22)	(23)	(24)	(25)
0.82	CM	179	1.43	SC	167	-0.08	SP	130	5.68	7.94	AP	*	1639+365
0.93	CM	172	1.26	SC	167	0.07	SP	130	6.77	9.44	AP	*	1644-018
0.84	*	*	0.75:	SC	167	-0.76	SP	130	9.36	8.31	AP	*	1645+476
1.17	CM	139	0.9:	D	133	-0.58	SP	66	9.99	9.87	AP	*	1650-220
1.06	CM	97	1.38	SC	167	-0.15	SP	130	6.55	8.78	AP	*	1654-040
1.68	CM	13	0.83	D	215	-0.42	e	215	11.29:	11.78:	AP	23	1656-370
0.99:	CM	99	0.73	SC	99	0.13	SC	99	14.2	(15.98)	Σ	99	1657-004
1.32:	CM	6	1.02	SC	117	-0.61	*	117	6.68	6.63	AP	*	1658-300
0.98:	CM	97	1.24	SC	166	-0.30	SP	130	6.73	7.95	AP	*	1659-262
1.13	*	*	1.00	*	133	-0.79	SP	130	8.85	8.07	AP	*	1701-246
1.27	*	*	0.98	*	133	-0.48	SP	130	9.31	9.69	AP	*	1702-226
1.08	*	*	1.17	*	133	-0.78	SP	130	8.15	7.75	AP	*	1707-265
			-0.22::	D	215	-1.11:	e	215					1708-271
1.56	CM	106	1.10	*	133	-0.58	SP	130	8.36	8.59	AP	*	1711-294
1.54	*	*	1.16	*	133	-0.76	SP	130	8.75	8.51	AP	*	1713-280
1.55:	CM	85	1.00:	*	133	-0.61	SP	130	10.56	10.30	AP	*	1714-237
			-0.29::	D	215	-1.34	e	205	16.3::	11.7::	AP	205	1715-277
			-0.22::	D	215	-1.01:	e	215					1715-262
1.80	BV	*	-0.05::	D	215	-1.27	e	205	16.58:	12.56::	AP	203	1715-278
0.74	CM	27	1.22	SC	167	-0.51	SP	130	6.39	6.81	AP	*	1715+432
1.10	*	*	1.19	SC	166	-0.38	SP	130	7.61	8.33	AP	*	1716-184
1.45	*	*	0.94	*	133	-0.88	SP	130	9.84	9.03	AP	*	1718-195
1.19:	CM	187	1.06	*	133	-0.53	SP	130	8.18	8.29	AP	*	1720-177
1.58	*	*	0.88:	*	133	-0.72	SP	130	9.68	9.56	AP	*	1720-263
1.29	CM	101	1.08	*	133	-0.17	AP	168	8.13	9.93	AP	*	1721-484
2.35:	Q	148	0.49	D	133	-0.97	SP	81	14.29:	12.12:	AP	149	1724-307
1.65	CM	171	1.3:	D	133	0.34	SC	167	8.88:	12.47	AP	*	1725-050
1.08	CM	2	1.22	SC	167	0.20	SC	167	7.52	10.20	AP	*	1726-670
2.65:	Q	148	0.22:	D	133	≤-1.00	e	215	16.0:	(≤13.34)	R	215	1727-315
2.18:	BV	67	0.67	D	133	-0.68	e	133	12.49	(11.62)	IR	132	1727-299
			0.94::	D	215	-0.27:	e	215	17.66:	18.64:	SP	80	1728-338
4.01:	Q	148	0.52	D	215	-1.23	SP	81	15.84:	13.42:	AP	*	1730-333
2.24	*	*	0.75	D	133	-0.46	AP	168	11.12:	11.14	AP	168	1731-390
2.71:	Q	148	0.59	D	133	-1.01	e	215	15.9	(13.74)	R	215	1732-304
1.33	CM	10	0.92	SC	117	-0.83	*	117	6.73	5.71	AP	*	1732-447
			0.71	D	133	-0.14	e	215	12.24	(13.27)	IR	132	1733-390
1.23:	CM	126	1.00	SC	167	-0.10	SP	130	7.57	9.52	AP	*	1735-032
1.69	*	*	1.12	SP	130	-0.63	SP	130	9.45:	9.76	AP	*	1735-238
0.87	CM	33	1.585	SC	49	-0.11	AP	49	5.75	8.14	AP	*	1736-536
2.67:	*	*	0.87	D	133	-0.17	e	133	11.55:	12.77	AP	168	1740-262
			0.04::	D	215	-0.75:	e	215					1741-328
1.08	*	*	0.87	*	133	-0.45	SP	130	11.13	11.76	AP	*	1742+031
3.14:	*	*	0.61	D	133	-1.32:	e	133	13.85:	10.70:	AP	175	1745-247
2.15	*	*	1.08	*	133	-0.94	SP	65	9.05	8.13	AP	*	1746-203
1.51	CM	105	0.88:	SC	117	-0.86	SP	81	7.19	6.26	AP	*	1746-370
1.42	*	*	0.7:	D	133	-0.79	SP	130	9.78	8.74	AP	*	1748-346
			0.18:	D	133	-1.08	e	215	13.85:	(10.85)	IR	132	1747-312
3.95:	Q	148	0.77	D	147	-0.66	AP	149	17.29:	16.63:	AP	149	1751-241
1.10	*	*	1.0:	D	133	0.25	AP	168	8.48:	11.18	AP	*	1755-442
2.77:	Q	148	0.43:	D	133	-0.91:	e	215	16.0:	(13.96)	R	215	1758-268
1.91	*	*	1.02:	*	133	-0.99	SP	130	10.30	9.09	AP	*	1759-089
			0.18:	D	202	-0.94:	e	215	14.9:	(12.39)	R	215	1800-260
1.29	CM	18	1.00	SP	130	-0.75	SP	130	8.35	7.63	AP	*	1800-300
1.13	CM	*	1.0:	*	133	-0.70	AP	168	10.48	10.27	AP	*	1801-003
1.56:	CM	213	0.67	SP	130	-0.89	SP	130	9.51	8.30	AP	*	1801-300

TABLE Ia. (cont'd.)

Number	Name		α_{1950}	δ_{1950}	V_{HB}			E_{B-V}		
(1)	(2)	(3)	(4)	(5)	(6)	(7)	(8)	(9)	(10)	(11)
1802-075	NGC 6539		18 02 07	-07 35.4	16.6	IR	90	1.10	CS	*
1804-250	NGC 6544		18 04 15	-25 00.3	15.0:	CM	12	0.73	CS	*
1804-437	NGC 6541		18 04 25	-43 43.4	15.20	CM	7	0.12	CS	*
1806-259	NGC 6553		18 06 11	-25 55.1	16.95	CM	100	0.80	CS	*
1807-317	NGC 6558		18 07 03	-31 46.4	16.7	*	90	0.43	CS	*
1808-072	IC 1276	Pal 7	18 08 03	-07 13.2	18.5	*	90	0.92	CS	86
1809-227	Ter 11		18 09 14	-22 45.3	22.5::	BG	215	1.57	CX	215
1810-318	NGC 6569		18 10 23	-31 50.5	17.1	*	90	0.55	CS	*
1812-121	Kod 1		18 12 19	-12 04.2				3.8:	IR	125
1814-522	NGC 6584		18 14 38	-52 14.1	16.8	*	90	0.10	CS	*
1820-303	NGC 6624		18 20 28	-30 23.3	16.05	CM	144	0.29	CS	*
1821-249	NGC 6626		18 21 28	-24 53.8	15.60	CM	11	0.37	GB	11
1827-255	NGC 6638		18 27 51	-25 32.0	15.92	CM	14	0.37	CS	*
1828-323	NGC 6637	M 69	18 28 07	-32 23.1	16.20	CM	87	0.17	CS	*
1828-235	NGC 6642		18 28 52	-23 30.8	15.5	*	90	0.37	CS	*
1832-330	NGC 6652		18 32 29	-33 01.9	16.7	BG	90	0.10	CS	*
1833-239	NGC 6656		18 33 21	-23 56.9	14.20	CM	5	0.35	uy	104
1838-198	Pal 8		18 38 32	-19 52.5	18.9	*	37	0.33	CS	*
1840-323	NGC 6681	M 70	18 39 57	-32 20.5	15.60	CM	*	0.05	CS	*
1850-087	NGC 6712		18 50 20	-08 46.2	16.11	CM	185	0.48	*	185
1851-305	NGC 6715	M 54	18 51 51	-30 32.6	17.71	CM	85	0.14	CS	*
1852-227	NGC 6717	Pal 9	18 52 05	-22 46.0	15.7	CM	76	0.20	CS	*
1856-367	NGC 6723		18 56 11	-36 41.9	15.48	CM	153	0.02	UB	153
1902+017	NGC 6749		19 02 43	+01 49.5	19.2	*	37	0.96:	CS	96
1906-600	NGC 6752		19 06 28	-60 03.9	13.80	CM	36	0.05	HB	*
1908+009	NGC 6760		19 08 39	+00 56.8	16.5	*	90	0.88	CS	*
1914-347	Ter 7		19 14 25	-34 44.8	18.6	BG	34	0.06	HI	29
1914+300	NGC 6779		19 14 39	+30 05.5	16.20	CM	21	0.20	Ψ	190
1916+184	Pal 10		19 15 49	+18 28.9	19.4	BG	37	1.15	CX	215
1925-304	Arp 2		19 25 34	-30 27.7	18.21	CM	192	0.11	HI	29
1936-310	NGC 6809	M 55	19 36 50	-31 04.6	14.40	CM	138	0.08	UB	138
1938-341	Ter 8		19 38 29	-34 07.1	19.4	BG	34	0.12	HI	29
1942-081	Pal 11		19 42 32	-08 07.7	17.38	CM	47	0.34	UB	38
1951+186	NGC 6838	M 71	19 51 32	+18 38.7	14.42	CM	46	0.19	XY	119
2003-220	NGC 6864	M 75	20 03 08	-22 03.9	17.45	CM	85	0.16	CS	*
2031+072	NGC 6934		20 31 45	+07 14.0	16.82	CM	94	0.11	Ψ	190
2050-127	NGC 6981	M 72	20 50 43	-12 43.6	16.85	CM	62	0.03	*	*
2059+160	NGC 7006		20 59 09	+15 59.4	18.72	CM	188	0.05	Ψ	190
2127+119	NGC 7078	M 15	21 27 33	+11 56.8	15.86	CM	179	0.10	*	*
2130-010	NGC 7089	M 2	21 30 53	-01 02.8	16.05	CM	85	0.02	Ψ	190
2137-234	NGC 7099	M 30	21 37 32	-23 24.4	15.09	CM	63	0.06	UB	63
2143-214	Pal 12	Cap	21 43 50	-21 29.0	17.10	CM	92	0.02	HI	29
2304+124	Pal 13	Peg	23 04 14	+12 30.2	17.70	CM	41	0.05	CS	86
2305-159	NGC 7492		23 05 49	-15 52.9	17.00	CM	22	0.00	CS	86

TABLE Ia. (cont'd.)

$(B-V)_g$			$\log\theta_t$			$\log\theta_c$			V_t	σ_c			Number
(12)	(13)	(14)	(15)	(16)	(17)	(18)	(19)	(20)	(21)	(22)	(23)	(24)	(25)
2.02	*	*	1.12:	*	133	-0.28	SP	130	9.77	11.16	AP	*	1802-975
1.39:	CM	12	1.07:	*	133	-0.33	SP	130	8.07	9.07	AP	*	1804-250
0.81	CM	7	1.50:	SC	167	-0.55	SP	218	6.08	6.86	AP	*	1804-437
1.87:	CM	100	1.02:	*	133	-0.17	SP	130	8.06	9.27	AP	*	1806-259
1.24	*	*	0.8:	D	133	-0.81	AP	168	9.82	8.79	AP	*	1807-317
1.89:	UB	175	1.1:	D	133	0.10	*	167	10.34:	12.83	AP	175	1808-072
			0.30:	D	133	-0.71:	e	215	16.4:	(14.86)	R	215	1809-227
1.48	*	*	0.85:	*	133	-0.46	SP	130	8.71	9.09	AP	*	1810-318
									20.1::		AP	125	1812-121
0.90	*	*	0.94	*	133	-0.40	AP	167	8.63	9.14	AP	*	1814-522
1.26	CM	144	1.06	*	133	-1.06	SP	81	7.99	6.63	AP	*	1820-303
1.11	CM	11	1.18	*	133	-0.42	*	167	6.83	7.29	AP	*	1821-249
1.32	CM	14	0.77	*	166	-0.63	SP	130	9.05	8.34	AP	*	1827-255
1.12	CM	87	1.02:	*	133	-0.48	SP	130	7.56	7.93	AP	*	1828-323
1.23	*	*	0.9:	D	133	-1.13:	SP	66	9.36	7.66	AP	*	1828-235
1.05	*	*	0.89	*	133	-0.82	SP	130	8.76	7.70	AP	*	1832-330
1.09	CM	5	1.52	SC	167	0.08	SP	130	5.07	8.35	AP	*	1833-239
1.38	*	*	0.35	SC	37	-0.40	SC	37	11.15	10.78	AP	*	1838-198
1.00:	CM	85	1.04	*	133	-1.03	SP	65	7.95	6.85	AP	*	1840-323
1.35	CM	185	1.04:	*	133	-0.17	SP	130	8.17	9.57	AP	*	1850-087
0.87:	CM	85	0.87	SC	117	-0.98	SP	130	7.57	6.00	AP	*	1851-305
0.85:	CM	76	0.9:	D	133	-1.24	SP	66	9.18	7.29	AP	*	1852-227
0.93	CM	153	1.10	SC	167	-0.08	SP	130	7.17	8.85	AP	*	1856-367
2.04	UB	*	0.43	SC	37	-0.32	SC	37	12.44	12.32	AP	*	1902+017
0.83	CM	36	1.49	SC	49	-0.30	*	49	5.48	6.85	AP	*	1906-600
1.85	*	*	1.08	*	133	-0.36	SP	130	9.08	10.03	AP	*	1908+009
			0.54	D	133	-0.44	e	215	12.0	(11.76)	R	215	1914-347
0.82	CM	21	1.08	*	133	-0.36	SP	130	8.26	9.33	AP	*	1914+300
			0.75	SC	37	-0.49	SC	37	13.22	(13.14)	IR	132	1916+184
0.84	CM	192	0.75	D	133	0.30	SC	166	12.3	(14.46)	R	215	1925-304
0.88	CM	138	1.27	SC	167	0.24	SC	167	6.36	9.97	AP	*	1936-310
			0.94	D	*	-0.11	e	215	12.4	(13.93)	R	215	1938-341
1.29	CM	47	0.87	SC	167	0.12	SC	167	9.80::	11.92	AP	168	1942-081
1.25	CM	46	1.14	*	133	-0.08	SP	130	8.00	9.80	AP	*	1951+186
0.93	CM	85	0.78	SC	117	-1.02	SP	130	8.53	6.68	AP	*	2003-220
0.97	CM	94	0.95	SC	167	-0.69	SP	130	8.72	8.31	AP	*	2031+072
0.83	CM	62	0.94	SC	167	-0.44	SP	130	9.32	9.91	AP	*	2050-127
0.81	CM	188	0.80	SC	167	-0.75	SP	130	10.46	9.44	AP	*	2059+160
0.78	CM	27	1.32	SC	167	-1.04	SP	81	6.02	5.90	AP	*	2127+119
0.76	CM	85	1.21	SC	167	-0.47	SP	130	6.36	7.18	AP	*	2130-010
0.71	CM	63	1.20	SC	167	-1.13	SP	65	7.32	6.33	AP	*	2137-234
0.92	CM	92	1.03	SC	166	-0.32	SC	166	11.71	12.85	AP	*	2143-214
0.99:	CM	41	0.58	SC	163	-0.42	SC	163	13.80:	13.52:	AP	175	2304+124
0.74	CM	22	0.88	SC	167	-0.09	SP	130	11.43	13.08	AP	*	2305-159

FOOTNOTES TO TABLE Ia

0021-723 **(24)**: 24,49,72,82,117,127,128,168,207
0050-268 **(24)**: 24,82,120,168
0100-711 **(19)**: SP,AP **(24)**: 24,82,117,128,168,207
0310-554 **(11)**: 24,84,86,221 **(24)**: 24,82,128,168,212
0443+313 **(7)**: BG,IR **(11)**: 43,44,89 **(14)**: 43,89
0444-840 **(14)**: 71,77,168,210 **(24)**: 77,168,210
0512-400 **(24)**: 24,82,117,128,168,209,212
0522-245 **(24)**: 24,82,128,168,177
0647-359 **(11)**: 24,86,129,206,221 **(16)**: AP,D **(24)**: 24,82,98,128,168
0734+390 **(24)**: 53,98,131,177
0911-646 **(19)**: SP,AP **(24)**: 24,40,98,117,128,168,207
0921-770 **(8)**: 107,211 **(14)**: 107,211
1003+003 **(8)**: 78,191
1015-461 **(19)**: SC,AP **(24)**: 24,128,168,207
1207+188 **(24)**: 24,131,168,175,177,219
1223-724 **(24)**: 24,168,175
1236-264 **(24)**: 23,24,43,131,168,207,220
1256-706 **(16)**: AP,D **(24)**: 24,82,128,168,207
1310+184 **(24)**: 24,122,131,168,177,207,220
1313+179 **(10)**: Ψ;UB **(11)**: 190;184 **(24)**: 43,131,175,177
1323-472 **(10)**: NB;rr **(11)**: 158;30,199 **(19)**: SP,AP **(24)**: 24,49,72,82,83,128,168,207
1339+286 **(24)**: 24,122,131,177,207,219,220
1343-511 **(11)**: 24,84,129,173,221 **(24)**: 24,82,98,128,168,207
1403+287 **(24)**: 43,131,175
1427-057 **(11)**: 24,84,86,129,145,221 **(13)**: ΔP;ΔS;UB;Q **(14)**: 180;193;24,82,83,98,131, 168,212,219;221 **(24)**: 24,82,98,131,168,212,219
1436-363 **(11)**: 24,86,129,145,221 **(13)**: UB;Q **(14)**: 24,82,83,98,131,168,212;223 **(16)**: AP,D **(24)**: 24,82,98,131,168,212
1452-820 **(24)**: 168,175
1500-328 **(11)**: 24,86,129,145,220,221 **(24)**: 24,82,98,131,168,220
1513+000 **(10)**: ΔU;Ψ **(11)**: 181;190
1514-208 **(11)**: 24,84,145,220 **(24)**: 24,82,131,168,219,220
1516+022 **(24)**: 24,40,82,120,122,131,168,177,207,219,220
1524-505 **(16)**: AP,D **(24)**: 24,82,128,168,207
1542-376 **(11)**: 24,86,129,145,221 **(24)**: 24,82,131,168,207
1608+150 **(8)**: 52,99 **(14)**: 52,99
1614-228 **(24)**: 24,82,83,117,131,168,207,220
1620-720 **(11)**: 24,86,223 **(13)**: UB;Q **(14)**: 24,82,168,175;223 **(24)**: 24,82,168,175
1620-264 **(10)**: HB,rr,RR **(24)**: 24,82,131,168,207,220
1624-259 **(11)**: 24,86,145,223 **(16)**: AP,D **(24)**: 24,82,131,168,175
1624-387 **(7)**: BG,IR **(11)**: 24,86,129,221 **(13)**: UB;Q **(14)**: 24,82,83,98,128, 168;221 **(24)**: 24,82,98,128,168
1625-352 **(17)**: 34,201
1629-129 **(24)**: 24,82,131,168,177,219,220
1639+365 **(24)**: 24,120,122,131,219,220
1644-018 **(24)**: 24,82,120,122,131,168,207,219,220
1645+476 **(13)**: ΔP;UB;Q **(14)**: 180;24,98,131,177,219;221 **(24)**: 24,131,177,219
1650-220 **(24)**: 24,82,168,175
1654-040 **(24)**: 24,120,122,131,168,207,219,220
1658-300 **(8)**: 6,85 **(10)**: GB,GR **(19)**: SP,AP,SC **(24)**: 24,82,117,131,168,207,220
1659-262 **(11)**: 24,84,86,129,145,220,221 **(24)**: 24,82,120,131,168,207,220
1701-246 **(7)**: RR,BG,IR **(11)**: 24,86,129,145,221 **(13)**: ΔS,UB;Q **(14)**: 193;24,67,82,83, 98,131,168;221 **(16)**: AP,D **(24)**: 24,82,131,168
1702-226 **(7)**: RR,BG,IR **(11)**: 24,86,145,221 **(13)**: UB;Q **(14)**: 24,82,83,98,131, 168;221 **(16)**: AP,D **(24)**: 24,82,98,131,168
1707-265 **(7)**: RR,BG,IR **(11)**: 24,86,129,145,220,221 **(13)**: UB;Q **(14)**: 24,82,83,120, 131,168,207;223 **(16)**: AP,D **(24)**: 24,82,120,131,168,207,220
1711-294 **(11)**: 24,84,129,145,173,220,221 **(16)**: AP,D **(24)**: 16,23,24,82,120,131,168, 207,220
1713-280 **(7)**: BG,IR **(11)**: 24,86,145,221 **(13)**: UB;Q **(14)**: 24,67,82,120,168, 175;223 **(16)**: AP,D **(24)**: 24,131,168,175
1714-237 **(11)**: 24,86,145,221 **(16)**: AP,D **(24)**: 24,131,168,175
1715-278 **(14)**: 203,204
1715+432 **(24)**: 24,120,122,131,177,219,220

FOOTNOTES TO TABLE Ia (cont'd.)

1716-184 **(7)**: RR,BG,IR **(11)**: 24,86,129,145,220,221 **(13)**: ΔP;UB;Q **(14)**: 180;24, 82,83,131,168,177,207;221 **(24)**: 24,82,131,168,177,207,220
1718-195 **(7)**: BG,IR **(11)** 24,86,145,221 **(13)**: UB;Q **(14)**: 24,82,98,131,168,175;223 **(16)**: AP,D **(24)**: 24,82,131,168,175
1720-177 **(11)**: 24,84,96,121,129,145,220,221 **(16)**: AP,D **(24)**: 16,24,82,131,168,177, 207,220
1720-263 **(7)**: BG,IR **(11)**: 24,86,145,221 **(13)**: UB;Q **(14)**: 24,82,83,131,168,175,223 **(16)**: AP,D **(24)**: 24,82,131,168,175
1721-484 **(10)**: uy,UB **(16)**: AP,D **(24)**: 24,82,128,168,175,207
1725-050 **(24)**: 168,175,219
1726-670 **(10)**: GB,GR **(24)**: 24,82,128,168,207
1730-333 **(24)**: 124,149
1731-390 **(8)**: 4,34 **(13)**: BV;Q **(14)**: 168;148
1732-447 **(11)**: 24,86,206,221 **(19)**: SP,AP **(24)**: 24,82,98,117,128,168,207
1735-032 **(11)**: 24,84,86,121,129,145,220,221 **(24)**: 24,82,120,131,168,207,219,220
1735-238 **(11)**: 86,145,223 **(13)**: UB;Q **(14)**: 23,82,168,175;223 **(24)**: 23,82,131,168,175
1736-536 **(24)**: 24,49,82,128,168,207
1740-262 **(13)**: BV;Q **(14)**: 168;148
1742+031 **(11)**: 86,145,221 **(13)**: ΔP;UB;Q **(14)**: 180;43,98,131,168,219;221 **(16)**: AP,D **(24)**: 43,131,168,219
1745-247 **(8)**: 175,194 **(13)**: BV;Q **(14)**: 175;148
1746-203 **(11)**: 24,96,129,145,221 **(13)**: UB;Q **(14)**: 24,67,82,83,98,131,168,175,209;223 **(24)**: 24,82,98,131,168,175,209
1746-370 **(24)**: 24,82,117,128,168,207,209
1748-346 **(7)**: RR,BG,IR **(11)**: 24,96,145,223 **(13)**: UB;Q **(14)**: 24,168,175;223 **(24)**: 24,131,168,175
1755-442 **(7)**: BG,IR **(11)**: 24,86,223 **(13)**: BV;Q **(14)**: 24,168,175;223 **(24)**: 24,168,175
1759-089 **(11)**: 24,86,145,223 **(13)**: UB;Q **(14)**: 24,67,98,120,131,175,219;223 **(16)**: AP,D **(24)**: 24,120,131,168,175,219
1800-300 **(11)**: 18,208 **(24)**: 24,82,120,131,168,207,220
1801-003 **(8)**: 15,140 **(14)**: 15,140 **(16)**: AP,D **(24)**: 24,120,168,177,219
1801-300 **(11)**: 24,84,96,129,145,221 **(24)**: 24,82,131,168,175
1802-075 **(11)**: 86,145,223 **(13)**: UB;Q **(14)**: 131,168,175,177,219;223 **(16)**: AP,D **(24)**: 131,168,177,219
1804-250 **(11)**: 24,86,129,145,220,221 **(16)**: AP,D **(24)**: 24,82,131,168,175,220
1804-437 **(11)**: 24,84,86,206,221 **(24)**: 24,82,128,168,207
1806-259 **(11)**: 24,84,86,129,145,220,221 **(16)**: AP,D **(24)**: 24,82,120,131,168,175,220
1807-317 **(7)**: RR,BG **(11)**: 24,86,221 **(13)**: UB;Q **(14)**: 24,82,83,120,131,168,175;223 **(16)**: AP,D **(24)**: 24,82,120,131,168,175
1814-522 **(7)**: BG,IR **(11)**: 24,84,86,221 **(13)**: UB;Q **(14)**: 24,82,83,128,168,207;221 **(16)**: AP,D **(24)**: 24,82,128,168,207
1820-303 **(11)**: 24,86,129,145,206,220,221 **(16)**: AP,D **(24)**: 19,24,32,82,103,120,131, 168,207,209,220
1821-249 **(16)**: AP,D **(19)**: AP,e **(24)**: 24,82,120,131,168,207,220
1827-255 **(11)**: 24,84,86,129,145,221 **(16)**: SC,AP **(24)**: 23,24,82,120,131,168,175
1828-323 **(11)**: 24,84,86,129,145,206,220,221 **(16)**: AP,D **(24)**: 24,82,131,168,207,220
1828-235 **(7)**: RR,BG **(11)**: 24,86,129,220,223 **(13)**: UB;Q **(14)**: 24,82,83,168,175;223 **(24)**: 24,82,168,175,220
1832-330 **(11)**: 24,84,86,129,145,220,221 **(13)**: UB;Q **(14)**: 24,82,83,131,168,207;223 **(16)**: AP,D **(24)**: 24,82,131,168,207,220
1833-239 **(24)**: 24,82,120,131,168,207,220
1838-198 **(7)**: BG,IR **(10)**: 86,223 **(13)**: UB;Q **(14)**: 168,175; 223 **(24)**: 168,175
1840-323 **(8)**: 85,143 **(11)**: 24,129,145,173,206,220,221 **(16)**: AP,D **(24)**: 24,82,131, 168,207,220
1850-087 **(10)**: B,UB,GB,GR **(16)**: AP,D **(24)**: 24,82,131,168,177,207,209,219,220
1851-305 **(11)**: 24,86,129,145,206,220,221 **(24)**: 24,82,117,131,168,207,220
1852-227 **(11)**: 86,223 **(24)**: 76,82,168,175
1856-367 **(24)**: 24,74,82,131,168,207
1902+017 **(7)**: BG,IR **(14)**: 168,175,219 **(24)**: 168,175,219
1906-600 **(11)**: 39,159 **(19)**: SP,AP **(24)**: 24,49,74,82,128,168,207
1908+009 **(7)**: RR,BG,IR **(11)**: 24,84,86,145,221 **(13)**: UB;Q **(14)**: 24,67,82,83,98, 131,168,177,219;223 **(16)**: AP,D **(24)**: 24,82,131,168,177,219
1914+300 **(16)**: AP,D **(24)**: 24,131,177,220

FOOTNOTES TO TABLE Ia (cont'd.)

1936-310 **(24)**: 24,82,120,131,168,207,220
1938-341 **(17)**: 34,201
1951+186 **(16)**: AP,D **(24)**: 24,131,168,207,219,220
2003-220 **(11)**: 24,86,129,145,206,220,221 **(24)**: 24,82,117,131,168,207,220
2031+072 **(24)**: 24,43,82,131,168,177,219
2050-127 **(10)**: Ψ;UB **(11)**: 190;62 **(24)**: 24,43,82,131,168,177,219
2059+160 **(24)**: 24,43,120,131,168,177,219
2127+119 **(10)**: HB;rr **(11)**: 27;25 **(24)**: 24,120,122,131,168,207,219,220
2130-010 **(24)**: 24,82,120,122,131,168,207,219,220
2137-234 **(24)**: 24,82,120,131,168,207,220
2143-214 **(24)**: 168,175
2305-159 **(24)**: 120,168,175

TABLE Ib. OBSERVED PARAMETERS OF GALACTIC DWARF SPHEROIDALS

Number	Name		α_{1950}	δ_{1950}	V_{HB}			E_{B-V}		
(1)	(2)	(3)	(4)	(5)	(6)	(7)	(8)	(9)	(10)	(11)
0057-340	Scl		00 57 44	-34 00.4	20.13	CM	134	0.02	*	*
0237-347	For		02 37 50	-34 44.4	21.4:	CM	214	0.00	HI	29
0640-509	Car		06 40 24	-50 55.0	20.52	CM	155	0.03	HI	29
1005+126	Leo I	DDO 74	10 05 46	+12 33.2	22.3	BG	115	0.02	HI	29
1110+224	Leo II	DDO 93	11 10 50	+22 26.1	22.45:	CM	59	0.01	HI	29
1508+674	UMi	DDO 199	15 08 12	+67 23.0	20.00	CM	189	0.06	*	189
1719+580	Dra	DDO 208	17 19 13	+57 57.5	20.07	CM	195	0.03	uy	196

FOOTNOTES TO TABLE Ib

0057-340 **(10)**: uy,UB **(20)**: 170;68
1110+224 **(22)**: 56,113 **(25)**: 56,113
1508+674 **(10)**: UB,GB

TABLE Ic. OBSERVED PARAMETERS OF THE FORNAX GLOBULAR CLUSTERS

Number	Name		α_{1950}	δ_{1950}	V_{HB}			E_{B-V}		
(1)	(2)	(3)	(4)	(5)	(6)	(7)	(8)	(9)	(10)	(11)
0235-344	For 1		02 34 57	-34 24.0	21.4:	CM	214	0.00	HI	29
0236-350	For 2		02 36 40	-35 01.5	21.4:	CM	214	0.00	HI	29
0237-345	For 3	NGC 1049	02 37 44	-34 28.3	21.4:	CM	214	0.00	HI	29
0238-348	For 4		02 38 05	-34 45.2	21.4:	CM	214	0.00	HI	29
0240-343	For 5		02 40 17	-34 18.9	21.4:	CM	214	0.00	HI	29
0238-346	For 6		02 38 06	-34 38.8	21.4:	CM	214	0.00	HI	29

FOOTNOTES TO TABLE Ic

0235-344 **(14)**: 110,112 **(24)**: 110,112
0236-350 **(14)**: 55,57,110,112 **(24)**: 55,57,77,110,112,222
0237-345 **(14)**: 55,57,73,77,110,112,131 **(17)**: 77,112,131
(24): 55,57,77,110,112,222
0238-348 **(14)**: 55,57,73,77,110,112 **(17)**: 55,77,112 **(24)**: 55,57,77,110,112,222
0240-343 **(14)**: 55,57,110,112 **(17)**: 55,112 **(24)**: 55,57,77,110,112,222

TABLE Ib. OBSERVED PARAMETERS OF GALACTIC DWARF SPHEROIDALS

$(B-V)_g$			$\log \theta_t$	$\log \theta_c$	ε			V_t			σ_c			Number
(12)	(13)	(14)	(15)	(16)	(17)	(18)	(19)	(20)	(21)	(22)	(23)	(24)	(25)	(26)
0.79	CM	134	1.76	1.01	0.35	SC	75	8.13	SP	111	14.55	SP	111	0057-340
0.83:	CM	214	1.76	1.10	0.35	SC	75	8.41	SP	54	14.42	SP	114	0237-347
0.78	CM	155	1.67	0.98	0.31	SC	75	10.74	Σ	58	17.13	Σ	155	0640-509
			1.11	0.64	0.31	SC	75	10.46	SP	116	(13.98)			1005+126
0.84:	CM	59	1.00	0.36	0.01	SC	75	12.18:	AP	*	15.18	AP	*	1110+224
0.75	CM	189	1.67	1.12	0.55	SC	75	10.69	R	108	(16.41)			1508+674
0.80	CM	195	1.54	0.97	0.29	SC	75	10.78	R	109	(16.31)			1719+580

TABLE Ic. OBSERVED PARAMETERS OF THE FORNAX GLOBULAR CLUSTERS

$(B-V)_g$			$\log \theta_t$			$\log \theta_c$			V_t	σ_c			Number
(12)	(13)	(14)	(15)	(16)	(17)	(18)	(19)	(20)	(21)	(22)	(23)	(24)	(25)
0.65	UB	*	0.18	D	215	-0.62	e	215	15.57	13.89	AP	*	0235-344
0.75	UB	*	0.28	AP	55	-0.84	SP	55	13.50	11.42:	AP	*	0236-350
0.73	UB	*	0.20	AP	*	-1.63:	e	77	12.61	7.72:	AP	*	0237-345
0.83	UB	*	0.04	AP	*	-1.79:	e	77	13.57	7.87:	AP	*	0238-348
0.74	UB	*	0.10	AP	*	-1.16	e	77	13.42	10.01:	AP	*	0240-343
0.7::	CM	214	-0.24:	D	215	-1.15:	e	215	14.34	10.43:	AP	112	0238-346

REFERENCES FOR TABLE I

1 Aaronson, M., Schommer, R. A., and Olszewski, E. W. 1984, Ap. J., **276**, 221.
2 Alcaino, G. 1972, Astr. Ap., **16**, 220.
3 _____. 1974, Astr. Ap. Suppl., **13**, 55.
4 _____. 1977a, Astr. Ap. Suppl., **27**, 255.
5 _____. 1977b, Astr. Ap. Suppl., **29**, 383.
6 _____. 1978, Astr. Ap. Suppl., **32**, 379.
7 _____. 1979a, Astr. Ap. Suppl., **35**, 233.
8 _____. 1979b, Astr. Ap. Suppl., **38**, 61.
9 _____. 1980, Astr. Ap. Suppl., **39**, 316.
10 _____. 1981a, Astr. Ap. Suppl., **44**, 33.
11 _____. 1981b, Astr. Ap. Suppl., **44**, 191.
12 _____. 1983a, Astr. Ap. Suppl., **52**, 105.
13 _____. 1983b, Astr. Ap. Suppl., **53**, 47.
14 Alcaino, G., and Liller, W. 1983, A. J., **88**, 1166.
15 Anthony-Twarog, B. J., and Twarog, B. A. 1984, preprint.
16 Arp, H. C. 1958, A. J., **63**, 118.
17 _____. 1962, Ap. J., **135**, 311.
18 _____. 1965, Ap. J., **141**, 43.
19 Bahcall, N. A. 1976, Ap. J. (Lett.), **204**, L83.
20 Bahcall, N. A., and Hausman, M. A. 1977, Ap. J., **213**, 93.
21 Barbon, R. 1965, Contr. Oss. astrofis. Univ. Padova, Asiago, No. 175.
22 Barnes, S. A. 1968, A. J., **73**, 579.
23 Bernard, A. 1976, Astr. Ap. Suppl., **25**, 281.
24 Bica, E. L. D., and Pastoriza, M. G. 1983, Ap. Space Sci., **91**, 99.
25 Bingham, E. A., Cacciari, C., Dickens, R. J., and Fusi Pecci, F. 1984, M.N.R.A.S., **209**, 765.
26 Buonanno, R., Buscema, G., Corsi, C. E., Iannicola, G., and Fusi Pecci, F. 1984, Astr. Ap. Suppl., **56**, 79.
27 Buonanno, R., Corsi, C. E., and Fusi Pecci, F. 1984, Astr. Ap., submitted.
28 Burbidge, E. M., and Sandage, A. R. 1958, Ap. J., **127**, 527.
29 Burstein, D., and Heiles, C. 1982, A. J., **87**, 1165.
30 Butler, D., Dickens, R. J., and Epps, E. 1978, Ap. J., **225**, 148.
31 Cacciari, C. 1979, A. J., **84**, 1542.
32 Canizares, C. R., Grindlay, J. E., Hiltner, W. A., Liller, W., and McClintock, J. E. 1978, Ap. J., **224**, 39.
33 Cannon, R. D. 1974, M.N.R.A.S., **167**, 551.
34 Cannon, R. D., Hawarden, T. G., and Tritton, S. B. 1978, unpublished.
35 Cannon, R. D., and Stobie, R. S. 1973a, M.N.R.A.S., **162**, 207.
36 _____. 1973b, M.N.R.A.S., **162**, 227.
37 Canterna, R., and Rosino, L. 1981, Astr. Ap. Suppl., **45**, 53.
38 Canterna, R., and Schommer, R. 1978, Ap. J. (Lett.), **219**, L119.
39 Carney, B. W. 1979, A. J., **84**, 515.
40 Chun, M. S., and Freeman, K. C. 1979, Ap. J., **227**, 93.
41 Ciatti, F., Rosino, L., and Sussi, M. G. 1965, Kleine Veröff. Remeis-Sternw., Bamberg, **4**, Nr. 40, p. 228.
42 Clement, C. C., Dickens, R. J., and Bingham, E. E. 1979, A. J., **84**, 217.
43 Corwin, H. G., Jr. 1977, A. J., **82**, 193.
44 Cowley, A. P., Hartwick, F. D. A., and Sargent, W. L. W. 1978, Ap. J., **220**, 453.
45 Crawford, D. L., and Barnes, J. V. 1975, P.A.S.P., **87**, 65.
46 Cudworth, K. M. 1984, A. J., submitted.
47 Cudworth, K. M., Schommer, R. A., and Canterna, R. 1984, in preparation.
48 Cuffey, J. 1965, A. J., **70**, 732.
49 Da Costa, G. S. 1979, A. J., **84**, 505.
50 _____. 1984, Ap. J., submitted.
51 Da Costa, G. S., Mould, J., and Kristian, J. 1984, in preparation.
52 Da Costa, G. S., Ortolani, S., and Mould, J. 1982, Ap. J., **257**, 633.
53 de Vaucouleurs, G. 1959, Lowell Obs. Bull., **4**, 105.
54 de Vaucouleurs, G., and Ables, H. D. 1968, Ap. J., **151**, 105.
55 _____. 1970, Ap. J., **159**, 425.
56 de Vaucouleurs, G., de Vaucouleurs, A., and Buta, R. 1981, A. J., **86**, 1429.
57 Demers, S. 1969, Ap. Lett., **3**, 175.
58 Demers, S., Beland, S., and Kunkel, W. E. 1983, P.A.S.P., **95**, 354.
59 Demers, S., and Harris, W. E. 1983, A. J., **88**, 329.
60 Demers, S., and Kunkel, W. E. 1976, Ap. J., **208**, 932.
61 Demers, S., Kunkel, W. E., and Hardy, E. 1979, Ap. J., **232**, 84.

REFERENCES FOR TABLE I (cont'd.)

62 Dickens, R. J. 1972a, M.N.R.A.S., **157**, 281.
63 ____. 1972b, M.N.R.A.S., **157**, 299.
64 Dickens, R. J., and Rolland, A. 1972, M.N.R.A.S., **160**, 37.
65 Djorgovski, S., and King, I. R. 1984, Ap. J. (Lett.), **277**, L49.
66 Djorgovski, S., and Penner, H. 1985, in Dynamics of Star Clusters, I.A.U. Symp. No. 113, eds. J. Goodman and P. Hut (Dordrecht: D. Reidel), this volume.
67 Dufay, J., and Bigay, J. H. 1959, C. R. Acad. Sci. Paris, **248**, 2162.
68 Eggen, O. J. 1970, Ap. J. Suppl., **22**, 289.
69 Fourcade, C. R. 1974, Perem. Zvezdy Pril., **2**, 18.
70 Fourcade, C. R., Laborde, J. R., and Arias, J. C. 1974, Perem. Zvezdy Pril., **2**, 3.
71 Gascoigne, S. C. B. 1966, M.N.R.A.S., **134**, 59.
72 Gascoigne, S. C. B., and Burr, E. J. 1956, M.N.R.A.S., **116**, 570.
73 Gascoigne, S. C. B., and Kron, G. E. 1952, P.A.S.P., **64**, 196.
74 Gascoigne, S. C. B., and Ogston, F. A. 1963, Observatory, **83**, 64.
75 Godwin, P. J. 1985, in Dynamics of Star Clusters, I.A.U. Symp. No. 113, eds. J. Goodman and P. Hut (Dordrecht: D. Reidel), this volume.
76 Goranskii, V. P. 1979, Astr. Zh., **56**, 510 [English transl.: Soviet Astr., **23**, 284].
77 Gordon, K. C., and Kron, G. E. 1983, P.A.S.P., **95**, 461.
78 Gratton, R. G., and Ortolani, S. 1984, Astr. Ap. Suppl., **57**, 177.
79 Grindlay, J. E. 1978, Ap. J. (Lett.), **224**, L107.
80 Grindlay, J. E., and Hertz, P. 1981, Ap. J. (Lett.), **247**, L17.
81 Grindlay, J. E., Hertz, P., Steiner, J. E., Murray, S. S., and Lightman, A. P. 1984, Ap. J. (Lett.), **282**, L13.
82 Hamuy, M. 1984, Astr. Ap. Suppl., **57**, 91.
83 Hanes, D. A., and Brodie, J. P. 1984, M.N.R.A.S., submitted.
84 Harris, H. C., and Canterna, R. 1977, A. J., **82**, 798.
85 Harris, W. E. 1975, Ap. J. Suppl., **29**, 397.
86 ____. 1976, A. J., **81**, 1095.
87 ____. 1977, P.A.S.P., **89**, 482.
88 ____. 1978, P.A.S.P., **90**, 45.
89 ____. 1980a, P.A.S.P., **92**, 43.
90 ____. 1980b, in Star Clusters, I.A.U. Symp. No. 85, ed. J. E. Hesser (Dordrecht: D. Reidel), p. 81.
91 ____. 1982, Ap. J. Suppl., **50**, 573.
92 Harris, W. E., and Canterna, R. 1980, Ap. J., **239**, 815.
93 Harris, W. E., and Hesser, J. 1976, P.A.S.P., **88**, 377.
94 Harris, W. E., and Racine, R. 1973, A. J., **78**, 242.
95 ____. 1974, A. J., **79**, 472.
96 ____. 1979, Ann. Rev. Astr. Ap., **17**, 241.
97 Harris, W. E., Racine, R., and de Roux, J. 1976, Ap. J. Suppl., **31**, 13.
98 Harris, W. E., and van den Bergh, S. 1974, A. J., **79**, 31.
99 ____. 1984, preprint.
100 Hartwick, F. D. A. 1975, P.A.S.P., **87**, 77.
101 Hartwick, F. D. A., and Hesser, J. E. 1972, Ap. J., **175**, 77.
102 ____. 1973, Ap. J., **186**, 1171.
103 Harvel, C. A., and Martins, D. H. 1977, Ap. J. (Lett.), **213**, L49.
104 Hesser, J. E. 1976, P.A.S.P., **88**, 849.
105 Hesser, J. E., and Hartwick, F. D. A. 1976a, Ap. J., **203**, 97.
106 ____. 1976b, Ap. J., **203**, 113.
107 Hesser, J. E., McClure, R. D., Hawarden, T. G., Cannon, R. D., von Rudloff, R., Kruger, B., and Egles, D. 1984, P.A.S.P., **96**, 406.
108 Hodge, P. W. 1964a, A. J., **69**, 438.
109 ____. 1964b, A. J., **69**, 853.
110 ____. 1965, Ap. J., **141**, 308.
111 ____. 1966, A. J., **71**, 204.
112 ____. 1969, P.A.S.P., **81**, 875.
113 ____. 1982, A. J., **87**, 1668.
114 Hodge, P. W., and Smith, D. W. 1974, Ap. J., **188**, 19.
115 Hodge, P. W., and Wright, F. W. 1978, A. J., **83**, 228.
116 Holmberg, E. B. 1958, Medd. Lunds Astr. Obs., ser. 2, No. 136.
117 Illingworth, G., and Illingworth, W. 1976, Ap. J. Suppl., **30**, 227.
118 Inman, R. T., and Carney, B. W. 1984, in preparation.
119 Janulis, R., and Straižys, V. 1984, Ap. Space Sci., **100**, 95.
120 Johnson, H. L. 1959, Lowell Obs. Bull., **4**, 117.
121 Johnson, S. L., and McNamara, D. H. 1969, P.A.S.P., **81**, 415.

REFERENCES FOR TABLE I (cont'd.)

122 King, I. R. 1966, A. J., **71**, 276.
123 Kinman, T. D., Stryker, L. L., and Hesser, J. E. 1976, P.A.S.P., **88**, 393.
124 Kleinmann, D. E., Kleinmann, S. G., and Wright, E. L. 1977, Ap. J. (Lett.), **210**, L83.
125 Kodaira, K. 1983, I.A.U. Circ., No. 3846.
126 Kogon, C. S., Wehlau, A., and Demers, S. 1974, A. J., **79**, 387.
127 Kron, G. E. 1966, P.A.S.P., **78**, 143.
128 Kron, G. E., and Gordon, K. C. 1984, in preparation.
129 Kron, G. E., and Guetter, H. H. 1976, A. J., **81**, 817.
130 Kron, G. E., Hewitt, A. V., and Wasserman, L. H. 1984, P.A.S.P., **96**, 198.
131 Kron, G. E., and Mayall, N. U. 1960, A. J., **65**, 581.
132 Kukarkin, B. V. 1974, Globular Star Clusters (Moscow: Nauka) [English transl.: 1977, NASA TT F-16, 157].
133 Kukarkin, B. V., and Kireeva, N. N. 1979, Astr. Zh., **56**, 465 [English transl.: Soviet Astr., **23**, 261].
134 Kunkel, W. E., and Demers, S. 1977, Ap. J., **214**, 21.
135 Lee, S.-W. 1977a, Astr. Ap. Suppl., **27**, 367.
136 _____. 1977b, Astr. Ap. Suppl., **27**, 381.
137 _____. 1977c, Astr. Ap. Suppl., **28**, 409.
138 _____. 1977d, Astr. Ap. Suppl., **29**, 1.
139 Liller, M. H. 1980a, A. J., **85**, 673.
140 _____. 1980b, A. J., **85**, 1480.
141 _____. 1981, A. J., **86**, 1204.
142 _____. 1983a, A. J., **88**, 104.
143 _____. 1983b, A. J., **88**, 1463.
144 Liller, M. H., and Carney, B. W. 1978, Ap. J., **224**, 383.
145 Lohmann, W. 1963, Z. Ap., **57**, 288.
146 Madore, B. F., and Arp, H. C. 1979, Ap. J. (Lett.), **227**, L103.
147 Malkan, M. 1978, paper presented at the NATO Advanced Study Institute on Globular Clusters, Cambridge, England.
148 _____. 1981, in Astrophysical Parameters for Globular Clusters, I.A.U. Colloq. No. 68, eds. A. G. D. Philip and D. S. Hayes (Schenectady: L. Davis Press), p. 533.
149 Malkan, M., Kleinmann, D. E., and Apt, J. 1980, Ap. J., **237**, 432.
150 Martins, D. H., Harvel, C. A., and Miller, D. H. 1980, A. J., **85**, 521.
151 McNamara, D. H., and Langford, R. 1969, P.A.S.P., **81**, 141.
152 Menzies, J. W. 1972, M.N.R.A.S., **156**, 207.
153 _____. 1974a, M.N.R.A.S., **168**, 177.
154 _____. 1974b, M.N.R.A.S., **169**, 79.
155 Mould, J., and Aaronson, M. 1983, Ap. J., **273**, 530.
156 Mould, J., Cannon, R. D., Aaronson, M., and Frogel, J. A. 1982, Ap. J., **254**, 500.
157 Neckel, Th., and Klare, G. 1980, Astr. Ap. Suppl., **42**, 251.
158 Newell, E. B., Rodgers, A. W., and Searle, L. 1969, Ap. J., **158**, 699.
159 Newell, E. B., and Sadler, E. M. 1978, Ap. J., **221**, 825.
160 Olszewski, E. W., Canterna, R., and Harris, W. E. 1984, Ap. J., **281**, 158.
161 Ortolani, S. 1984, Astr. Ap., in press.
162 Ortolani, S., and Gratton, R. 1984, Astr. Ap., in press.
163 Ortolani, S., Rosino, L., and Sandage, A. 1984, in preparation.
164 Penny, A. J. 1975, M.N.R.A.S., **172**, 65P.
165 Persson, S. E., Aaronson, M., Cohen, J. G., Frogel, J. A., and Matthews, K. 1983, Ap. J., **266**, 105.
166 Peterson, C. J. 1976, A. J., **81**, 617.
167 Peterson, C. J., and King, I. R. 1975, A. J., **80**, 427.
168 Peterson, C. J., and Knipp, D. 1985, in Dynamics of Star Clusters, I.A.U. Symp. No. 113, eds. J. Goodman and P. Hut (Dordrecht: D. Reidel), this volume.
169 Peterson, C. J., and Kunkel, W. E. 1977, P.A.S.P., **89**, 634.
170 Philip, A. G. D. 1974, Ap. J., **190**, 573.
171 Pike, C. P. 1976, M.N.R.A.S., **177**, 257.
172 Racine, R. 1971, A. J., **76**, 331.
173 _____. 1973, A. J., **78**, 180.
174 _____. 1974, unpublished (see Harris 1976).
175 _____. 1975, A. J., **80**, 1031.
176 Racine, R., and Harris, W. E. 1975, Ap. J., **196**, 413.
177 Rousseau, J. 1964, Ann. Ap., **27**, 681.
178 Sandage, A. 1969, Ap. J., **157**, 515.
179 _____. 1970, Ap. J., **162**, 841.
180 _____. 1982, Ap. J., **252**, 553.

REFERENCES FOR TABLE I (cont'd.)

181 Sandage, A., and Hartwick, F. D. A. 1977, A. J., **82**, 459.
182 Sandage, A., and Katem, B. 1968, Ap. J., **153**, 569.
183 ____. 1982, A. J., **87**, 537.
184 Sandage, A., Katem, B., and Johnson, H. L. 1977, A. J., **82**, 389.
185 Sandage, A., and Smith, L. L. 1966, Ap. J., **144**, 886.
186 Sandage, A., and Walker, M. F. 1955, A. J., **60**, 230.
187 Sandage, A., and Wallerstein, G. 1960, Ap. J., **131**, 598.
188 Sandage, A., and Wildey, R. 1967, Ap. J., **150**, 469.
189 Schommer, R. A., Olszewski, E. W., and Kunkel, W. E. 1978, in The HR Diagram, I.A.U. Symp. No. 80, eds. A. G. D. Philip and D. S. Hayes (Dordrecht: D. Reidel), p. 269.
190 Searle, L., and Zinn, R. 1978, Ap. J., **225**, 357.
191 Seitzer, P., Da Costa, G. S., and Mould, J. 1984, in preparation.
192 Seitzer, P., and Phillips, M. 1984, in preparation.
193 Smith, H. A., and Perkins, G. J. 1982, Ap. J., **261**, 576.
194 Spinrad, H., Smith, M. G., and Harlan, E. 1974, Ap. J., **192**, 405.
195 Stetson, P. B. 1979a, A. J., **84**, 1149.
196 ____. 1979b, A. J., **84**, 1167.
197 ____. 1981, A. J., **86**, 687.
198 Stetson, P. B., and Harris, W. E. 1977, A. J., **82**, 954.
199 Sturch, C. R. 1978, P.A.S.P., **90**, 264.
200 Terzan, A. 1965, Ann. Ap., **28**, 935.
201 ____. 1968, C. R. Acad. Sci. Paris, **267**, 1245.
202 ____. 1971, Astr. Ap., **12**, 477.
203 Terzan, A., and Bernard, A. 1978, Messenger, No. 15, p. 14.
204 Terzan, A., Bernard, A., and Ju, K. H. 1978a, C. R. Acad. Sci. Paris, **287**, sér. B, 157.
205 ____. 1978b, C. R. Acad. Sci. Paris, **287**, sér. B, 235.
206 van Albada, T. S., de Boer, K. S., and Dickens, R. J. 1981, M.N.R.A.S., **195**, 591.
207 van den Bergh, S. 1967, A. J., **72**, 70 [Erratum: 1970, A. J., **75**, 131].
208 ____. 1971, A. J., **76**, 1082.
209 ____. 1977, A. J., **82**, 796.
210 ____. 1981, Astr. Ap. Suppl., **46**, 79.
211 van den Bergh, S., Demers, S., and Kunkel, W. E. 1980, Ap. J., **239**, 112.
212 van den Bergh, S., and Hagen, G. L. 1968, A. J., **73**, 569.
213 van den Bergh, S., and Younger, F. 1979, A. J., **84**, 1305.
214 Verner, G., Demers, S., Hardy, E., and Kunkel, W. E. 1981, A. J., **86**, 357.
215 Webbink, R. F. 1985, this paper.
216 Webbink, R. F., and Hunter, D. A. 1981, unpublished.
217 West, R. M., and Bartaya, R. A. 1979, Astr. Ap. Suppl., **38**, 69.
218 Williams, T. B., and Bahcall, N. A. 1979, Ap. J., **232**, 754.
219 Zaitseva, G. V., Lyutyi, V. M., and Kukarkin, B. V. 1974, Astr. Zh., **51**, 438 [English transl.: Soviet Astr., **18**, 257].
220 Zdanavičius, K. V. 1983, Astr. Zh., **60**, 44 [English transl.: Soviet Astr., **27**, 26].
221 Zinn, R. 1980, Ap. J. Suppl., **42**, 19.
222 Zinn, R., and Persson, S. E. 1981, Ap. J., **247**, 849.
223 Zinn, R., and West, M. J. 1984, Ap. J. Suppl., **55**, 45.

TABLE IIa. DERIVED PARAMETERS OF GALACTIC GLOBULAR CLUSTERS

Number	Name		ℓ	b	R_o	X	Y	Z	R_{GC}
(1)	(2)	(3)	(4)	(5)	(6)	(7)	(8)	(9)	(10)
0021-723	NGC 104	47 Tuc	305.895	-44.889	4.6	-6.9	-2.7	-3.3	8.1
0050-268	NGC 288		151.147	-89.377	8.2	-8.9	0.0	-8.2	12.1
0100-711	NGC 362		301.533	-46.247	8.7	-5.6	-5.1	-6.3	9.9
0310-554	NGC 1261		270.539	-52.127	16.1	-8.7	-9.9	-12.7	18.3
0325+794	Pal 1		130.067	+19.023	13.7	-17.1	9.9	4.5	20.3
0344-718	NGC 1466	SL 1	286.700	-39.537	39.4	-0.1	-29.1	-25.1	38.4
0354-498	AM 1	E 1	258.360	-48.472	116.4	-24.4	-75.6	-87.2	117.9
0422-213	Eri		218.108	-41.331	84.7	-58.9	-39.3	-56.0	90.2
0435-590	Ret	Sé 40/3	268.664	-40.269	50.4	-9.7	-38.5	-32.6	51.3
0443+313	Pal 2		170.532	-09.070	13.6	-22.0	2.2	-2.1	22.2
0444-840	NGC 1841		297.016	-30.147	40.9	7.2	-31.5	-20.5	38.3
0512-400	NGC 1851		244.512	-35.036	12.0	-13.0	-8.9	-6.9	17.2
0522-245	NGC 1904	M 79	227.231	-29.350	13.0	-16.5	-8.3	-6.4	19.5
0647-359	NGC 2298		245.629	-16.007	10.6	-13.0	-9.3	-2.9	16.2
0734+390	NGC 2419		180.370	+25.242	91.4	-91.4	-0.5	39.0	99.4
0737-337	AM 2		248.126	-05.876	57.7	-30.2	-53.2	-5.9	61.5
0911-646	NGC 2808		282.193	-11.252	9.5	-6.8	-9.1	-1.9	11.6
0921-770	E 3		292.269	-19.018	8.3	-5.8	-7.2	-2.7	9.7
0923-545	UKS 2		276.003	-03.008	9.0	-7.9	-9.0	-0.5	11.9
1003+003	Pal 3	Sex C	240.142	+41.866	87.9	-41.4	-56.8	58.7	91.5
1015-461	NGC 3201		277.229	+08.641	5.0	-8.2	-4.9	0.7	9.5
1117-649	ESO 093-SC?08		293.508	-04.041	59.5	14.9	-54.4	-4.2	56.6
1126+292	Pal 4	UMa	202.293	+71.801	93.3	-35.8	-11.1	88.7	96.2
1207+188	NGC 4147		252.848	+77.189	17.3	-9.9	-3.7	16.8	19.9
1223-724	NGC 4372		300.995	-09.881	4.9	-6.3	-4.2	-0.8	7.6
1235-509	Rup 106		300.888	+11.670	26.7	4.6	-22.4	5.4	23.5
1236-264	NGC 4590	M 68	299.625	+36.051	9.6	-5.0	-6.7	5.6	10.1
1256-706	NGC 4833		303.604	-08.014	5.8	-5.6	-4.8	-0.8	7.4
1310+184	NGC 5024	M 53	332.965	+79.764	18.5	-5.9	-1.5	18.2	19.2
1313+179	NGC 5053		335.675	+78.946	15.8	-6.0	-1.3	15.5	16.7
1323-472	NGC 5139	ω Cen	309.100	+14.971	5.2	-5.6	-3.9	1.3	7.0
1339+286	NGC 5272		042.218	+78.707	10.4	-7.3	1.4	10.2	12.6
1343-511	NGC 5286		311.614	+10.568	9.7	-2.5	-7.1	1.8	7.7
1353-269	AM 4		320.280	+33.506	30.3	10.6	-16.2	16.7	25.6
1403+287	NGC 5466		042.137	+73.593	15.8	-5.5	3.0	15.2	16.4
1427-057	NGC 5634		342.210	+49.260	25.0	6.7	-5.0	18.9	20.7
1436-263	NGC 5694		331.056	+30.360	31.3	14.9	-13.1	15.8	25.4
1452-820	IC 4499		307.354	-20.473	18.0	1.5	-13.4	-6.3	14.9
1500-328	NGC 5824		332.555	+22.071	24.6	11.4	-10.5	9.2	18.0
1513+000	Pal 5	Ser	000.847	+45.853	21.4	6.1	0.2	15.4	16.5
1514-208	NGC 5897		342.948	+30.294	11.8	1.0	-3.0	6.0	6.7
1516+022	NGC 5904	M 5	003.860	+46.797	7.6	-3.6	0.4	5.6	6.6
1524-505	NGC 5927		326.605	+04.859	8.8	-1.5	-4.8	0.7	5.1
1531-504	NGC 5946		327.582	+04.192	9.2	-1.1	-4.9	0.7	5.1
1535-499	BH 176		328.417	+04.344	85.7	64.0	-44.7	6.5	78.3
1542-376	NGC 5986		337.028	+13.273	10.5	0.6	-4.0	2.4	4.7
1608+150	Pal 14	AvdB	028.755	+42.177	75.3	40.1	26.8	50.6	69.9
1614-228	NGC 6093	M 80	352.674	+19.462	8.0	-1.3	-1.0	2.7	3.1
1620-720	NGC 6101		317.751	-15.828	16.1	2.7	-10.4	-4.4	11.6
1620-264	NGC 6121	M 4	350.975	+15.974	2.1	-6.8	-0.3	0.6	6.8
1624-259	NGC 6144		351.929	+15.702	9.5	0.3	-1.3	2.6	2.9
1624-387	NGC 6139		342.365	+06.939	8.1	-1.2	-2.4	1.0	2.9
1625-352	Ter 3		345.083	+09.190	27.2	17.2	-6.9	4.4	19.0
1629-129	NGC 6171	M 107	003.371	+23.012	6.2	-3.1	0.3	2.4	3.9
1636-283	ESO 452-SC 11		351.912	+12.097	10.3	1.2	-1.4	2.2	2.8

TABLE IIa. DERIVED PARAMETERS OF GALACTIC GLOBULAR CLUSTERS

[m/H]	M_V	r_c	r_t	c_{der}	log T_r	log ρ_o	σ_o	v_{esc}	Number
(11)	(12)	(13)	(14)	(15)	(16)	(17)	(18)	(19)	(20)
-0.75	-9.63	0.52	60.3	2.08	7.870	5.024	13.15	56.8	0021-723
-1.39	-6.59	3.96	37.0	0.89	9.037	2.017	2.84	10.0	0050-268
-1.39	-8.37	0.52	25.9	1.71	7.774	4.829	10.29	41.8	0100-711
-1.17	-7.75	1.86	37.1	1.28	8.618	3.186	5.46	20.7	0310-554
-1.01	-2.54	0.63	19.9		7.097	2.373	0.74	2.9	0325+794
-2.15	-7.54	2.81	73.9		8.818	2.477	3.69	14.3	0344-718
-1.68	-4.49	9.54	81.2	0.93	9.270	-0.001	0.68	2.4	0354-498
-1.22	-4.94	6.19	85.5	1.14	8.997	0.584	0.90	3.3	0422-213
-2.01	-5.87	14.66	146.6		9.740	-0.064	0.98	3.5	0435-590
-1.68	-7.26	0.31	12.5	1.63	7.290	5.090	8.39	33.7	0443+313
-1.56	-7.38	7.85	94.4	1.08	9.548	1.292	2.54	9.3	0444-840
-1.25	-8.30	0.23	28.3	2.16	7.086	5.480	9.77	42.9	0512-400
-1.47	-7.79	0.66	40.5	1.89	7.786	4.160	6.06	25.3	0522-245
-2.06	-6.40	1.34	24.4	1.20	8.209	3.120	3.62	13.6	0647-359
-1.98	-9.58	11.08	284.8	1.38	10.053	1.528	4.86	18.8	0734+390
	-6.50	25.97	98.7		10.401	-0.087	1.38	4.6	0737-337
-1.47	-9.47	0.70	39.2	1.76	8.132	4.845	14.13	57.9	0911-646
-0.96	-4.20	4.48	25.2		8.803	1.049	0.99	3.4	0921-770
-0.37	-4.15	2.04	9.3		8.326	2.167	1.55	5.2	0923-545
-1.68	-5.96	11.96	109.1	0.96	9.634	0.269	1.17	4.2	1003+003
-1.60	-7.47	1.58	52.4	1.55	8.400	3.118	4.36	17.3	1015-461
	-6.90	1.07	75.5		7.974	3.198	3.26	13.5	1117-649
-1.30	-5.65	14.58	83.9		9.790	0.081	1.06	3.7	1126+292
-1.68	-6.03	0.65	36.4	1.80	7.529	3.536	2.91	12.0	1207+188
-1.77	-7.61	3.76	45.3	1.09	9.103	2.332	4.05	14.9	1223-724
	-7.00	7.42	89.2		9.451	1.216	2.20	8.1	1235-509
-1.85	-7.26	1.72	82.1	1.63	8.400	2.878	3.60	14.4	1236-264
-1.98	-7.82	1.95	33.8	1.13	8.696	3.249	6.05	22.4	1256-706
-1.89	-8.86	2.20	118.0	1.66	8.809	3.175	6.50	26.2	1310+184
-2.02	-6.09	11.31	63.6	0.77	9.690	0.577	1.47	5.1	1313+179
-1.60	-10.40	3.96	65.7	1.15	9.583	3.347	13.79	51.2	1323-472
-1.30	-9.16	1.26	117.4	1.89	8.428	3.856	8.19	34.3	1339+286
-1.60	-8.42	0.60	33.8	1.80	7.854	4.587	9.07	37.4	1343-511
-2.23	-1.70	3.43	35.1		8.196	0.152	0.30	1.1	1353-269
-1.85	-7.05	7.31	96.3	1.08	9.448	1.253	2.27	8.3	1403+287
-1.77	-7.77	1.71	55.2	1.45	8.520	3.200	5.14	20.1	1427-057
-1.89	-8.63	0.46	138.0	2.40	7.479	4.453	5.95	27.5	1436-263
-1.77	-7.63	6.61	83.2	1.06	9.482	1.630	3.15	11.5	1452-820
-1.98	-9.56	0.39	142.6	2.39	7.532	5.034	9.98	46.0	1500-328
-1.43	-5.00	17.97	113.4		9.808	-0.495	0.69	2.4	1513+000
-1.47	-7.01	4.43	39.5	1.19	9.087	1.818	2.67	10.0	1514-208
-1.60	-8.82	0.97	64.1	1.83	8.223	4.109	8.44	34.9	1516+022
-0.67	-8.08	1.23	35.4	1.46	8.352	3.745	6.93	27.1	1524-505
-1.34	-7.12	0.44	26.6	1.79	7.452	4.482	5.90	24.3	1531-504
	-8.00	11.13	153.6		9.860	1.044	2.74	10.1	1535-499
-1.72	-8.42	1.46	38.3	1.43	8.529	3.676	7.61	29.6	1542-376
-1.34	-4.80	16.62	102.5		9.730	-0.463	0.66	2.3	1608+150
-2.15	-7.89	0.24	21.7	2.04	7.105	5.360	8.99	38.6	1614-228
-1.68	-7.00	3.72	74.2	1.26	8.953	1.997	2.77	10.5	1620-720
-1.09	-6.99	0.75	26.5	1.53	7.839	3.912	5.14	20.3	1620-264
-1.81	-6.85	3.42	32.6	1.09	8.918	2.154	3.00	11.0	1624-259
-1.60	-7.99	0.52	29.5	1.58	7.756	4.742	9.40	37.5	1624-387
	-6.20	6.59	60.1		9.283	1.142	1.75	6.3	1625-352
-0.88	-6.93	1.31	41.5	1.34	8.244	3.272	4.28	16.4	1629-129
-1.01	-4.06	1.17	10.4		7.833	2.549	1.57	5.6	1636-283

TABLE IIa. (cont'd.)

Number	Name		ℓ	b	R_o	X	Y	Z	R_{GC}
(1)	(2)	(3)	(4)	(5)	(6)	(7)	(8)	(9)	(10)
1639+365	NGC 6205	M 13	059.006	+40.914	7.1	-6.0	4.6	4.6	8.9
1644-018	NGC 6218	M 12	015.715	+26.313	5.3	-4.2	1.3	2.4	5.0
1645+476	NGC 6229		073.638	+40.306	31.6	-2.0	23.1	20.5	30.9
1650-220	NGC 6235		358.918	+13.520	9.5	0.4	-0.2	2.2	2.3
1654-040	NGC 6254	M 10	015.138	+23.074	4.5	-4.8	1.1	1.8	5.3
1656-370	NGC 6256	Ter 12	347.791	+03.307	9.1	0.0	-1.9	0.5	2.0
1657-004	Pal 15		018.873	+24.293	69.7	51.3	20.5	28.7	62.3
1658-300	NGC 6266	M 62	353.575	+07.317	6.1	-2.8	-0.7	0.8	3.0
1659-262	NGC 6273	M 19	356.869	+09.381	10.6	1.7	-0.6	1.7	2.5
1701-246	NGC 6284		358.347	+09.939	10.3	1.3	-0.3	1.8	2.2
1702-226	NGC 6287		000.132	+11.023	7.2	-1.7	0.0	1.4	2.2
1707-265	NGC 6293		357.620	+07.834	7.7	-1.2	-0.3	1.0	1.6
1708-271	TJ 5		357.261	+07.283					
1711-294	NGC 6304		355.825	+05.374	6.0	-2.9	-0.4	0.6	2.9
1713-280	NGC 6316		357.175	+05.765	12.8	3.9	-0.6	1.3	4.2
1714-237	NGC 6325		000.973	+08.003	6.2	-2.7	0.1	0.9	2.8
1715-277	TJ 16	TBJ 2	357.713	+05.636					
1715-262	TJ 15		358.939	+06.468					
1715-278	TJ 17	TBJ 1	357.688	+05.564					
1715+432	NGC 6341	M 92	068.339	+34.858	7.7	-6.5	5.9	4.4	9.8
1716-184	NGC 6333	M 9	005.544	+10.705	7.5	-1.5	0.7	1.4	2.2
1718-195	NGC 6342		004.899	+09.725	11.6	2.6	1.0	2.0	3.4
1720-177	NGC 6356		006.723	+10.220	16.7	7.5	1.9	3.0	8.3
1720-263	NGC 6355		359.585	+05.428	7.1	-1.7	-0.1	0.7	1.8
1721-484	NGC 6352		341.421	-07.164	6.6	-2.6	-2.1	-0.8	3.4
1724-307	Ter 2	HP 3	356.320	+02.298	10.0	1.2	-0.6	0.4	1.4
1725-050	NGC 6366		018.411	+16.041	4.0	-5.1	1.2	1.1	5.4
1726-670	NGC 6362		325.555	-17.569	7.7	-2.7	-4.2	-2.3	5.5
1727-315	Ter 4	HP 4	356.024	+01.308	16.1	7.3	-1.1	0.4	7.4
1727-299	HP 1		357.423	+02.113	9.5	0.7	-0.4	0.4	0.9
1728-338	Gri 1		354.304	-00.151	11.8	2.9	-1.2	0.0	3.2
1730-333	Lil 1		354.841	-00.161	7.9	-0.9	-0.7	0.0	1.2
1731-390	NGC 6380	Ton 1	350.182	-03.414	4.0	-4.9	-0.7	-0.2	5.0
1732-304	Ter 1	HP 2	357.558	+00.992	10.6	1.8	-0.5	0.2	1.9
1732-447	NGC 6388		345.557	-06.738	13.5	4.2	-3.3	-1.6	5.6
1733-390	Ton 2	Pis 26	350.797	-03.419	8.7	-0.3	-1.4	-0.5	1.5
1735-032	NGC 6402	M 14	021.322	+14.803	10.2	0.4	3.6	2.6	4.5
1735-238	NGC 6401		003.451	+03.978	7.1	-1.7	0.4	0.5	1.8
1736-536	NGC 6397		338.165	-11.959	2.2	-6.8	-0.8	-0.5	6.9
1740-262	Pal 6		002.092	+01.779	5.9	-2.9	0.2	0.2	2.9
1741-328	TJ 23		356.676	-01.916					
1742+031	NGC 6426		028.088	+16.233	17.5	6.0	7.9	4.9	11.1
1745-247	Ter 5		003.838	+01.687	7.1	-1.7	0.5	0.2	1.8
1746-203	NGC 6440		007.729	+03.800	7.1	-1.8	0.9	0.5	2.1
1746-370	NGC 6441		353.532	-05.006	11.7	2.8	-1.3	-1.0	3.3
1748-346	NGC 6453		355.717	-03.873	10.7	1.9	-0.8	-0.7	2.1
1747-312	Ter 6	HP 5	358.572	-02.163	12.8	3.9	-0.3	-0.5	4.0
1751-241	UKS 1		005.125	+00.764	10.4	1.5	0.9	0.1	1.8
1755-442	NGC 6496		348.026	-10.012	6.3	-2.8	-1.3	-1.1	3.2
1758-268	Ter 9		003.603	-01.988	7.0	-1.8	0.4	-0.2	1.9
1759-089	NGC 6517		019.225	+06.762	6.1	-3.1	2.0	0.7	3.8
1800-260	Ter 10		004.421	-01.864	14.6	5.8	1.1	-0.5	5.9
1800-300	NGC 6522		001.026	-03.929	6.6	-2.3	0.1	-0.4	2.3
1801-003	NGC 6535		027.176	+10.435	6.9	-2.8	3.1	1.2	4.3
1801-300	NGC 6528		001.138	-04.175	6.8	-2.0	0.1	-0.5	2.1

TABLE IIa. (cont'd.)

[m/H]	M_V	r_c	r_t	c_{der}	log T_r	log ρ_o	σ_o	v_{esc}	Number
(11)	(12)	(13)	(14)	(15)	(16)	(17)	(18)	(19)	(20)
-1.60	-8.67	1.72	55.5	1.44	8.672	3.556	7.79	30.4	1639+365
-1.89	-7.53	1.82	28.1	1.21	8.584	3.173	5.22	19.6	1644-018
-1.39	-8.14	1.60	51.7	1.51	8.524	3.398	6.07	23.9	1645+476
-1.60	-6.11	0.73	21.9	1.53	7.685	3.603	3.49	13.8	1650-220
-1.51	-7.50	0.92	31.2	1.66	8.023	3.765	5.37	21.6	1654-040
-1.56	-6.31	1.00	17.8	1.39	7.957	3.343	3.55	13.8	1656-370
-1.26	-5.40	27.35	108.9		10.245	-0.625	0.79	2.6	1657-004
-1.26	-8.67	0.43	18.4	1.67	7.716	5.209	13.29	53.7	1658-300
-2.40	-9.62	1.55	53.8	1.49	8.752	4.042	12.36	48.5	1659-262
-1.34	-7.10	0.48	29.8	1.77	7.511	4.364	5.64	23.1	1701-246
-1.72	-6.61	0.69	20.0	1.53	7.733	3.860	4.49	17.8	1702-226
-1.85	-7.39	0.37	32.9	1.95	7.328	4.695	6.33	26.7	1707-265
									1708-271
-0.54	-7.19	0.46	21.9	1.75	7.496	4.488	6.15	25.2	1711-294
-0.62	-8.45	0.65	53.8	1.98	7.850	4.365	7.58	32.2	1713-280
-2.02	-6.14	0.44	17.9	1.53	7.362	4.262	4.53	17.9	1714-237
									1715-277
									1715-262
≤-0.07									1715-278
-1.89	-8.11	0.69	37.2	1.65	7.938	4.381	8.22	33.1	1715+432
-1.77	-7.88	0.91	33.7	1.42	8.136	4.087	7.59	29.5	1716-184
-0.75	-6.96	0.45	29.5	2.00	7.370	4.242	4.53	19.3	1718-195
-1.17	-8.89	1.43	55.7	1.51	8.573	3.838	9.02	35.6	1720-177
-1.34	-6.92	0.39	15.7	1.94	7.302	4.429	4.98	21.0	1720-263
-0.07	-6.42	1.30	23.1	1.44	8.134	3.021	3.19	12.4	1721-484
-0.54	-4.91	0.31	9.0		6.973	4.261	3.20	12.5	1724-307
-0.71	-6.22	2.56	23.3	0.94	8.678	2.402	2.89	10.3	1725-050
-0.71	-7.18	3.57	37.4	0.83	9.089	2.454	4.13	14.4	1726-670
-0.29	-5.00	≤0.47	7.8		≤7.306	≥3.915	≥3.18	≥11.9	1727-315
-1.68	-6.91	0.58	12.9		7.703	4.329	6.35	24.4	1727-299
	-7.94	1.84	29.9	1.21	8.659	3.313	6.23	23.4	1728-338
-0.29	-7.96	0.14	7.6	2.06	6.726	6.139	12.23	52.6	1730-333
-1.30	-6.28	0.40	6.5	1.21	7.398	4.649	6.27	23.5	1731-390
+0.10	-4.10	0.30	12.0		6.804	3.891	2.04	8.1	1732-304
-0.62	-9.91	0.58	32.6	1.75	8.089	5.268	19.13	78.2	1732-447
	-5.36	1.83	12.9		8.354	2.577	2.45	8.6	1733-390
-2.19	-9.33	2.36	29.7	1.30	9.025	3.496	9.91	37.8	1735-032
-1.01	-7.25	0.49	27.4	1.93	7.492	4.297	5.27	22.2	1735-238
-2.02	-6.55	0.50	24.7	1.63	7.484	4.198	4.79	19.2	1736-536
+0.22	-6.95	1.16	12.8		8.246	3.635	5.56	20.3	1740-262
									1741-328
-1.94	-6.27	1.81	37.8	1.59	8.288	2.435	2.27	9.1	1742+031
-0.71	-7.25	0.10	8.4		6.454	6.380	11.76	49.4	1745-247
-0.54	-8.75	0.24	24.7	2.08	7.206	5.708	13.01	56.2	1746-203
-0.07	-9.31	0.47	25.9	1.89	7.814	5.200	14.41	60.2	1746-370
-1.51	-7.32	0.51	15.6	1.61	7.616	4.504	6.88	27.6	1748-346
	-6.35	0.31	5.6		7.229	4.977	7.11	26.9	1747-312
-1.22	-7.61	0.66	17.7		7.881	4.387	7.79	30.3	1751-241
-0.71	-5.82	3.23	18.2	0.72	8.854	2.156	2.52	8.6	1755-442
-0.45	-3.70	0.25	5.5		6.682	4.137	2.21	8.5	1758-268
-1.47	-7.10	0.18	18.5	2.06	6.778	5.416	7.10	30.6	1759-089
	-6.40	0.49	6.4		7.573	4.490	6.33	23.4	1800-260
-1.56	-7.30	0.34	19.1	1.69	7.333	4.965	7.88	31.9	1800-300
-1.56	-4.77	0.40	20.1	1.84	7.022	3.629	2.01	8.3	1801-003
-0.96	-6.64	0.26	9.3	1.81	7.012	4.984	6.08	25.1	1801-300

TABLE IIa. (cont'd.)

Number	Name		ℓ	b	R_o	X	Y	Z	R_{GC}
(1)	(2)	(3)	(4)	(5)	(6)	(7)	(8)	(9)	(10)
1802-075	NGC 6539		020.795	+06.775	3.1	-5.9	1.1	0.4	6.0
1804-250	NGC 6544		005.837	-02.201	2.6	-6.2	0.3	-0.1	6.2
1804-437	NGC 6541		349.286	-11.189	7.0	-2.1	-1.3	-1.4	2.8
1806-259	NGC 6553		005.253	-03.029	5.7	-3.1	0.5	-0.3	3.2
1807-317	NGC 6558		000.200	-06.024	8.8	0.0	0.0	-0.9	0.9
1808-072	IC 1276	Pal 7	021.832	+05.666	9.8	0.3	3.6	1.0	3.8
1809-227	Ter 11		008.357	-02.100	23.7	14.7	3.4	-0.9	15.1
1810-318	NGC 6569		000.481	-06.681	8.9	0.0	0.1	-1.0	1.0
1812-121	Kod 1		018.072	+02.415					
1814-522	NGC 6584		342.144	-16.413	15.0	4.9	-4.4	-4.2	7.8
1820-303	NGC 6624		002.788	-07.913	8.0	-0.9	0.4	-1.1	1.5
1821-249	NGC 6626		007.799	-05.580	5.8	-3.1	0.8	-0.6	3.2
1827-255	NGC 6638		007.897	-07.153	6.7	-2.2	0.9	-0.8	2.5
1828-323	NGC 6637	M 69	001.722	-10.269	10.3	1.3	0.3	-1.8	2.3
1828-235	NGC 6642		009.814	-06.439	5.5	-3.4	0.9	-0.6	3.6
1832-330	NGC 6652		001.535	-11.377	14.3	5.2	0.4	-2.8	6.0
1833-239	NGC 6656		009.890	-07.552	3.1	-5.7	0.5	-0.4	5.8
1838-198	Pal 8		014.103	-06.797	28.1	18.3	6.8	-3.3	19.8
1840-323	NGC 6681	M 70	002.853	-12.510	9.3	0.3	0.5	-2.0	2.1
1850-087	NGC 6712		025.353	-04.318	6.2	-3.2	2.7	-0.5	4.2
1851-305	NGC 6715	M 54	005.607	-14.088	21.5	12.0	2.0	-5.2	13.2
1852-227	NGC 6717	Pal 9	012.876	-10.901	7.8	-1.3	1.7	-1.5	2.6
1856-367	NGC 6723		000.072	-17.298	9.2	0.0	0.0	-2.7	2.7
1902+017	NGC 6749		036.201	-02.204	12.8	1.5	7.5	-0.5	7.7
1906-600	NGC 6752		336.495	-25.628	4.1	-5.4	-1.5	-1.8	5.9
1908+009	NGC 6760		036.108	-03.924	4.1	-5.5	2.4	-0.3	6.0
1914-347	Ter 7		003.387	-20.063	36.4	25.4	2.0	-12.5	28.4
1914+300	NGC 6779		062.659	+08.336	9.8	-4.3	8.6	1.4	9.8
1916+184	Pal 10		052.437	+02.726	10.6	-2.4	8.4	0.5	8.7
1925-304	Arp 2		008.543	-20.787	28.3	17.4	3.9	-10.0	20.4
1936-310	NGC 6809	M 55	008.798	-23.272	5.1	-4.2	0.7	-2.0	4.7
1938-341	Ter 8		005.758	-24.558	48.2	34.8	4.4	-20.0	40.4
1942-081	Pal 11		031.806	-15.577	13.8	2.5	7.0	-3.7	8.3
1951+186	NGC 6838	M 71	056.742	-04.562	4.4	-6.4	3.7	-0.3	7.4
2003-220	NGC 6864	M 75	020.304	-25.748	18.5	6.8	5.8	-8.0	12.0
2031+072	NGC 6934		052.105	-18.894	14.9	-0.1	11.1	-4.8	12.1
2050-127	NGC 6981	M 72	035.163	-32.683	17.0	2.9	8.2	-9.2	12.7
2059+160	NGC 7006		063.769	-19.407	39.1	7.5	33.1	-13.0	36.3
2127+119	NGC 7078	M 15	065.013	-27.313	9.7	-5.1	7.8	-4.5	10.4
2130-010	NGC 7089	M 2	053.371	-35.770	11.9	-3.0	7.8	-7.0	10.9
2137-234	NGC 7099	M 30	027.179	-46.835	7.2	-4.4	2.3	-5.3	7.2
2143-214	Pal 12	Cap	030.510	-47.680	19.4	2.4	6.6	-14.3	16.0
2304+124	Pal 13	Peg	087.104	-42.699	24.4	-7.9	17.9	-16.6	25.7
2305-159	NGC 7492		053.392	-63.479	19.1	-3.7	6.8	-17.0	18.7

TABLE IIa. (cont'd.)

[m/H]	M_V	r_c	r_t	c_{der}	log T_r	log ρ_o	σ_o	v_{esc}	Number
(11)	(12)	(13)	(14)	(15)	(16)	(17)	(18)	(19)	(20)
-1.05	-6.23	0.48	12.0	1.53	7.430	4.189	4.52	17.9	1802-075
-2.15	-6.33	0.35	8.8	1.44	7.269	4.688	5.89	22.9	1804-250
-2.02	-8.52	0.57	64.1	1.97	7.786	4.567	8.45	35.8	1804-437
-0.41	-8.29	1.13	17.4	1.04	8.445	4.218	10.52	38.3	1806-259
-1.51	-6.28	0.40	16.2	1.68	7.279	4.356	4.57	18.5	1807-317
-0.84	-7.56	3.59	35.9		9.094	2.449	4.34	15.7	1808-072
	-5.50	1.35	13.8		8.124	2.892	2.72	9.8	1809-227
-1.01	-7.79	0.89	18.3	1.46	8.102	4.043	7.13	27.9	1810-318
									1812-121
-1.56	-7.57	1.74	38.0	1.34	8.528	3.164	5.00	19.2	1814-522
-0.84	-7.46	0.20	26.8	2.15	6.880	5.323	7.19	31.5	1820-303
-1.81	-8.17	0.64	25.5	1.37	7.967	4.681	10.60	40.9	1821-249
-0.92	-6.27	0.46	11.5	1.29	7.466	4.411	5.52	21.0	1827-255
-0.92	-8.04	0.99	31.3	1.52	8.192	3.976	7.31	28.9	1828-323
-1.30	-5.54	0.12	12.8	2.15	6.247	5.244	3.86	16.9	1828-235
-0.92	-7.34	0.63	32.3	1.70	7.742	4.168	5.86	23.7	1832-330
-1.81	-8.53	1.10	30.2	1.59	8.319	3.991	8.27	33.0	1833-239
-0.50	-7.15	3.25	18.3	0.80	9.035	2.587	4.34	15.1	1838-198
-0.92	-7.05	0.25	29.6	2.19	6.941	4.833	5.08	22.4	1840-323
-1.26	-7.34	1.23	19.9	1.16	8.310	3.641	6.00	22.3	1850-087
-1.85	-9.54	0.65	46.4	1.86	8.073	4.880	13.85	57.6	1851-305
-2.19	-5.92	0.13	18.0	2.28	6.308	5.134	3.72	16.8	1852-227
-1.09	-7.71	2.22	33.6	1.05	8.789	3.091	5.69	20.7	1856-367
-0.37	-6.16	1.78	10.0	0.73	8.511	3.058	3.94	13.5	1902+017
-1.64	-7.72	0.59	36.5	1.59	7.789	4.475	7.79	31.1	1906-600
-0.84	-6.82	0.53	14.5	1.51	7.590	4.319	5.77	22.7	1908+009
	-6.00	3.85	36.8		8.895	1.746	2.06	7.4	1914-347
-2.32	-7.34	1.25	34.3	1.59	8.214	3.351	4.50	18.0	1914+300
	-5.58	0.99	17.3		7.877	3.157	2.81	10.6	1916+184
-1.85	-5.31	16.42	46.3		10.004	0.269	1.13	3.7	1925-304
-1.56	-7.44	2.59	27.7	1.27	8.784	2.637	4.03	15.3	1936-310
	-6.40	10.89	122.2		9.614	0.497	1.41	5.1	1938-341
-0.92	-6.98	5.27	29.7	0.75	9.344	1.947	3.29	11.3	1942-081
-0.45	-5.82	1.06	17.6	1.13	7.986	3.242	3.27	12.1	1951+186
-1.68	-8.32	0.51	32.5	1.82	7.731	4.739	9.24	38.2	2003-220
-1.30	-7.50	0.89	38.7	1.70	7.989	3.790	5.32	21.6	2031+072
-1.56	-6.93	1.80	43.1	1.53	8.401	2.744	3.22	12.7	2050-127
-1.72	-7.66	2.02	71.7	1.49	8.604	2.908	4.37	17.2	2059+160
-2.06	-9.24	0.26	59.1	2.54	7.133	5.263	8.55	40.9	2127+119
-1.81	-9.09	1.18	56.4	1.77	8.408	4.001	9.04	37.1	2130-010
-2.19	-7.17	0.16	33.4	2.40	6.554	5.260	5.15	23.8	2137-234
-1.13	-4.79	2.70	60.4	1.50	8.350	1.380	1.00	4.0	2143-214
-0.96	-3.30	2.70	27.0	0.90	8.285	1.196	0.75	2.7	2304+124
-1.81	-4.97	4.51	42.0	1.06	8.815	1.065	1.12	4.1	2305-159

TABLE IIb. DERIVED PARAMETERS OF GALACTIC DWARF SPHEROIDALS

Number	Name		ℓ	b	R_o	X	Y	Z	R_{GC}
(1)	(2)	(3)	(4)	(5)	(6)	(7)	(8)	(9)	(10)
0057-340	Scl		287.685	-83.133	78.	-6.	-9.	-78.	78.
0237-347	For		237.294	-65.654	145.	-41.	-50.	-132.	147.
0640-509	Car		260.113	-22.223	92.	-23.	-84.	-35.	94.
1005+126	Leo I	DDO 74	225.980	+49.109	212.	-105.	-100.	161.	217.
1110+224	Leo II	DDO 93	220.143	+67.236	231.	-77.	-58.	213.	234.
1508+674	UMi	DDO 199	104.969	+44.843	69.	-22.	48.	49.	72.
1719+580	Dra	DDO 208	086.363	+34.746	75.	-5.	61.	43.	75.

TABLE IIc. DERIVED PARAMETERS OF THE FORNAX GLOBULAR CLUSTERS

Number	Name		ℓ	b	ρ	θ	X	Y	R_F
(1)	(2)	(3)	(4)	(5)	(6)	(7)	(8)	(9)	(10)
0235-344	For 1		236.724	-66.298	41.1	299.5	-343.	1694.	1729.
0236-350	For 2		238.082	-65.838	22.4	220.1	-941.	-15.	941.
0237-345	For 3	NGC 1049	236.660	-65.721	16.2	355.3	475.	486.	679.
0238-348	For 4		237.303	-65.609	3.2	104.7	59.	-119.	133.
0240-343	For 5		236.089	-65.226	39.5	50.0	1641.	-260.	1662.
0238-346	For 6		237.021	-65.629	6.4	29.3	264.	55.	270.

TABLE IIb. DERIVED PARAMETERS OF GALACTIC DWARF SPHEROIDALS

[m/H]	M_v	r_c	r_t	c_{der}	log T_r	log ρ_o	σ_o	v_{esc}	Number
(11)	(12)	(13)	(14)	(15)	(16)	(17)	(18)	(19)	(20)
-1.68	-11.40	232.	1309.	0.91	12.346	-1.098	3.74	13.3	0057-340
-1.43	-12.39	529.	2420.	0.55	13.248	-1.323	5.28	17.4	0237-347
-1.77	-9.18	256.	1255.	0.94	12.039	-2.176	1.24	4.4	0640-509
	-11.24	270.	796.		12.690	-0.801	4.62	15.1	1005+126
-1.43	-9.67	154.	672.	0.63	12.055	-1.203	2.34	7.8	1110+224
-2.02	-8.71	266.	945.		12.044	-1.665	1.49	4.9	1508+674
-1.68	-8.69	203.	756.		11.999	-1.656	1.48	4.9	1719+580

TABLE IIc. DERIVED PARAMETERS OF THE FORNAX GLOBULAR CLUSTERS

[m/H]	M_v	r_c	r_t	c_{der}	log T_r	log ρ_o	σ_o	v_{esc}	Number
(11)	(12)	(13)	(14)	(15)	(16)	(17)	(18)	(19)	(20)
-2.19	-5.23	10.09	63.6	0.71	9.508	0.454	1.10	3.7	0235-344
-1.77	-7.30	6.08	80.1	1.08	9.370	1.598	2.80	10.3	0236-350
-1.85	-8.19	0.99	66.6	1.83	8.132	3.836	6.26	25.9	0237-345
-1.43	-7.23	0.68	46.1	1.82	7.742	3.936	4.86	20.1	0238-348
-1.81	-7.38	2.91	52.9	1.26	8.854	2.469	3.73	14.1	0240-343
-1.98	-6.46	2.98	24.2		8.824	2.324	3.05	10.8	0238-346

PROPER MOTION STUDIES OF STARS IN AND AROUND OPEN CLUSTERS

Floor van Leeuwen

Royal Greenwich Observatory
Herstmonceux Castle
Hailsham, East Sussex, BN27 1RP
United Kingdom

Abstract. A compilation of proper motion studies of stars in and around open clusters is presented. It can serve as a reference to cluster member selections, studies of cluster dynamics or as a guide to further improvement of the data presently available. The present paper is only a preliminary version; it is hoped for that reactions from the astronomical community will lead to a more complete and correct study in the near future.

Introduction. Compilations of data on open clusters have been made by Alter and Ruprecht, who still collect all references relevant to the research on open clusters, associations and globular clusters. For a general reference this is the most complete set of publications. Also in the present paper extensive use has been made of these compilations. Lynga (IAU Symp.85, 123) keeps an up to date record of the characteristic properties of open clusters, such as their size, content, age, distanc etc. Mermilliod prepared compilations of photometric and radial velocity data on open clusters (ref 171, 172 and IAU Symp.85, 129). Some of these studies are available at the CDS in Strasbourg. At least four other studies also provide many references to different open clusters: a paper by Van Schewick on absolute proper motions of 61 open clusters (ref 169), a paper by Janes and Adler on open clusters and the galactic distance scale (ref 125), a paper by Nicolet on distances of open clusters (ref 013) and a paper by Harris on colour magnitude diagrams of star clusters (ref 124). Each of these papers refers to earlier important reviews. Vasilevskis presented in 1962 an excelent review paper on proper motion studies of open clusters (ref 082). It describes extensively the why and how of this type of work. In his paper Vasilevskis counted some 90 individual studies. At present, 22 years later, this number has doubled. At least 81 clusters have now been investigated, and some of these several times. Improvements can still be obtained for most of the presented work, either by employing more of the photographic material available over the world, or by improving the measuring and reduction techniques. The present study also includes references to some of the old proper motion studies, not

J. Goodman and P. Hut (eds.), Dynamics of Star Clusters, 579–606.

only to do them justice, but also to show todays astronomers where they might find old plate material. This is, however, no historically complete compilation: the earliest references to studies of the Pleiades, the Hyades and probably some other clusters, have been omitted. Van Leeuwen (ref 377) and Van Bueren (ref 152) give reviews of the historical work on the Pleiades and the Hyades respectively.

The present paper is a compilation concentrating mainly on proper motion studies of stars in and around open clusters. Proper motions are the strongest selection criterion for cluster stars available. They do not depend on any other astrophysical assumption then the coincidence of the velocities of cluster members. (Apart from a time basis shorter than 5 years, when a possible disturbance due to orbital motions may influence the results.) This compilation will provide references to the selection of cluster members, which is the first step into the study of cluster dynamics: membership selection is needed to derive density profiles, which are reflections of the energy and angular momentum distribution in the cluster. In very few cases the proper motions are accurate enough to study the internal velocity dispersion in the cluster, providing additional data on the dynamics of the clusters involved.

The paper has been organized as follows. The first section gives a short review of how proper motions are derived and what this means to their interpretation. Some attention will be paid to intercomparing different studies and to deriving internal proper motion and velocity dispersions. In order to faciliate the use of the major review-table (II), we present in section two brief reviews of the most important centres of open cluster proper motion studies. The reference Table II , presented in section three, shows the characteristics of each individual study on proper motions in an open cluster. Table IV presents a listing of the literature which has been consulted.

The time given to prepare the present study has been short. It is therefore likely that it will not be complete and will contain a number of errors. A study like this should in addition contain references to work in progress. However, in this respect only a very limited picture could be obtained. My intention is, to publish this compilation once more in the supplements of one of the main journals. Hopefully then it will be more complete. That presentation could possibly also include some references to the vast amount of useful plate material scattered all over the world. The communication of any corrections and/or additions to the current version would be most appreciated.

Proper motions. The present paper will deal primarily with proper motions derived from comparing positions on photographic plates. Some studies do, however, also include meridian data and/or triangularization in the focal plane by means of micrometers or heliometers. However, most of these, older, studies ceased to be of value. Photographic proper motions in open clusters have been obtained with a multitude of instruments: initially with refractors only, but more recently also with different types of reflectors. A still very important contribution was made during the end of the 19th century and the first decades of the 20th century, using the Astrographs, of which many were build for the first sky surveys, the Astrographic Catalogue and the Carte du Ciel. Often these astrographs are refered to as Double Astrographs for having two tubes, one with a visually and the other with a photographically corrected objective. Their field size of 2.5 by 2.5 degrees and scale of 60"/mm make them still useful for the research of the nearby clusters. The second generation, with scales of 40" and 30" per mm (such as the Leiden, Bonn, Greenwich, Moscow and Zo-Se refractors) provide useful plate material on intermediate distance clusters. The range of large US telescopes with scales from 15" to 8" per mm has been used extensively, in particularly for the more distant clusters, that require a higher resolving power. Most of the instruments for which good plate files exist are still in working order, allowing one to take second epoch exposures with the same equipment as up to almost 100 years ago. Table I presents data on the telescopes that were found to be used in proper motion studies of stars in and around open clusters. The first part identifies the Astrographic Catalogue telescopes and the second part all others.

Proper motions are derived by means of comparing positions recorded on photographic plates at different epochs. In the past this has been done by means of plate pairs. A second epoch exposure was taken through the glass at the same hour angle and (almost) the same time of the year. The two exposures were superimposed and measured for differences in positions of the stars. The accuracies of the manual measurements vary from 1.5 to possibly 4 micron. More recent methods simply measure positions on the plates in a relative coordinate system of either a manual or automatic measuring machine. This method allows one to use plates taken at different hour angles and to combine plates at varies epochs. In particularly the automatic measuring machines brought a significant improvement in the measuring accuracy, to a repeatability of better than 0.8 micron. This means that the intrinsic positional noise on a photographic plate (roughly 0.8 to 1.8 micron) is no longer much increased by a measuring noise. The improved accuracy and availability of computers provide 2 to 4 times more accurate results. The contribution from the computer is mainly in the complexity of the mathematical model and the number of reference stars that can be used.

The reduction of positions to proper motions and the combination of different sets of proper motions involve transformations that inevitably remove some of the information contained in them. These transformations are needed in order to remove external influences on the positions of stars on a photographic plate. A linear transformation, which will always be needed, removes all the linear proper motion information from the stars used as reference points and defines by means of these stars a reference system for all the proper motions in the field. It is therefore virtually impossible to derive from such relative proper motions a linear rotation around the line of sight or a linear contraction/expansion in the plane of projection. This also means that by using different sets of reference stars the results of different studies of the same object will be different. They can only be compared in detail after applying a transformation that brings one system in accordance with the other.

The least squares solution used in the transformation of positions do not use the measuring errors as residuals, but instead the actual proper motions. Only in an iterative process or simultaneous plate-star solution this can be avoided (see e.g. papers by H.Eichhorn such as his study on the Pleiades, Mem.Roy.Aston.Soc., 73, part 2, 125). Nevertheless, the selection of stars used to perform the transformation or to constrain the all-in solution define a non-ideal reference system, which contains relative to an absolute system remnants of linear rotation and expansion, and has an arbitrary zero point. The linear deviations will be smaller when the intrinsic proper motion dispersion of the reference stars is smaller. In the study of open clusters the cluster stars themselves can provide such a very low intrinsic dispersion reference frame. It is therefore advisable always to use a selection of cluster stars as final reference points in the definition of the proper motion system.

The same type of complications disturb the picture of the internal proper motion dispersions. Here the level of the dispersions is less than 0".001/annum, with only a few stars to define the reference system. Values derived for internal proper motion dispersions are therefore sensitive to the actual definition of the proper motion system. In order to derive the information about the internal velocity dispersion, one needs in addition to correct for the influences of the parallactic motions using the space density distribution. In comparing results obtained by different authors it is therefore best to use the initial proper motions rather than the interpretated results, which are unlikely to be unique.

Still one other type of proper motion determination has been in use. In order to study the proper motions of members of the very nearby Hyades and Coma Berenice clusters, the use of many plates covering a large area of the sky is needed. These plates are often reduced to star positions from catalogues such as the AGK2. This introduces, however, the systematic errors of the catalogues in the proper motion field, which has made it always difficult to compare different Hyades studies and which caused a lot of confusion about the existence of the Coma Berenice cluster.

Centres of Activity. A major contribution to the proper motion studies of stars in and around open clusters has come from a few groups of astronomers. Although their definition is somewhat arbitrary, their brief descriptions given below may still provide some useful background information that would otherwise be difficult to trace back.

1. Bonn and Vienna. Uses mainly the Bonn refractor (40"/mm, 16x16cm plates) but also occasionally the AG refractors (100"/mm) of Bonn and Bergedorf. Studies of more than 20 open clusters were performed from 1953 to 1967, mainly by Van Schewick and Meurers. Plates have always been measured manually. First epoch plates from the period 1900-1920, second epoch plates taken between 1950 and 1965. Publications in Bonn Veroffentlichungen and Wien Mitteilungen.

2. Greenwich-MtWilson. Uses mainly the Greenwich 26 inch refractor (30"/mm 16x16 cm plates) and MtWilson 60 inch reflector (8".24/mm, 20x20 cm plates). Studies of a number of open clusters by Van Maanen, Murray, Cannon and others. Plates were all manually measured. Publications in the Royal Observatory Bulletin, Ap.J. and MNRAS.

3. Lick-Allegheny-Univ.New Mex.. Uses the astrographs of Lick (55"/mm) and Allegheny (14".7/mm). Extensive use of the Allegheny plate file on open clusters, initiated by Vasilevskis, and continued by Sanders, Jones, Cudworth, McNamara and others. Close collaboration with Yerkes. Most of the measurements done by automatic machine. Publications in A.J. and Aston.Astroph.Suppl.

4. Moscow Sternberg. Uses mainly the wide angle astrograph (90"/mm, 24x24 cm plates) and the double astrograph (30"/mm, 16x16cm plates). The first has been used for cluster halo surveys, in comparison with published positions in Astrographic Catalogues. The second is used for differential proper motions of higher accuracies in more distant clusters. Plates are measured manually. Most of the work has been performed by Arthiukhina from 1954 to 1976. Publications in Sternberg Trudy.

5. Pulkovo. Uses the 33 cm astrograph (60"/mm, 16x16cm plates). Studies of a large number of open clusters, which are still going on. Old epoch plates date back to around 1910 and are exposed at least till 14th magnitude. All measurements have been performed manually. The studies have been performed mainly by Bronnikova, Lavdovskij and Frolov. New plates are still being taken. Publications in Pulkovo Bulletin and -Trudy.

6. Sydney. Uses the 33 cm Astrograph (60"/mm, 16x16 cm plates). Unique studies of southern open clusters. Old epoch plates from last decenium of 19th century, exposed till 12th to 13th magnitude. Most of the work has been performed by D.S.King. All measurements were done manually. Publications from 1977 on in Sydney Observatory Papers.

7. Tashkent. Uses the 33 cm astrograph (60"/mm, 16x16 cm plates) with probably one of the finest plate files on open clusters in existence. Several deep studies (16th magn) have been performed by Ishmukhamedov, Latypov and Muminov. Many of the first epoch

exposures (taken around 1920) are on 8x8 cm plates. A study of NGC7654 covered five different field centres, and was performed from 1927 to 1954 by Subbotin and Savitskij. All measurements are performed manually. Publications mainly in Tashkent Circulars and Trudy.

8. <u>Yerkes</u>. Uses the Yerkes refractor (10"/mm) which provides on a short time interval good proper motions. First studies by Ebbighausen, later work by Van Altena and others, also in collaboration with Lick Observatory. Publications in A.J.

9. <u>Zô-Sé</u>. Uses the 40 cm Zô-Sé refractor (30"/mm). This former Jesuit missionary observatory has taken up the task of measuring proper motions again, and presented so far some of the most extensive studies available. All measurements are performed manually in as far as I could find out. Early publications in the Annals of the Zo-Se observatory (in French), later publications in the annals of Shanghai Observatory (in Chinese with English abstracts).

In addition significant work has been performed at a number of other observatories, but non of this has resulted yet in a long lasting and systematic research for proper motions in and around open clusters.

<u>The tables</u>. The present compilation is presented in the form of 4 tables. The first, which was presented already above, contains a review of telescopes that have been used. Table II presents a compilation of 200 studies, that were at least to a great extend aiming at and successful in measuring proper motions of stars in and around open clusters. It presents for each of these studies a reference number, the year(-1900) in which it was presented, the cluster identification, the investigated field, the limiting magnitude, total number of stars, spread in epochs of exposures, number of plates or exposures, telescopes that have been used (see table I), highest accuracy of proper motions obtained (in arcsec/1000 years), the highest likelihood for cluster members and an indication on whether the study has been selective on members only (indicated by s). When a value is followed by ":", it applies only to part of the study. An asterisk is used if part of the data has come from existing catalogues, e.g. the indication "ihg*" means that the zones i, h and g (see Table I) of the Astrographic Catalogue have been used for reference positions.

References to papers presenting also some proper motion information are presented in Table III (underlined numbers), as well as references to some papers presenting discussions of the studies shown in Table II.

In addition to the papers presented above, there are some studies under way. In this respect only a very limited picture can be presented. The following studies I know about from private communications: 1. J.Stauffer et al, a proper motion study of 4000 stars in a Per, using plates from (PS, L1) to a limiting magnitude of

V=17. 2. Zhao Jun-liang, a study on NGC1718 has been completed and a study on NGC2286 is well on its way at Shanghai Observatory (Zô-Sè). 3. B.Green (Mt.Stromlo), a proper motion study of Melotte 66, based amongst others on plates obtained in the Newtonian focus of the SAAO Radcliffe reflector. 4. F.van Leeuwen et al, proper motion studies of NGC2204, NGC2243 and Melotte 66. 5. B.McNamara, study of the internal motions of NGC6705. 6. F.van Leeuwen, study of internal motions in a 3x3degr field centred on the Pleiades and a survey for faint members in the outer halo of the Pleiades (see present volume). This list is hopefully far from complete, as many more proper motion studies need to be performed.

References to photometry, radial velocities, distance moduli and other relevant data can be found in the papers mentioned in the introduction. Roughly half of the clusters for which proper motions have been obtained have also been investigated photometrically. This applies also to radial velocities, although here often only very few stars were investigated. Table IV presents the list of references, sorted on Journal in order to faciliate the reference to less commonly known journals and to keep studies with the same origin together. This type of presentation also allowed me to indicate the content of the study, the clusters investigated or whether it concerns a review or catalogue of interest, and the type of study. All those indicated with "p" are shown in Table II, the others in Table III.

Some clusters are mentioned in Table IV, but not in Table II. Most of these are insignificant and fall within the field of a more obvious cluster. This applies to the following clusters: Tr.1: see NGC 581, NGC1907: see NGC1912, NGC2158: see NGC2168, Cr.401: see NGC6611, NGC6882: see NGC6885, NGC7790: see NGC7788. There may also be some confusion between the cluster NGC6530, often by mistake refered to as M8, and the nearby situated NGC6523, which is the true M8 and not an open cluster. The exact borders of this compilation were also not always clear. The cluster M71, NGC6838 has been refered to as an old open cluster as well as a globular cluster, while a Per and IC 348 are also regarded as parts of associations rather than individual open clusters.

Acknowledgements. The present compilation was made possible mainly due to two factors: the very extensive library of the Royal Greenwich Observatory, with the much appreciated assistence of J.Hutchins, and the assistence of my wife Małgorzata in translating the many Russian references. In addition I enjoyed discussions with R.D.Mathieu and C.A.Murray, and would like to thank those who bothered answering my enquiries.

Table I

data on telescopes used in open cluster proper motion studies

Cd	site and telescope	altitude (m)	latitude	scale ("/mm)
a	Greenwich Astrograph	47	+51.5	60.
b	Vatican Astrograph	100	+41.9	60.
c	Catania Astrograph	47	+37.5	60.
d	Helsinki Astrograph	33	+60.2	60.
e	Hyderabat Astrograph	554	+17.4	60.
f	Potsdam Astrograph	97	+52.4	60.
g	Oxford Astrograph	64	+51.8	60.
h	Paris Astrograph	57	+48.8	60.
i	Bordeaux Astrograph	73	+44.8	60.
j	Toulouse Astrograph	195	+43.6	60.
k	Alger Astrograph	345	+36.8	60.
p	Perth Astrograph	60	-32.0	60.
q	Cape Astrograph	8	-33.9	60.
r	Sydney Astrograph	44	-33.9	60.
A1	Allegheny Thaw	370	+40.5	15.
A2	Allegheny Doublet	370	+40.5	127.
B	Bonn Astrograph	62	+50.7	40.
BA	Bergedorf AG astrograph	41	+53.5	100.
C	Camebridge Sheepshanks equ.	28	+52.2	36.
Co	Copenhagen Astrograph	14	+55.7	43.
CV	Cape, Victoria triplet	10	-33.9	30.
G	Greenwich 26 inch	47	+51.5	30.
GA	Gottingen Astrograph	161	+51.5	50.
Go	Goloseevo Astrograph	184	+50.5	40.
H	Harvard Boyden	1387	-29.0	43.
L1	Lick Carnegie	1283	+37.3	55.
L2	Lick 36 inch	1283	+37.3	12.
LA	Lippert Astrograph	41	+52.5	60.
Le	Leiden Astrograph	6	+52.1	40.
Lo	Lowell	2210	+52.1	122.
Lp	Leipzig	119	+52.3	60.
M1	Moscow wide angle	130	+55.6	90.
M2	Moscow Astrograph	130	+55.6	30.
MC	McCormic 26 inch	259	+38.0	21.
P	Pulkovo Astrograph	75	+59.8	60.
PS	Palomar Schmidt	1706	+33.4	67.
R	Oxford, Radcliffe	65	+51.8	30.
Ru	Rutherfurd Astrograph	25	+40.6	53.
S	Swarthmore, Sproul	63	+39.9	19.
St	Stockholm Astrograph	44	+59.3	25.

Table I, continued

Cd	site and telescope	altitude (m)	latitude	scale ("/mm)
T	Tashkent Astrograph	479	+41.3	60.
To	Cerro Tololo 1m	2399	-30.2	20.
U	Uppsala Astrograph	21	+59.9	47.
W1	MtWilson 60 inch	1742	+34.2	8.
W2	MtWilson 100 inch	1742	+34.2	16.
Wi	Wien Astrograph	240	+48.2	60.
Y	Yerkes refractor	334	+42.6	11.
Zo	Zô-Sé Astrograph	100	+31.1	30.
Zu	Zürich Astrograph	469	+47.4	60.

Table III

cross-references to papers not incorporated in Table II

Hyades: 044, 048, 140, 150, 151, 177, 209, 366, 373, 378; Orion: 033; a Per: 211, 301; h & X Per: 205, 223, 230, 335, 375, 376; Pleiades: 136, 137, 220, 358, 377; Praesepe: 049, 063; NGC 129: 130, 228; NGC 457: 228; NGC 581: 228, 345; NGC 654: 139; NGC1513: 227; NGC1907: 230; NGC1912: 230; NGC1960: 227, 229; NGC2099: 227, 229, 328, 374; NGC2168: 230, 309, 374, 378; NGC2287: 119; NGC2506: 122; NGC2682: 308, 378; NGC6494: 014; NGC6530: 145; NGC6705: 010, 055, 060, 229, 378; NGC6750: 227; NGC6882: 230; NGC6885: 056, 230; NGC6910: 040, 059; NGC6913: 038, 039; NGC6940: 005; NGC7092: 230, 300; NGC7209: 228; NGC7654: 057, 058, 310; NGC7788: 061, 233; NGC7789: 302; NGC7790: 233; IC1805: 005; IC2391: 221; Tr.1: 145, 228; Berk.58: 062, 233;

Notes to Table II

1) ref 182, tel: P, a, St(87"/mm), d, r,Lick(39"/mm), f, y, A1 and Dearborn(30"/mm)
2) ref 135, tel: k, B, i, Co, a, Harvard, d, Le, Oxford, R, h, f, P, T, b

Table II, review of open cluster proper motion studies

ref	yr	cluster	field	range	stars	ep.	tel.	exp.	acc	ctr	sel
052	56	Coma B	5.x5.d	<14	99	50	R	4	9.		
178	04	Hyades	7.x5.d	<9.5	395	5	d	12x2	8.		
181	24	Hyades	7.x5.d	<12	328	16	d	12x4	3.		
152	52	Hyades	30x30d	<9.0	152		BA,Y	*	10.		s
050	54	Hyades	5.x5.d	<13.5	1359	21	BA,B,LA	6	7.		
191	62	Hyades	11x14d	<16.5	1000	30	Lo	2	26.		s
086	66	Hyades	6.5x6.5d	<18	778	14	L1	2	19.0	.91	s
367	66	Hyades	17.7sq d	<15.5	70	46:	W1	39x2	7.0		s
372	71	Hyades	6.5x6.5d	<21	4800	18	PS	2	20.		s
009	75	Hyades	30x30d	<12.5	112	55	ghi*,Le	*	6.0	-	s
094	75	Hyades	6.5x6.5d	<18	>600	24:	L1	5	5.0	.95	s
257	75	Hyades	3x(5x5d)	<13	3481	60:	ij*,M2	*,3	7.		
258	76	Hyades	6x(5x5d)	<13	6728	60:	hi*,M2	*,6	7.		
317	30	Orion	100'x100'		705	21	Zo	6	5:		
250	54	Orion	0.8x3.0d	<15	1058	54:	M1,T	30	3.8		
120	58	Orion	0.5x0.5d	<14.5	224	52	Y	19	0.6		
095	76	Orion	1.x1.d	<14	222	84	H,L1	8	0.7		
141	77	Orion	1.5x1.5d	<12.5	136	50	*	04			
054	55	a Per	5.5x5.d	<12.2	1379	50	cd*,BA,	10	4.0		
							GA,LA	4,6	6.0		
096	80	a Per	5.x5.d	<12	1200	25?	PS	8*	4.0	1.00	
043	09	h & X Per	2.x1.5d	<13	17	12	P	2			s
354	11	h & X Per	2.x2.d	<13.5	1418	13	d,P	16	8.1		
117	44	h & X Per	20'x20'	<16.0	800	29	W1	4	0.7		
053	55	h & X Per	6.x 6.d		203	81	BA	*,8	0.8		
164	60	h & X Per	1.5x1.5d	<14	800	43	B	8	2.0		
241	61	h & X Per	80'x80'	<14.6	2914	60	P	6	1.2		
283	83	h & X Per	1.0x1.0d	<17	1323	80	T	4	2.0		
045	19	Pleiades	1.5x1.5d	< 9.0	70	75:	*,Lp	*			
185	21	Pleiades	6.x6.d:	<15::	246	*	hg*,*,A1,				
							A2,Zu	*,23			s
204	21	Pleiades	1.5x1.5d	8-10	130	22	C	10	80		
148	24	Pleiades	2.5x2.5	<14.5	112	29	a,d,f	10	4.0		s
072	38	Pleiades	1.5x1.5d	<9.5	51	67	Ru	52	0.6		
115	40	Pleiades	40'x32'	<17.8	137	19	W2	4	2.0		s
118	45	Pleiades	.9x.9 d	<15.9	409	25	W1	10			s
135	47	Pleiades	2.5x2.5d	<16	2920	56:	(2	161	0.8		
089	70	Pleiades	1.6x1.6d	<15:	98	47:	Y,L2	66	0.3		s
255	70	Pleiades	9.x9.d	<14	4228	60:	gh*,M2	*,5	6.0		
023	73	Pleiades	5.x5.d	12-18	*	21	L1,PS	13	2.-8.	-	s
088	73	Pleiades	6.5x6.5d	10-17	125	16	L2	2	14.0		s
030	79	Pleiades	1.5x1.5d	<12.5	146	77:	Le	88	0.2		
097	81	Pleiades	6.5x6.5d	14-18	232	16	L1	7	1.0	.97	s

Table II, continued

ref	yr	cluster	field	range	stars	ep.	tel.	exp.	acc	ctr	sel
179	16	Praesepe	2.x2.d	<13.6	531	25	f	4	4.0		
149	24	Praesepe	2.2x2.2d	<13.5	173	29	P,a,R	10	3.5		s
046	25	Praesepe	1.5x1.5d	<10	45	64:	*,B	*,2			
182	27	Praesepe	2.0x2.0d	<14.	577	29:	(1	42	1.5		
110	34	Praesepe	50'x30'	<10.5	52	20	MC	20	0.8		
071	38	Praesepe	1.5x1.5d	6.3-8.7	37	65	Ru	32	0.8		
210	39	Praesepe		<13.2	179	30	R	4	4.0		
156	53	Praesepe	1.5x1.5d	<13	290	29	B	4	2.0		
254	66	Praesepe	5.x5.d	<13.0	1978	60:	h,M2	1*	5.0		
215	67	Praesepe	30'x50'	<10.5	35	51	MC	25	0.9		
245	71	Praesepe	1.5x1.5d	<16.6	784	30	M1	3	4.0		
090	71	Praesepe		<13	91	67:	Y,L2	68	0.2		s
099	83	Praesepe	2.5x2.d	<16.5	3000	23	PS	8	2.0	.99	
241	61	NGC 129	80'x80'	<14.6	1025	50	P	6	2.0		s
080	61	NGC 129	40'x40'	<12.4	70	42	Y	4	1.1		
232	75	NGC 129	2.5x2.5d	<14.7	326	72	P	6	2.1		
092	72	NGC 188	50'x40'	<15.6	228	40	A1	6	1.8	.97	
208	26	NGC 225	44'x36'	<12.3	62	22	Y	2			
241	61	NGC 457	80'x80'	<15.	905	51	P	4	2.0		s
275	72	NGC 457	1.4x1.4d	<14.5	659	43	T	8	1.7		
356	79	NGC 559	2.x2.d	<15.5	811	35	b	2			
241	61	NGC 581	80'x80'	<14.	1155	56	P	4	1.8		
346	65	NGC 581	80'x80'	<15.5	1618	49	U	8	0.5		
011	77	NGC 654	40'x40'	<14.5	186	55	Y	4	0.7	.93	
070	19	NGC 663			209	13	Y	4			
362	60	NGC 663	30'x30'	<14	168	54	Wi	5	2.8		
276	73	NGC 663	1.0x1.0d		1086	44	T	6	1.8		
111	39	NGC 752	50'x50'	<12.5	125	21	A1	10	1.1		
241	61	NGC 752	80'x80'	<15.	883	46	P	2			
051	54	NGC1039	50'x43'	<10.5?	136	50:	d,BA	6	3.0		
277	73	NGC1039	1.0x1.0d	<13.8	696	46	T	4	2.0		
138	75	NGC1039			700	50:	S,A1	90:	1.0		
361	58	NGC1502	1.x1.d	<13.3	146	49	b,Wi	6	5.0		
274	66	NGC1502	2.0x2.0d	<13.4	296	40	T	2	1.9		
240	58	NGC1513	80'x80'	<14.6	664	55	P	2			

Table II, continued

ref	yr	cluster	field	range	stars	ep.	tel.	exp.	acc	ctr	sel
093	73	NGC1664	50'x40'	<13.4	222	40	A1	4	1.3	.91	
318	53	NGC1750	100'x100'	<14	2461	12	Zo	4	10:		
318	53	NGC1817	100'x100'	<14	2103	27	Zo	4	5:		
241	61	NGC1912	80'x80'	<14.6	1252	58	P	8	1.3		
330	67	NGC1912	1.5x1.5d	<15	964	72	h	2	1.0	.79	
206	23	NGC1960		<11	170	20	C	4			
155	24	NGC1960	1.5x1.5d	<11.4	34	20:	f,B	*,3	6.0		
161	58	NGC1960	0.5x0.5d	<15	550	36	B	6	3.0		
240	58	NGC1960	80'x80'	<14.8	1634	50:	P	4	5.4		
319	66	NGC1960	100'x100'	<12.7	360	51	Zo	2	2.2		
069	16	NGC2099	0.5x0.5d	<13.5	292	12	Y	3			
327	53	NGC2099	1.x1.d	<15	691	15	St	4	4.:		
240	58	NGC2099	80'x80'	<14.9	2532	59	P	2			
163	60	NGC2099	1.5x1.5d	<15.8	693	53	B	3	2.3		
081	62	NGC2099	40'x40'	?	243	50	*,MC	6	3.0		
087	66	NGC2099	20'x20'	<14	489	60	Y	11	1.0		
207	25	NGC2168	1.3x1.3d	<14.	200	21	C	6	8.0		
073	42	NGC2168	26'x26'	<12.0	150	30:	Y	10	0.9		
163	60	NGC2168	1.5x1.5d	<16	853	40	B	4	2.3		
241	61	NGC2168	80'x80'	<15.5	2784	60	P	8	1.1		
091	71	NGC2168	40'x40'	<14.5	763	54:	A1,L2	5	1.1	.97	
162	58	NGC2244	25'x25'	<14.2	161	39	B	6	2.0		
098	82	NGC2244	40'x40'	<14.4	287	60:	Y,A1	17	0.6	.99	
162	58	NGC2252	6'x6'	<14.2	33	39	B	6	2.0		s
083	65	NGC2264	40'x40'	<15	245	53:	A1,W1,Lo	18	1.0		
079	59	NGC2281	50'x40'	<13.0	127	40	A1	4	0.9		
318	53	NGC2286	100'x100'	<14	2242	18	Zo	2	7:		
213	70	NGC2420	21'x19'	<15	208	42	W1	3	0.6		
001	70	NGC2420	40'x40'	<13.6	85	52	Y	4	0.7	.91	
167	66	NGC2422	1.1x0.9	<11.5	138	62	B	6	1.9		
273	66	NGC2422	1.4x1.4d	<14.0	727	42	T	2			
167	66	NGC2423	1.1x0.7	<12.5	74	57	B	4	2.:		

Table II, continued

ref	yr	cluster	field	range	stars	ep.	tel.	exp.	acc	ctr	sel
165	62	NGC2437	1.x1.d	<12.5	226	58	B	3	1.0		
121	81	NGC2506	10'x10'	<18	724	57	W1,To	3	0.6	.95	
263	78	NGC2516	2.5x2.5	<12.0	119	84	r	10	0.7	.99	
201	60	NGC2547		<11	67	62	q		3.0		s
112	39	NGC2548	30'x30'	<12.5	115	28:	Y	8	1.3		
318	53	NGC2548	100'x100'	<14	2451	14	Zo	2	10:		
265	79	NGC2669	2.5x2.5d	<12.5	226	86	r	28	0.9	.90	
180	22	NGC2682	2.0x2.0d	<14.	555	27	f	2			
116	42	NGC2682	20'x20'	12-15	343	21	W1	4	0.8		
113	42	NGC2682	30'x30'	<13.0	158	35:	Y	8	1.2		
365	65	NGC2682	27'x27'	<16.5	487	62	G,W1	4	2.0		
368	68	NGC2682	1.x1.d	<16.5	671	65	G	2	2.3		
277	73	NGC2682	1.0x1.0d	<14.0	550	44	T	6	1.9		
026	77	NGC2682	1.5x1.5d	<17.0	1866	67:	B,Y	12	1.0	.96	
324	82	NGC2682	100'x100'	<15.5	1067	69	Zo	8	2.7	.93	
336	30	NGC3064	24'x30'	<15.1	133	11	W1	4	2.7		
153	59	NGC3532	1.x1.d	<11.4	255	52	CV	4	1.9		
262	78	NGC3532	2.5x2.5	<12.5	647	84	r	6	1.3	.90	
264	79	NGC4103	2.5x2.5d	<13.0	171	84	r	24	0.8	.85	
267	81	NGC4755	2.5x2.5d	<11.9	164	88	r	10	0.6	.93	
266	80	NGC5662	2.5x2.5d	<12.6	188	87	r	14	0.9	.89	
261	77	NGC6025	2.5x2.5d	<12.5	132	78	r	6	1.5		
268	82	NGC6087	2.5x2.5	<12	157	89	r	10	1.5	.96	
350	69	NGC6475	1.5x1.5d	<13.5	746	75:	p,q	8			
012	80	NGC6494	40'x40'	<13.5	304	34	Y	13	0.5	.98	
319	66	NGC6530	100'x100'	<11.5	160	48	Zo	2	2.2		
008	72	NGC6530	60'x30'	<13.6	363	34	Y	12	0.6	.86	
323	81	NGC6530	100'x100'		303	66:	Zo	6	1.0		
166	62	NGC6611	1.3x1.3d	<13.0	231	40	B	10	2.1		
024	74	NGC6611	40'x40'	<13.2	142	47	Y	3	0.7	.94	

Table II, continued

ref	yr	cluster	field	range	stars	ep.	tel.	exp.	acc	ctr	sel
078	58	NGC6633	50'x40'	<13.5	207	40	A1	6	1.0		
021	73	NGC6633	40'x40'	<14.0	497	33	Y	10	0.7	.93	
278	73	NGC6633	1.0x1.0d	<15.5	1000	42	T	4	1.3		
280	75	NGC6694	1.0x1.0d	<14.7	627	42	T	4	2.6		
157	54	NGC6705	1.5x1.5d	<13.5	426:	51:	B	5	1.6		
240	58	NGC6705	80'x80'	<14.2	1097	53	P	2			
280	75	NGC6705	1.0x1.0d	<13.6	762	68	T	8	1.2		
027	77	NGC6705	30'x30'	<16.5	1890	70:	Y	30	0.4	.99	
281	77	NGC6709	1.2x1.2d	<14.6	1139	43	T	4	1.5		
032	83	NGC6709	30'x30'		502	57:	Y	10	0.4	.90	
279	73	NGC6755	1.0x1.0d	<15.7	706	43	T	6	2.2		
004	71	NGC6811	40'x40'	<14.1	296	53:	Y	14	0.7	.97	
007	72	NGC6819	40'x40'	<14.5	189	60	Y	2	1.2	.91	
003	71	NGC6823	40'x40'	<13.6	146	49	Y	4	0.8	.98	
251	56	NGC6838	1.5x1.1d	<14.5	1372	43	M1	2	3.0		
253	61	NGC6866	50'x50'	<14.1	392	25	M1	4	7.0		
231	71	NGC6866	2.5x2.5d	<15.	2036	45:	P	8	1.3		
224	30	NGC6885	2.0x2.0d	<14.7	1433	29	P	4	5.7		
288	54	NGC6885	1.x1.d	<13.5	431	46	T	2	2.		
161	58	NGC6885	1.4x1.4d	<13.5	374	36	B	2?	3.2		
241	61	NGC6885	80'x80'	<14.0	544	56	P	10	1.2		
313	56	NGC6910	40'x40'	<14.3	129	33	P	4	5.3		
022	73	NGC6913	40'x40'	<13.9	228	22	Y	2	2.5	.89	
160	57	NGC6939	1.x1.d	<15.	739	40	B	6	2.4		
370	70	NGC6939	16'x16'	<15.5	232	41	W1	3	0.9		
077	57	NGC6940	50'x40'	<13.0	216	40	A1	4	1.0	.94	
282	77	NGC6940	1.0x1.0d	<15.0	1351	42	T	4	1.0		
002	70	NGC7062	40'x40'	<14.5	334	48	Y	4	0.7	.87	

Table II, continued

ref	yr	cluster	field	range	stars	ep.	tel.	exp.	acc	ctr	sel
114	42	NGC7092	30'x30'	<13.0	50	23	Y	4	1.3		
159	57	NGC7092	0.6x0.6d	<13.2	128	40	B	8	2.5		
241	61	NGC7092	80'x80'	<14.8	1079	53	P	8	1.3		s
256	70	NGC7092	5.x5.d	<12.8	2964	67:	d*,M2	1*	6.0		
028	77	NGC7092	1.5x1.5d	<14.5	1710	52:	B,Y,A1	18	1.8		
047	30	NGC7209	2.0x2.0d	<13.5	35	32:	P	4			s
241	61	NGC7209	80'x80'	<14.5	1264	56	P	8	1.1		
252	61	NGC7209	1.5x1.5d	<12.9	332	62	d*,M1	1*	9.0		
167	66	NGC7209	1.1x1.0	<13.2	191	40	B	6	2.1		
225	37	NGC7243	1.6x1.3d	<13.5	814	34	P	4	2.0		
159	57	NGC7243	0.5x0.5d	<13.2	133	32	B	6	2.8		
318	53	NGC7380	100'x100'	<15	4443	16	Zo	2	8:		
319	66	NGC7380	100'x100'	<13.2	63	41:	Zo	8	1.8		
274	66	NGC7380	2.0x2.0d	<14.3	831	41	T	2	1.8		
285	27	NGC7654	2.0x2.0d	<13.8	1186	28	T	2			
287	28	NGC7654	2.0x2.0d	<13.4	1168	28	T	2			
286	30	NGC7654	2.0x2.0d	<13.8	860	30	T	2	3.3		
074	42	NGC7654	14'x14'	<13.2	84	33:	Y	16	0.7		
340	46	NGC7654	1.7x1.7d	<15.0	1937	42	U	6	2.3		
289	54	NGC7654	2.0x2.0d	<13.9	820	54:	T	6	3.7		
272	66	NGC7788	1.0x1.0d	<14.0	1088	42	T	6			
061	74	NGC7788	1.0x1.0d	<16.5	2170	49	T	2			
270	41	NGC7789	1.0x1.0d	<14.5	1183	40	T	6			
158	56	NGC7789	1.8x1.8d	<15	920	37	B	3	3.0		
031	81	NGC7789	30'x30'	<15.5	1387	68:	Y,A1	20	0.4	.98	
076	56	IC 348	45'x20'	<13.4	31	36:	S,	75	0.5		
					27	32	MC	60	0.4		
084	65	IC1805	50'x40'	<14	354	40	A1	4	1.6	.84	
075	55	IC4665	2.x2.d	<11.0	125	50:	jk*,*,L1	*,2	3.3		
006	72	IC4665	40'x40'	<13.8	275	33	Y	10	0.9	.84	
025	75	IC4756	8'x10'	<14	464	51:	A1	18	0.9	.97	
360	56	IC4996	1.4x1.4d	<15.5	157	59:	f,g,B,Wi	10	0.5		
168	67	IC4996	17'x17'	<14.2	79	43	B	6	2.3		
167	66	an.Bar.1	1.5x1.5d	<12.5	134	41:	B	4	2.1		

Journal: <u>Astronomy and Astrophysics</u>

ref	year	Vol	page	subjects	type	Authors
001	1970	8	112	NGC2420	p	W.F.van Altena, B.F.Jones
002	1970	9	86	NGC7062	p	B.F.Jones, W.F.van Altena
003	1971	10	270	NGC6823	p	R.R.Erickson
004	1971	15	368	NGC6811	p	W.L.Sanders
005	1972	16	58	NGC6940,IC1805	r	W.L.Sanders
006	1972	17	193	IC4665	p	W.L.Sanders,W.F.van Altena
007	1972	19	155	NGC6819	p	W.L.Sanders
008	1972	20	425	NGC6530	p	W.F.van Altena, B.F.Jones
009	1975	43	423	Hyades	p	G.Pels, J.H.Oort, H.A.Pels-Kluyver
010	1977	54	569	NGC6705	r	B.J.McNamara, W.L.Sanders
011	1977	54	803	NGC 654	p	R.C.Stone
012	1980	88	102	NGC6494	p	W.L.Sanders, R.Schroeder
013	1981	104	185	distance OCl	r	B.Nicolet
014	1983	118	361	NGC6494	r	B.J.McNamara, W.L.Sanders

Journal: <u>Astronomy and Astrophysics Supplement Series</u>

ref	year	Vol	page	subjects	type	Authors
021	1973	9	213	NGC6633	p	W.L.Sanders
022	1973	9	221	NGC6913	p	W.L.Sanders
023	1973	9	313	Pleiades	p	B.F.Jones
024	1974	16	1	NGC6611	p	L.W.Kamp
025	1975	19	211	IC4756	p	A.D.Herzog, W.L.Sanders, W.Seggewiss
026	1977	27	89	NGC2682	p	W.L.Sanders
027	1977	27	117	NGC6705	p	B.J.McNamara, N.M.Pratt, W.L.Sanders
028	1977	30	45	NGC7092	p	B.J.McNamara, W.L.Sanders
029	1979	36	163	Cross references of numbering in 50 OCl	r	J.-C.Mermilliod
030	1979	37	333	Pleiades	p	S.Vasilevskis, F.van Leeuwen, W.Nicholson, / C.A.Murray
031	1981	43	337	NGC7789	p	B.J.McNamara, S.Solomon
032	1983	51	541	NGC6709	p	J.Hakkila, W.L.Sanders, R.Schroeder
033	1983	54	221	Orion	r	B.J.McNamara, S.Huels

Journal: <u>Astrometriya i Astrofizika</u>

ref	year	Vol	page	subjects	type	Authors
038	1981	45	56	NGC6913	r	E.A.Herts
039	1982	47	58	NGC6913	r	E.A.Herts
040	1983	50	52	NGC6910	r(p)	E.A.Herts

Journal: Astronomische Nachrichte

ref	year	Vol	page	subjects	type	Authors
043	1909	182	369	h & X Per	p	S.Kostinskij
044	1912	194	3	Hyades	r	A.Hnatek
045	1919	209	355	Pleiades	p	F.Hayn
046	1925	225	49	Praesepe	p	O.Heckmann
047	1930	238	245	NGC7209	p	S.Kostinskij
048	1940	269	303	Hyades	r	W.Losert
049	1947	275	73	Praesepe	r(p)	O.Heckmann,W.Kruse
050	1954	281	193	Hyades	p	V.V.Osvalds
051	1954	282	25	NGC1039	p	W.Dieckvoss
052	1956	283	1	Coma B.	p	A.König
053	1955	283	67	h & X Per	p	W.Dieckvoss
054	1955	283	109	a Per	p	O.Heckmann, W.Dieckvoss, H.Kox

Journal: Astronomisheskij Tsirkulyar

ref	year	No	page	subjects	type	Authors
055	1949	87	2	NGC6705	r	P.A.Savitskij
056	1950	108	1	NGC6885	r	P.A.Savitskij
057	1953	134	8	NGC7654	r	P.A.Savitskij
058	1953	136	16	NGC7654	r	P.A.Savitskij
059	1954	153	13	NGC6910	r	A.B.Onegina
060	1955	165	10	NGC6705	r	P.A.Savitskij
061	1974	845	5	NGC7788,NGC7790	r(p)	V.N.Frolov
062	1974	848	1	Berk.58,Anon.1	r(p)	V.N.Frolov
063	1975	863	3	Praesepe	r(p)	I.A.Zykov

Journal: Astronomical Journal

ref	year	Vol	page	subjects	type	Authors
069	1916	29	101	NGC2099	p	A.H.Joy
070	1919	32	117	NGC 663	p	V.M.Gushee
071	1938	46	197	Praesepe	p	J.Schilt, J.Titus
072	1938	47	25	Pleiades	p	J.Titus
073	1942	50	1	NGC2168	p	E.G.Ebbighausen
074	1942	50	91	NGC7654	p	E.G.Ebbighausen
075	1955	60	384	IC4665	p	S.Vasilevskis
076	1956	61	437	IC 348	p	L.W.Fredrick

077	1957	62	175	NGC6940	p	S.Vasilevskis, R.A.Rach
078	1958	63	387	NGC6633	p	S.Vasilevskis, A.Klemola, G.Preston
079	1959	64	170	NGC2281	p	S.Vasilevskis, A.G.A.Balz Jr
080	1961	66	16	NGC 129	p	A.P.Lenham, O.G.Franz
081	1962	67	532	NGC2099	p	W.H.Jefferys III
082	1962	67	699	** review **	r	S.Vasilevskis
083	1965	70	797	NGC2264	p	S.Vasilevskis, W.L.Sanders, A.G.A.Balz
084	1965	70	806	IC1805	p	S.Vasilevskis, W.L.Sanders, W.F.van Altena
085	1966	71	385	program US-naval Obs	r	H.H.Guetter, A.R.Upgren
086	1966	71	482	Hyades	p	W.F.van Altena
087	1966	71	736	NGC2099	p	A.R.Upgren
088	1968	73	44	Pleiades	p	W.F.van Altena
089	1970	75	563	Pleiades	p	B.F.Jones
090	1971	76	470	Praesepe	p	B.F.Jones
091	1971	76	475	NGC2168	p	K.M.Cudworth
092	1972	77	74	NGC 188	p	A.R.Upgren, W.S.Mesrobian, S.J.Kerridge
093	1973	78	53	NGC1664	p	S.J.Kerridge, R.M.Nelson, W.S.Mesrobian
094	1795	80	379	Hyades	p	R.B.Hanson
095	1976	81	375	Orion	p	B.J.McNamara
096	1980	85	66	a Per	p	A.Fresneau
097	1981	86	290	Pleiades	p	B.F.Jones
098	1982	87	1497	NGC2244	p	L.A.Marschall,W.F.van Altena, Liang-Tai G. Chiu
099	1983	88	215	Praesepe	p	B.F.Jones, K.M.Cudworth

Journal: <u>Astrophysical Journal</u>

ref	year	Vol	page	subjects	type	Authors
110	1934	81	297	Praesepe	p	P.van de Kamp
111	1939	89	431	NGC 752	p	E.G.Ebbighausen
112	1939	90	689	NGC2548	p	E.G.Ebbighausen
113	1942	91	244	NGC2682	p	E.G.Ebbighausen
114	1942	92	434	NGC7092	p	E.G.Ebbighausen
115	1942	94	399	Pleiades	p	A.van Maanen
116	1942	96	382	NGC2682	p	A.van Maanen
117	1944	100	31	h & X Per	p	A.van Maanen
118	1945	102	26	Pleiades	p	A.van Maanen
119	1955	119	188	NGC2287	r(p)	A.N.Cox
120	1958	128	14	Orion	p	K.A.Strand
121	1981	243	827	NGC2506	p	L.T.G.Chiu, W.F.van Altena
122	1981	243	841	NGC2506	r	R.D.McClure, B.A.Twarg, W.T.Forrester

Journal: <u>Astrophysical Journal Supplement Series</u>

ref	year	Vol	page	subjects	type	Authors
124	1976	30	451	evolved stars in OCl	r	G.L.H.Harris
125	1982	49	425	catalogue of OCl	r	K.Janes, D.Adler

Journal: <u>Pis'ma v Astronomisheskij Zhurnal</u>

ref	year	Vol	page	subjects	type	Authors
130	1975	1	3	NGC 129	r	V.N.Frolov

Journal: <u>Annalen van de Sterrewacht te Leiden</u>

ref	year	Vol	part	subjects	type	Authors
135	1947	19	1,A	Pleiades,catalogue	p	E.Hertzsprung
136	1949	19	1,B	Pleiades,measurements	r(p)	E.Hertzsprung
137	1946	19	2	Pleiades	r	L.Binnendijk

Journal: <u>Bulletin of the American Astronomical Society</u>

ref	year	Vol	page	subjects	type	Authors
138	1975	7	435	NGC1039	p	P.A.Ianna, H.A.Alister
139	1975	7	509	NGC 654	r	R.C.Stone
140	1976	8	524	Hyades	r	S.C.Morris,G.Hill, W.J.Luyten
141	1977	9	333	Orion	p	F.W.Fallow

Journal: <u>Bulletin of the Astronomical Society of India</u>

ref	year	Vol	page	subjects	type	Authors
145	1976	4	82	NGC6530	r	R.Sagar, U.C.Joshi
146	1976	4	82	Tr.1	r	R.Sagar, U.C.Joshi

Journal: <u>Bulletin of the Astronomical Institutes of the Netherlands</u>

ref	year	Vol	page	subjects	type	Authors
147	1923	1	218	h & X Per	r	E.Hertzsprung
148	1924	2	55	Pleiades	p	E.A.Kreiken
149	1924	2	183	Praesepe	p	W.J.Kleyn Wassink
150	1934	7	168	Hyades	r(p)	W.Chr.Martin
151	1936	7	305	Hyades	r(p)	G.van Herk
152	1952	11	385	Hyades	p	H.G.van Bueren
153	1959	14	265	NGC3532	p	D.Koelbloed

Journal: <u>Veroffentligungen der Astronomischen Institute Bonn</u>

ref	year	No	subjects	type	Authors
155	1924	19	NGC1960	p	J.Hopmann
156	1953	40	Praesepe	p	K.-W.Schrick
157	1954	41	NGC6705	p	J.Meurers, W.Richels
158	1956	43	NGC7789	p	J.Meurers, O.Bahr,H.H.Thomas
159	1957	47	NGC7092, NGC7243	p	H.van Schewick
160	1957	48	NGC6939	p	J.Meurers, K.J.Scharf
161	1958	49	NGC6885, NGC1960	p	J.Meurers, K.Zentgraf, H.F.Mohn
162	1958	51	NGC2244, NGC2252	p	H.van Schewick
163	1960	53	NGC2168, NGC2158, NGC2099	p	J.Meurers, A.Schwarz
164	1960	56	h & X Per	p	J.Meurers, A.Acksoy
165	1962	61	NGC2437	p	J.Meurers, H.J.Grandjean
166	1962	62	NGC6611, Cr.401	p	H.van Schewick
167	1966	74	NGC2422, NGC2423, NGC7209, An.Barkh.1	p	H.van Schewick, H.S.Haase, G.Heidrich, H.Nentwig
168	1967	76	NGC2301 (r), IC4996	p	H.van Schewick, H.Bergeder, M.Kohl, R.Ober, G.Vogler
169	1972	84	abs.prop.mot.61 OCl	r	H.van Schewick

Journal: <u>Bulletin d'information du Centre des Donnees Stellaires</u>

ref	year	No	page	subjects	type	Authors
171	1979	16	2	Bibl.rad.vel. OCl	r	J.-C.Mermilliod
172	1984	26	9	Bibl.rad.vel. OCl	r	J.-C.Mermilliod

Journal: Groningen, Publications of the Kapteyn Astronomical Laboratory

ref	year	No	subjects	type	Authors
177	1904	13	Hyades	r(p)	H.A.Weersma
178	1904	14	Hyades	p	A.Donner, W.de Sitter, J.C.Kapteyn
179	1916	26	Praesepe	p	P.J.van Rhijn
180	1922	33	NGC2682	p	P.J.van Rhijn
181	1924	35	Hyades	p	A.Donner, W.J.Kleyn Wassink, P.J.van Rhijn
182	1927	41	Praesepe	p	W.J.Klein Wassink
183	1955	56	NGC2301	r(p)	P.J.van Rhijn, L.Plaut

Journal: Lick Observatory Bulletin

ref	year	Vol	page	subjects	type	Authors
185	1921	333	110	Pleiades	p	R.Trumpler
186	1938	494	167	Coma	r(p)	R.Trumpler

Journal: Lowell Observatory Bulletin

ref	year	Vol	page	subjects	type	Authors
191	1962	5	257	Hyades	p	H.L.Giclas, R.Burnham, N.G.Thomas

Journal: Mitteilungen des Astronomisches Geselschaft

ref	year	Vol	page	subjects	type	Authors
195	1954		44	NGC6705	r	J.Meurers
196	1966	21	123	prop.mot. OCl	r	H.van Schewick
197	1968	25	172	IC4756	r	W.Seggewiss

Journal: Monthly Notices of the Astronomical Society of South Africa

ref	year	Vol	page	subjects	type	Authors
201	1960	19	120	NGC2547	p	J.D.Fernie

Journal: <u>Monthly Notices of the Royal Astronomical Society</u>

ref	year	Vol	page	subjects	type	Authors
203	1921	81	213	NGC6633	r(p)	W.J.Luyten
204	1921	81	536	Pleiades	p	W.M.Smart
205	1922	83	79	h & X Per	r	W.M.Smart
206	1923	83	334	NGC1960	p	C.H.Payne
207	1925	85	257	NGC2168	p	W.M.Smart
208	1926	86	645	NGC 225	p	O.J.Lee
209	1939	99	168	Hyades	r	W.M.Smart
210	1939	100	378	Praesepe	p	P.C.Chaudhuri
211	1940	100	560	a Per	r	W.M.Smart, A.Ali
212	1969	146	479	Coma B	r(p)	A.N.Argue, C.M.Kenworthy
213	1970	150	279	NGC2420	p	R.D.Cannon, C.Lloyd

Journal: <u>Publications of the Leander McCormic Observatory</u>

ref	year	Vol	page	subjects	type	Authors
215	1967	11	191	Praesepe	p	J.L.Hershey

Journal: <u>Publications of the Astronomical Society of the Pacific</u>

ref	year	Vol	page	subjects	type	Authors
220	1920	32	43	Pleiades	r(p)	R.Trumpler
221	1960	72	85	IC2391	r(p)	A.R.Hogg

Journal: <u>Bulletin of the Pulkovo Astronomical Observatory</u>

ref	year	Vol	Nr	page	subjects	type	Authors
223	1924	9	92	277	h & X Per	r	J.Balanovskij
224	1930	12	108	1	NGC6885	p	I.Lehmann-Balanovskaja
225	1937	15	126	1	NGC7243	p	G.G.Lengauer
226	1938	16	131	1	Coma B.	r	G.Shajn
227	1958	21	161	144	NGC1513, NGC1960, NGC2099, NGC6750	r	N.M.Bronnikova

228	1962	23	171	121	NGC 129, NGC 457, NGC 581, NGC7209 Tr.1	r	V.V.Lavdovskij
229	1964	23	174	144	NGC1960, NGC2099, NGC6705	r	N.M.Bronnikova
230	1965	23	176	138	NGC1907, NGC1912, NGC2168, NGC6882, NGC6885, NGC7092, h & X Per	r	V.V.Lavdovskij
231	1971		189	207	NGC6866	p	L.S.Koroleva
232	1975		193	80	NGC 129	p	V.N.Frolov
233	1979		196	69	NGC7788, NGC7790, Berk.58	r	V.N.Frolov

Journal: Transactions of the Pulkovo Astronomical Observatory

ref	year	Vol	page	subjects	type	Authors
240	1958	72	77	NGC1513, NGC1960, NGC2099, NGC6705	p	N.M.Bronnikova
241	1961	73	5	NGC 129, NGC 457, NGC 581, NGC 752, NGC1907, NGC1912, NGC2168, NGC6882, NGC6885, NGC7092, NGC7209, h & X Per	p	V.V.Lavdovskij

Journal: Communications from the Sternberg Astronomical Observatory

ref	year	Vol	page	subjects	type	Authors
245	1971	172	3	Praesepe	p	N.M.Artiukhina

Journal: Transactions from the Sternberg Astronomical Observatory

ref	year	Vol	page	subjects	type	Authors
250	1954	25		Orion	p	P.P.Parenago
251	1956	27	3	NGC6838, Har.20	p	N.M.Arthyukhina
252	1961	30	196	NGC7209	p	N.M.Arthyukhina
253	1961	30	219	NGC6866	p	N.M.Arthyukhina
254	1966	34	181	Praesepe	p	N.M.Arthyukhina

255	1970	39	111	Pleiades	p	N.M.Arthyukhina, E.P.Kalinina
256	1970	40	3	NGC7092	p	N.M.Arthyukhina, E.P.Kalinina
257	1975	46	57	Hyades	p	N.M.Arthyukhina, P.N.Kholopov
258	1976	47	105	Hyades	p	N.M.Arthyukhina, P.N.Kholopov

Journal: <u>Sydney Observatory Papers</u>

ref	year	No	subjects	type	Authors
261	1977	75	NGC6025	p	W.H.Robertson
262	1978	79	NGC3532	p	D.S.King
263	1978	81	NGC2516	p	D.S.King
264	1979	83	NGC4103	p	D.S.King
265	1979	86	NGC2669, IC2391	p	D.S.King
266	1980	87	NGC5662	p	D.S.King
267	1981	89	NGC4755	p	D.S.King
268	1982	93	NGC6087	p	D.S.King

Journal: <u>Bulletin of the Tashkent Astronomical Observatory</u>

ref	year	Vol	No	page	subjects	type	Authors
270	1941	2	5(15)	201	NGC7789	p	J.P.Tsukervanik

Journal: <u>Circulars of the Astronomical Observatory in Tashkent</u>

ref	year	Nol	No	page	subjects	type	Authors
272	1966	345		1	NGC7788	p	Kh.Ishmukhamedov
273	1966	346		16	NGC2422	p	Kh.Ishmukhamedov
274	1966	347		1	NGC1502	p	Kh.Ishmukhamedov
275	1972	(384)	37	11	NGC 457	p	A.A.Latypov
276	1973	(386)	39	12	NGC 663	p	A.A.Latypov
277	1973	(388)	41	1	NGC1039, NGC2682	p	A.A.Latypov
278	1973	(389)	42	11	NGC6633	p	A.A.Latypov
279	1973	(390)	43	10	NGC6755	p	A.A.Latypov
280	1975	(405)	58	1	NGC6705, NGC6694	p	A.A.Latypov
281	1977	(421)	74	1	NGC6709, NGC6940	p	A.A.Latypov
282	1977	(423)	76	1	NGC6940	p	A.A.Latypov
283	1982	(445)	98	3	h & X Per	p	M.Muminov

Journal: Publications de l'Observatoire de Tashkent

ref	year	Vol	page	subjects	type	Authors
285	1927	1	3	NGC7654 (1)	p	M.T.Subbotin
286	1930	3	3	NGC7654 (3)	p	P.A.Savitskij

Journal: Transactions of the Tashkent Astronomical Observatory

ref	year	Ser	Vol	page	subjects	type	Authors
287	1928	1	1	3	NGC7654 (2)	p	P.A.Savitskij
288	1954	2	3	3	NGC6885	p	P.A.Savitskij
289	1954	2	4	3	NGC7654 (4)	p	P.A.Savitskij
290	1978	3	2	3	catalogues of Tashkent p.m. studies	r(p)	A.A.Latypov

Journal: Astronomical Society of the Pacific, Leaflets

ref	year	No	subjects	type	Authors
295	1949	240	Table of Messier's nebulae and st.cl	r	J.C.Duncan

Journal: Sovjet Astronomy, Astronomical Journal

ref	year	Vol	No	page	subjects	type	Authors
300	1970	14	1	130	NGC7092	r	N.M.Arthyukhina
301	1972	16	2	317	a Per	r	N.M.Arthyukhina
302	1975	18	5	584	NGC7789	r	L.S.Koroleva

Journal: Publications of the American Astronomical Society

ref	year	Vol	page	subjects	type	Authors
308	1940	10	11	NGC2682	r	E.G.Ebbighausen
309	1941	10	124	NGC2168	r	E.G.Ebbighausen
310	1942	10	161	NGC7654	r	E.G.Ebbighausen

Journal: Ukrainian Academy of Sciences, Bulletin of the Main Astronomical Observatory

ref	year	Vol	No	page	subjects	type	Authors
313	1956	1	2	61	NGC6910	p	A.B.Onegina

Journal: Annals of the Sheshan (Zo-Se) Section of Shanghai Observatory

ref	year	Vol	page	subjects	type	Authors
317	1930	18	1	Orion	p	S.Chevalier, S.J.
318	1953	23	1	NGC1750, NGC1817, NGC2286, NGC2548, NGC7380	p	Li Khen
319	1966	26	63	NGC1960, NGC6530, NGC7380	p	Chian Bay-tzen, Zhu Guo-liang

Journal: Annals of Shanghai Observatory, Academia Sinica

ref	year	Vol	page	subjects	type	Authors
323	1981	3	151	NGC6530	p	Mi Liang-liang, Jiang Pei-fang, Qian Bo-chen
324	1982	4	17	NGC2682	p	Tian Kai-ping, Yin Ming-guan, Xu Zong-hai, Zhao Jun-liang

Journal: Stockholm Observatorium Annaler

ref	year	Band	No	subjects	type	Authors
327	1954	18	1	NGC2099	p	P.O.Lindblad
328	1956	19	6	NGC2099	r(p)	P.O.Lindblad

Journal: Journal des Observateurs

ref	year	Vol	page	subjects	type	Authors
330	1967	50	179	NGC1912	p	G.A.Mills

Journal: <u>Contributions from the Mount Wilson Observatory</u>

ref	year	Vol	page	subjects	type	Authors
335	1921	205		h & X Per	r(p)	A.van Maanen
336	1930	405		NGC2264	p	A.van Maanen

Journal: <u>Uppsala Astronomiska Observatoriums Annaler</u>

ref	year	Band	No	subjects	type	Authors
340	1946	1	10	NGC7654	p	A.Lundby

Journal: <u>Meddelanden fran Astronomiska Observatorium Uppsala</u>

ref	year	No	subjects	type	Authors
345	1939	77	NGC 581	r(p)	Å.Wallenquist
346	1965	153	NGC 581, Tr.1	p	T.Oja

Journal: <u>Proceedings of the Astronomical Society of Australia</u>

ref	year	Vol	page	subjects	type	Authors
350	1969	1	207	NGC6475	p	S.M.Constantine, B.J.Harris, I.Nikoloff

Journal: <u>Recherches Astronomiques de l'Observatoire d'Utrecht</u>

ref	year	No	subjects	type	Authors
354	1911	5	h & X Per	p	A.van Maanen

Journal: <u>Vatican Observatory Publications</u>

ref	year	Vol	No	subjects	type	Authors
356	1979	1	16	NGC 559	p	E.de Graeve, S.J.

Journal: Specola Vaticana, Ricerche Astronomiche

ref	year	Vol	No	subjects	type	Authors
358	1968	7	12	Pleiades	r(p)	M.F.McCarthy, S.J., P.J.Treanor, S.J.

Journal: Mitteilungen der Universitats-Sternwarte Wien

ref	year	Vol	No	page	subjects	type	Authors
360	1956	9	3		IC4996	p	J.Hopmann, K.Haidrich
361	1958	9	13		NGC1502	p	J.Hopmann
362	1960	10		129	NGC 663	p	J.Hopmann, K.Haidrich

Journal: Royal Observatory Bulletin

ref	year	No	subjects	type	Authors
365	1965	91	NGC2682	p	C.A.Murray, P.M.Corben, M.R.Allchorn
366	1965	98	Hyades	r	P.A.Wayman, L.S.T.Symon, K.C.Blackwell
367	1966	108	Hyades	p	C.A.Murray, C.M.Lowne, E.D.Clements
368	1968	139	NGC2682	p	C.A.Murray, E.D.Clements
369	1968	141	NGC2682	p	C.A.Murray
370	1970	158	NGC6939	p	R.D.Cannon, A.G.Purcell

Journal: others

ref	year		subjects	type	Authors
372	1971	Minnesota	Hyades	p	W.J.Luyten
373	1936	Leipzig Diss	Hyades	r(p)	G.Beulig
374	1960	Bonn Diss.	NGC2168, NGC2099	r(p)	A.Schwarz
375	1960	Bonn Diss.	h & X Per	r(p)	A.Aksoy
376	1967	Bonn Diss.	h & X Per	r(p)	E.Voosholz
377	1983	Leiden Diss.	Pleiades	r(p)	F.van Leeuwen
378	1983	Berk. Diss.	NGC2168, NGC2682, NGC6705, Hyades	r	R.D.Mathieu

INDEX OF NAMES

INDEX OF SUBJECTS

GPSR Compliance

The European Union's (EU) General Product Safety Regulation (GPSR) is a set of rules that requires consumer products to be safe and our obligations to ensure this.

If you have any concerns about our products, you can contact us on ProductSafety@springernature.com

In case Publisher is established outside the EU, the EU authorized representative is:

Springer Nature Customer Service Center GmbH
Europaplatz 3
69115 Heidelberg, Germany

Batch number: 10684222

Printed by Printforce, the Netherlands